28th International Conference on Advanced Ceramics and Composites: A

Ceramic Engineering & Science Proceedings | Volume 25, Issue 3, 2004

Series Editor: Greg Geiger
Production Manager: John Wilson
Director, Technical Publications: Mark Mecklenborg

Editorial and Circulation Offices

PO Box 6136
Westerville, Ohio 43086-6136

Contact Information

Editorial: (614) 794-5858
Customer Service: (614) 794-5890
Fax: (614) 794-5892
E-Mail: info@ceramics.org
Website: www.ceramics.org/cesp

Ceramic Engineering & Science Proceedings (CESP) (ISSN 0196-6219) is published five times a year by The American Ceramic Society, PO Box 6136, Westerville, Ohio 43086-6136; www.ceramics.org. Periodicals postage paid at Westerville, Ohio, and additional mailing offices.

The American Ceramic Society assumes no responsibility for the statements and opinions advanced by the contributors to its publications. Papers for this issue were submitted as camera-ready by the authors. Any errors or omissions are the responsibility of the authors.

Change of Address: Please send address changes to *Ceramic Engineering and Science Proceedings*, PO Box 6136, Westerville, Ohio 43086-6136, or by e-mail to info@ceramics.org.

Subscription rates: One year $220 (ACerS member $176) in North America. Add $40 for subscriptions outside North America. In Canada, add GST (registration number R123994618).

Single Issues: Single issues may be purchased online at www.ceramics.org or by calling Customer Service at (614) 794-5890.

Back Issues: When available, back issues may be purchased online at www.ceramics.org or by calling Customer Service at (614) 794-5890.

Copies: For a fee, photocopies of papers are available through Customer Service. Authorization to photocopy items for internal or personal use beyond the limits of Sections 107 or 108 of the U.S. Copyright Law is granted by The American Ceramic Society, ISSN 0196-6219, provided that the appropriate fee is paid directly to Copyright Clearance Center, Inc., 222 Rosewood Dr., Danvers, MA 01923, USA; (978) 750-8400; www.copyright.com. Prior to photocopying items for educational classroom use, please contact Copyright Clearance Center, Inc.

This consent does not extend to copying items for general distribution, or for advertising or promotional purposes, or to republishing items in whole or in part in any work in any format. Please direct republication or special copying permission requests to the Director, Technical Publications, The American Ceramic Society, P.O. Box 6136, Westerville, Ohio 43086-6136, USA.

Indexing: An index of each issue appears at www.ceramics.org/ctindex.asp.

Contributors: Each issue contains a collection of technical papers in a general area of interest. These papers are of practical value for the ceramic industries and the general public. The issues are based on the proceedings of a conference. Both The American Ceramic Society and non-Society conferences provide these technical papers. Each issue is organized by an editor, who selects and edits material from the conference proceedings. The opinions expressed are entirely those of the presenters. There is no other review prior to publication. Author guidelines are available on request.

Postmaster: Please send address changes to *Ceramic Engineering and Science Proceedings*, P.O. Box 6136, Westerville, Ohio 43086-6136. Form 3579 requested.

Ceramic Engineering & Science Proceedings   Volume 25, Issue 3, 2004

28th International Conference on Advanced Ceramics and Composites: A

A collection of Papers Presented at the 28th International Conference and Exposition on Advanced Ceramics and Composites held in conjunction with the 8th International Symposium on Ceramics in Energy Storage and Power Conversion Systems

Edgar Lara-Curzio
Michael J. Readey
Editors

January 25-30, 2004
Cocoa Beach, Florida

Published by
The American Ceramic Society
735 Ceramic Place
Westerville, OH 43081
www.ceramics.org

Contents

28th International Conference on Advanced Ceramics and Composites: A

Ceramics and Components in Energy Conversion Systems

vi

Solid Oxide Fuel Cells

Ceramics in Environment Applications

xii

Ceramic Armor

Preface

The 28th International Conference and Exposition on Advanced Ceramics and Composites was held in Cocoa Beach, Florida, January 25-30, 2004. The conference was organized in conjunction with the 8th International Symposium on Ceramics in Energy Storage and Power Conversion Systems. More than 600 participants from 23 countries registered to attend the meeting, which had a record-breaking number of presentations with 447 oral and 93 poster presentations.

Dr. Jithendra P. Singh of Argonne National Laboratory presented the James I. Mueller Memorial lecture, and received the James I. Mueller award, which is the most prestigious award granted by the Engineering Ceramics Division of The American Ceramic Society. The title of his presentation was Residual Stresses in Composites and Coatings. Professor Roger Naslain of the University of Bordeaux, France was the recipient of the 2004 Bridge Building Award, which recognizes researchers outside the United States who have made significant contributions to engineering ceramics. The title of his Bridge Building Award lecture was SiC-Matrix Composites: Non-Brittle Ceramics for Thermostructural Applications.

The success of this conference, which has become one of the premier international meetings on ceramic materials and composites, is the result of the quality of the papers presented by the participants and the dedication of session chairs and volunteers who champion the organization of symposia and focused sessions. The success of this conference is also in great part due to the contributions of the staff of The American Ceramic Society. In particular, we are indebted to Chris Schnitzer, Greg Geiger and Marilyn Stoltz for their professionalism and dedication.

This proceedings contains 186 peer-reviewed manuscripts on the topics of solid oxide fuel cells; ceramics in environmental applications; mechanical properties; advanced ceramic coatings; biomaterials and biomedical applications; ceramic armor; nanomaterials and biomimetics; and ceramics and components in energy conversion.

We hope you will find these manuscripts both interesting and useful. We look forward to seeing you in Cocoa Beach next year.

Edgar Lara-Curzio
Michael J. Readey

Ceramics and Components in Energy Conversion Systems

CERAMIC COMPONENTS IN GAS TURBINE ENGINES: WHY HAS IT TAKEN SO LONG?

David W. Richerson, Richerson and Associates
and the University of Utah, Salt Lake City, Utah

ABSTRACT

Extensive efforts have been in progress worldwide since the late 1960s to develop ceramic turbine components, particularly since the introduction of high-strength silicon nitride and silicon carbide materials. This paper identifies the difficult challenges that have been encountered and highlights some of the key milestones in overcoming these challenges. Issues of improvements in material properties (especially long term durability at high temperature), property measurement and standards, component fabrication, quality assurance, design methodology for brittle materials, life prediction codes, and accumulation of rig and engine testing are all discussed.

INTRODUCTION

Remarkable progress has been made since the invention of the gas turbine engine, especially during the past 20 years, to increase thermal efficiency. Early engines had well under 30% efficiency. Modern utility scale gas turbines are now being commercially introduced with thermal efficiency greater than 60%. This has been accomplished through a combination of design and advanced materials to increase turbine inlet temperature, to increase pressure ratio, and in some cases to reclaim waste heat by use of heat exchangers. Ceramics have already played important roles as thermal barrier coatings [1] and as cores, molds, molten metal filters, and other refractories for investment casting of sophisticated superalloy blades and vanes with intricate internal cooling passages [2]. But the big hope has been to take advantage of the high temperature properties of ceramics to replace the complex cooled metal components with simple uncooled ceramic components.

Gas turbine engine designers explored the use of ceramic components in turbine engines intermittently since the middle of the 1940s [3]. The thermal and mechanical properties of oxides, carbides, cermets, and other compositions were studied and some prototype components were tested in engines [4]. None of these ceramic-based materials survived because they did not have the right combination of properties to withstand the severe conditions inside a gas turbine engine. The oxides and other monolithic (non-composite) compositions could not tolerate the thermal shock conditions, while the cermets exhibited too high creep at the turbine engine operation temperature (which was much lower than modern turbine engines).

Indeed, the hot section of a gas turbine engine is a challenging environment for materials to survive. The materials must withstand high mechanical and/or thermal stress at high temperature for thousands of hours, resist oxidation and corrosion by high velocity gases (often entrained with particles or condensed phases), avoid chemical reaction with adjacent components, be stable in high-cycle and low-cycle vibration conditions, and have acceptable creep and stress rupture life. They must also withstand handling during inspection and assembly, localized contact stress concentrations at points of attachment or contact with adjacent components, and occasional events of impact.

During the late 1960s some new silicon nitride and silicon carbide materials were developed that appeared to have the right combination of properties to justify a renewed effort to implement ceramics in turbine engines [5-11]. This led to considerable enthusiasm, optimism and hype [12,13] worldwide and initiation of a series of major well-funded programs during the 1970s and 1980s and continued effort to the present time [14,15]. But in spite of all the effort and enthusiasm, ceramic gas turbine engine components have not yet achieved broad commercial success. Why? Because the hot section of a gas turbine engine is such a severe environment and because we have been forced to address a wide spectrum of challenges:

- generating a detailed property database for each new ceramic material modification to support design and guide further material improvement
- improving material properties (especially long term durability at high temperature)
- learning to design with brittle materials (establishing and experimentally verifying design and life prediction methods and codes)
- developing fabrication technology to reproducibly fabricate at reasonable cost the advanced ceramics into the required complex shaped turbine engine components,
- establishing quality assurance procedures (including dimensional measurements, non-destructive evaluation techniques, and proof test protocols)
- conducting iterative rig, engine and field test trials to identify problems such as impact, contact stress, enhanced rate of oxidation in a moist high pressure turbine hot section, and hot corrosion

The remainder of this paper briefly discusses each of these challenges and the evolution of technology to progress in solving each challenge. A more detailed discussion is available in references [16] and [17], from which this was extracted.

Generating a Detailed Property Database

Very little property data were available for silicon nitride and silicon carbide candidate turbine materials in 1970, and there was little understanding of the relationships of chemistry, processing, and microstructure to the properties. Scanning electron microscopy, energy dispersive x-ray analysis, transmission electron microscopy, and Aüger spectroscopy were just emerging as analysis tools available to industry, but had not yet been applied to the new silicon nitride and silicon carbide materials. Strength testing for ceramics was done primarily in 3-point bending, but there were no standards for sample preparation (such as surface grinding procedure and surface finish), sample size, test fixtures, or methods of data analysis or fracture surface analysis. Use of Weibull statistics had not yet been established or validated for brittle material design. The significance of fracture toughness was just beginning to be realized. In addition to all of these deficiencies and limitations in knowledge, we didn't even have computers on our desks or user-friendly software.

All of these things needed to be developed and put in place concurrent to and often as a prerequisite to making progress in development of ceramics for gas turbine engines. As shown in Figure 1, materials characterization became a center point of virtually every program. Industry, government organizations, research institutes, and universities collaborated to develop standardized tests for each key property, to design and build reliable test fixtures and equipment, and to gather and validate a wide range of property data [18-40]. Strength testing evolved from 3-point bending to 4-point bending to uniaxial tension (which was considered necessary by engine designers). Test methodologies were established and implemented for creep, stress rupture, fracture toughness, thermal conductivity, thermal shock resistance, and other properties.

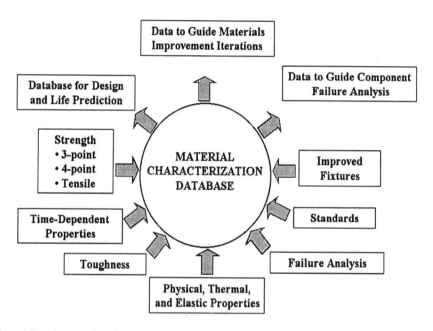

Figure 1 Development Requirements for Materials Characterization to Support Design and Life Prediction, Guide Materials Improvement, and Guide Component Failure Analysis [17]

Silicon nitride and silicon carbide materials available in the early 1970s proved to not have adequate properties (especially at high temperature) to provide acceptable life and reliability. As data was generated (from characterization as well as component rig or engine testing) that identified deficiencies, companies that developed the materials worked diligently to develop improved versions. Every new version or modification required extensive characterization before the material could be considered for fabrication into turbine engine components. Each iteration of development, characterization, and testing took years. As a result, technology of ceramics for turbines progressed slowly from one plateau to another. One plateau would be reached that solved one challenge and allowed more extensive engine testing, but this testing then would identify a further limiting challenge that would require more material development to reach a new plateau. The subsequent section identifies some of these plateaus.

Improving Properties

By 1970 four types of ceramic materials had been identified as important candidates for turbine components:

1. Reaction bonded silicon nitride (RBSN)
2. Reaction sintered silicon carbide (RSSC)
3. Hot pressed silicon nitride (HPSN)
4. Lithium aluminum silicate (LAS)

As gas turbine engine companies began to evaluate these 1970-vintage materials, key deficiencies were identified that limited the stress and temperature at which they could be used and the time that they would remain functional. Initial RBSN materials had about 30% porosity and were thus low strength and vulnerable to oxidation degradation. Early HPSN materials had excellent room temperature strength, but the strength, creep resistance, and stress rupture life were inadequate above about 1000° C. Furthermore, HPSN could not be fabricated directly to the complex shapes required for turbines, necessitating extensive and costly diamond grinding. !970-vintage RSSC looked promising for low stress combustor components, but did not have adequate strength for other high stress components. LAS performed well as a honeycomb rotary regenerator in laboratory tests, but failed during on-road tests due to cracking caused by ion exchange of lithium ions with sodium ions (from road salt) and hydrogen ions (from sulfuric acid condensate from sulfur impurities in the fuel).

In spite of the deficiencies, the early 1970-vintage ceramic materials were good enough to demonstrate that ceramic turbine components could be designed to survive at least for a short time under gas turbine start, steady state, and shutdown conditions. However, it was clear that improvements in properties and reduction in fabrication cost would be required before ceramics would be suitable for significant application in gas turbine engines. Since 1970 extensive worldwide efforts have been conducted that have resulted in large improvements in the baseline ceramic materials and in development of new categories of silicon nitride and silicon carbide ceramics that are fabricated by net-shape fabrication processes such as pressureless sintering, overpressure sintering, and hot isostatic pressing (HIP). This progression is illustrated in Figure 2 and reviewed in subsequent paragraphs.

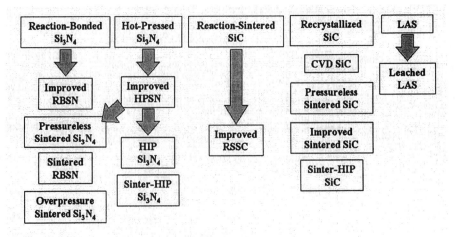

Figure 2. Progression of Ceramic Materials Development for Turbine Components [17]

Reaction-bonded Silicon Nitride

RBSN is fabricated by forming a compact of silicon particles into the desired shape and reacting with nitrogen in a high temperature furnace (about 1400° C) to convert to silicon nitride [6,41-48]. Because this requires a gas phase reactant, the resulting RBSN contains interconnected porosity. 1970-vintage RBSN typically had a density of about 2.2 g/cm^3, which means that the material contained about 30% porosity. This porosity limited the strength and also allowed ingress of oxygen to cause material degradation at high temperature. The typical strength of the 2.2 g/cm^3 RBSN in 3-point bending was under 100 MPa.

Through process development iterations during the 1970s, the density and strength of RBSN were progressively increased. By 1979 several RBSN materials were commercially available with a density of 2.7-2.8 g/cm^3 and an average strength in 4-point bending greater than 350 MPa [49-52]. These materials, along with hot pressed silicon nitride and reaction-sintered silicon carbide, allowed engine companies to reach a first plateau of rig and engine testing. However, the properties were not adequate for reliability or long-term life.

Reaction-sintered Silicon Carbide

The second category of ceramics available in 1970 was RSSC, which was also referred to as siliconized silicon carbide. RSSC was fabricated by preparing a shaped perform from a mixture of silicon carbide powder and a source of carbon such as graphite powder. This preform was then infiltrated with molten silicon in a vacuum above 1405° C. The molten silicon reacted with the carbon to form additional SiC and also filled all of the remaining pores. The resulting RSSC had near-zero porosity, strength comparable or better to RBSN, high thermal conductivity, and excellent oxidation resistance [53-58].

1970-vintage RSSC typically had flexure strength under 200 MPa. Improvements in density, uniformity, microstructure, and processing resulted in substantial increase in strength [59-64] during the early 1970s and 1980s as illustrated in Figure 3.

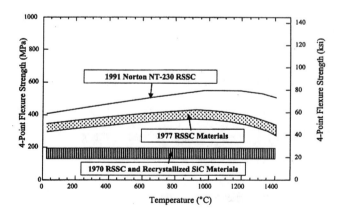

Figure 3. Progress in Improvement of RSSC Materials [17]

7

Hot Pressed Silicon Nitride

A second category of silicon nitride that was just becoming available by 1970 was HPSN [4,65-68]. HPSN was fabricated by adding an oxide such as MgO to silicon nitride powder and applying pressure through graphite punch and die tooling at about 1750° C. This resulted in a multiphase ceramic with near-zero porosity and very high strength (>700 MPa as measured in 3-point flexure) at room temperature, as illustrated by the Lucas HS-110 HPSN in Figure 4. In spite of the excellent room temperature strength, though, the strength dropped rapidly as the use temperature was increased above 1000-1100° C.

Studies during the early 1970s determined that the high temperature strength and rupture life were controlled by the chemistry and crystallinity of the grain boundary phase (typically a complex silicate). A non-crystalline (glass) phase would soften above a critical temperature and lead to deformation and slow crack growth by grain boundary sliding. Extensive "grain boundary engineering" studies and process control studies were conducted worldwide that resulted in substantial improvements in the properties of HPSN materials [69-92]. Figure 4 shows some examples of the level of improvements that were achieved.

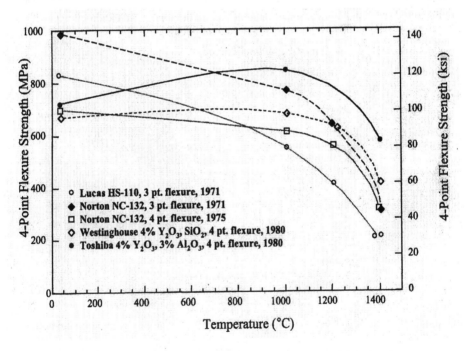

Figure 4. Improvements in the Strength of Hot Pressed Silicon Nitride Materials [17]

Lithium Aluminum Silicate (LAS)

The fourth material available in the early 1970s was lithium aluminum silicate. LAS had near-zero thermal expansion and was studied extensively as a honeycomb rotary regenerator core, as a thick-walled flow separator housing, and later as a matrix for SiC fibers. The honeycomb regenerator was tested successfully under a laboratory environment, but proved not to be successful during on-road testing. Ion exchange occurred between the lithium and sodium (from road salt) and hydrogen (from sulfuric acid condensate from sulfur impurities in the fuel) and resulted in cracking. A technique was developed to leach out the lithium from the LAS to form an aluminosilicate that was more stable. Considerable development also was conducted with cordierite-based compositions similar to those later developed for catalytic converter honeycomb catalyst substrates.

Hot Isostatic Pressing (HIP)

Hot pressed silicon nitride was an important material during early turbine engine developments, especially for various rotor configurations. However, the high temperature properties, especially creep and stress rupture life, were still limiting. Also, complex shapes could not be achieved without extensive, expensive diamond grinding. Asea in Sweden began in about 1971 to explore an alternative, hot isostatic pressing (HIP) of silicon nitride. By 1977 Larker and his co-workers at Asea reported direct fabrication of complex shapes of silicon nitride using a glass encapsulation process and achieving 3-point flexure strength of about 550 MPa at 1370° C [93-95]. Other organizations also conducted development of HIP silicon nitride, and some companies ultimately licensed the Asea technology for complex shape fabrication [96-112].

Figure 5 compares the strength versus temperature of some of the HIP silicon nitride materials with 1975-vintage (NC-132) hot pressed silicon nitride. The CSN 101 was an Asea material, the NT154 and NT164 were Norton materials, and the GN10 was an Allied-Signal Ceramic Components material. Even more dramatic improvements are illustrated in Figure 6 for the static fatigue life based on tensile creep tests at 1370 °C and 100 MPa. The property improvements were made possible by the much higher pressure that could be applied during HIP than by uniaxial hot pressing. The higher pressure allowed full densification to be achieved with decreased sintering aid or alternate sintering aid (such as Ytterbium) and thus resulted in increased control of the composition and characteristics of the grain boundary phase.

Although HIP silicon nitride fabricated using glass encapsulation had excellent bulk properties, chemical interaction during the HIP cycle between the silicon nitride and the glass resulted in lesser properties at the surface. The overall probability of survival for candidate turbine components was decreased unless the surface material could be removed by machining. In most cases this was considered not to be cost effective, so the silicon nitride HIPed with glass encapsulation lost favor during the mid 1990s, especially for long life applications. However, "sinter-HIP" was still used whereby the ceramic was first densified by sintering to the point that the porosity was all internal (closed porosity) rather than connected to the surface. The component could then be HIPed without glass encapsulation to remove much of the closed porosity and achieve significant increase in strength. Even in this case, though, the HIP step represented an undesirable added cost.

9

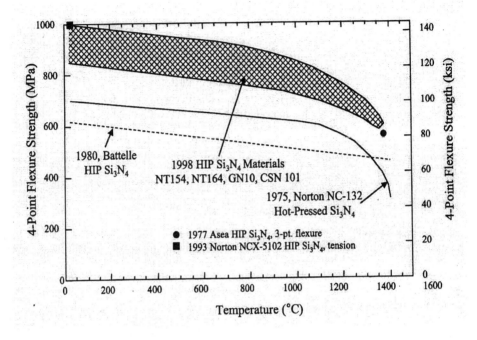

Figure 5. Evolution of Hot Isostatically Pressed Silicon Nitride [17]

Sintered Silicon Nitride

The ideal low cost approach for fabrication of ceramics is conventional pressureless sintering. Pressureless sintering was demonstrated for silicon nitride by the mid-1970s, but the properties were not much better than those of RBSN or RSSC. For example, the first reported sintered silicon nitride by Terwilliger [130] in 1974 was less than 77% of theoretical density. This material, which contained MgO as the sintering aid, exhibited high weight loss due to decomposition of silicon nitride and volatilization of MgO.

Many developments conducted over many years would be required to achieve properties in sintered silicon nitride that would approach those of HIP silicon nitride. Some of the key directions of development included (1) control of silicon nitride decomposition, (2) development of processes for synthesis of high quality, fine particle, reactive silicon nitride powders, (3) refinement of sintering aids and the resulting grain boundary chemistry, (4) increasing fracture toughness by achieving elongated, fibrous microstructure, (5) sintered RBSN. International efforts resulted in progress in each of these areas [114-129,52]. Examples illustrating the evolution of improvement in room temperature and high temperature strength are shown in Figure 6. The strength improvements were accompanied by large improvements in static fatigue life (Figure 7) and fracture toughness.

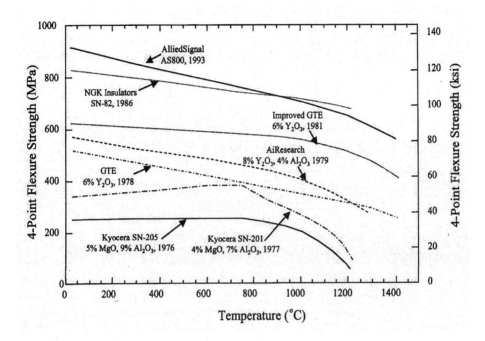

Figure 6. Improvements in Strength of Sintered and Over-pressure Sintered silicon nitride [17]

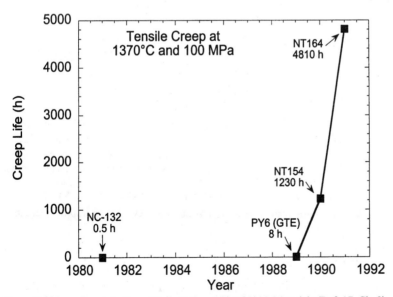

Figure 7. Improvements in Creep/Static Fatigue Life of Si_3N_4 Materials (Ref. 17, Ch. 3)

Learning to sinter in a moderate overpressure atmosphere of nitrogen (about 1-4 atmospheres) to minimize silicon nitride decomposition during sintering was one especially key advance. This allowed sintering at higher temperature to achieve close to theoretical density (typically less than 2% porosity), but also resulted in enhanced growth of elongated beta silicon nitride grains during sintering, as illustrated for AS 800 silicon nitride in Figure 8. These grains intertwined in such a way that fracture toughness was substantially increased [119-121]. 1975-vintage hot pressed silicon nitride had fracture toughness of about 4.5-5.0 MPa.m$^{1/2}$. Some of the overpressure sintered silicon nitride materials by 1990 had fracture toughness greater than 8.0 MPa.m$^{1/2}$. This unusually high toughness for a ceramic has been an important factor in the reliability and successful use of silicon nitride for turbine components and for other applications.

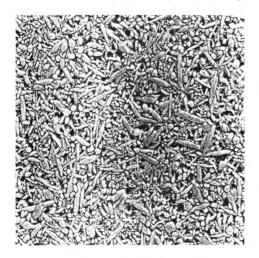

Figure 8. Microstructure of Over-Pressure Sintered AS 800 Silicon Nitride Showing the Elongated Grains That Result in High Fracture Toughness (Photo courtesy of Honeywell Engines, Systems, and Services)

Two other important technologies evolved. One was referred to as "sintered RBSN" and the other was "sinter-HIP". For sintered RBSN, sintering aids were mixed with the initial silicon powder, the silicon was nitrided to achieve partial densification, and then the RBSN containing the sintering aid was overpressure sintered. This allowed fabrication of dense silicon nitride parts using a less expensive starting powder (silicon versus silicon nitride) and also reduced the total shrinkage during sintering. For sinter-HIP, overpressure sintering would be conducted with silicon nitride powder or sintered RBSN to achieve a high enough density that only closed porosity remained. Then the part could be placed in a HIP unit without requiring glass encapsulation to reduce the porosity to close to zero and thus achieve further improvement in strength.

Sintered Silicon Carbide

SiC evolved similarly to silicon nitride. RSSC and recrystallized SiC were the primary SiC materials available up to the mid 1970s, followed sequentially by hot pressed SiC [130-132], pressureless sintered SiC [133-145], and HIP and sinter-HIP SiC. Hot pressed SiC became commercially available in the early 1970s, but never really caught on. In tests that provided a direct comparison, it did not have comparable thermal shock resistance to hot pressed silicon nitride. By the mid to late 1970s sintered SiC, especially from Carborundum Company, became an important candidate for turbine components and was evaluated extensively throughout the 1980s. 1977-vintage Hexoloy SA (sintered alpha-SiC from Carborundum) had 4-point flexure strength of 300 MPa at room temperature and about 350 MPa at 1450° C. By 1980 several sintered SiC materials were available with 4-point flexure strength between 400 and 500 MPa.

Ceramic Matrix Composites (CMCs)

An alternate approach to monolithic SiC was pursued starting in the late 1970s when high-strength SiC-based fibers became available. SEP in France developed a composite consisting of SiC-based fibers (Nicalon fibers from Japan) in a chemical vapor infiltration (CVI) matrix [146]. This general category of composites became known as CVI SiC/SiC. Later General Electric Company developed a melt-infiltrated reaction sintered composite material (MI-SiC/SiC) and Dow Corning developed a polymer infiltrated and pyrolyzed (PIP-SiC/SiC) composite. Extensive development has been conducted on these ceramic matrix composite materials particularly since 1990 [147-156].

The advantage of the continuous fiber reinforced CMC materials has been their fracture behavior. Rather than failing in a brittle mode like the monolithic ceramics, these CMC materials are able to tolerate much larger material defects and to sustain substantial accumulation of structural damage prior to fracture. The result is increased strain-to-failure and a non-catastrophic failure mode. However, studies during the 1980s and 1990s determined that the SiC/SiC ceramic matrix composites were susceptible to degradation due to interaction with the application environment (especially oxygen and water vapor) at high temperature [157]. Extensive developments on interface layers, fiber improvements, matrix modification, and surface coatings have been conducted during the 1990s and are continuing. Parallel efforts have focused on oxide/oxide ceramic matrix composites [158], although current versions have lower strength and lower use temperature than the SiC/SiC composites.

Learning to Design with Brittle Materials

Design methodology for brittle materials was at a very early stage prior to about 1970. Ford Motor Company initiated a program in 1967 to design an automotive gas turbine engine with an all-ceramic hot section (their 820 ceramic engine design). At that time computers were still cumbersome and there were no established design tools and guidelines for the complex 3D analyses that would be required for brittle materials. Ford and other early pioneers adopted by necessity a "learn-by-doing" approach that required substantial data gathering, many iterations, and experimental verification at each step of the way [159]. This involved a systems approach that became the model for future programs. It consisted of time-consuming iterations of the following sequence of activities:
- Define a platform for development, e.g. a specific engine or a generic engine concept.
- Select one or more conceptual component designs.
- Gather a material property database suitable to support analytical design requirements.

- Conduct detailed design analysis using finite element and finite difference analysis methodology.
- Conduct probabilistic life prediction for the components using materials properties and component boundary conditions as inputs.
- Select one or more configurations to fabricate.
- Select structural materials and suppliers for component fabrication.
- Fabricate prototype hardware.
- Inspect the hardware using property certification, dimensional inspection techniques, and nondestructive evaluation (NDE) technology.
- Test the prototype hardware in a test rig to simulate the thermal and mechanical stresses expected in an engine.
- Conduct failure analysis to determine whether failure resulted from a design deficiency or material deficiency.
- Modify the component designs, materials, and/or fabrication process steps. Generate additional material property data as needed for guiding material improvement and for design analysis input.
- Repeat the complete design/development sequence or a specific loop within the sequence.
- When successfully demonstrated in rigs, proceed to the definitive test of ceramic components in an engine.
- Accumulate engine test time and ultimately conduct long term engine demonstration in a field test (engine operating in real service at an on-site application.

The deterministic approach used for metals was found not to be suitable for ceramics. Instead, a probabilistic approach evolved [160,161] that required a statistically significant Weibull (weakest link) representation of the strength of the candidate ceramic material that distinguished both volumetric and surface strength-controlling flaw distributions.

The initial design codes that evolved were for fast fracture reliability. These were verified by short term rig and engine tests, and the viability of the overall probabilistic design methodology was clearly demonstrated by the mid 1970s, i.e., it was clearly demonstrated that ceramic gas turbine components could be analytically designed and could survive the severe start-up, steady state, and shut-down condition in a gas turbine engine [19,20]. However, most of the property data up to this time was fast fracture data, and all of the rig/engine tests were of short duration. It was recognized that long term uses in a turbine engine would involve changes in the material properties over time due to factors such as oxidation, slow crack growth and creep. As a result, efforts during the late 1970s and thereafter focused on gathering a time-dependent property database [23-27] (up to thousands of hours) and establishing design codes that would calculate reliability both for fast fracture and time-dependent properties [162-176]. Since each new improved version of candidate material required statistically-significant database, the evolution of design and reliability codes and validation of component designs stretched out over many years and is still continuing today.

Learning to Fabricate Complex Turbine Engine Components

Learning to design with ceramics and achieving acceptable material properties were both difficult challenges. A third major challenge was to cost-effectively fabricate the required complex shaped, close tolerance turbine components. Fabrication of a ceramic component

14

involves many steps, and each step requires a high level of reproducibility and careful controls to achieve adequate properties for successful turbine use [177,178]. If all of the process parameters in each step are not done properly, the resulting component either will not have acceptable properties or will not meet dimensional tolerances. The degree of process understanding and control necessary to produce acceptable ceramic gas turbine components did not exist in 1970. Each process for each material iteration required extensive long-term development and refinement.

Ceramic turbine components during the 1970s were primarily fabricated from RBSN, RSSC, and HPSN. RBSN and RSSC could be fabricated by conventional shape forming processes, but had the favorable characteristic that very little dimensional change (typically less than 1%) occurred during densification (compared to conventional pressureless sintering that typically exhibited 15-18% linear shrinkage). The raw materials also were commercially available and relatively inexpensive. In contrast, only simple blocks and flat or slightly contoured plates had been made by hot pressing, and high quality silicon nitride powder was not commercially available.

Extensive fabrication development of RBSN and RSSC was conducted during the 1970s on three approaches: injection molding, slip casting, and cold isostatic pressing (CIP)/green machining. Ford Motor Company in the US and AME in England conducted early development of injection molding [49,179,180], followed later by AiResearch Casting Company [51,181]. Ford and AiResearch also conducted much development on slip casting of RBSN during the mid-1970s [51,182,183]. Norton Company focused on CIP/green machining [184]. Figure 9 illustrates some of the turbine components fabricated at Ford by injection molding. Extensive RBSN and RSSC development also was conducted in Germany during the 1970s, some of which is described in references [38,73,185].

Figure 9 Silicon Nitride Parts Fabricated by Ford Motor Co. in the 1970s by Injection Molding (from reference 17, Chapter 2)

15

Much was learned about injection molding, slip casting, and CIP/green machining of RBSN and also of RSSC. However, these materials did not have adequate strength for rotors. Efforts during the 1970s explored ways to make rotors using hot pressing. Ford demonstrated a "duo-density" integral axial rotor consisting of injection molded RBSN blades and a hot pressed silicon nitride hub [186,187]. Däimler-Benz prepared an integral axial rotor completely of hot pressed silicon nitride by a combination of ultrasonic grinding and profile grinding [188,189]. Garrett AiResearch prepared individual blades by diamond grinding and ultrasonic machining [190]. All of these methods were very expensive and had little chance of becoming commercially viable. However, they did provide high quality rotors of 1975-vintage hot pressed silicon nitride for rig and engine testing.

As HIP, pressureless sintering, and overpressure sintering emerged as potential affordable near-net-shape fabrication approaches, a whole new set of challenges needed to be resolved. Control of shrinkage and distortion during sintering became major issues. Also, new fine particle size high purity powders became available that were important for achieving acceptable properties and density, but behaved rheologically much differently than the silicon and silicon carbide powders that had been used for RBSN and RSSC. As a result, extensive study of powder processing and rheology was conducted during the 1980s and 1990s for both silicon nitride and silicon carbide material systems. Many material and fabrication iterations were required, each of which had to be accompanied by detailed property characterization.

Many companies worldwide demonstrated the technology to fabricate prototype engine-quality ceramic hardware [14,15,191-198] of monolithic ceramics using injection molding, slip casting, pressure casting, CIP/green machining, and gel casting followed by HIP, overpressure sintering or sinter-HIP. During the 1990s additional efforts were conducted to fabricate prototype engine components from continuous fiber reinforced ceramic matrix composites [151-158], especially combustor liners and tip shrouds. In spite of the progress in monolithic ceramics and ceramic matrix composites, though, cost of fabrication and assurance of component reliability are still concerns that are impeding significant commercial introduction.

Quality Assurance

Quality assurance has been a concern and challenge from the earliest programs. A combination of certification samples fabricated (or cut from) each lot of components, non-destructive evaluation (NDE), proof testing, and rig or engine testing have been implemented to qualify ceramic turbine components.

Many efforts have focused on NDE. NDE that was available in 1970 included dye penetrants, conventional X-ray radiography, and low frequency ultrasonic c-scan [199,200]. Microfocus X-ray radiography and image enhancement of digitized images were introduced during the 1970s [201]. A variety of different ultrasonic methods including high frequency and surface waves also were evaluated in the 1970s [202,203]. CT was explored in the 1980s and refined throughout the 1990s [204,205]. Thermal imaging and laser scattering were also developed during the 1990s [206]. Although good progress has been made in NDE development and assessment, turbine companies still do not trust NDE exclusively for qualification of ceramic components. Virtually all of the turbine development programs have used a proof test for final accept/reject of highly stressed ceramic turbine components such as rotors and stators.

Conducting Rig, Engine and Field Testing

Since about 1970 many programs have been conducted worldwide to develop ceramic components for various sizes of gas turbine engines [14,207-218]. Table 1 lists some of the major programs that involved significant levels of rig and engine testing of ceramic components.

Table 1. Examples of Worldwide Programs to Develop Turbine Engines with Hot Section Ceramic Components

Program	Approximate Duration	Engine Companies
Brittle Material Design, High Temperature Turbine	1971-1979	Ford Motor Co., : 100 kW automotive. Westinghouse: 30+ MW stator vanes.
Ceramic Gas Turbine Engine Demonstration Program	1976-1981	AiResearch (later became Garrett Turbine Engine Co., then Allied-Signal Engines, then Honeywell Engines, Systems, and Services)
BMFT Ceramic Components for Vehicular Gas Turbines; Structural Ceramics Program	1974-1986	Daimler-Benz, MTU, Volkswagen
Ceramic Applications for Turbine Engines	1978-1983	Detroit Diesel Allison (later became Allison Gas Turbine Division of General Motors, then Allison Engine Co., then Rolls-Royce)
100 MW Reheat Cycle Gas Turbine	1978-1987	Ishikawajima-Harima Heavy Industries (IHI), Kawasaki Heavy Industries (KHI)
Advanced Gas Turbine (AGT) Programs	1979-1987	Detroit Diesel Allison, Garrett Turbine Engine Co.
300 kW Ceramic Gas Turbine Engine Programs	1984-1999	IHI, KHI, Yanmar Diesel
High Temperature Nozzle	1985-1989	MTU
15-20 MW Gas Turbines	1985-1996	CRIEPI, TEPCO, Toshiba, Hitachi, Mitsubishi Heavy Industries
Advanced Turbine Technology Applications Projects (ATTAP)	1987-1993	Allison Allied-Signal Engines
AGATA Program	1992-1998	United Turbines, Volvo Aero, ONERA C&C
APU Ceramic Turbine Engine Demonstration Projects	1992-present	Honeywell Engines, Systems, and Services
Cooled Silicon Nitride Vane	Early 1990s to present	UTRC/P&W
Advanced Turbine Systems Projects	1995-present	Rolls-Royce Allison General Electric (CMC components)
8000 kW Cogeneration	1999-present	KHI
Ceramic Matrix Composites for Advanced Engine Components	1999-2002	Siemens Westinghouse Power Corp.

Even though there have been many programs and substantial funding, the accumulation of rig and engine running time with ceramic hot section components has been painfully slow, as illustrated by the rough approximation in Figure 10.

17

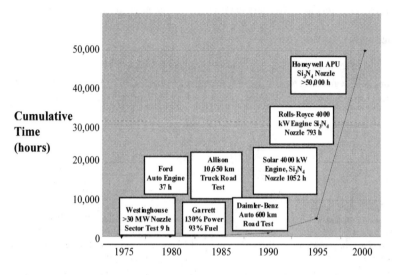

Figure 10. Curve Illustrating the Slow Rate of Accumulation of Engine Test Time for Monolithic Ceramics and Giving a Few Examples from Specific Programs

Many reasons can be cited for this for the slow accumulation of engine test time:

1. Initial materials had inadequate properties and required many development iterations before reliability was high enough to sustain engine tests.
2. Most engine development programs encountered design problems that required either modification or re-design, often requiring procurement of new tooling before the next set of ceramic hardware could be fabricated.
3. Initially there were no commercial suppliers with experience at fabrication of complex shaped ceramic engine components, and much time was required to establish the technology and experience.
4. Prototype turbine components were developmental and thus very expensive, so only very small quantity could be afforded by most programs.
5. Delivery times were very long because multiple development/fabrication iterations were invariably necessary to achieve ceramic prototype components, and these were often of marginal quality and failed during proof testing or rig testing.
6. Unexpected problems and failure modes were frequently encountered that required major material changes or design changes.

Because of these reasons and because the baseline technologies (design methods/codes, life prediction methods/codes, database generation, material improvement and new materials, fabrication development, quality assurance, understanding of failure modes) all had to be developed in parallel, progress occurred in plateaus. One technology would reach a level to allow engine testing in a certain temperature range and for a limited duration, but engine testing could not progress to the next plateau until another technology area caught up. Just when it appeared that one problem was resolved, longer test time or higher temperature testing would

18

identify a new problem that would require another plateau of time to study and try to resolve. Table 2 gives a rough idea of the nature of the plateaus, although in reality the path seemed more like a maze.

Table 2. Rough Approximation of the Plateaus of Materials and Technology Development

Technology Category	Plateau # 1: ~1968-1980	Plateau #2: ~1980-1992	Plateau #3: ~1992-1998	Plateau #4: ~1998-present
Materials	RBSN, HPSN, RSSC	Sinter, HIP, Over-pressure sinter	Same, but later generations; CMCs	Same with EBCs
Design and Life Prediction	Emergence of probabilistic; Fast fracture, minor time-dependent	Focus on time-dependent and contact stress; Experimental verification	Same plus oxidation and recession; CMCs	Same plus EBCs
Properties	3-pt. and 4-pt. flexure; Flexural stress rupture; elastic and thermal properties	Tensile strength, creep, and static fatigue; Standards	Long term (>7000 hours) tensile creep and static fatigue; CMCs	Same plus environmental degradation
Fabrication	Injection molding; slip casting; CIP/green machining; diamond grinding	Near-net-shape dense ceramics	Gel casting; cast and green machine; composite lay-up	CIP/green machine; Lay-up
NDE	Microfocus X-ray radiography with image enhancement; dye-penetrant; ultrasonic c-scan	Same with improvements	CT; thermal imaging; inspection of CMCs	Refinements; ability to inspect larger parts
Engine Testing	Short time (~2 h cycles); limited time accumulation	Longer time, greater time accumulation	>1000 h engine tests, including engines in field	Large increase in field test time
Key Failure Modes	Contact stress; impact	Same plus concerns about long term reliability	Surface recession; oxidation degradation	Same plus hot corrosion

Key Failure Modes

Of particular importance in Table 2 are the key failure modes that were encountered during rig and engine testing: contact stress, impact, inadequate time-dependent properties, oxidation, surface recession, and hot corrosion. These are discussed in detail in references [14] and [15] and in the literature cited in those references, but are reviewed briefly in subsequent paragraphs.

Contact Stress

Contact stress has proven to be a major problem in nearly every program. Ceramic materials have very high compressive strength and are thus very resistant to high Hertzian loads where the stress is 100% perpendicular (normal) to the surfaces. However, it is very difficult to design turbine components to achieve this simple normal loading; there always seems to be some tangential stress vector due to thermal expansion mismatch, temperature gradients in the engine, tolerance mismatches, friction, or relative movement (such as between a ceramic blade inserted into a metal disk slot). Whenever there is a biaxial stress condition, there is a localized increase in the tensile stress concentration that is often high enough to result in a surface crack in the ceramic [176].

Contact stress induced failures have occurred in most programs. The Ford Motor Company 820 automotive engine developed during the 1970s successfully operated for about 25 hours, but failed during shutdown when high friction at the curvic coupling interface of the ceramic rotor with the metal shaft resulted in a contact stress failure [19]. In the same Brittle Materials Design, High Temperature Turbine Program, Westinghouse encountered fractures at the interface between stator vanes and end caps in the 30 MW engine components rig testing [211]. Honeywell (and their predecessors) in their many programs encountered contact stress failures in various components: (1) points of attachment of static structure components, (2) attachment dovetail of ceramic blades inserted in metal disks, and (3) shrink fit attachment of a radial rotor to a metal shaft [20,163]. Solar Turbines had similar experiences with larger ceramic engine components [210].

These few examples are intended to illustrate that contact stress is a major issue that has been a substantial source of delay in successful implementation of ceramic components in gas turbine engines. Enormous effort has been expended trying to come up with designs that minimize contact stress, and good progress has been made. However, even a change in friction (increase) over time at a critical interface can result in unexpected failure. Contact stress still remains a reliability concern that will require extensive field testing of engines before engine companies are comfortable that it is no longer a critical failure mode.

Impact

Impact (foreign object damage --- FOD) is a special case of contact stress that has resulted in catastrophic engine failures [219]. Designers initially focused on configurations of ceramic airfoils that had thin cross sections and trailing edges similar to metal components. These were susceptible to damage by foreign objects and even by small pieces of carbon formed in the combustor. Impact resistant designs evolved with much thicker airfoils and trailing edges and generally with fewer blades or vanes.

Creep and Static Fatigue

Dense versions of silicon nitride (other than CVD) have a grain boundary phase that can soften at high temperature and limit creep and static fatigue (stress rupture) life. Early hot pressed and sintered versions could not be used very long at temperatures above 1000 °C. As shown earlier in Figure 7, many years were required to develop improved materials to meet the needs of higher turbine inlet temperatures (>1300 °C) and longer engine life. Early automotive engines only required 3500 hours life and only operated about 200 hours at peak operating temperature and stress. Industrial turbines under development in the 1990s required much longer life with a substantial percentage of the operation at near-peak conditions.

Oxidation/Corrosion/Recession

Silicon nitride and silicon carbide normally form a layer of silicon dioxide at the surface when exposed to an oxidizing atmosphere at high temperature. Based on furnace tests and early short time rig and engine tests, the silica layer appeared to provide an adequate barrier to further oxidation. Even during long term (but relatively low turbine inlet temperature) field testing of APUs during the 1990s, oxidation did not appear to be a problem for monolithic ceramics. Everyone's confidence was growing. However, high velocity burner rig studies at NASA and elsewhere suggested that the silica layers were not stable at higher temperature in turbine-simulating atmospheres [220]. Of special concern was the presence of water vapor in the combustion gases. Water vapor could have three detrimental effects: (1) enhance transport of impurities, (2) increase the oxidation rate due to solubility of water molecules in the silica, and (3) react with the silica to form a volatile silicon hydroxide product.

The APUs and many of the other smaller engines did not have high water vapor content in the combustion gas. However, when efforts began in the 1990s to integrate ceramics into an industrial engine at Solar Turbines, ceramic components were exposed for long time in a combustion environment at high pressure and with high water vapor content. The first of these components that were exposed for long time were SiC/SiC CMC materials for the combustor liners. Substantial oxidation and surface recession occurred during both a 948 hour field test and a 5028 hour field test, but did not result in component failure [157,210].

Evaluations during the late 1990s determined that a protective "environmental barrier coating" (EBC) would be required to have any chance of meeting engine company goals of 30,000 hours component life. Fortunately coating development was already well under way at United Technologies Research Center (UTRC) under the NASA High Speed Civil Transport (HSCT) program [218,221]. By November 2000, a 13,937 hour field test was conducted with a Solar engine containing SiC/SiC combustor liners coated with an experimental UTRC EBC [210]. Additional engine field tests are in progress with combustor liners with improved EBCs. Boroscope inspections indicate that the improved coated combustor liners are performing very well and should easily exceed the life of the earlier ones. Two alternate technologies, TBCs on metal combustor liners and oxide/oxide CMC combustor liners, also are in field test evaluation and are performing very well.

During 1999 a field test was conducted by Rolls-Royce Allison of monolithic silicon nitride first stage vanes in an industrial turbine engine [162]. The vanes were inspected after 815 hours of total engine operation and found to have a significant level of surface recession consistent with the silicon hydroxide volatilization mechanism. The consensus now is that silicon nitride will require an EBC for long life industrial turbine applications and probably for many other turbine applications.

One final challenge has emerged within the past year. Honeywell Engines, Systems, and Services has pursued an aggressive APU development effort with silicon nitride components since about 1990. APUs with silicon nitride nozzle vanes have accumulated thousands of hours in ground cart service and in commercial aircraft. They appeared to be performing well until problems arose for aircraft in commercial service in the Middle East. A combination of dust and sea salt contamination caused rapid and severe hot corrosion degradation of the silicon nitride vanes. This is under study, but may be another case where an EBC is required.

Technology Status and Prognosis

Achieving improvements in gas turbine engine efficiency through the use of hot section ceramic components has been a difficult challenge and has not yet reached commercial success. Much progress has been accomplished in design, life prediction, material property improvement, material characterization and standards, quality assurance, and resolution of key problems such as contact stress, impact, and thermochemical stability. But each time we seem to be close to success, a new problem or failure mechanism is encountered during field testing that requires new R&D to solve.

Another area of major advancement has been in fabrication of both monolithic and composite ceramic engine components. Monolithic silicon nitride components can now be produced consistently to turbine engine standards and are projected to be cost competitive as quantity production increases. There are still a couple important concerns, though: (1) can the existing manufacturers maintain their effort until a sustaining level of business is reached, or will they be forced to drop out like many earlier ceramic developers did and (2) can EBCs be successfully developed and cost-effective.

Ceramic matrix composites have been under development for many years less than the monolithic ceramics, but have progressed to successful field testing much more quickly. Since May of 1997 Solar Turbines has accumulated over 50,000 hours of industrial engine field testing with CMC combustor liners, with the high time liner exceeding 15,000 hours. Although the CMC liners have exhibited degradation, they have not catastrophically failed, thus allowing field testing to be used as a development tool.

The success of the Solar field tests has encouraged other companies to progress more quickly into field tests with CMC materials. General Electric Company has now successfully (during 2003) conducted first field tests of CMC tip shrouds in an F-class utility size turbine engine. The Solar and GE field tests are very encouraging, but there is still much material development and fabrication work that needs to be done to assure reliability and long life of CMCs with EBC coatings.

So, after all of these years of extensive funding and effort, we still have not reached our goal. Has the effort been worthwhile? Based only on the turbine results, one would probably say no. However, the efforts have greatly enhanced the technology of covalent silicon-based ceramics and have led to spin off of many other applications and markets as shown in Table 3. The impact and market growth of silicon nitride cutting tools and bearings has been especially significant.

Table 3. Applications That Emerged from Materials Technology Developed in Support of the Gas Turbine Engine Programs

Applications	Commercial Introduction
Silicon nitride and sialon cutting tool inserts	Late 1970s
Sintered SiC water pump seals for trucks	1980s
Sintered SiC for numerous other wear and corrosion resistance parts	1980s and 1990s
Silicon nitride diesel glow plugs	1981
Silicon nitride rocker arm wear pad for alternate fuel automobile engines	1984
Silicon nitride diesel swirl chamber	1986
Silicon nitride turbocharger rotors	1980s
Silicon nitride bearings	Late 1980s
Sintered SiC and RSSC for the electronics industry (integrated circuit fabrication)	1990s
Silicon nitride diesel cam follower rollers	1990s
Wide variety of silicon nitride parts for wear resistance applications	1990s
Silicon nitride for aluminum casting applications	1990s
Silicon nitride "seal runner" propulsion turbine engine ring seals for the LP and HP turbine shaft, the HP compressor shaft, and the fan gearbox.	Late 1990s

One item in Table 3, the silicon nitride "seal runner", is a demanding turbine engine component application and deserves further comment [222]. The ring seals range in diameter from 7.9-9.4 cm diameter and rotate at 30,100 rpm. They were introduced into Honeywell TFE731 aircraft propulsion engines in 1996 and have now been retrofitted into >90% of the fleet. They have accumulated over 20 million seal hours and have reduced unscheduled engine removal by >90%.

Conclusions

Applying ceramic materials to the hot section of gas turbine engines has been a very difficult challenge, and not all of the problems have yet been solved. Furthermore, we have faced a moving target. Metals engines were thought during the early 1970s to be reaching their temperature limits, but advances in cooling, ceramic thermal barrier coatings, and metals (single crystal for example) have increased turbine inlet temperature and engine efficiency even above the original goals thought only achievable with ceramics. Ceramics can still provide improvement, but significant challenges still remain to reach the desired high temperatures, reliability comparable to metal turbines, and component cost and life. On a positive note, much progress has been accomplished and impressive spinoff applications have resulted over the past 30 years.

Acknowledgements

The author gratefully acknowledges ASME Press for allowing generous extraction from references [16] and [17], including many figures.

References

1. Padture, N. P., Gell, M., and Jordan, E. H., "Thermal Barrier Coatings for Gas Turbine Engine Applications, *Science*, Vol. **296**, Apr. 12, 2002.
2. Richerson, D.W., *The Magic of Ceramics*, The American Ceramic Society, Westerville, OH, 2000.
3. Conway, H. M., *The Possible Use of Ceramic Materials in Aircraft Propulsion Systems*, NACA CB No. 4D10, 1944.
4. Morgan, W. C., and Deutsch, G. C., *Experimental Investigation of Cermet Turbine Blades in an Axial-Flow Turbojet Engine*, NACA TN-4030, Oct. 1957.
5. Collins, J. G., and Gerby, R. W., "New Refractory Uses for Silicon Nitride Reported," *J. Metals*, **7**, 612-615, 1955.
6. Parr, N. L., "Silicon Nitride, a New Ceramic for High Temperature Engineering and Other Applications," *Res.* (Lon.) **13**, 261-269, 1960.
7. Popper, P., and Ruddlesden, S. N., "The Preparation, Properties and Structure of Silicon Nitride," *Trans. Brit. Ceram. Soc.*, **60**, 603-626, 1961.
8. Deely, G. G., Herbert, J. M., and Moore, N. C., "Dense Silicon Nitride," *Powder Met.*, **8**, 145-151, 1961.
9. Popper, P., "The Preparation of Dense Self-Bonded Silicon Carbide," pp. 209-219 in *special Ceramics*, P. Popper ed., Academic Press, New York, 1960.
10. Godfrey, D. J., and Taylor, P. G., "Designing with Brittle Materials," *Eng. Mater. Design*, **12** [9], 1339-1342, 1969.
11. Godfrey, D.J., "The Use of Ceramics in Engines," *Prod. Brit. Ceram. Soc.*, **26**, 1-15, 1978.
12. Mueller, J. I., "Handicapping the World's Derby for Advanced Ceramics," *Am. Ceram. Soc. Bull.*, **61** [5], 588-591, 1982.
13. Kenney, G. B., and Bowen, H. K., "High Tech Ceramics in Japan: Current and Future Markets," *Am. Ceram. Soc. Bull.*, **62** [5], 590-596, 1983.
14. van Roode, M., Ferber, M.K., and Richerson, D.W., eds, *Ceramic Gas Turbine Design and Test Experience, Progress in Ceramic Gas Turbine Development, Volume I*, ASME Press, New York, 2002.
15. van Roode, M., Ferber, M.K., and Richerson, D.W., eds., *Ceramic Gas Turbine Component Development and Characterization, Progress in Ceramic Gas Turbine Development Volume 2*, ASME Press, New York, 2003.
16. Richerson, D.W., Iseki, T, Soudarev, A.V., and van Roode, M., "An Overview of Ceramic Materials Development and Other Supporting Technologies," Chapter 1 in reference 15, pp 1-30.
17. Richerson, D.W., Ferber, M.K., and van Roode, M., "The Ceramic Gas Turbine—Retrospective, Current Status and Prognosis," Chapter 29 in reference 15, pp 693-741.
18. Kossowsky, R., "Creep and Fatigue of Silicon Nitride as Related to Microstructure," in *Ceramics for High Performance Applications*, J.J. Burke, A.E. Gorum, and R.N. Katz, eds., Brook Hill Publishing, Chestnut Hill, MA, USA,1974. pp. 347-372.
19. McLean, A. F., and Fisher, E. A., *Brittle Materials Design, High Temperature Gas Turbine*, AMMRC TR 81-14, Final Report, March 1981.
20. Richerson, D. W., and Johansen, K. M., *Ceramic Gas Turbine Engine Demonstration Program*, Final Report DARPA/Navy Contract N00024-76-C-5352, May 1982.
21. Fairbanks, J.W. and Rice, R.W., *Proc. 1977 DARPA/NAVSEA Ceramic Gas Turbine Demonstration Engine Program Review*, MCIC Report MCIC-78-36, March 1978.
22. Evans, A. G. and Wiederhorn, S. M., "Crack Propagation and Failure Prediction in Silicon Nitride at Elevated Temperatures," *J. Mater. Sci.*, **9**, 270-278, 1974.
23. Evans, A.G., Russell, L.R., and Richerson, D.W., "Slow Crack Growth in Ceramic Materials at Elevated Temperatures," *Met. Trans. A.*, **6**A, 707-716, 1975.
24. Quinn, G. D., *Characterization of Turbine Ceramics after Long-Term Environmental Exposure*, AMMRC-TR-80-15, Final Report, April 1980.
25. Larsen, D. C., and Adams, J. W., *Property Screening and Evaluation of Ceramic Turbine Materials*, Final Technical Report AFWAL-TR-83-4141, April 1984.

26. Chuck, L., Goodrich, S. M., Hecht, N. L., and McCullum, D. E., "High-Temperature Tensile Strength and Tensile Stress Rupture Behavior of Norton/TRW NT-154 Silicon Nitride," *Ceram. Eng. Sci. Proc.*, **11**[7-8],1007-1027, 1990.

27. Ferber, M. K., Jenkins, M. G., and Tennery, V. J., "Comparison of Tension, Compression, and Flexure Creep for Alumina and Silicon Nitride Ceramics," *Ceram. Eng. Sci. Proc.*, **11**[7-8],1028-1045, 1990.

28. Cranmer, D.C. and Richerson, D.W., eds., *Mechanical Testing Methodology for Ceramic Design and Reliability*, Marcel Dekker, New York, 1998.

29. Sanders, W.A., and Probst, H.B., "High Gas Velocity Burner Tests on Silicon Carbide and Silicon Nitride at 1200 °C,: pp. 493-531 in *Ceramics for High Performance Applications*, J.J. Burke, A.E. Gorum, and R.N. Katz, eds., Brook Hill Publishing Co., Chestnut Hill, MA, 1974.

30. Carruthers, W.D., Richerson, D.W., and Benn, K.W., "Combustion Rig Durability Testing of Turbine Ceramics," pp. 571-595 in *Ceramics for High Performance Applications II*, E.M. Lenoe, R.N. Katz, and J.J. Burke, eds., Plenum Press, New York, 1983.

31. Schienle, J.L., *Durability Testing of Commercial Ceramic Materials*, Final Report DOE/NASA/0027-1, NASA CR-198497, Jan. 1996.

32. Liu, K.C. and Brinkman, C.R., "High Temperature Tensile Strength and Fatigue of Silicon Nitride," *Proc. 1989 Automotive Tech. Devt. Contractors Coord. Meeting*, Dearborn, Michigan, Oct. 23-26, 1989.

33. Quinn, G.D., and Wiederhorn, S.M., "Structural Reliability of Ceramics at Elevated Temperatures," *ASTM Symposium on Life Prediction Methodologies for Ceramic Materials in Advanced Applications—A Basis for Standards*, Cocoa Beach, Florida, January 11-13, 1993; *Ceram. Eng. and Sci. Proc.*, **14**[7-8], 305, 1993.

34. Williams, R.M. and Uy, J.C., "Ceramic Material Characterization," in *Ceramics for High Performance Applications-II*, J.J. Burke, E.N. Lenoe, and R.N. Katz, eds., Brook Hill Publishing, Chestnut Hill, MA, USA,1978. pp. 151-176.

35. Thümmler, F. and Grathwolh, G., "Creep of Ceramic Materials for Gas Turbine Application," in *Mechanical Properties of Ceramics for High Temperature Applications*, AGARD AD-A034 262, December 1976. pp. 1-26.

36. Davidge, R., Tappin, G., and McLaren, J., "Strength Parameters Relevant to Engineering Applications for Reaction bonded Silicon Nitride and REFEL Silicon Carbide," *Powder Metall. Int.*, **8**[3], 110-114, 1976.

37. Quinn, G.D., "Review of Static Fatigue in Silicon Nitride and Silicon Carbide," *Ceram. Eng. and Sci. Proc.*, **3**[1-2], 77-98, 1982.

38. Bunk, W., Gugel, E., and Walzer, P., "Overview of the German Ceramic Gas Turbine Program," in *Ceramics for High Performance Applications III Reliability*, E.M. Lenoe, R.N. Katz, and J.J. Burke, eds., Plenum Press, New York, NY, USA,1983. pp. 29-50.

39. Oda, I., Matsue, M., Soma, T., Masuda, M. and Yamada, N., "Fracture Behavior of Sintered Silicon Nitride under Multiaxial Stress States," *J. Ceram. Soc. Jpn. Inter. Ed.*, **96**, 523-530, 1988.

40. Lara-Curzio, E., "Properties of Continuous Fiber-Reinforced Ceramic Matrix Composites for Gas Turbine Applications," Chapter 22 in Reference 15, pp. 441-491.

41. Parr, N. L., Sands, R., Pratt, P.L., May, E. R. W., Shakespeare, C. R., and Thompson, D. S., "Structural Aspects of Silicon Nitride," *Powder Met.*, **8**, 152-163, 1961.

42. Parr, N. L., Martin, G. F., and May, E. R. W., "Preparation, Microstructure, and Mechanical Properties of Silicon Nitride," in *Special Ceramics*, P. Popper, ed., Academic Press, New York, NY, USA, 1960. pp 102-135.

43. Pratt, P. L., "The Microstructure and Mechanical Properties of Silicon Nitride," in *Mechanical Properties of Engineering Ceramics*, W. W. Kriegel and H. Palmour, eds. III, Interscience Publishers, New York, NY, USA, 1961. pp 507-519.

44. Noakes, P. B., and Pratt, P. L., "High-Temperature Mechanical Properties of Reaction-Sintered Silicon Nitride," in *Special Ceramics*, Vol. 5, Proceedings of the 5th International Symposium on Special Ceramics, held by the BCRA, Stoke-on-Trent, England, UK, 14-16 July 1970, P. Popper, ed., British Ceramic Research Association, Manchester, England, UK, 1972. pp. 299-310.

45. Cutler, I. B., and Croft, W. J., "Powder Met. Review 7: Silicon Nitride (Part 1), "*Powder Met. Int.* **6**, 92-96, 1974; "(Part 2)," *Powder Met. Int.* **6**, 144-146,1974.

46. Davidge, R. W., Evans, A. G., Gilling, D., and Wilyman, P. R., "Oxidation of Reaction-Sintered Silicon Nitride and Effects on Strength," in *Special Ceramics*, Vol. 5, *Proc. 5th Int. Symp. on Special Ceramics*, BCRA, Stoke-on-Trent, England, UK, 14-16 July 1970, P. Popper, ed., British Ceramic Research Association, Manchester, England, UK, 1972. pp 329-344.

47. F. L. Riley, ed., *Nitrogen Ceramics*, Noordhoff Int. Publishing, Leyden, The Netherlands, 1977.

48. Davidge, R. W., "Mechanical Properties of Reaction Bonded Silicon Nitride," in *Nitrogen Ceramics*, F.L.Riley, ed., Noordhoff Int. Publishing, Leyden, The Netherlands, 1977. pp 541-559.

49. Mangels, J. A., "Development of Injection Molded Reaction Bonded Si_3N_4," in *Ceramics for High Performance Applications II*, Proceedings of the Fifth Army Materials Conference, Newport , RI, USA, March 21-25, 1977, J. J. Burke, E. N. Lenoe, and R. N. Katz, eds., Brook Hill Publ., Chestnut Hill, MA, USA, 1978. pp. 113-130.

25

50.   Washburn, M. E. and Baumgartner, H. R., "High-Temperature Properties of Reaction-Bonded Silicon Nitride," in *Ceramics for High Performance Applications*, Proceedings of the Second Army Materials Conference, Hyannis, Mass, November 13-16, 1973, J. J. Burke, A. E. Gorum, and R. N. Katz, eds., Brook Hill Publishing Company, Chestnut Hill, MA, USA, 1974. pp. 479-491.

51.   Gersch, H., Mann, D., Rorabaugh, M., and Schuldies, J. J., "Slip-Cast and Injection-Molding Process Development of Reaction-Bonded Silicon Nitride at AiResearch Casting Company," pp. 313-340 in reference 21.

52.   Richerson, D. W., Smyth, J. R., and Styhr, K., "Material Improvement Through Iterative Process Development," *Ceram. Eng. and Sci. Proc.*, 4 [9-10], 841-852, 1983.

53.   Andersen, J.C., "Method of Making Refractory Body," U.S. Patent 2,938,807, May 31, 1960.

54.   Taylor, K. M., "Cold Molded Dense Silicon Carbide Articles and Method of Making the Same," U.S. Patent 3,205,043, Sept. 7, 1965.

55.   Fredricksson, J.I., "Process of Making Recrystallized Silicon Carbide Articles," U.S. Patent 2,964,823, Dec. 20, 1960.

56.   Forrest, C.W., "Manufacture of Dense Bodies of Silicon Carbide (Refel)," U.S. Patent 3,495,939, Feb. 17, 1970.

57.   Popper, P., US Patent 3,275,722, "Production of Dense Bodies of Silicon Carbide," Sept. 27, 1966. .

58.   Sawyer, G.R., and Page, T.F., "Microstructural Characterization of 'Refel' (Reaction –Bonded) Silicon Carbides," *J. Mater. Sci.*, 13, 885-904, 1978.

59.   Alliegro, R. A., and Coes, S. H., "Reaction bonded Silicon Carbide and Silicon Nitride for Gas Turbine Applications," ASME paper 72-GT-20, presented at the Gas Turbine Conference, San Franciso, CA, USA, March 26-30, 1972.

60.   Lucek, J. W., Torti, M.L., Weaver, G. Q., and Olson, B. A., "Cast Densified Silicon Carbides," SAE Paper 790253, 1979.

61.   Whalen, T. J., Noakes, J. E., and Terner, L. L., "Progress on Injection-Molded, Reaction-Bonded SiC," in *Ceramics for High Performance Applications II*, Proceedings of the Fifth Army Materials Conference, Newport, RI, USA, March 21-25, 1977, J. J. Burke, E. N. Lenoe, and R. N. Katz, eds., Brook Hill Publ., Chestnut Hill, MA, USA, 1978. pp. 179-189.

62.   Hucke, E.E., Process Development for Silicon Carbide Based Structural Ceramics, AMMRC TR 82-10, February 1982.

63.   Weaver, G.Q. and Olson, B.A., "Process for Fabricating Silicon Carbide Articles," U.S. Patent 4,019,913, April 26, 1977.

64.   Weaver, G.Q. and Logan, J.C., "Process for Forming Dense Silicon Carbide Bodies," U.S. Patent 4,127,629, Nov. 28, 1978.

65.   Lumby, R. J. and Coe, R. F., "The Influence of Some Process Variables on the Mechanical Properties of Hot-Pressed Silicon Nitride," *Proc. Brit. Ceram. Soc.*, 15, 91-101, 1970.

66.   Wild, S., Grieveson, P., Jack, K. H., and Latimer, M. J., "The role of Magnesia in Hot-Pressed Silicon Nitride," in *Special Ceramics*, Vol. 5, Proc. 5th Int. Symp. on Special Ceramics, BCRA, Stoke-on-Trent, England, 14-16 July 1970, P. Popper, ed., British Ceramic Research Association, Manchester, England , UK, 1972. pp. 377-384.

67.   Wild, S., Grieveson, P., and Jack, K. H., "The Crystal Chemistry of New Metal-Silicon-Nitrogen Ceramic Phases," ibid., pp. 289-298.

68.   Bowen, L. J., Weston, R. J., Carruthers, T. G., and Brook, R. J., "Mechanisms of Densification During the Pressure Sintering of Alpha-Silicon Nitride," *Ceramurgia Int.*, 2[4], 173-176, 1976.

69.   Richerson, D. W., "Effects of Impurities on the High Temperature Properties of Hot-Pressed Silicon Nitride," *Am. Ceram. Soc. Bull.*, 52 [7], 560-62, 1973.

70.   Richerson, D.W. and Washburn, M.E., "Hot Pressed Silicon Nitride," U.S. Patent 3,836,374, Sept. 17, 1974.

71.   Gazza, G. E., "Hot-Pressed Silicon Nitride," *J. Am. Ceram. Soc.*, 56[12], 662, 1973.

72.   Weaver, G.Q. and Lucek, J.W., "Optimization of Hot-Pressed Si$_3$N$_4$-Y$_2$O$_3$ Materials," *Am. Ceram. Soc. Bull.*, 57[12], 1131-1134, 1136,1978.

73.   van Roode, M., "The German National Program for Ceramic Materials and Component Development," Chapter 9 in Reference 15, pp. 175-198.

74.   Tsuge, A. and Nishida, K., "High Strength Hot Pressed Silicon Nitride with Concurrent Y$_2$O$_3$ and Al$_2$O$_3$ Additions," *Amer. Ceram. Soc. Bull.* 57[4], 424-426, 431, 1978.

75.   Tsuge, A., Nishida, K., and Komatsu, M., "Effect of Crystallizing the Grain-Boundary Glass Phase on the High-Temperature Strength of Hot-Pressed Silicon Nitride Containing Yttria," *J. Am. Ceram. Soc.*, 58, 323-326, 1975.

76.   Tsuge, A., Nishida, K., and Komatsu, M., "High-Temperature Strength of Hot-Pressed Si$_3$N$_4$ Containing Y$_2$O$_3$," *J. Amer. Ceram. Soc.* 58[3], 323-326,1975.

77.   Baumgartner, H. R. and Richerson, D. W., "Inclusion Effects on the Strength of Hot-Pressed Si$_3$N$_4$," in *Fracture Mechanics of Ceramics*, Vol. 1, R. C. Bradt, D. P. H. Hasselman, and F. F. Lange, eds.,Plenum Press, New York, NY, USA, 1974. pp. 367-386.

78.   Mitomo, M., "Pressure Sintering of Silicon Nitride," *J. Mater. Sci.*, 11,1103-1107, 1976.

26

79. Osipova, I.I. and Pogorelova, D.A., "Recrystallization of Silicon Nitride During Hot Pressing," *Sov. Powder Met. Metal Ceram.*, **14**, 1015-1017,1975.

80. Iskoe, J.L., Lange, F.F., and Diaz, E.S., "Effect of Selected Impurities on the High Temperature Mechanical Properties of Hot_Pressed Silicon Nitride," *J. Mater. Sci.*, **11**, 908-912, 1976.

81. Colquhoun, I., Thompson, D.P., Wilson, W.I., Grieveson, P., and Jack, K.H., "The Determination of Surface Silica and Its Effect on the Hot-Pressing Behavior of Alpha-Silicon Nitride Powder," *Proc. Brit. Ceram. Soc.*, **22**, 181-195,1973.

82. Knoch, H. and Ziegler, G., "Influence of MgO Content and Temperature on Transformation Kinetics, Grain Structure and Mechanical Properties of Hot-Pressed Silicon Nitride," *Science of Ceramics 9*, 494-501,1977.

83. Weston, R.J. and Carruthers, T.G., "Kinetics of Hot-Pressing of Alpha-Silicon Nitride Powder with Additives, " *Proc. Br. Ceram. Soc.*, **22**, 197-206,1973.

84. Fickel, A.F. and Kessel, H., "Technology and Applications of Dense Silicon Nitride," *Glas. Email. Keram. Tech.*, **25**, 193-196, 1974.

85. Gugel, E., Fickel, A.F., and Kessel, H., "Developments in the Production of Hot-Pressed Silicon Nitride," *Powder Met. Int.* **6**, 136-140, 1974.

86. Oyama, Y. and Kamigaito, O., "Sintered Silicon Nitride – Magnesium Oxide System," *Yogyo Kyokai Shi*, **81**, 290-293,1973.

87. Bowen, L.J., Carruthers, T.G., and Brook, R.J., "Hot-Pressing of Si_3N_4 with Y_2O_3 and Li_2O Additives," *J. Amer. Ceram. Soc.*, **61** [7-8] 335-339, 1978.

88. Tsuge, A., Kudo, H., and Komeya, K., "Reaction of Silicon Nitride and Yttrium Oxide in Hot-Pressing," *J. Amer. Ceram. Soc.* **57** [6] 269-270, 1974.

89. Rice, R.W. and McDonough, W.J., "Hot Pressed Silicon Nitride with Zr Based Additives," *J. Amer. Ceram. Soc.* **58** [5-6] 264,1975.

90. Huseby, I.C. and Petzow, G., "Influence of Various Densifying Additives on Hot-Pressed Silicon Nitride," *Powder Met. Int.*, **6**, 17-19, 1974.

91. Weston, J.E., Pratt, P.L., and Steele, B.C.H., " Crystallization of Grain boundary Phases in Hot-Pressed Silicon Nitride Materials, Parts 1 and 2," *J. Mat. Sci.*, **13**, 2137-2156,1978.

92. Lange, F.F., "Relation Between Strength, Fracture Energy and Microstructure of Hot-Pressed Silicon Nitride," *J. Am. Ceram. Soc.*, **56**[10], 518-522, 1973.

93. Adlerborn, J., and Larker, H.T., "Method of Manufacturing Bodies of Silicon Nitride," US Patent 4,455,275, 1974

94. Larker, H., Adlerborn, J. and Bohman, H., "Fabricating of Dense Silicon Nitride Parts by Hot Isostatic Pressing," SAE Paper 770335, 1977.

95. Larker, H.T., "Hot Isostatic Pressing of Ceramic Turbine Components to Shape—Experience at Asea High Pressure Laboratory, Asea Cerama AB, ABB Cerama AB and AC Cerama AB," Chapter 10 in Reference 15, pp. 199-210.

96. Wills, R., Brockway, M.C., McCoy, L., and Niesz, D., "Preliminary Observations on the Hot Isostatic Pressing of Silicon Nitride," *Ceram. Eng. Sci. Proc.*, 1[7-8B], 534-539, 1980.

97. Richerson, D.W. and Wimmer, J.M., "Properties of Isostatically Hot-Pressed Silicon Nitride," *J. Am. Ceram. Soc.*, **66**[9], C-173 – C-176, 1983.

98. Heinrich, J. and Böhmer, M., in *Science of Ceramics*, Vol. 11, R. Carlsson and S.K. Carlsson, eds., Swedish Ceramic Soc., Göteborg, Sweden, 1981. pp. 439-446.

99. Moritoki, M., "HIP Activities in Japan," *Metal Powder Report*, 38[7],393-396, 1983.

100. Ishii, T., Tsuzuki, H., and Inoue, Y., "Modular-type, Hot-loading-type and Compact Hot Isostatic Pressing Systems," *Proc. 2nd Int. Conf. On Isostatic Pressing*, Vol. 2, 1982. pp. 20.1-20.24

101. Fujikawa, T., Moritoki, M. and Homma, K., "The High Temperature HIP Furnace and Its Application in Densification of Ceramics," ibid., Vol. 1, 1982. pp.12.1-12.25.

102. Homma, K., Tatuno, T, Okada, H., Kawai, N., and Nishihara, M., "HIP Treatment of Sintered Silicon Nitride,"*Zairyo-Journal of the Soc. Of Mater. Sci. Japan*, **30**, 1005-1011, 1981 (in Japanese with English summary).

103. Larker, H.T., "On Hot Isostatic Pressing of Shaped Ceramic Parts," in *Ceramic Components for Engines*, S. Sōmiya, E. Kanai and K. Ando, eds., Elsevier Science Publishing, New York, NY, USA, 1986. pp. 304-310.

104. Fujikawa, T., Moritoki, M., Kando, T., Homma, K. and Okada, H., "Hot Isostatic Pressing: Its Application in High Performance Ceramics," ibid., pp. 425-433.

105. Hirota, K., Ichikizaki, T., Hasegawa, Y.T., and Suzuki, H., "Densification of Silicon Nitride by Hot Isostatic Press in N_2 Atmosphere," ibid, pp. 434-441.

106. Ziegler, G. and Wötting, "Post-treatment of Pre-sintered Silicon Nitride by Hot Isostatic Pressing," *Int. J. High Tech. Ceram.*, **1**, 31-58, 1985.

27

107. Richerson, D.W., Bright, E., and Licht, R., "Ceramic Component Development at Saint-Gobain/Norton Advanced Ceramics," Chapter 3 in Reference 15, pp. 55-75.
108. Pollinger, J.P., "The Development of Monolithic Silicon Nitride Structural Ceramics at Honeywell," Chapter 5 in Reference 15, pp. 97-123.
109. McEntire, B.J., Taglialavore, A.P., Heichel, D.N., Johnson, J.W., Bright, E., and Yeckley, R.L., "ATTAP Ceramic Component Development Using Taguchi Methods," *Proc. 26th Auto. Tech. Dev't Contractors Coord. Meeting*, Dearborn, MI, USA, Oct. 24-27, 1988, SAE Publication P-219, April 1989. pp. 289-300.
110. Hengst, R.R., Heichel, D.N., Holowczak, J.E., Taglialavore, A.P., and McEntire, B.J., "A Comparison of Forming Technologies for Ceramic Gas-Turbine Engine Components," ASME paper 90-GT-184, presented at the Gas Turbine and Aeroengine Congress and Exposition, Brussels, Belgium, June 11-14, 1990.
111. Pujari, V.K., Amin, K.E., and Tewari, P.H., "Development of Improved Processing and Evaluation of Silicon Nitride," ASME paper 91-GT-317, presented at the Int. Gas Turbine and Aeroengines Congress and Exposition, Orlando, FL, USA, June 3-6, 1991.
112. Pujari, V.K., and Tracey, D.M., "Processing Methods for High Reliability Silicon Nitride Heat Engine Components," ASME paper 93-GT-319, presented at the Int. Gas Turbine and Aeroengine Congress and Exposition, Cincinnati, OH, USA, May 24-27, 1993.
113. Terwilliger, G. R., "Properties of Sintered Silicon Nitride," *J. Am. Ceram. Soc.*, **57**[1], 48-49, 1974.
114. Mitomo, M., Tsutsumi, M., Bannai, E., and Tanaka, T., "Sintering of Silicon Nitride," *Am. Ceram. Soc. Bull.* **55**[3], 313, 1976.
115. Masaki, H., and Kamigaito, O., "Pressureless Sintering of Silicon Nitride with Additions of MgO, Alumina, and/or Spinel," *Yogyo Kyokai Zasshi* (J. Ceram. Soc. Japan), **84**[10], 508-512, 1976 (in Japanese).
116. Oda, I., Kaneno, M., and Yamamoto, N., "Pressureless Sintered Silicon Nitride, in *Progress in Nitrogen Ceramics*, *Proc. NATO-ASI Nitrogen Ceramics Conf.*, Canterbury, England, UK, Aug. 16-27, 1976. pp. 359-365.
117. Greskovich, C., and Rosolowski, J. H., "Sintering of Covalent Solids," *J. Am. Ceram. Soc.*, **59**[7-8], 336-343, 1976.
118. Priest, H. F., Priest, G. L., and Gazza, G. E., "Sintering of Si_3N_4 under High Nitrogen Pressure," *J. Am. Ceram. Soc.*, **60**[1-2], 81, 1977.
119. Rowcliffe, D. J., and Jorgensen, P. J., "Sintering of Silicon Nitride, in *Proc. Workshop on Ceramics for Heat Engines*, Energy Res. & Dev. Adm., CONRT, Orlando, FL, USA, Jan. 24-26, 1977. pp. 191-196.
120. Tani, E., Umebayashi, S., Kishi, K., Kobayashi, K., and Nishijima, M., "Gas-pressure Sintering of Silicon Nitride with Concurrent Addition of Alumina and 5 wt.% of Rare Earth Oxide: High Fracture Toughness Silicon Nitride with Fiber-like Structure," *Am. Ceram. Soc. Bull.*, **65**[9], 1311-1315, 1986.
121. Matsuhiro, K. and Takahashi, T., "The Effect of Grain Size on the Toughness of Sintered Silicon Nitride," Ceram. Eng. and Sci. Proc., **10**[7-8], 807-816, 1989.
122. Li, C. W. and Yamanis, J., "Super-tough Silicon Nitride with R-Curve Behavior," *Ceram. Eng. and Sci. Proc.*, **10**[7-8], 632-645, 1989.
123. Smith, J. T. and Quackenbush, C. L., "A Study of Sintered Silicon Nitride Compositions with Yttria and Alumina Densification Additives," in *Proc. Int. Symp. on Factors in Densification and Sintering of Oxide and Nonoxide Ceramics*, Hakone, Japan, Oct. 3-6, 1978, S. Sōmiya and S. Sayto, eds., Association of Science Documents Information, Tokyo Institute of Technology, 1979. pp. 426-443.
124. Mangels J. A., "Sintered Reaction-Bonded Silicon Nitride," *Ceram. Eng. Sci. Proc.*, **2** [7-8] 589-593, 1981.
125. Giachello, A., and Popper, P., "Post-Sintering of Reaction-Bonded Silicon Nitride," *Ceramurgia Int.* **5** [3],110, 1979.
126. Takahashi, T., "Sintering Character of Yttria-Magnesia-Zirconia Doped and Yttria-Ytterbia Doped Silicon Nitrides," *Proc. of the Symposium on the Sintering of Advanced Ceramics*, May 3, 1988, American Ceramic Society, Westerville, OH, 1989.
127. Hampshire, S., *The Sintering of Nitrogen Ceramics*, Parthenon Press, Lancashire, England, 1986.
128. Lumby, R.J., North, B., and Taylor, A.J., "Properties of Sintered Sialons," in *Ceramics for High Performance Applications II*, J.J. Burke, E.N. Lenoe, and R.N. Katz, eds., Brook Hill Publishing, Chestnut Hill, MA, USA, 1978. pp. 893-906.
129. Lumby, R.J., "The Preparation, Structure and Properties of commercial Sialon Ceramic Materials," *Ceram. Eng. and Sci. Proc.*, **3**[1-2], 50-66, 1982.
130. Alliegro, R.A., Coffin, L.B., and Tinklepaugh, J.R., "Pressure-Sintered Silicon Carbide," *J. Am. Ceram. Soc.*, **39**[11], 386-389, 1956.
131. Marshall, R.C., Faust, J.W., and Ryan, C.E., "Silicon Carbide – 1973," Univ. of South Carolina Press, Columbia, SC, 1974.
132. Lange, F.F., "Hot-Pressing Behavior of Silicon Carbide Powders with Additions of Aluminum Oxide," *J. Mater. Sci.*, **10**, 314-320, 1975.
133. Prochazka, S., "Dense Polycrystalline Silicon Carbide," U.S. Patent 3,954,483, May 4, 1976.

28

134. Prochazka, S., "Silicon Carbide Sintered Body; Boron Additive," U.S. Patent 4,041,117, Aug. 8, 1977.
135. Coppola, J.A. and Smoak, R.H., "Method of Producing High Density Silicon Carbide Product; Sintering Silicon Carbide with Boron or Boron Compounds Present as Densification Aids," U.S. Patent 4,080,415, March 21, 1978.
136. Stroke, F., "Submicron Beta Silicon Carbide Powder and Sintered Articles of High Density Prepared Therefrom," U.S. Patent 4,133,689, Jan. 9, 1979.
137. Lipp, A. and Schwetz, K.A., "Dense Sintered Shaped Articles of Polycrystalline Alpha-Silicon Carbide and Process for Their Manufacture," U.S. Patent 4,230,497, Oct. 28, 1980.
138. Coppola, J.a., Hailey, L.N. and McMurtry, C.N., "Process for Pruducing Sintered Silicon Carbide Ceramic Body," U.S. Patent 4,124,667, Nov. 7, 1978.
139. Boecker, W., Landfermann, H. and Hausner, H., "Sintering of Alpha Silicon Carbide with Additions of aluminum," *Powder Met Int.*, **11**[2], 83-85, 1979.
140. Tanaka, H. and Inomata, Y., "Normal sintering of Al-doped Beta SiC," *J. Mater. Sci. Letters*, **4**, 315-317, 1985.
141. Omori, M. and Takei, H., "Pressureless Sintering of SiC," *J. Am. Ceram. Soc.*, **65** [6], C-92, 1982.
142. Omori, M. and Takei, H., "Method for Preparing Sintered Shapes of Silicon Carbide," U.S. Patent 4,564,490, Jan. 14, 1986.
143. Negita, K., "Effective Sintering Aids for Silicon Carbide Ceramics: Reactivities of Silicon Carbide with Various Additives," *J. Am. Ceram. Soc.*, **69**[12] C-308 – C-310, 1986.
144. Prochazka, S., "The Role of Boron and Carbon in the Sintering of Silicon Carbide," *Special Ceramics* **6**, 171-181, 1974.
145. Schwetz, K.A. and Lipp, A., "The Effect of Boron and Aluminum Sintering Additives on the Properties of Dense Sintered Alpha Silicon Carbide," *Science of Ceramics*, **10**, 149-158, 1980.
146. Lamicq, P.J. and Jamet, J.F., "Thermostructural CMCs: an Overview of the French Experience," in *High-Temperature Ceramic-Matrix Composites I: Design, Durability, and Performance*, A.G. Evans and R. Naslain, eds., American Ceramic Society, Westerville, OH, USA,1995. pp. 1-11.
147. Johnson, D.D. and Sowman, H.G., "Ceramic Fibers," in *Composites, Engineered Materials Handbook* Vol 1, ASM Int., Metals Park, OH, 1987. pp. 60-65.
148. Vasilos, T., "Structural Ceramic Composites," ibid, pp. 925-932.
149. Richerson, D.W., "Ceramic Matrix Composites,", Chapter 19 in *Composites Engineering Handbook*, P.K. Mallick, ed., Marcel Dekker, New York, NY, USA, 1997. pp. 983-1038.
150. Belitskus, D., *Fiber and Whisker Reinforced Ceramics for Structural Applications*, Marcel Dekker, New York, 1993.
151. *High Temperature / High Performance Composites*, F.D. Lemkey, S.G. Fishman, A.G. Evans, and J.R. Strife, eds., *MRS Proc.*, Vol. **120**, Mater. Res. Soc., Pittsburgh, PA, USA, 1988.
152. *High-Temperature Ceramic-Matrix Composites, I and II*, A.G. Evans, and R. Naslain, eds., American Ceramic Society, Westerville, OH, USA, 1995.
153. Landini, D.J., Fareed, A.S., Wang, H., Craig, P.A. and Hemstad, S., "Ceramic Matrix Composites Development at GE Power Systems Composites, LLC," Chapter 14 in Reference 15, pp. 277-289.
154. Corman, G.S., Luthra, K.L. and Brun, M.K., "Silicon Melt Infiltrated Ceramic Composites – Processes and Properties," Chapter 16 in Reference 15, pp. 291-312.
155. Hiromatsu, M, Nishide, S., Nishio, K., Igashira, K., Matsubara, G. and Suemitsu, T., "Development of Hot Section Components Composed of Ceramic Matrix Composites (CMC) at the Research Institute of Advanced Materials Gas Generator (AMG)," Chapter 17 in Reference 15, pp. 313-329.
156. Kameda, T., Suyama, S., Itoh, Y., Nishida, K., Ikeda, I., Hijikata, T. and Okamura, T., "Development of continuous fiber-Reinforced Reaction Sintered Silicon Carbide Matrix Composites for Gas Turbine Hot Parts Applications," Chapter 18 in Reference 15, pp. 331-351.
157. More, K. L., Tortorelli, P. F., Ferber, M. K., Walker, L. R., Keiser, J. R., Brentnall, W. D., Miriyala, N., and Price, J. R., "Exposure of Ceramics and Ceramic Matrix Composites in Simulated and Actual Combustor Environments," ASME paper 99-GT-292, presented at the International Gas Turbine and Aeroengine Congress and Exposition, Indianapolis, Indiana, USA, June 7-10, 1999.
158. Szweda, A., Easler, T.E., Jurf, R.A. and Butner, S.C., "Ceramic Matrix Composites for Gas Turbine Applications," Chapter 15 in Reference 15, pp.277-289
159. Hartsock, D.L., "Ford's Development of the 820 High Temperature Ceramic Gas Turbine Engine," Chapter 3 in Reference 14, pp. 17-75.
160. Paluszny, A., and Wu, W., "Probabilistic Aspects of Designing with Ceramics," *Transactions of the ASME, Journal of Engineering for Power*, **99**[4], 617-630, 1977.
161. McClean, A. F., and Hartsock, D. L., "Design with Structural Ceramics," in *Structural Ceramics*, J. B. Wachtman, Jr., ed., *Treatise of Material Science and Technology*, Academic Press, Inc., Boston, MA, USA, Vol. 29, 27-97, 1989.

29

162. Khandelwal, P and Heitman, P., "Ceramic Gas Turbine Development at Rolls-Royce in Indianapolis," Chapter 5 in Reference 14, pp. 111-132.

163. Schenk, B., Easley, M.L., and Richerson, D.W., "Evolution of Ceramic Turbine Engine Technology at Honeywell Engines, Systems & Services," Chapter 4 in Reference 14, pp. 77-109.

164. Gyekenyesi, J.P., "SCARE: a Post Processor Program to MSC/NASTRAN for Reliability Analysis of Structural Ceramic Components," *Transactions of the ASME, Journal of Engineering for Gas Turbine Power*, **108**[3], 540-546, 1986.

165. Nemeth, N.N., Powers, L.M., and Janosik, L.A., "Time-dependent Reliability Analysis of Monolithic Ceramic Components Using the CARES/LIFE Integrated Design Program, *ASTM Symposium on Life Prediction Methodologies for Ceramic Materials in Advanced Applications—A Basis for Standards*, Cocoa Beach, Florida, Jan. 11-13, 1993; *Cer. Eng. and Sci. Proc.*, **14**[7-8], 318-319, 1993.

166. Comfort, A. M., and Cuccio, J. S., "Life Prediction Methodology for Ceramic components of Advanced Heat Engines," in *Proc. 27th Automotive Tech. Devt. Contractors' Coord. Meeting*, Dearborn, MI, USA, Oct. 23-26, 1989, SAE P-230, Society of Automotive Engineers, Warrenville, PA, USA, 1990. pp. 275-284.

167. Schenk, B., Brehm, P.G., Menon, M.N., Tucker, W.T., and Peralta, A.D., "Status of the CERAMIC/ERICA Probabilistic Life Prediction Codes Development for Structural Ceramic Applications," ASME paper 99-GT-318, presented at the International Gas Turbine and Aeroengine Congress and Exposition, Indianapolis, IN, USA, June 7-10, 1999.

168. Schenk, B., Brehm, P.G., Menon, M.N., Tucker, W.T., and Peralta, A.D., "A New Probabilistic Approach for Accurate Fatigue Data Analysis of Ceramic Materials," *Transactions of the ASME, Journal of Engineering for Gas Turbines and Power*, **122**[4], 637-645, 2000.

169. Bornemisza, T and Saith, A., "SPSLIFE: A User Friendly Approach to the Structural Design and Life Assessment of Ceramic Components," ASME paper 94-GT-486, presented at the International Gas Turbine and Aeroengine Congress and Exposition, The Hague, The Netherlands, June 13-16, 1994. *Transactions of the ASME, Journal of Engineering for Gas Turbines and Power*, **118**[1], 179-183, 1996.

170. Bornemisza, T. and Jones, A., "Ceramic Gas Turbine Development at Hamilton Sundstrand Power Systems," Chapter 6 in Reference 14, pp. 133-154.

171. Teramae, T., Hamada, S., and Hamanaka, J., "Probabilistic Structural analyses of Ceramic Gas Turbine Components," *IUTAM Symposium on Probabilistic Structural Mechanics—Advances in Structural Reliability Methods*, 1994. pp. 518-533.

172. Furuse, Y., Teramae, T., Tsuchiya, T., Maeda, F., Tsukuda, Y., and Wada, K., "Application of Ceramics to a Power Generating Gas Turbine," Chapter 17 in Reference 14, pp. 381-402.

173. Stürmer, G., Schultz, A., Wittig, S., "Life Time Prediction for Ceramic Gas Turbine Components," ASME paper 91-GT-96, presented at the 36th ASME International Gas Turbine & Aeroengine Congress and Exposition, Orlando, FL, USA, June 3-6, 1991.

174. Schulz, A., "New Design Principles for Ceramic Hot Gas Components of Gas Turbines," Chapter 25 in Reference 14, pp. 555-582.

175. Duffy, S.F., Janosik, L.A., Thomas D.J., Wereszczak, A.A., and Lamon, J., "Life Prediction of Structural Components," Chapter 25 in Reference 15, pp. 553-606.

176. Richerson, D.W., Finger, D.G., and Wimmer, J. M., "Analytical and Experimental Evaluation of Biaxial Contact Stress," in *Fracture Mechanics of Ceramics, Vol. 5, Surface Flaws, Statistics and Microcracking*, R. C. Bradt, A.G. Evans, D.P.H. Hasselman, and F.F. Lange, eds., Plenum Press, New York, NY, USA, 1983. pp. 163-184.

177. Reed, J.S., *Principles of Ceramic Processing*, Wiley, New York, 1988.

178. Richerson, D.W., *Modern Ceramic Engineering: Properties, Processing, and Use in Design, 2nd Ed.*, Chapters 9-12, Marcel Dekker, Inc., New York, 1992.

179. Mangels, J.A., "Ceramic Technology Development at Ford Motor Company from 1967 to 1986," Chapter 2 in Reference 15, pp. 31-53.

180. Johnson, C.R. and Mohr, T.G., "Injection Molding 2.7 g/cc Silicon Nitride Turbine Rotor Blade Rings Utilizing Automatic Control," in *Ceramics for High Performance Applications-II*, J.J. Burke, E.N. Lenoe and R.N. Katz, eds., Brook Hill Publishing, Chestnut Hill, MA, USA, 1978. pp. 193-206.

181. Richerson, D.W., Smyth, J.R., and Styhr, K, *Low-Cost Net-Shape Ceramic Radial Turbine Program*, Final Report AMMRC TR 85-2 under contract DAAG46-81-C-0006, May 1985.

182. Ezis, A., "The Fabrication and Properties of Slip-Cast Silicon Nitride," in *Ceramics for High Performance Applications*, J.J. Burke, A.E. Gorum and R.N. Katz, eds., Brook Hill Publishing, Chestnut Hill, MA, 1974. pp. 207-222.

183. Ezis, A and Neil, J. T., Fabrication and Properties of Fugitive Mold Slip-cast Si_3N_4, *Am. Ceram. Soc. Bull.*, **58**[9], 883 (1979).

184. Torti, M.L., Baker, S.H., Tuffs, S and Olson, B.A., "Norton Process Optimization Studies for Hot Pressed Silicon Nitride Blades and Reaction Sintered Silicon Nitride Stator and Static Components," pp. 341-363 in reference 21.

185. Burke, J.J., Lenoe, E.N., and Katz, R.N., Eds., *Ceramics for High Performance Applications—II*, Brook Hill Publishing, Chestnut Hill, MA and the Metals and Ceramics Information Center, Columbus, OH, 1978, pp. 445-547.

186. Uy, J.C., Williams, R.M. and Goodyear, M.U., "Processing of Hot Pressed Silicon Nitride," in *Ceramics for High Performance Applications – II*, J.J. Burke, E.N. Lenoe and R.N. Katz, eds., Brook Hill Publishing, Chestnut Hill, MA, USA, 1978. pp. 131-149.

187. Baker, R.R., Ezis, A., Hartsock, D.L. and Goodyear, M.U., "Developments in Press bonding of Duo-Density Rotors," ibid., pp. 207-230.

188. Bunk, W., "Overview on the German Program on Ceramic Componoents for Vehicular Gas Turbines," in *Ceramic Components for Engines*, Proc. 1st Int. Symp. 1983, Japan, S. Sōmiya, E. Kanai and K. Ando, eds., Elsevier Applied Science, London, 1986. pp. 9-20.

189. Mörgenthaler, K.D. and Neubrand, F., "Thermal Shock Testing of a Hot-Pressed Silicon Nitride Turbine Rotor," in *Ceramics for High Performance Applications III Reliability*, E.M. Lenoe, R. N. Katz and J.J. Burke, eds., Plenum Press, New York, NY, USA, 1983. pp. 805-810.

190. Robare, M.W. and Richerson, D.W., "Rotor Blade Machining Development," pp. 291-311 in Reference 21.

191. McEntire, B. J., Taglialavore, A. P., Heichel, D. N., Bright, E., Yeckley, R. L., Holowczak, J. E., and Quackenbush, C. L., "Silicon Nitride Component Development for Advanced Gas Turbine Engines," in *Proc. 27th Automotive Tech. Devt. Contractors' Coord. Meeting*, Dearborn, MI, USA, Oct. 23-26, 1989, SAE P-230, Society of Automotive Engineers, Warrendale, PA, USA, 1990. pp. 341-356.

192. Pollinger, J. P., and Busovne, B. J., "Development of Pressure Slip Cast Silicon Nitride Rotors for ATTAP," ibid., pp. 357-364.

193. Funk, J. E., Lawler, H. A., Ohnsorg, R. W., and Storm, R. S., "ATTAP SiC Ceramic Component Development Year 2," ibid., pp. 319-326.

194. Neil, J., Bandyopadhyay, G., Sordelet, D., and Mahoney, M., "Fabrication of Silicon Nitride ATTAP Components at GTE Laboratories," ibid., pp. 303-310.

195. Pasto, A. E. and Natansohn, S., "Advanced Processing of High-Performance Silicon Nitride Ceramics," ibid., pp. 197-206.

196. McEntire, B.J., Hengst, R.R., Collins, W.T., Taglialavore, A.P., and Yeckley, R.L., "Ceramic Component Processing Development for Advanced Gas Turbine Engines," ASME paper 91-GT-120 presented at the Int. Gas Turbine and Aeroengine Congress and Exposition, Orlando, FL, USA, June 3-6, 1991. *J. Engineering for Gas Turbines and Power* **115**[1], 1-8, 1993.

197. Pollinger, J. P., "Improved Silicon Nitride Materials and Component Fabrication Processes for Aerospace and Industrial Gas Turbine Applications," ASME paper 95-GT-159, presented a the International Gas Turbine and Aeroengine Congress and Exposition, Houston, TX, USA, June 5-8, 1995.

198. Bright, E., Burleson, R., Dynan, S.A., and Collins, W.T., "NT 164 Silicon Nitride Gas-Turbine Engine Turbine Blade Manufacturing Development," ASME Paper 95-GT-74, presented at the Int. Gas Turbine and Aeroengine Congress and Exposition, Houston, TX, USA, June 5-8, 1995.

199. Cassidy, D. J., "NDE Techniques Used for Ceramic Turbine Rotors," pp. 231-242 in reference 185.

200. Schuldies, J. J., and Richerson, D. W., "NDE Approach, Philosophy, and Standards for the DARPA/NAVSEA Ceramic Turbine Program," pp. 381-402 in reference 21.

201. Schuldies, J. J., and Spaulding, W. H., "Radiography and Image Enhancement of Ceramics," ibid., pp. 403-428.

202. Schuldies, J. J., and Derkacs, T., "Ultrasonic NDE of Ceramic Components," ibid., pp. 429-448.

203. Kino, G. S., Khuri-Yakub, B. T., Murakami, Y., and Yu, K. H., "Defect Characterization in Ceramics Using High Frequency Ultrasonics," in *Proceedings of the ARPA/AFML Review of Progress in Quantitative Nondestructive Evaluation*, July 17-21, 1978, La Jolla, California, Scripps Institute of Oceanography, CA, USA, 1979. pp. 242-245.

204. Ellingson, W. A., and Vannier, M. W., *X-Ray Computed Tomography for Nondestructive Evaluation of Advanced Structural Ceramics*, Argonne National Laboratory Report ANL-87-52, Argonne National Laboratory, Argonne, IL, USA, 1987.

205. Ellingson, W.A., Ikeda, Y. and Goebbels, J., "Nondestructive Evaluation/Characterization," Chapter 23 in reference 15, pp.493-519.

206. Sun, J. G., Shirber, M., and Ellingson, W. A., "Laser-Based Optical Scattering Detection of Surface and Subsurface Defects in Machined Silicon Nitride Components," *Ceram. Eng. Sci. Proc.* ,**18**[4], 273-280, 1997.

207. Richerson, D.W., and Anson, D., "Evolution of Ceramic Gas Turbine Development Programs at Engine Manufacturers in the United States," Chapter 2 in reference 14, pp. 11-16.

208. Itoh, T, and Sugawara, A., "Evolution of Ceramic Gas Turbine Development Programs at Engine Manufacturers in Japan," Chapter 12 in reference 14, pp. 277-281.

209. van Roode, M., "Evolution of Ceramic Gas Turbine Development Programs at Engine Manufacturers in Western Europe," Chapter 20 in reference 14, pp. 445-452.

210. Brentnall, W.D., van Roode, M., Norton, P.F., Gates, S., Price, J.R., Jimenez, O., and Miriyala, N., "Ceramic Gas Turbine Development at Solar Turbines Incorporated," Chapter 7 in reference 14, pp. 155-192.

211. Morrison, J., Lane, J., and Burke, M., "Progress in Ceramic Gas Turbine Development at Westinghouse/Siemens Westinghouse," Chapter 8 in reference 14, pp. 193-223.

212. Itoh, T., Yoshida, Y., Sasaki, S., Sasaki, M., and Ogita, H., "Japanese Automotive Ceramic Gas Turbine Development," Chapter 13 in reference 14, pp. 283-303.

213. Sasa, T., Tanaka, S., and Mikami, T., "Development of 300 kW-Class Regenerated Single-Shaft Ceramic Gas Turbine CGT301," Chapter 14 in reference 14, pp. 305-329.

214. Tatsumi, T., "Development of the 300 kW Ceramic Gas Turbine CGT302," Chapter 15 in reference 14, pp. 331-360.

215. Arakawa, S., and I. Ohashi, "Development of 300 kW Class Ceramic Gas Turbine CGT303," Chapter 16 in reference 14, pp. 361-380.

216. Helms, E.H., Lindgren, L.C., Heitman, P.W., and Thrasher, S.R., *Ceramic Applications in Turbine Engines*, Noyes Publications, Park Ridge, NJ, USA, 1986.

217. Lundberg, R. and Gabrielsson, R., "Ceramic Components for Automotive Gas Turbines – Experience at Volvo," Chapter 22 in reference 14, pp.499-512.

218. Holowczak, J., Jarmon, D.C., Mendelson, M. I., and Linsey, G.D., "United Technologies Pratt & Whitney/United Technologies Research Center Experiences with Ceramics and CMCs in Turbine Engines," Chapter 10 in reference 14 pp. 237-260.

219. Peralta, A.D., and Yoshida, H., "Design of Impact-Resistant Ceramic Structural Components," Chapter 28 in reference 15, pp. 665-692.

220. Jacobson, N.S., Fox, D.S., Smialek, J.L., Opila, E.J., Tortorelli, P.F., More, K.L., Nickel, K.G., Hirata, T., Yoshida, M., and Yuri, I., "Corrosion Issues for Ceramics in Gas Turbines," Chapter 26 in reference 15, pp. 607-640.

221. Lee, K.N., Fritze, H., and Ogura, Y., "Coatings for Engineering Ceramics," Chapter 27 in reference 15, pp. 641-664.

222. Boyd, G.L., Moy, J., and Fuller, F., "Hybrid Ceramic Circumferential Carbon Ring Seal," SAE Paper 2002-01-2956, Society of Automotive Engineers, Warrendale, PA, 2002.

DEVELOPMENT OF THE 8000KW CLASS HYBRID GAS TURBINE

Takao Sugimoto, Yoshihiro Ichikawa,
Hitoshi Nagata, and Kenichiroh Igashira
Kawasaki Heavy Industries, Ltd.
1-1 Kawasaki-cho
Akashi, 673-8666 Japan

Sazo Tsuruzono, and Takero Fukudome

Kyocera Corporation
1-4 Yamashita-cho
Kokubu, 899-4396 Japan

ABSTRACT

Based on the successful result of the Japanese national project for 300 kW class ceramic gas turbine development (CGT: 1989-1999FY)[1-7], the New Energy and Industrial Technology Development Organization (NEDO) started "Research and Development on Practical Industrial Co-generation Technology" program (NEDO-HGT: 1999-2003FY)[8-10], funded by the Ministry of Economy, Trade and Industry (METI). The objective of this project was to investigate a high efficient co-generation system that employs "Hybrid Gas Turbine (HGT)". The HGT means a new gas turbine using both metallic and ceramic parts for its high-temperature section.

In this R&D activity, ceramic material evaluation test and long-term engine operation test of the HGT have been demonstrated. This paper gives some results from the R&D of HGT.

INTRODUCTION

HGT was developed by applying ceramic components to the existing commercial 7000 kW class gas turbine (KAWASAKI M7A-02). The ceramic components were used for its high temperature stationary parts, such as combustor liners, transition ducts, and the first stage turbine nozzles.

The target performance of HGT is shown in Table 1. The turbine inlet temperature (TIT) of HGT is higher than the M7A-02, so the thermal efficiency and output are higher.

Long-term engine test was started from 2002FY at the Kawasaki Heavy Industries Akashi Works.

Figure 1 shows the development organization of NEDO-HGT program. Kawasaki Heavy Industries, Ltd. (KHI) takes charge of the research on the reliability of HGT engine. Kyocera corporation takes charge of the development and evaluation of the heat-resistant ceramic component.

Table 1   Target performance of HGT

	Target
Output Power	8000 kW class
Thermal Efficiency	34 % or higher
Turbine Inlet Temp.	1523 K class

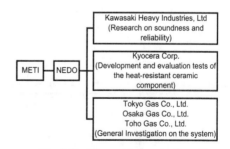

Fig. 1 Development organization of NEDO-HGT program.

Figure 2 shows the overview of the M7A-02 and the ceramic components for HGT. The combustor liners, the transition ducts, and the first stage turbine nozzles are replaced with the ceramic components. Silicon nitride SN282 of Kyocera was selected as the material for the ceramic components; the material was used in the CGT-302 type ceramic gas turbine in CGT project and proved its high temperature durability.

The development schedule is shown in Figure 3. The basic design, detail design and engine manufacturing were carried from 1999FY to 2001FY. Long-term engine tests were started from 2002FY. Partial-load (-5000 kW) tests and full-load (8000 kW) test were carried out.

PRESENT STATUS OF THE DEVELOPMENT
Material Characteristic
1) Static Strength

Kyocera SN282 was selected for the ceramic components such as the combustor liners, the transition ducts, and the first stage turbine nozzles; the material was used in the CGT302 project and proved its high temperature durability. Figure 4 shows flexure strengths of SN282 test sample from room temperature to high temperatures. High temperature strengths of test sample that were cut out from actual components were about 500 MPa as shown in Table 2, and were equivalent to those of standard test sample.

2) Creep Strength

To find the dominant fracture mechanism of SN282 at high temperature, creep rupture tests were carried out at 1523 K, where is the turbine inlet temperature of HGT. The longest lifetime in

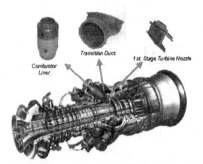

Fig. 2   Overview of the Hybrid Gas-turbine(HGT) and its ceramic component.

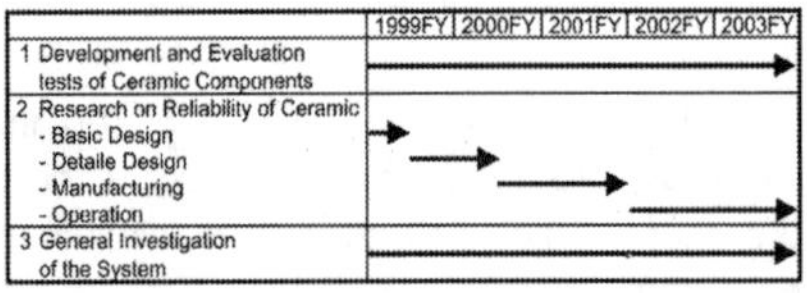

Fig. 3   Development schedule of NEDO-HGT.

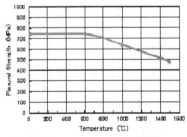

Fig. 4   Flexure strength of SN282.

Table 2   Flexure strength of cut out sample at 1673 K

Combustor Liner (MPa)	1st Turbine Nozzle (MPa)
511	498

34

these tests was 11482 hours under 385 MPa stress, but no inclusions or defects were observed in the fracture origin. There was almost no creep strain until the sample was ruptured and the fracture surface did not indicate creep damage. So, it was considered that the dominant fracture mechanism of SN282 at 1523 K is not the creep, but sub-critical crack growth (SCG).

3) Cyclic Fatigue Strength

Figure 5 shows the S-N diagram of SN282 at 1523 K, 1623 K and 1723 K. The cyclic fatigue tests were carried out by sine wave of 10 Hz, and trapezoidal wave with 1 sec. and 100 sec. hold time. Fatigue strength at 1523 K is in the variation of the static strength, but at higher temperatures, SN282 shows the time dependent strength degradation. The fatigue strength of the sample with 100 s. hold time was lower than that with 1 sec. hold time at 1623 K and 1723 K. Employing the effective loading time (T_e) as shown in Figure 6, the strength of the sample with 1 sec. and 100 sec. hold time were almost the same at 1623 K. However, at higher temperature as 1723 K, the strengths of the sample with 1 sec. and 100 sec. hold time seemed to be different. This result suggests that the fracture mode change from the sub-critical crack growth(SCG) to the creep damage failure.

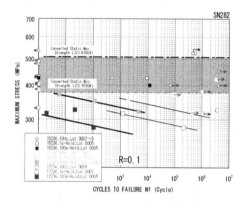

Fig. 5   S-N diagram of SN282.

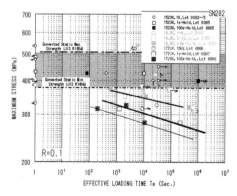

Fig. 6   S-T_e diagram of SN282.

35

4) Sub-critical Crack Growth

To extrapolate the long-term time dependent strength caused by SCG, we tried to evaluate the SCG rate of SN282 by the observation of actual crack tip before and after loading test. At 1473 K and 1523 K in Ar, a constant flexural load was given to the test sample with two pre-cracks that were introduced by Vickers indentor. Failure should occur from one pre-crack, then SCG could be seen at the tips of survived pre-crack. SCG rate (a_{SCG}/t_f) against the mean stress intensity factor $K_{Imean} \left(= \frac{K_I(a_0 + a_{SCG}) - K_I(a_0)}{2} \right)$, both are calculated from the SCG length from SEM observation of the crack tip, are plotted in Figure 7.

5) Exposure Test[11]

The conditions of exposure test simulate the combustion gas of HGT are shown in Table 3. The exposure tests were made for atmost 30 hours. For SN282 without EBC, recession rate was estimated about 0.5 μm/h from the result of the exposure tests as shown in Figure 8. It is a serious problem for the industrial gas turbine applications which are required long-term operation as several thousand hours at least. On the other hand, EBCed SN282 exhibited a good resistance against the recession. This type of EBC was employed to the actual SN282 parts in HGT.

Design of Engine[12]
1) Insulation Structures for Mounting Ceramic Parts

The thermal expansion of the ceramic material is much smaller than that of metallic material, and the fracture toughness of it is low, so that ceramic parts can not be fixed tightly to the metallic supporting parts. The elastic-supporting structure using coil spring and the independent-supporting structure were employed based on the experience in CGT302. On the other hand, the HGT has horizontally-separated casing, so that the seal structure must be changed from that of CGT302. The ceramic segment seals were employed for HGT.

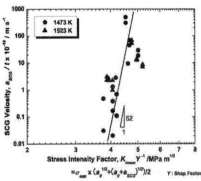

Fig. 7 SCG rate at 1473 K and 1523 K in Ar.

Table 3   Condition of exposure test

Temperature	1523 K
Gas Flow Velocity	110 m/s
Gas Pressure	1.6 MPa
H₂O Partial Pressure	120 kPa
Fuel	Methane

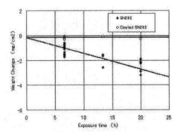

Fig. 8 Weight changes of SN282 after exposure to combustion gas.

2) Dry Low Emission Combustor

Dry Low Emission Combustor (DLE) using ceramic combustor liner was designed for HGT. The dimension of the ceramic combustor liner was similar to existing metallic one for M7-02A, but it did not have cooling air holes. To reduce NOx emissions, lean pre-mix combustion and supplemental combustion system was employed. To accommodate the difference of thermal expansion between the ceramic liner and the metal casing, the elastic-supporting structure using coil springs was employed.

Operation Study
1) Shakedown Operations

HGT was run in non-load or partial-load operations for about 400 minutes prior to full-load operation.

At the 1st operation in non-load condition, cracking was occurred in outer shroud of the ceramic turbine nozzles. This cracking was made by the irregular contact between metallic supporting parts and ceramic turbine nozzles. The irregular contact was prevented by improving the mounting structure and optimizing the initial clearance between the ceramic turbine nozzles and the metallic supporting parts.

In a partial-load operation, small chipping was occurred in the tab of the ceramic turbine nozzle as shown in Figure 9. Each ceramic nozzle was elastically fixed by clamping the single tab, so the gas force in operation could incline the nozzles. Therefore the contact manner that was originally designed in area contact between the ceramic tab and metallic supporting parts was changed; in this case, it was changed to point contact at the edge of the tab. So the grasping force given by coil springs was increased and the metallic supporting part was redesigned not to give the point contact even when the nozzles were inclined in operating condition.

2) Full-load Operations

In a full-load operation, HGT exhibited the target performance as shown in Table 4. HGT had already run for 1000 hours in full-load operation.

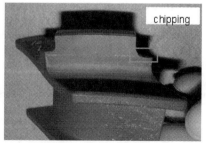

Fig. 9 Cracked ceramic turbine nozzle
after operation.

Table 4   Performance of HGT in full-load
operation

	Target	Result
Output Power	8000 kW class	8333 kW
Thermal Efficiency	34 % or higher	34.1 %
Turbine Inlet Temp.	1523 K class	1514 K

37

SUMMARY

1. HGT is a 8000 kW class gas turbine using ceramics for high temperature static component as combustor liners, transition ducts, and 1st stage turbine nozzles.
2. Elastic supporting structure were employed for the ceramic components from the experience of CGT302.
3. Considering a long-term operation, data of time dependent strength, such as creep and low cyclic fatigue properties of SN282 are accumulating. To supporting these creep and LCF data, SCG rate was measured by the actual observation of crack tip.
4. Exposure test that simulates the HGT condition gave a severe recession for SN282, and developed EBC could improve the recession resistance of it.
5. HGT could exhibit the target performans as 34 % in thermal efficiency.
6. Imporving the supporting structure for ceramic parts, HGT had run for 1000 hours.

ACKNOWLEDGMENTS

This study is proceeding under the Japanese HGT R&D Program conducted by NEDO. The authors wish to express their gratitude to METI and to NEDO for making this study possible and permitting this paper to be published.

REFERENCES

[1] I. Takehara, et al, 1996,"Research and Development of Ceramic Gas Turbine (CGT302)", ASME96-GT-477.
[2] A. Okuto, et al, 1998,"Development of a Low NOx Combustor for 300kW-Class Ceramic Gas Turbine (CGT302)", ASME98-GT-1272.
[3] "Research and Development of Ceramic Gas Turbine (300kW class) 1998 Annual Report", 1999, NEDO Japan.
[4] T. Tatumi, et al, 1999, Development Summary of the 300kW Ceramic Gas Turbine CGT302", ASME1999-GT-105.
[5] I. Takehara, 1999, "Research and Development of CGT302 Ceramic Gas turbine", JSME No.994-2pp.9-17.
[6] I. Takehara, et al, 1999, Development summary of CGT302 Ceramic Gas Turbine", IGTC '99 in KOBE, Vol.1pp.57-64.
[7] I. Takehara, et. al, 2000, "Summary of CGT302 Gas Turbine Research and Development Program", ASME2000-GT-644.
[8] "Research and Development on Practical Industrial Cogeneration Technology 1999 Annual Report", 2000, NEDO Japan.
[9] "Research and Development on Practical Industrial Cogeneration Technology 2000 Annual Report", 2001, NEDO Japan.
[10] "Research and Development on Practical Industrial Cogeneration Technology 2001 Annual Report", 2002, NEDO Japan.
[11] T. Fukudome, et al, 2002, "Development and Evaluation of Ceramic Components for Gas Turbine", ASME2002-GT-30627.
[12] R. Tanaka, et al, 2001, "Development of the Hybrid Gas Turbine", ASME2001-GT-0515.

DEVELOPMENT AND EVALUATION OF CMC VANE FOR NGSST ENGINE

Akira KAJIWARA*, Takeshi NAKAMURA*, Takahito ARAKI* and Hiroshige MURATA**

Ishikawajima-Harima Heavy Industries Co., Ltd.
*3-5-1, Mukodai-cho, Nishitokyo-shi, Tokyo 188-8555 JAPAN
**1, Shinnakahara-cho, Isogo-ku, Yokohama-shi, Kanagawa 235-8501 JAPAN

ABSTRACT

Ceramic matrix composites (CMCs) exit guide vane (EGV) has been investigated and evaluated in ESPR (Research and Development of Environmentally Compatible Propulsion System for Next-Generation Supersonic Transport) program in Japan. CMCs are attractive materials for hot section parts in jet engine because they are lightweight and have excellent high temperature capabilities. However, the durability of conventional CMCs has not been enough at elevated temperature and the parts life has been limited. The material durability represented by creep rupture property has been modified in this study and the strength was improved 70% at intermediate temperature (1273K), comparing with the CMCs that were developed in previous work. Near net shape process technology is also needed in order to fabricate complex vane shape. Near net manufacturing of CMC vane has been developed and the fabrication technology was evaluated by tensile strength using cut up specimens from the prototype vane. The results of cut up specimens test showed appropriate strength that was estimated from flat plate specimens.

INTRODUCTION

ESPR project started in 1999JFY has developed the basic technology for a practical, environmental friendly, economical propulsion system for civil supersonic transport. Aiming at reduce environmental pollution, one of these main objectives in ESPR project is reduction CO_2. In order to reduce CO_2, improvement of engine efficiency is necessary. Therefore reduction of engine weight, increasing of turbine inlet temperature and reduction of the amount of turbine cooling air are needed.

Application of CMCs to engine part is a very efficient approach on improvement in engine performance because CMCs have significant advantage in lightweight and high heat resistance, compared to Ni based superalloys. SiC fiber reinforced SiC composites (SiC/SiC) have the high potential of thermal stability, but there is a technical issue which is improvement in oxidation resistant at elevated temperature. Development of preform structure is also a major issue in order to apply SiC/SiC to engine part. The purpose of this study is to improve the high temperature durability of SiC/SiC and to develop fabrication technology. While acquiring of the material characteristic data

100mm

Fig. 1 Schematic view of CMC vane

and modifying of the matrix manufacturing process, development of ESPR engine scaled CMC parts has been carried out. [1] Modification of matrix forming process and development of CMC parts, especially CMC vane, are provided in this paper. A Schematic view of CMC vane is shown in Fig. 1.

MODIFICATION OF MATRIX MANUFACTURING PROCESS

When CMC vane is exposed to hot exhaust gas, its temperature reaches 1273K to 1473K. The strength of SiC/SiC such as creep strength could be degraded by oxidation of its interface layer in such high temperature. In order to reduce of strength degradation, modification of matrix manufacturing process has been studied. The modification of process was to add inhibitor which is boron compound and transforms to borate glass at high temperature. Borate glass could seal matrix crack and intercept oxygen. High temperature durability of modified SiC/SiC was evaluated by comparing with non-modified SiC/SiC in creep strength.

TEST SPECIMEN

The architecture of the specimen is shown in Table 1. The specimen has a 3-D orthogonal fabric to it described as X:Y:Z = 1:1:0.4 and fiber content is 40 volume percent. The first step in manufacturing of the specimen, carbon interface layer was deposited by chemically vapor infiltration (C-CVI). Then, SiC matrix was achieved in two steps that consist of a sequential process of CVI (SiC-CVI) followed by a polymer impregnation and pyrolysis (PIP). Inhibitor was added before SiC-CVI, on PIP and after PIP to modify of manufacturing process.

Table 1 The architecture of the specimen

Fiber	Weave structure	Fiber ratio and volume fraction
Tyrano-ZMI[TM]	3D orthogonal	X:Y:Z=1.0:1.0:0.1
		Volume fiber: 40%

CREEP RUPTURE TEST

Creep rupture test was performed at 1473K in uniform heating condition and local heating condition in order to confirm the creep strength at uniform and nonuniform temperature. Figure 2 shows the creep rupture test results. The results in uniform heating condition are described the circle symbols, and the results in local heating condition are described the triangular symbols. The creep rupture strength at 1473K for 100Hr was improved 50MPa as compared with SiC/SiC without inhibitor addition in uniform heating condition (square symbols). [1]

While the creep strength in local heating condition showed 40% less as compared with the results in uniform heating condition. The tested specimen after testing in local heating condition is shown in Fig. 3. The specimen has fractured outside gage section and it was confirmed that the fracture position was kept at 1223K to1273K during test. This temperature range is equivalent to the severest intermediate temperature that CMC vane will be exposed in ESPR engine operation. These results means that crack sealing has not effected because inhibitor has not transformed to glass (or inhibitor had high viscosity unless transforming to glass) at intermediate temperature. The temperature distribution of the specimen was confirmed with thermal paint. The specimen is shown in Fig. 4, which was exposed at 1473K in local heating condition.

From the test results, the creep strength was improved by inhibitor but prevention of the strength degradation at intermediate temperature was necessary. Therefore, borosilicate glass was selected for external coating candidate because of its crack sealing effect at intermediate temperature. Then, the coated specimens were tested at 1473K in local heating condition again. The test results are described the diamond symbols in Fig. 2. It revealed that the performance of the coated specimen is equivalent as that of the non-coated specimen in uniform heating condition. The specimen after testing is shown in Fig. 5. The specimen was fractured within gage section which was kept at 1473K.

From these results, creep strength degradation at intermediate temperature was prevented by borosilicate glass coating and it was confirmed that the creep strength was improved 70%, comparing with only inhibitor added SiC/SiC.

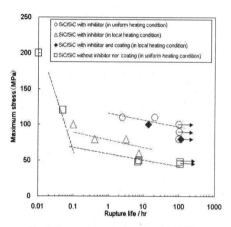

Fig. 2 Creep rupture test results at 1473K

Fig. 3 Non-coated specimen after testing
(Fractured position is out of gage section)

Fig. 4 Thermal painted specimen after exposing
at 1473K in local heating condition

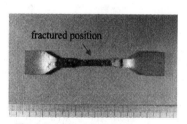

Fig. 5 Coated specimen after testing
(Fractured position is within gage section)

DEVELOPMENT OF CMC VANE

CMC vane that has been developed in this study consists of three sections; hollow airfoil, inner band and outer platforms. The fabric of CMC vane was tried to manufacture with following procedure; braided airfoil was integrated to cloth plied inner and outer platforms by stitching. Densification process was the same as flat plate specimen manufacturing process. Prototype CMC vane of ESPR engine form is shown in Fig. 6. It was confirmed in previous works that CVI and PIP matrices were well infiltrated in tows by microstructure observation. [2] Ultimate tensile strength (UTS) was confirmed with the specimens cut up from CMC vane in order to evaluate the strength at braiding and cloth plying combined section.

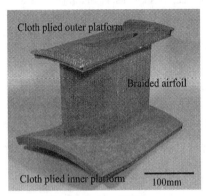

Fig. 6 Proto type CMC vane

EVALUATION OF UTS AT INNER BAND

The UTS at braiding and cloth plying combined section of inner platform was evaluated with two kinds of the specimens that dimensions are 70x10x6tmm. Schematic view of the specimens cut up location is shown in Fig. 7 and the cross sections of each specimen are shown in Fig. 8. The specimen 'A' has the structure that braiding is sandwiched with cloth plying. The specimen 'B' has the structure that braiding and cloth plying are sandwiched with cloth plying and has discontinuity at braiding and cloth plying steped-lap section. Tensile test was performed by loading in parallel direction of engine axis at room temperature.

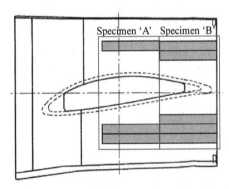

Fig. 7 Schematic view of specimens cut up location of inner band for tensile strength

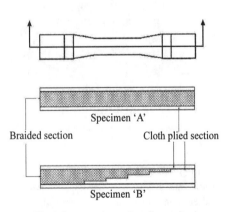

Fig. 8 Cross sections of tensile specimens

42

The specimens 'A' and 'B' after testing are shown in Fig. 9. The specimen 'A' was fractured at middle of the specimen and the specimen 'B' was fractured along the steped-lap. The typical load-displacement curves are shown in Fig. 10. The estimated strength that was calculated from fiber volume fraction in tensile direction is 185MPa and it is also shown in Fig. 10. The average strength of the specimen 'A' was 170MPa and this result was almost equivalent as the estimated strength. Therefore, it was confirmed that the inner platform was well infiltrated and have expected strength. While the average strength of specimen 'B' was 90MPa and this results showed to be about half strength of the specimen 'A'. Therefore, it was confirmed that the strength of steped-lap section was the half because of its fiber discontinuity, so that sufficient consideration will be needed in specifying the criteria for part design.

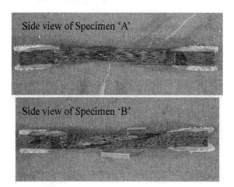

Fig. 9 Cut up specimens after testing

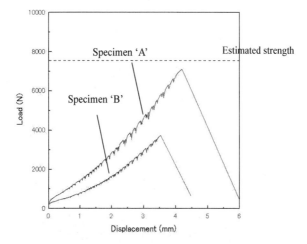

Fig. 10 Load-displacement curve of tensile

SUMMARY

The creep strength was improved by inhibitor and glass coating. And the strength degradation at intermediate temperature was not appeared by glass coating.

The fabrication technology of CMC vane was evaluated by tensile strength using two types of cut up specimens from the prototype vane. The test results of specimen 'A' that has the estimable strength showed the expected strength. And the test results of specimen 'B' that has fiber discontinuity showed to be about half strength of specimen 'A'.

ACKNOWLEDGEMENTS

The authors would like to express their thanks to the New Energy and Industrial Technology Development Organization (NEDO) and the Ministry of Economy, Trade and Industry (METI), who gave them opportunity to conduct "Research and Development of Environmentally Compatible Propulsion System for Next-Generation Supersonic Transport (ESPR) project".

REFERENCES

[1] K. Watanabe, et al., Research of CMC material application to HPT components; The First International Symposium of Environmentally Compatible Propulsion System for Next-Generation Supersonic Transport, (2002) C1

[2] K. Watanabe, et al., Development of CMC vane for gas turbine engine; 27th Annual Cocoa Beach Conference and Exposition on Advanced Ceramics and Composites, (2003)

44

CERAMIC COMBUSTOR DESIGN FOR ST5+ MICROTURBINE ENGINE

Jun Shi, Venkata Vedula, Ellen Sun, David Bombara, John Holowczak, William Tredway,
Alex Chen, Catalin Fotache
United Technologies Research Center
411 Silver Lane, East Hartford, CT 06108

ABSTRACT
Ceramic combustor liners require no film cooling on the hot-side and minimal backside cooling, due to their high temperature capability. The hotter combustor walls lead to minimal "wall quenching" and better carbon monoxide (CO) oxidation because less cooling is applied. A ceramic combustor can was designed for the ST5+ microturbine engine to improve the engine performance and emissions. The can was designed with special attention to attachment methods that minimize thermal stresses due to the large difference in thermal expansion coefficients between metallic and ceramic materials. Detailed thermal and stress analyses were performed to guide design decisions that lead to optimal component reliability and manufacturability. This paper gives a detailed account of the design and analysis associated with the ceramic combustor can.

INTRODUCTION
World population growth and the rise in living standards have increased the demand for electricity substantially over the last two decades. The increased demand for electric power has put on a strain over the electricity grids, most of which are dated and over-stretched for their capacity. This is best exemplified by recent massive blackout in North America highlighted the vulnerability of the current electric grid system for reliable distribution of electric power[1]. The blackout on August 14, 2003, affected 50 million people in eight U.S. states and eastern Canada. Billed as the worst blackout in America's history, it cost at least $6 billion in economic and other losses.

The U.S. Department of Energy (DOE) had foreseen the potential problems with antiquated power distribution systems and initiated an Advanced Microturbine Systems program[2]. Under this program, electric generators driven by high efficiency microturbine engines are developed to provide electric power local to power demands, thereby relieving the reliance on electricity grids that are already congested and are costly to maintain and upgrade. United Technologies Research Center and Pratt and Whitney recognized the market potential for microturbine engines for distributed power generation and were awarded a DOE contract to develop a high thermal efficiency (>40%) and low installed cost (<$500/kW) ST5+ microturbine with a power output of 400kW.

A key improvement to the advanced microturbine is application of structural ceramics to hot gas path components: combustor, first stage turbine vanes, integrally bladed rotor (IBR), and turbine tip shroud. Together, these ceramic components account for 3% increase in engine thermal efficiency. A picture of the ST5+ microturbine with its major components is shown in Figure 1. Engine specifications and design of turbine components have been published previously[3]. This paper describes the design and analysis of the ceramic combustor can.

Combustors in gas turbine engines are subjected to severe thermal loading generated by the combustion process. The current metal combustor can in gas turbines are either film cooled or impingement cooled (more recent engines) to maintain structural integrity under severe thermal loading. In both cases, the cooled combustor wall leads to local quenching of CO burnout, thereby resulting in high CO emissions. In addition, active film cooling of the combustor wall introduces flow losses (pressure loss) that reduce the thermal efficiency of the engine. These effects become more pronounced in a lean premixed combustor system that operates at a lower temperature and has very limited air available for liner cooling. By using high temperature ceramic materials, CO wall quenching due to film or impingement cooling, as well as associated flow losses can be eliminated[4]. Since a ceramic combustor wall operates at a higher temperature than a metal combustor wall, it helps to stabilize combustion process, especially at part-load conditions for lean premixed combustion, which is prone to combustion instability and flameout.

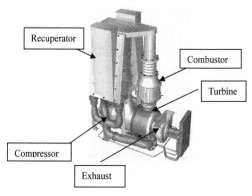

Figure 1: ST5+ Microturbine

MATERIAL SELECTION

In the current program, six ceramic materials were considered for the microturbine combustor, four monolithic ceramics; silicon carbide (SiC), silicon nitride (Si_3N_4), siliconized silicon carbide (SiSiC), alumina (Al_2O_3), and two ceramic matrix composites (CMCs)-Oxide/Oxide and SiC/SiC. Although CMC combustors have been successfully demonstrated for industrial gas turbine engines[5,6,7,8], their cost were found to be prohibitive for the cost sensitive microturbine. In addition, the small size and low stress expected in the microturbine combustor warranted monolithic ceramics as past engine experience has illustrated[9,10,11]. As a result, CMCs were not pursued for the ST5+ microturbine combustor. Alumina was not pursued due to low thermal shock resistance.

ATTACHMENT DESIGN AND ANALYSIS

The ST5+ microturbine engine has a single silo type combustor (see Figure 1). The current metal combustor in the baseline configuration is coated with thermal barrier coating (TBC) and is impingement cooled to withstand the high temperature and corrosive combustion environment. The silo combustor is connected to a combustor transition duct (CTD) that turns a vertical combustion gas flow into horizontal flow that goes into the turbine. Such a turning leads to high thermal loading on the CTD and since ST5+ engine is recuperated, there is concern that there may be insufficient cooling for the CTD. A use of ceramic combustor is advantageous since the combustor would need less cooling air (compared to a metal one) and if required, more cooling could be diverted to the CTD to ensure its mechanical integrity.

The current metal can is supported at the fuel nozzle and is free to slide at its exit in and out of the combustion transition duct to minimize thermal stress. The ceramic combustor adopted the same arrangement at exit, but special attention was paid to the attachment method at the fuel nozzle end in order to minimize thermal stress resulting from the thermal expansion difference between the metal support and the ceramic combustor.

A number of concepts were proposed during the concept design phase. Two ceramic combustor cans were down-selected for preliminary and detail designs. Both ceramic cans are deigned with a flange at the fuel nozzle end for attachment (see Figure 2). One design has a right angle flange, while the other design has a 45-degree flange. The ceramic cans are clamped down to a metal support through a wave spring that purports to absorb the thermal expansion mismatch between the can and the metallic support. At the interface between the ceramic can and the metal supports, ceramic seals with metal overbraid are inserted to avoid stress concentration by point contact and allow some relative movement.

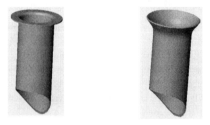

Figure 2: Two Ceramic Combustor Attachment Methods

Extensive thermal and stress analyses were performed to ascertain the temperature and thermal/mechanical stress. The temperature and maximum principal stress distributions are shown in Figure 3. The temperature is lower at the flange area, necessary for attachment, but higher at the main combustor can body. The temperature difference between metal and ceramic at attachment generates substantial thermal stress and insulation material was added to minimize heat flow from the ceramic can to the metal support, thereby raising temperature at the ceramic can attachment area while still maintaining a low temperature for the metal support.

The final results for the candidate ceramic materials for both designs are summarized in Table 1. The ceramic can design with a 45 degree flange has a lower thermal stress than the 90 degree flange design because of its lower constraint on the thermal expansion of the hotter main ceramic can body.

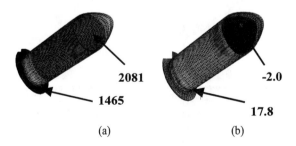

$$2081$$
$$1465$$
$$-2.0$$
$$17.8$$

(a)   (b)

Figure 3: (a) Temperature distribution (F) and (b) maximum principal stress (ksi) in SiC ceramic combustor can

Table 1: Maximum Temperature and Stress in Two Ceramic Combustor Designs

Material	Design			
	Flange		Taper	
	Max./Min. Temp (F)	Max. Stress (Ksi)	Max./Min. Temp (F)	Max. Stress (ksi)
Honeywell AS800	2103/1330	14.6	2103/1320	7.9
Hexoloy SA SiC	2081/1465	17.8	2080/1457	7.8
CoorsTek SCRB-210	2090/1391	16.6	2090/1392	8.1
Kyocera SN282	2103/1330	11.3	2103/1320	6.1

The monolithic ceramics were ranked according to their strength, resistance to thermal shock, and cost of manufacturing. Si-SiC was eliminated because of its low strength at room and elevated temperatures. In addition, SiSiC was also found to be prone to creep at stress and temperature regime expected for the ceramic combustor can. In-situ toughened Si_3N_4 offers the highest strength and toughness, but it is more costly than SiC. Considering that the ceramic combustor is not as highly stressed as ceramic turbine vanes and blades, SiC was selected as the primary candidate material for combustor can.

Eigen value frequency analysis was also performed to determine if the thermo-acoustic and structural frequencies might lead to resonance. Baseline combustor configuration test was performed and the measured frequency-dependent impedance was incorporated in a thermo-acoustic model, which predicted that at 100% power, system instability most likely exists near 480 Hz. The analysis predicted that the vibration frequencies for the first three bulk modes for combustor were 1605, 2731, and 3648 Hz due primarily to the fact that CMC's have high stiffness and low density. The vibration modes are shown in Figure 4. The first two frequencies are lower than 3000 Hz where thermo-acoustic frequencies are active, indicating that there is adequate dynamic excitation margin and resonance will not be an issue.

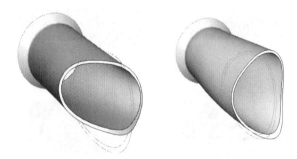

Figure 4: Vibration modes: Eigen value analysis (amplified displacements)

In addition to the combustor can material selection and design, there are several other elements under development to ensure that the overall combustor module can achieve the required life and performance. One such key element is a wave spring in the attachment design. The wave spring is used to mitigate the thermal stress that may built-up between the ceramic combustor can and the metal support under engine operating conditions when both expand thermally, but by different amount. The major concern was that the spring may not be able to maintain the clamping load due to creep, mainly due to high recuperated air temperature (~1130

^{0}F). To verify the creep strength, wave springs were placed in a furnace and a constant load was applied to simulate service condition. The first set of wave springs were made of INCONEL X750 and they showed substantial creep deformation under load and temperature. A second set of wave springs were made using a more creep resistant Nickel based superalloy (IN718) and tested under same conditions. Unfortunately, the springs showed unacceptable creep deformation again. The springs were then redesigned to reduce stress and more temperature resistance materials were also considered. Ceramic rope seal is another important element of the low stress attachment design. To ensure its integrity under engine conditions, a special testing rig was set up to simulate actual thermal and mechanical loadings. Fatigue tests were run up to 1 million cycles and no visible damage was observed on the seals.

The other key element in the use of silicon carbide based materials in the gas turbine environments is the development and use of protective environmental barrier coatings (EBC's). EBC's are required because of the accelerated oxidation of SiC and subsequent volatilization of silica in the high temperature high-pressure steam environment. EBC systems for silicon carbide fiber reinforced silicon carbide composites (SiC/SiC CMC's) were first developed under NASA HSCT Program[12]. The multi-layer coating system typically consists of a silicon bond layer, a mullite containing intermediate layer and a top layer of a celsian-based complex silicate (such as barium strontium aluminum silicate – BSAS). The effectiveness of such multi-layer coating systems have been demonstrated via > 50,000-hour field tests in industrial gas turbine engines under different DOE programs. Coating trials were performed to apply the multi-layer EBC to the ST5+ combustor liner. The plasma spray parameters were optimised to ensure high quality EBC coatings. Preliminary coating tests showed that a modified EBC process could be applied successfully to the combustor liner. The coating fixtures and parameters that were established during coupon trials were refined. A dense BSAS based coating with desired and uniform thickness was demonstrated on the inner surface of a SiC Hexoloy cylinder with the same ID (*circa* 6 inches) and similar length (*circa* 13 inches) as that of the advanced microturbine combustor liner (Figure 4).

Figure 5: Cross-section view of EBC on the ID surface of a 6" SiC Hexoloy cylinder

SUMMARY

Significant progress has been made in designing the ceramic combustor for ST5+ microturbine engine. It is believed that the ceramic combustor can meet the thermal and stress design requirements. The ceramic combustor can was designed with special attention to attachment methods that minimize thermal stresses due to the large difference in thermal expansion coefficients between metallic and ceramic materials. Detailed thermal and stress analyses were performed to that lead to optimal component reliability and manufacturability.

ACKNOWLEDGEMENTS

The authors would like to thank DOE for its financial support and Debbie Haught of the DOE Office of Power Technologies and Stephen Waslo of the DOE Chicago Operations Office for their programmatic support. They also wish to acknowledge the technical support from their colleagues in Pratt & Whitney Canada.

REFERENCES

[1] Stauffer, J., 2003, August 14, 2003, Blackout: Updated Sequence of Events, Cambridge Energy Research Associates Report

[2] U.S. Department of Energy, http://www.eere.energy.gov/der/microturbines/microturbines.html #advanced

[3] Shi, J., V. Vedula, J. Holowczak, C. E. Bird, S. S. Ochs, L. Bertuccioli and D. J. Bombara, 2002, Preliminary Design of Ceramic Components for the ST5+ Advanced Microturbine Engine Paper ASME GT-2002-30547, ASME TURBOEXPO 2002, June 3-6, 2002, Amsterdam, The Netherlands

[4] Smith, K.O. and Fahme, A, 1996, Experimental assessment of the emissions benefits of a ceramic gas turbine combustor, ASME Turbo Expo 1996, Paper 96-GT-318

[5] Price, J., et al, 2001, Ceramic Stationary Gas Turbine Development Program-eighth annual summary, ASME Turbo Expo Paper, 2001-GT-0517

[6] Brewer,D., Ojard,G. and Gibler,M., 2000, Ceramic Matrix combustor liner rig test, ASME Turbo Expo 2000, Munich, ASME 2000-GT-0670

[7] Verrilli, M.J., D. Brewer, 2002, Characterisation of CMC fastners exposed in a combustor liner rig test, ASME GT-2002-30459

[8] Igashira, K., Matsuda, Y., 2001, Development of the advanced combustor liner composed of CMC/GMC hybrid composite material, ASME 2001-GT-0511

[9] Mikami, T., et al , 1996, Status of the development of the CGT301, a 300KW Class Ceramic Gas Turbine, ASME 96-GT-252

[10] Tatsumi, T., et al, 1999, Development summary of the 300kw ceramic gas turbine CGT302, ASME 99-GT-105

[11] Tanaka,R, T. Tatsumi, Y. Ichikawa, K. Sanbonsugi, 2001,Development of the Hybrid Gas Turbine (1st year summary),ASME 2001-GT-0515

[12] Eaton, H.E., Linsey, G.D., Sun, E.Y., More, K.L., Kimmel, J.B., Price, J.R., and Miriyala, N., "EBC Protection of SiC/SiC Composites in the Gas Turbine Combustion Environment - Continuing Evaluation and Refurbishment Considerations", 2001-GT-0513, 2001, ASME TURBOEXPO 2001, Land, Sea and Air, New Orleans, LA, 2001

CMC COMBUSTOR LINER DESIGN FOR A MODEL RAM JET ENGINE

Tetsuya Morimoto[1], Shinji Ogihara[2], Hideyuki Taguchi[1], Takayuki Kojima[1], Kazuo Shimodaira[1], Keiichi Okai[1], and Hisao Futamura[1]

1)Japan Aerospace Exploration Agency (JAXA)
7-44-1 Jindaiji Higashi-machi
Chofu-shi Tokyo 182-8522 JAPAN

2)Tokyo University of Science
2641 Yamazaki, Noda, Chiba 278-8510 JAPAN

ABSTRACT

The development of high temperature resistant ceramic fibers has led to potential applications of ceramic matrix composites (CMCs) for aerospace engine components such as combustion chamber liner, turbine blade, and nozzle. The superior temperature capability of CMCs may then enable in the near future the turbine based combined cycle (TBCC) be the practical thruster for the launch system of two-stages to orbit (TSTO). The current study presents an engineering approach to design the CMC combustion liner for a model RAM jet engine that will be tested in supersonic wind tunnel. FEM thermal analyses have revealed with the thermal parameters of Tyrannohex and SA-Tyrannohex SiC/SiC composites (UBE Industry Co., UBE Japan) the maximum heat flux through the combustor liner implying the limitation in the engine operation.

INTRODUCTION

Japan Aerospace Exploration Agency (JAXA) has been studying a turbine based combined cycle (TBCC) propulsion system, which consists of low-speed turbine engine propulsion and high-speed RAM jet engine propulsion, to achieve an access to space in the form of two-stages to orbit (TSTO)[1-3]. The first stage seeks the operational velocity up to Mach 6, which provides the RAM combustion temperature of 1900°C with hydrogen as the propellant and the oxygen of atmospheric air[4]. The temperature is too high for the most material to retain the performance, thus cooling is inevitable for the components. However, air-cooling is not practical at the high-speed RAM jet mode due to the aerodynamic heating on the air itself. Thus, the propellant, liquefied hydrogen (LH$_2$), needs to cool the combustion chamber liner in addition to other high temperature components such as intake strut vane and nozzle. Current metal-based design, however, needs huge capacity of cooling that may excess the practical amount of LH$_2$. Thus, high temperature resistant materials, which save LH$_2$ and the cooling channel weight, are needed to withstand the high temperature environment associated with the high Mach number flight and the combustion. The superior temperature capability of new materials such as CMCs is thus the key factor for realizing TBCC as the practical thruster for TSTOs[5]. However, new materials such as many CMCs at the experimental production stage often need around 10 years before the commercialization. In addition, not a little number of them remains experimental production level due to the lack of competency in the commercial market. Thus, selecting them may risk the future modification and components supply for the model RAM jet engine.

The current study presents an engineering approach on the CMC combustion chamber liner for a model RAM jet engine, which will be tested in supersonic wind tunnel. 'To be affordable in the commercial market during 2003' has been set as the condition to select the combustion chamber materials. This condition also leads to the reduction not only in the materials cost but also in the machining and non-destructive evaluation costs.

During the wind tunnel tests under the design condition, the pressure shell of the combustion chamber is cooled with liquefied hydrogen and the faces are film-cooled with gaseous hydrogen. Thus, the face is kept 1350°C, which is low enough for the two CMCs, Tyrannohex SA-grade and LOX-M grade, to keep the thermo-mechanical parameters. However, parameter mismatch between the model and the wind tunnel, malfunctioning of the wind tunnel, and so on may cause the heat spot formation of 1900°C, which is the maximum temperature of the combustion gas. The worst condition may empirically take 10 seconds before the shut down or resetting the parameters for the design condition operation. Then, the authors would like to estimate if the wind tunnel model survives from the failure such as pressure shell melt down. Therefore, intensive study on the CMC thermal conductivities and FEM analyses were performed to assess the maximum heat flux through the CMC heat shield tiles.

CMC THERMAL CONDUCTIVITIES

The specific heat capacity and the thermal diffusivity were measured on the 8-harnes satin type Tyrannohex LOX-M grade and SA grade specimens in the disc form of 10mm in the diameter and 2mm in the thickness. Following the supplier's data sheet, the fiber volume fraction (VF) of the LOX-M grade is 87% and the VF of the SA grade is 91%. The VFs are extremely high as CMCs, thus the thermal parameters may strongly reflect the fiber parameters including the orientation. Therefore, both in-plane and out-of-plane layer directions were sampled as depicted in Fig. I for the measurements in vacuum at room temperature, 600°C, 800°C, 1000°C, and 1200°C using ULVAC-RIKO TC-7000 laser flash thermal constants analyzer (ULVAC-RIKO Inc., Yokohama Japan). The specific heat capacity and the thermal diffusivity were then converted to the thermal conductivity, which was applied for the FEM thermal analyses.

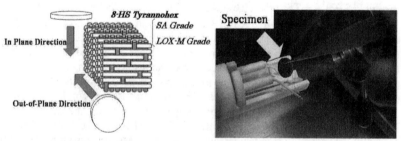

(a)Specimen Sampling Directions　　(b) An Example of Thermal Analysis Setup
Figure I　CMC specimen preparation for the thermal analyses

Figure II and III show the derived thermal conductivities of 8-HS SA-Tyrannohex and 8-HS Tyrannohex for the 10 samples of both in-plane and out-of-plane directions. Figure II (a) and (b) indicate that Tyrannohex SA-grade possesses larger thermal conductivity in the in-plane direction than the out-of-plane direction. Thus, the out-of-plane direction must be set parallel to the heat flux direction in the tile design, for the redistribution of localized heat and minimizing the heat through-flux.

Figure III indicates that the thermal conductivity of in-plane case is close to or equal to the out-of-plane case, thus Tyrannohex LOX-M grade may be acceptable to regard isotropic on the thermal conductivity.

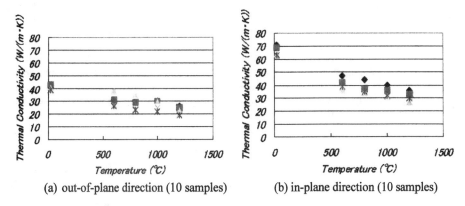

(a) out-of-plane direction (10 samples)   (b) in-plane direction (10 samples)

Figure II.   Thermal conductivity of 8-HS Tyrannohex SA grade

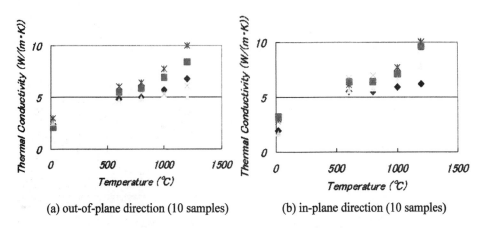

(a) out-of-plane direction (10 samples)   (b) in-plane direction (10 samples)

Figure III.   Thermal conductivity of 8-HS Tyrannohex LOX-M grade

DESIGN APPROACH

The model RAM jet engine is for the wind tunnel tests, or only for the ground-based operations in limited time span. Thus, the needs on the weight reduction and the long-term structural reliability may take lower priority in the design. However, "cost and the margin for the future modification" are the key factors as the model of preliminary experimental stage.

Figure IV shows a concept design of model RAM jet engine, which has two dimensional intake and nozzle. The combustion chamber has been set as an orthorhombic of 700mm in the length, 100mm in the width, and 80mm in the height.

During the wind tunnel tests under the planned condition, the outer shell is cooled with liquefied hydrogen and the inner faces are film-cooled with gaseous hydrogen to 1,350°C. However, the maximum combustion temperature has been set to 1,900 °C in the cycle analysis, thus the heat spots may reach to 1,900 °C during the wind tunnel operations. Considering the case as the worst one, the evaluation condition on the liner surface temperature has been set to 1,900 °C with the exposure time of empirically set 10 seconds before the emergency shut down. Note that this condition provides safe-side over estimation, thus the expected life-time leads to low-cost modifications of the wind tunnel model. In addition, the test completes in limited time such as maximum 120 seconds due to the wind tunnel specification.

CMCs are rather new for the pressure shell applications although hopeful ones are on the developments for the future applications. The pressure shell material is thus assumed INCO 718, which is a well-matured and stably affordable alloy for the long-term operations at the temperature around 700 °C. Therefore, CMCs work as load-free heat-shield tiles, which protect the metallic shell from the high temperature combustion atmosphere of 1,900 °C. The combustion chamber wall has been then designed as thermal barrier CMC liner tiles and the surrounding metallic pressure shell, as depicted in Fig.V and VI. The measured thermal conductivitiess have indicated that Tyrannohex SA grade, which keeps large thermal conductivity under high temperature condition, provides a potential for avoiding heat spot formation through the in-plane redistribution of localized high heat flux. Thus, Tyrannohex SA grade covers the inner wall of the combustion chamber in the form of expendable tiles, or "in-plane isotropic temperature tiles." Tyrannohex LOX-M grade, which keeps heat shield performance at high temperature, surrounds as "heat shield tiles" the inner SA grade tiles for protecting the metallic pressure shell from the emergency high heat flux.

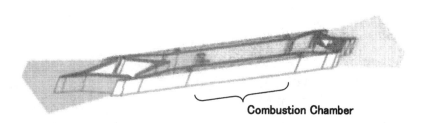

Combustion Chamber

Figure IV. A concept design of model RAM jet engine

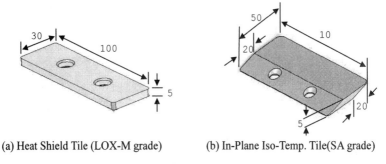

(a) Heat Shield Tile (LOX-M grade)   (b) In-Plane Iso-Temp. Tile(SA grade)

Figure V. Tyrannohex tile design plan of combustion chamber wall

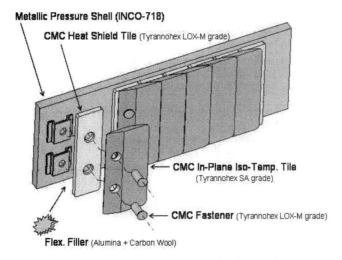

Figure VI. A design plan of combustion chamber wall for the model RAM jet engine

FEM THERMAL ANALYSES

A uniform temperature was set in the FEM analyses for the in-plane isotropic temperature tile surface of 1900°C and 1200°C was set as the practical lower limit for the heat shield tile face, which bounds to the flexible filler. Inter-tile is also filled with the flexible filler of 1kW/m^2·°C as the heat transfer coefficient. Note that the alumina and carbon wool filler may melt or sublimate during the "worst condition," equalizing the heat flux to the heat shield tiles.

Figure VII shows a result of the FEM analyses, which provided the heat flux of 1.2×10^5 W/m^2. The heat flux is too high for reusable structures to withstand, or the flux is within the order of reentry vehicle nose cap and the reentry capsule bottom that are protected with ablation cooling for limited time. Similarly, alumina and carbon wool flexible filler may need to work as ablator when the high heat flux lasts long before the emergency shut down.

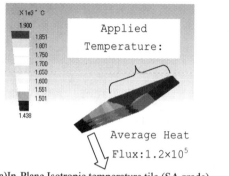
(a)In-Plane Isotropic temperature tile (SA grade)

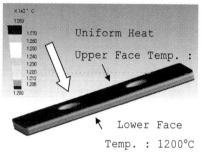
(b) Heat shield tile (LOX-M grade)

Figure VII. Temperature contour of the Tyrannohex CMC tiles for the "Worst Case"

SUMMARY

An engineering approach has been performed to design CMC tile wall for a model ramjet combustion chamber that operates at the combustion temperature of 1900°C. The tiles were assumed of Tyrannohex LOX-M grade and SA grade SiC/SiC composites surrounded by INCO 718 pressure shell. The design efforts involved the measurements of thermal conductivity on the two Tyrannohex and FEM thermal analyses on the heat flux through the tiles. The FEM analyses provided very high heat flux, which needs ablation cooling, through the tiles at the "worst case" of 1900°C on the faces. Thus, the wind tunnel test parameters must be carefully set for the design point operation to keep the tile face maximum temperature less than or equal to 1350°C.

References

[1] T. Kanda and K. Kudo, "Payload to Low Earth Orbit by Aerospace Plane with Scramjet Engine," Journal of Propulsion and Power, Vol. 13, No. 1, pp.164-166, 1997

[2] T. Kanda, T. Hiraiwa, T. Mitani, S. Tomioka, and N. Chinzei, "Mach 6 Testing of a Scramjet Engine Model," Journal of Propulsion and Power, Vol. 13, No. 4, pp.543-551, 1997

[3] T. Mitani, T. Hiraiwa, S. Sato, S. Tomioka, T. Kanda, and K. Tani, "Comparison of Scramjet Engine Performance in Mach 6 Vitiated and Storage-Heated Air," Journal of Propulsion and Power, Vol. 13, No.5, pp. 635-642, 1997

[4] H. Taguchi, H. Futamura, R. Yanagi, and M. Maita, "Analytical Study of Pre-Cooled Turbojet Engine for TSTO Spaceplane," AIAA 2001-1838, 2001

[5] T. Morimoto, "Design of CMC Inlet Vane for a Model Air Breathing Engine," Ceramic Engineering and Science Proceedings, Vol. 24, Issue 4, pp.269-274, 2003

BURNER RIG TEST OF SILICON NITRIDE GAS TURBINE NOZZLE

Masato Ishizaki, Tomohiro Suetsuna,
*Masahiro Asayama and Motohide Ando
Fine Ceramics Research Association
2268-1 Anagahora, Shimo-shidami
Moriyama-ku, Nagoya city
468-8687 JAPAN

Naoki Kondo and Tatsuki Ohji
Synergy Materials Research Laboratory
Agent of Industrial Science and Technology
2268-1 Anagahora, Shimo-shidami
Moriyama-ku, Nagoya city
468-8687 JAPAN

ABSTRACT

Cyclic burner rig test was performed with model parts of gas turbine nozzle made of silicon nitride with and without environmental barrier coating (EBC) based on lutetium silicate. A short cycle of burner rig test were repeated up to 100 minutes of holding time at 1400 °C, followed by 10 cycles of 1 minute holding each at every 50 °C increase of holding temperature up to 1600 °C. The model part showed no visible cracks under zyglo inspection after the test cycles. The model part without EBC showed a little weight increase after the test cycles, whereas the part with EBC showed a slight weight decrease. The thermal stress distribution analysis was carried out by FEM based on the temperature distribution recorded by ThermoVison infrared camera during the test cycles. The maximum stress after flame turned off was 175 MPa which is lower than the strength of the material (400 MPa at 1500 °C), suggesting the validity of the application of the material to real parts.

INTRODUCTION

Silicon Nitride is one of the most promising ceramic materials for high temperature structural use[1,2]. It is expected to enhance the energy efficiency of gas turbine generator for 10 % relatively when the turbine inlet temperature is elevated from 1300 °C to 1500 °C with application of nozzle vanes made of silicon nitride.

High-temperature durable silicon nitride has been developed by the Synergy Ceramics project which holds the mechanical properties up to 1500 °C. A turbine nozzle model part of this material was manufactured. A cyclic burner rig test was carried out with the model part to show its potential ability for gas turbine parts application. Environmental barrier coating for protection from water vapor erosion has been also developed in the project, which is coated on one of the model part to apply to the burner rig test.

EXPERIMENTAL

Starting powder containing 8 wt % of lutetium oxide and 2 wt % of silica was prepared. The CIPed green body was machined to net shape and sintered at 1920 °C under nitrogen pressure of 1 MPa for 6 hours. The sintered model part was applied to the cyclic burner rig test. The dimensions of the model part, shown in fig 1, are in the typical range of 20 to 40 MW turbine generators. The model part has a hollow structure to give a potential functionality of air cooling. A model part with environment barrier coating based on

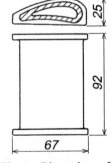

Fig. 1 Dimensions of the nozzle model part (dimensions in mm)

---

* Present affiliation: Toshiba co., ltd.

lutetium silicate was also prepared for the cyclic burner rig test.

The burner rig test was carried out with flame of a gas mixture of LNG and oxygen. The rig test for bear model part was a cyclic test of controlled flame profile for 1 minute. The highest temperature on the outer surface was set at 1400 °C for the first 100 cycles followed by 1450 °C, 1500 °C, 1550 °C and 1600 °C for 10 cycles each. The similar rig test was carried out with the model part with EBC, where the cycles of 1400 °C was 10-minute keeping for 10 cycles followed by 10 cycles for 1minute keeping at higher temperatures. Every test cycle consisted of 2 minutes of elevation followed by 1 (or 10) minute(s) of holding, and cooling duration with the flame turned off for 17 minutes. In the latter 7 minutes of the cooling duration, cooling air was applied on the metallic holder which holds the test piece from the outside.

Temperatures of 6 points on the surface of the test piece were measured with 3 pyrometers and 3 thermocouples. The weight change before and after the cyclic test was measured. Zyglo inspection was carried out before and after the test cycles for defect investigation.

Temperature distribution on the outer surface of the model part was measured by ThermoVision infrared camera from foreside and backside in typical test cycles. The thermal stress analysis was carried out based on the temperature distribution measurement.

RESULTS AND DISCUSSION

Fig. 2 shows outer views of the model part before and after sintering. Figs. 3 and 4 show a

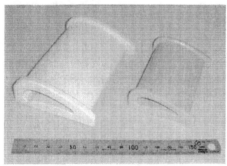

Fig. 2   Model parts before (left)
and after sintering

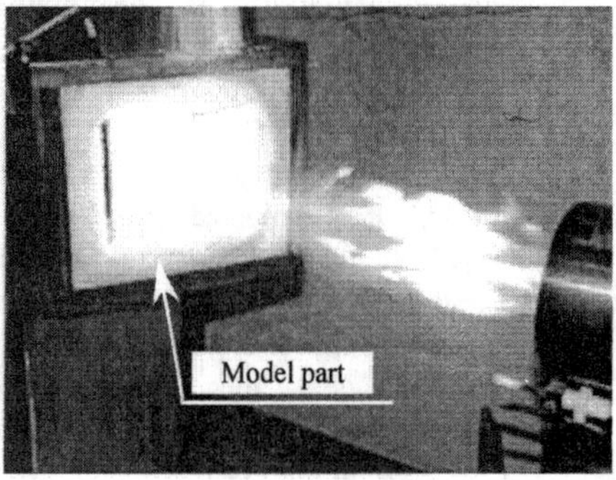

Fig. 3   A view of a test set-up

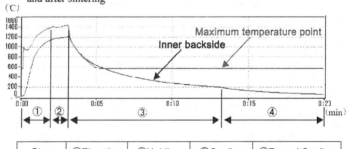

Fig. 4   Temperature profile of the model part during the test cycle

Step	①Elevation	②Holding	③Cooling	④Forced Cooling
Duration	2 minutes	1 minutes	10 minutes	7 minutes

typical view and temperature profile of the model part during the test cycle respectively. The temperature was successfully elevated to the holding temperature in 2 minutes in every cycle. Fig. 5 shows the measured temperatures at the end of the holding duration in a typical test cycle of 1400 °C. The temperature was the highest at the center of the flame exposure (Point #6 in Fig. 5) and the difference was as small as 10 °C at the center of the outer side (Point #5), and the difference through the thickness of some 5 mm was about 200 °C (Point #3). There was no significant change in the measured temperature distribution between test cycles for the same targeted temperature.

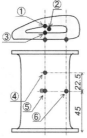

Measurement points

Point	*Medium	Temperature
☐	T/C	1070 °C
☐	T/C	1120 °C
☐	T/C	1210 °C
☐	P/M	1290 °C
☐	P/M	1400 °C
☐	P/M	1410 °C

*T/C: Thermocouple, P/M: Pyrometer

Fig. 5 Measured temperatures during a test

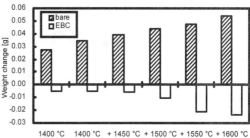

Fig. 6 Weight change of model parts

Fig. 6 shows weight change of the model parts with and without EBC coating during the test cycles. Neither of the model parts showed any defect after all the cycles of the burner rig test by zyglo inspection. The bear model part showed small weight increase due to the surface oxidation, whereas the model with EBC showed slight weight decrease during the test cycles. The EBC enhanced the anti-erosion property of the model part drastically. The weight of the model part with EBC showed no change after 50 and 100 minutes of exposure at 1400 °C.

Temperature distribution of a test cycle of 1400 °C measured by ThermoVision is shown in Fig. 7. An FEM analysis was carried out from the temperature distribution measurement. The temperature

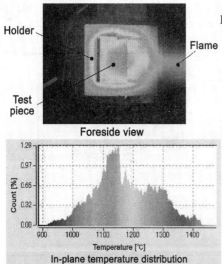

Fig. 7 Temperature distribution measured by ThermoVison

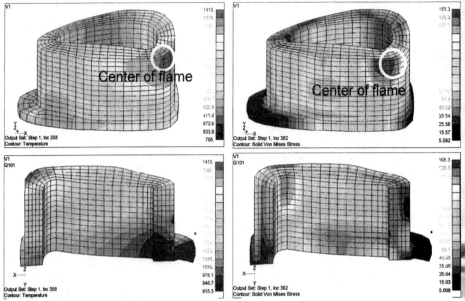

Fig. 8   Temperature distribution at 179 s   Fig. 9   Stress distribution at 181 s
(just before flame turned off) by FEM   (just after flame turned off) by FEM

distribution in the model part was assumed symmetrical between the two sides of equator section
for convenience of calculation.   The contours of temperature at 179 s and maximum principal
stress at 181 s are shown in Figs. 8 and 9, respectively.   The flame was turned off at 180 s thus
fig 7 represents when the temperature was the highest all over the body and fig. 8 represent
when the stress distribution was the most severe.   The maximum stress was found at the outer
surface of the center of the flame just after the flame turned off.   Fig. 10 shows the stress

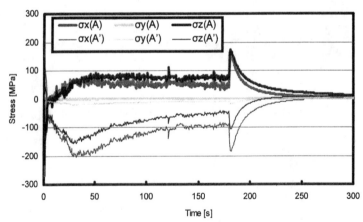

Fig.10   Stress transition at the flame center

transition at the outer surface of the center of the flame. The maximum stress was calculated as 175 MPa in tensile direction. The strength of the material of the model part is more than 400 MPa at 1500 °C,[3] which suggests the validity of applying the material to the real part for the gas turbine of the next generation. Further steps for real application would include the probability of failure analysis to consider the volume of the part.

CONCLUSIONS

The model parts of gas turbine nozzle manufactured of high-temperature durable silicon nitride was applied to a cyclic burner rig test. No defect was observed on the surface after the test cycles. The bear model part showed small weight increase after the test cycles and the model part with EBC showed slight weight decrease. The FEM analysis showed the maximum stress of 175 MPa which is smaller than the strength of the material of 400 MPa at 1500 °C, suggesting the validity of application of the material to the real part of the next generation gas turbine.

ACKNOWLEDGEMENT

This work has been supported by METI, Japan, as a part of the Synergy Ceramics Project. A part of the work has been supported by NEDO. The authors are members of the Joint Research Consortium of Synergy Ceramics.
REFERENCES

[1]M. van Roode et al.,"Ceramic Gas Turbine Design and Test Experience" (2002)

[2]M. van Roode et al., "Ceramic Gas Turbine Component Development and Characterization" (2003)

[3]N. Kondo et al., "Properties of gas-pressure sintered silicon nitride with lutetium oxide as additive", Proceedings of Annual Meeting of The Ceramic Society of Japan, 2003(2003)

MATERIALS FOR ADVANCED BATTERY AND ENERGY STORAGE SYSTEMS (BATTERIES, CAPACITORS, FUEL CELLS)

Alvin J. Salkind
Rutgers University, School of Engineering
98 Brett Road
Piscataway, NJ 08854

ABSTRACT

The annual production of Electrochemical Energy Storage and Conversion Devices including Batteries, Capacitors and Fuel Cells is approaching a $100 billion value (at manufacturers values). Approximately one half of this amount is the value of materials used in processing or product. These materials include metals, plastics, organic and polymer materials, fabrics, glass, ceramics, inorganic materials, and composites. They are utilized or processed in powders, pastes, films, sheets, fabrics, mats, electrolytes, seals, moldings, fibers, and composites. An overview of the presently used and advanced materials is presented and the needs for further innovations to meet emerging applications described. Some advanced materials studied in the Center for Energy Storage Materials and Engineering at Rutgers University are discussed. They include metal hydrides for the storage of hydrogen, bipolar-NiMH battery designs, manganese-bismuth mixed metal oxides for rechargeable alkaline and lithium cells, and super capacitor materials.

INTRODUCTION

The significant increase in overall battery manufacturing over the last decade was overshadowed by a dramatic increase in several technology/market segments. As shown in Table 1, the world level of production rose from approximately $23 billion in 1990 to an estimated $50 billion in 2003. Approximately one-half of production costs are materials. The electrical circuit demands for many newer and emerging application require higher power either sustained or in pulses. As the application base has shifted, secondary batteries have become an even more dominant segment of the industry. In 1990, the value ratio of secondary batteries to primaries was 1.2 (neglecting specials). It is estimated that it will be 1.8 in 2003 and even higher in 2005. Along with higher power more compact batteries, there has been an increased demand for advanced capacitors. The combination of a high power capacitor and high energy battery solves many electrical circuit needs. Capacitors are made by a wide variety of materials and technologies. The electrochemical capacitors and super-capacitors are similar to batteries in their fabrication technology and are themselves a fast growing market for advanced materials and are estimated to be about 1/3 the size of the battery production values.

The electrochemical energy storage systems, batteries, capacitors, and fuel cells are forecast to reach production values of $100 billion in the next few years.

Table 1- Battery Production by Technology/Market Segments
annual world production, *values in US dollars- billions*

System/Year	1990	2000	2003e	*2005 f*
1- Primary Cells[total]	10.0	13.1	15-16	*17-20*
Lechlanche and HD	*4.4*	*4.3*	*3.9*	*3.7*
Alkaline MnO2	*4.1*	*5.8*	*6.5-7*	*9-10*
Lithium(consumer/OEM)	*0.5*	*1.2*	*1.3*	*1.4*
Metal-Air	*0.4*	*0.8*	*0.9*	*1.1*
Silver/other	*0.7*	*0.7*	*0.6*	*0.6*
2-Secondary [total]	12.0	22.0	26-30	29-34
Lead-Acid SLI	*7.7*	*10.0*	*11.4*	*13* *36v begins*
Lead-Acid Ind.	*1.8*	*3.1*	*3.8*	*4*
Lead-Acid Cons.	*0.2*	*0.6*	*0.8*	*1*
subtotal lead-acid	*9.7*	*13.7*	*16.0*	*18*
Ni-Cd consumer	*1.6*	*1.6*	*1.1*	*1.1*
Ni-MH consumer	*-0-*	*1.6*	*1.1*	*1.1*
Ni-Cd/Ni-MH ind	*0.4*	*0.8*	*1.3*	*2.0***hybrid ev*
Other alkaline	*0.3*	*0.5*	*0.6*	*0.5*
MnO2-Zn, Ni-Fe, Metal-Air, Ni-Z etc				
subtotal alkaline	*2.3*	*4.5*	*4.1*	*4.8*
Lithium ion/other lithium	*–0-*	*3.1*	*4.2*	*4.5*
3- Special [total]	1.0	2.5	3.5	*4.0*
Reserve	*0.1*	*0.2*	*0.2*	*0.2*
Medical	*0.1*	*0.3*	*0.4*	*0.6*
Aerosp/military	*0.5*	*0.6*	*0.9*	*1.0*
Fuel Cells	*0.1*	*1.2*	*1.6*	*1.8*
Overall Totals	23	40	~47-50	~50-57

Note 1: Application/ technology changes reflected in ratio secondary/primary production
                     1.2   *1.7*   *1.8*
Note 2: Values are manufacturers values, Retail approx 2-3 times as great.
Note 3: Estimates by A.J.Salkind from ILZRO and Electrochem Soc data.

ELECTRICAL CIRCUIT REQUIREMENTS

Designs and types of battery systems are optimized for specific applications. Table 2 indicates the typical demand loads of present and emerging applications. Some newer high current pulse applications include: heart defibrillators, batteries for hybrid EV's, and special aircraft applications. The extreme power needed in some in the emerging applications has been met by going to much higher voltage designs, e.g., 240 V systems in some hybrid cars. Over the 150 years of commercial battery fabrication the requirements of circuits, motors, and electrical devices have dramatically changed.

Typical newer requirements include the following:
- Little or No Maintenance
- Higher Energy Density[weight /volume]
- Higher Power Density
- Longer Cycle Life/Longer Standby Life
- Use in Any Orientation\
- Undersea/Space Environments
- Biomedical Compatible and/or Implantable
- Stable in compact electrical circuits
- Minimal Environmental Problems in Manufacturing,
  Use, or disposal

To meet the new requirements materials studies have resulted in substantial changes in the nature and physical condition of the active and inactive materials used for electrochemical energy storage. Also, as part of the improvement in performance and minimization of environmental problems, new processing techniques, such as tape-casting, have evolved in high speed production.

Table 2- Energy/Power Profile of Electrical Devices

Power level	Micro <1 m W	Low >1 m W	Med. 1-10 W	High 10 W	Very High >100 W	Extreme >kW
Duty Cycle						
Low *< 1%*	*Appliance* *backup*	*Pager* *Camera*	*Flash* *light*	*Appliances* *Cam-* *corder lights*	*- Auto SLI-* *-Aircraft Power-*	
Medium *1-33%*	*Smart* *cards*	*Cordless* *phone*	*Cell* *phone* *GPS* *-Heart* *Defibrillator-*	*Power* *tools*		
High *>33%*	*Computer* *Heart-pacer*		*Portable* *PCS*	*EV*		*Torpedoes* *Hybrid EV*

BATTERY/ENERGY STORAGE MATERIALS

Background

The materials of construction and components for batteries encompass a great variety of chemical and physical properties and shapes. They fall into six main categories as shown in Table 3:
· Negative Active Materials
· Positive Active Materials
· Substrates and Conductivity Enhancers

65

· Electrolytes (solutes and solvents)
· Separators, Absorbers, and Membranes
· Seals and Packaging Materials

As shown in the table, almost all negative electrode materials are metallic elements, positive active materials are mainly metal oxides, substrates are metallic or carbon, electrolytes can be aqueous or non-aqueous, separator systems vary widely by chemical system but always must be inert in the electrolyte in which they function and include such functions as wicking, ionic barrier properties, and active materials separation. The separator must be penetrable by the electrolyte by means of absorption, swelling, or micro-porosity. Most packaging materials are plastic.

Table 3- Categories and Illustrations of Battery Materials

a) Negative a.m.
Hydrogen
Metal hydrides
Zinc, Cadmium
Iron, Lead
Lithium, lithium ion
Al, Mg
Sodium, Calcium

b) Positive a.m (oxidants)
Oxygen, Air $LiMn_2O_4$ SO_2 $(CFx)n$
Iodine $LiCoO_2$ $SOCl_2$ NiOOH
Bromine $LiNO_3$ V_2O_5 m-DNB
Sulfur MnO_2 Silver Vanadium-oxide
FeS, FeS_2 $NiCl_2$ TiS_2 $K_2Cr_2O_7$
CuS $CuO,CuCl$ WO_3 Organo-
AgO, AgO_2 AgCl PbO_2 Sulfides

c) Substrates and conductivity enhancers
Carbon, graphite,
Nickel, Lead ,copper
Al ,Tin, Tin Oxide
Stainless
Forms:powder.fiber,
foam,sinter,screen,
foil,film,sheet,cast

d)Electrolytes   (solutes and solvents)

Aqueous	Non-Aqueous	Solid
Sulfuric,	[solvents: Glymes,	LiI
KOH,	Propylene Carbonate,	βalumina
NaOH	Ethylene Carbonates	K_2CO_3molten
NH4Cl	mixed ethers]	
	[solutes:$LiClO_4,LiAsF_4$,	
	$LiBF_4,LiPF_6$]	

e)Separators, absorbers
glass fiber mat, wood,
fumed silica,paper
Microporous; PVC, PE,
 PP, rubber, ceramics
Films: Cellophane,
 grafted polyolefins
Fabrics: nylon,PP,cotton,

f) seals, packaging, cases
rubber, glass, ceramic, steel, stainless,
PVC, ABS, PP,PE, polysulfone, acrylates

seals: plastic, ceramic-metal, glass-metal,
 asphalt, rubber

COMPATABILITY is THE KEY

Materials Improvement Studies

A dramatic illustration of the improvement in a performance characteristic because of improvements in materials technology is the change in capacity of AA size nickel rechargeable cells over the last 40 years as shown in Table 4.

Table 4- Capacities of AA size consumer Nickel-Cad and Ni-MH cells: (Ah)

```
2.0                                                    x

                                           x   improved MH
1.5                                    x  [Ni-MH]  •

                                           •  •

1.0                            •           spherical nickel
                        •                  hydroxide introduced

0.5      •  [Ni-Cd]  •    Ni foam substrates


year   1970  75    80    85    90    95   2000   2005
```

The overall change in capacity from 0.5 to 2.0 A`h was due to 3 changes in materials technology: the development of nickel foams of >90% porosity which could be mechanically pasted, innovation in the preparation of nickel hydroxide by a processing techniques which eliminated unnecessary large pores (spherical nickel hydroxide), and the development of MH alloys with good stability and high density.

In the many projects in which the Energy Storage Materials group at Rutgers University is involved, 4 illustrate the importance of materials technology. We recently reported the development of bipolar Ni-MH batteries in collaboration with the project sponsor Electro Energy Inc. (Prime sponsors DOE and Navair). These have achieved over 20,000 cycles [1] and have been successfully operated over a wide temperature range. The design is suitable for extremely high current pulses and the energy density is over 60 W`h/kg. The design is compact and is less costly than conventional assembly methods. They key technology is extruded plastic sheet electrodes laminated to nickel foil.

Another study, reported by our group [2], is the effect of using various glass mat separator materials on VRLA Lead-Acid Battery performance. Wicking, separator compression, and mat density effect performance and ability to cycle. This study was unique in that the batteries were fabricated in our laboratory from commercial electrodes supplied by a major manufacturer the cells were cycled in 3 orientations to gravity. Since this study was reported, major changes have been made to glass fiber mat technology in order to eliminate over compression.

A 3rd development recently reported [3] is the synthesis of a manganese-bismuth mixed oxide with high energy storage capability for use in both alkaline cells and lithium cells. This material is also been evaluated for secondary batteries. The development was part of a CRADA arrangement with the US Army CECOM –Ft Monmouth and Rutgers and the inventors include Dr. Atwater and Prof. Salkind. A most important project area is the work on capacitors of the Energy Storage Technology group, which Rutgers acquired in May 2003 from Telcordia. Their innovations and recent Rutgers works in organic based capacitors were reported by Dr. Glenn Amatucci [4] (now a Research Associate Professor in the Department of Ceramic and Materials Engineering).

Of the many other studies for which space does not allow discussion, of note is the fabrication and testing of nano MnO_2 as an electrode material and catalyst [5], work on Ni-Zn battery technology, the fabrication of Nickel Hydroxide electrodes by electroplating techniques, the development of battery separators [5], work on room and high temperature fuel cells, and the development of ceramic materials and separators for lead-acid batteries.

Materials Development Needs for Emerging Applications

Many of the emerging applications have electrical and performance demands which require innovation in electrode materials, substrates, separators, and processing methods. These include:

Metal Hydride alloys with lower VP and higher hydrogen storage content
Lithium cell electrolytes with higher voltage and temperature stability
Better separators, absorbers, and membranes for many systems (including glass mats, cellophane, and micro-porous polypropylene)
Improved VRLA materials and assembly techniques
A safety shuttle mechanism for lithium cells
Higher valence nickel hydroxide crystal (better than 1 valence change)
Lower cost/higher energy substitute for CoO_2 in lithium-ion cells

SUMMARY AND CONCLUSIONS

The battery and energy storage business is a large but niche oriented industry with high materials utilization [e.g. over 70% of the world supply of lead is used in batteries]. The overall annual growth rate of the industry is approximately 8%, but it is many times this average in such rapidly growing segments such as Ni-MH in hybrid EV s, lithium-ion in consumer electronics and cell phones, portable tools, and medical devices external and implanted, fuel cells (from a small beginning), and capacitors. To achieve performance goals of emerging applications improvements in material science and materials processing will be required.

REFERENCES

1] M. Klein, M.Eskra, R. Plivelich, A.J. Salkind, and J.B. Ockerman, "Performance and Electrochemical Characterization Studies of Advanced High-Power Bipolar Ni-MH Batteries", Proc. 23rd IPSS, Sept 2003, Amsterdam, Netherlands.

2] R.V. Biagetti, L.C. Baeringer, F.J. Chiacchio, A.G. Cannone, J.J. Kelley, J.B.Ockerman, and A.J. Salkind, " Influence of Compression of Microfiber Glass Separator on Battery Performance", Proc. Intelec 94, Vancouver, B.C.

3] T. Atwater and A.J. Salkind, "Thermodynamic and Kinetic Study of the Li/MnO2 –BiO2 Couple", J. Electrochem. Soc. 145, L17-19, Mar 1998

4] G.Amatucci,, Private Communication

5] H.Lewis, C. Grun, and A.J. Salkind,"Cellulosic Separator Applications" J. Power Sources, 65, 29-38, (1997)

EFFECT OF Ni-Al PRECURSOR TYPE ON FABRICATION AND PROPERTIES OF TiC-Ni$_3$Al COMPOSITES*

T. N. Tiegs, F. C. Montgomery, and P. A. Menchhofer
Oak Ridge National Laboratory
Oak Ridge, TN 37831-6087

ABSTRACT

TiC-Ni$_3$Al composites containing about 50 vol. % metal are of interest for application in diesel engines. Processing studies have shown that the Ni-Al precursor type can have a considerable effect on the densification behavior and final mechanical properties. Several different metal powder combinations were examined that would react to produce Ni$_3$Al. These included: NiAl + Ni; Al-30 Ni + Ni; Ni+Al+NiAl + Ni; and Ni + Al. Sintering behavior was acceptable for all types except when elemental Al was used. Those particular samples exhibited macroscopic cracking due to differential heating and shrinkage resulting from the highly exothermic reaction between the Ni and Al. While the other samples densified well, only the composites made from the NiAl + Ni combination had mechanical properties that were acceptable.

INTRODUCTION

TiC-Ni$_3$Al composites (50:50 volume %) are under development for application in diesel engines because of desirable physical and mechanical properties [1-4]. The composites are densified by liquid phase sintering and most of the development of the composites has been done using Ni and NiAl powders (along with the TiC) to form Ni$_3$Al by an in-situ reaction during sintering. The in-situ reaction process was developed primarily because it produced high mechanical properties and developed a fine TiC grain size that is favored because of better wear resistance [5]. When the economics of producing the TiC-Ni$_3$Al composites was examined a significant cost was associated with the use of the NiAl precursor powder (~55% of the total raw material cost). The purpose of the study was to examine what effect alternate Ni-Al sources have on the behavior of TiC-Ni$_3$Al composites.

Elemental Ni and Al powders are produced in sufficient quantities to be relatively cost-effective and so tests were done to maximize their usage in the composites. Currently, Al-30% Ni powder (Raney-type nickel) is produced in commercial size quantities for use as a catalyst and consequently, is a cost-effective alternative for a Ni-Al alloy for fabricating TiC-Ni$_3$Al composites.

EXPERIMENTAL

The composites were fabricated by attritor milling appropriate amounts of TiC, Ni, Al, NiAl, and Al-30 Ni together so that Ni$_3$Al would form in-situ as a reaction product during sintering. The sample compositions are summarized in Table 1. The final compositions were calculated to be TiC-50 vol. % Ni$_3$Al. The same grade of TiC powder was used in each of the samples. Specimens of about 100g were uniaxially pressed at ~100 MPa (15 ksi) into compacts approximately 4.3 cm X 7 cm X 1 cm. Sintering was done in a graphite element furnace utilizing a vacuum/low pressure hot-isostatic pressure cycle (V-LPHIP) which consisted of a

* Research sponsored by the Heavy Vehicle Propulsion System Materials Program, DOE Office of FreedomCAR and Vehicle Technology Program, under contract DE-AC05-00OR22725 with UT-Battelle, LLC.

ramp of 10°C/min from room temperature to the final sintering temperature of 1400°C all under vacuum. The temperature was maintained at the sintering temperature for 0.5 h under vacuum followed by an argon gas pressurization to 1 MPa (150 psi) in 10 minutes and a hold under pressure for 10 minutes. The total time at the sintering temperature was 50 minutes. Selected samples of high density were machined into flexural strength specimens (nominal dimensions of 3x4x50 mm) and testing was done in four point bending (inner and outer spans of 20 mm and 40 mm, respectively). The fracture toughness was measured using the indentation and fracture technique using a 20 kg indent load. Hardness was measured on a polished surface with a Vickers indenter and a load of 20 kg.

Table 1 – Summary of compositions (wt.%). All compositions contained 39.1 % TiC (Kennametal Grade 1000, average particle size ~0.8 μm) and 0.8 % Mo (Alfa Aesar, Ward Hill, MA, 3-7 μm).

Sample ID	Ni[a]	Al[b]	NiAl[c]	Al-30Ni[d]
NiAl+Ni	34.7	--	25.4	--
Al-30Ni+Ni	44.1	--	--	16.0
NiAl+Ni+Al	43.5	4.0	12.7	--
Ni+Al	52.1	8.0	--	--

[a]Novamet, Wyckoff, NY, Type 123; [b] Alfa Aesar, Ward Hill, MA, 20 μm; [c] XForm, Cohoes, NY, 20 μm; [d] Activated Metals and Chemicals, Sevierville, TN, 12 μm.

DISCUSSION OF RESULTS

Differential thermal analysis (DTA) showed the reaction to form Ni_3Al from precursors is initially endothermic as the Al melts (~625-640°C) and then highly exothermic as the reaction with the Ni proceeds [6]. Initial sintering using a normal heating rate of 10°C/min produced varied results. The compositions containing NiAl+Ni or the Al-30 Ni product produced by the catalyst manufacturer showed good densification behavior and the densities achieved were >99.5 % T. D. However, the samples utilizing Ni+Al, or Ni+Al+NiAl showed severe macroscopic cracking caused by differential shrinkage during densification (Fig. 1). This behavior is most likely caused by non-uniform heating as a result of the exothermic reaction. Using the heat of formation for the reactions of Ni and Al, the adiabatic temperature rise for the 3Ni+Al→Ni_3Al reaction in the TiC-based composite is >1200°C (considering no heat loss to the surroundings and as such is a maximum condition). Since the reaction commences at ~650°C, the temperature increase is higher than the melting point of the Ni_3Al. In contrast, the adiabatic temperature rise in the composite for the NiAl+2Ni→Ni_3Al reaction is <300°C. Such an increase would not be sufficient to lead to significant melting of the Ni_3Al that is formed. Reducing the heating rate to 0.5°C/min in the temperature range of 600-700°C did not eliminate the cracking problem [6]. In addition, reducing the Al content (ie. the Ni+Al+NiAl sample) did not eliminate the cracking problem.

The exothermic reaction was recognized before the sintering run, but it was thought that the large fraction of TiC contained in the samples would act as a diluent to reduce the effects of any temperature rise. Previous studies of reaction sintering of Ni_3Al have been done where no sample cracking was observed [7,8]. However, the sample sizes used in both those studies were small (<4 g of Ni+Al per sample) in comparison to the samples in this study (~60 g of Ni+Al per sample). Heat dissipation from the smaller samples is better due to the increased surface area-to-mass ratios. Evidently, at the sample sizes in the current study, the heat loss from the exothermic reaction is not sufficient to alleviate the differential shrinkage.

The two compositions that did not crack were examined in greater detail for mechanical properties and microstructures. The fracture strength, toughness and hardness results are summarized in Figs. 2, 3, and 4, respectively. As shown, the strength measurements for the sample made by the reaction of Al-30 Ni (catalyst precursor) and Ni were very poor in comparison to the other composite. The fracture toughnesses were excellent for both of the composite compositions made with either NiAl or Al-30 Ni precursors. The main toughening mechanism in these composites is plastic deformation by the Ni_3Al binder phase and evidently it is effective for both composites. The hardness for the Al-30 Ni composite was lower than for the composite with NiAl. This is most likely due to the larger grain sizes and presence of a second phase in those materials.

Examination of the samples by optical microscopy revealed a significant amount of a second phase within the composite made with the Al-30 Ni (catalyst precursor) as shown in Figs 5 and 6. X-ray diffraction did not identify any secondary phases, however, elemental analysis in the SEM identified the inclusions to be composed primarily of silicon and oxygen. During production of the catalyst materials, silica contamination is introduced, which is acceptable for the catalyst application, but is detrimental for the structural composite application.

CONCLUSIONS

In the fabrication of $TiC-Ni_3Al$ composites it has been determined that the Ni-Al precursor type can have a significant effect on the densification behavior and final mechanical properties. Several different metal powder combinations were examined in the present study including: NiAl+Ni; Al-30 Ni+Ni; NiAl+Ni+Al; and Ni+Al. High densities were achieved for compositions utilizing NiAl or Al-30 Ni precursors. When elemental Al was used, the samples exhibited macroscopic cracking due to differential heating and shrinkage resulting from the exothermic reaction between the Ni and Al. Reducing the Al content and slowing the heating rate during the exothermic reaction did not eliminate the cracking problem.

Samples fabricated using the Al-30 Ni precursor exhibited a second phase in the microstructure that was not identified by x-ray diffraction. This phase is believed to be the cause of the low observed strength and hardness associated with these composites. Only the composites made from the NiAl + Ni combination had mechanical properties that were acceptable for engine applications.

REFERENCES
1. T. N. Tiegs, et al, Mater. Sci. Eng., Vol. A209, No. 1-2, 243-47 (1996).
2. T. N. Tiegs, et al, Int'l J. Powder Metall., 36 (7) 46-57 (2000).
3. K. P. Plucknett, et al, Ceram. Eng. Sci. Proc., 17 [3] 314-321 (1996).
4. T. N. Tiegs, et al, pp. 339-357 in Internat. Symp. Nickel and Iron Aluminides, ASM International, Metals Park, OH (1997).
5. T. N. Tiegs, et al, Ceram. Eng. Sci. Proc.,19 [3] 447-455 (1999).
6. T. N. Tiegs, et al, pp. 6.41-6.51 in Advances in Powder Metallurgy and Particulate Materials-2003, Metal Powder Industries Fed., Princeton, NJ (2003).
7. A. Bose, B. H. Rabin, and R. M. German, Powd. Metall. Internat., 20 [3] 25-29 (1988).
6. K. S. Hwang and Y. C. Lu, Powd. Metall. Internat., 24 [5] 279-282 (1992).

Fig. 1. Visual appearance of samples after sintering at heating rate of 10°C/min. Sample compositions: NiAl+Ni (left); Ni+Al (center); and NiAl+Ni+Al (right).

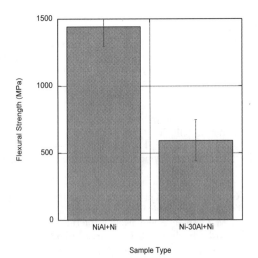

Fig. 2. Results on fracture strength of composites made with either NiAl+Ni or Al-30 Ni+Ni.

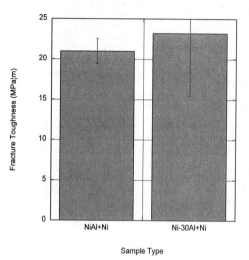

Fig. 3. Results on fracture toughness of composites made with either NiAl+Ni or Al-30 Ni+Ni.

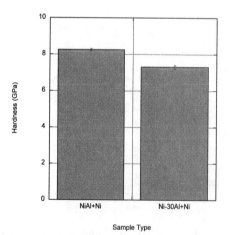

Fig. 4. Results on hardness of composites made with either NiAl+Ni or Al-30 Ni+Ni.

Fig. 5. Microstructure of sample made with NiAl+Ni showing uniform distribution of TiC grains in Ni₃Al matrix and no secondary phases.

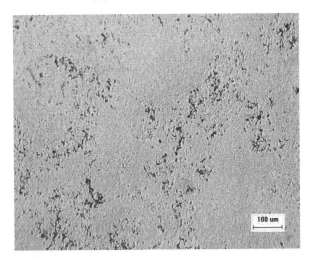

Fig. 6. Microstructure of sample made with Al-30 Ni+Ni showing a second phase (black spots) distributed within the TiC and Ni₃Al.

MMCs BY ACTIVATED MELT INFILTRATION: HIGH-MELTING ALLOYS AND OXIDE CERAMICS

J. Kuebler, K. Lemster, Ph. Gasser, U.E. Klotz and T. Graule
EMPA, Swiss Federal Laboratories for Materials Testing and Research
Ueberlandstrasse 129
8600 Duebendorf, Switzerland

ABSTRACT

For the fabrication of MMCs (Metal Matrix Composites), the infiltration behavior of metallic melts in ceramic substrates is decisive. The use of titanium in pressureless activated melt infiltration for some ceramic/alloy combinations is currently of considerable interest.

Since most ceramics are only poorly wetted by metal melts, an activating element such as titanium or chromium is needed. It is still not known whether the improvement of wetting, and thereby infiltration, is due to the formation of a reaction layer on the ceramic by the active element, or whether titanium is only necessary to promote the beginning of the infiltration process, which may then proceed further via gravitational and capillary forces.

In order to improve the understanding of the infiltration process, time-dependent infiltration experiments have been performed and the obtained microstructures and ceramic-metal interfaces characterized.

INTRODUCTION

For many applications in mechanical engineering and manufacturing of tools, materials with good wear and corrosion resistance are needed. Although ceramics show good corrosion resistance, their brittleness inhibits their use in monolithic form for many applications. MMCs ideally combine the good corrosion properties of ceramics with the ductility of metals.

Since metal melts only poorly wet most ceramics, an activating element, e.g. titanium, is needed to achieve infiltration. The effect of Ti or Cr as activators to enhance wetting on alumina has been studied in copper and nickel-base alloys used in brazing and soldering technology [1, 2]. The causes for the enhancement of the wettability through the addition of titanium have been reviewed by Li [3]. The formation of a TiO_x activated layer, but also modification of fluid flow, active metal adsorption and surface roughness have been recognized as wettability-enhancing processes. The stoichiometry of TiO_x, which is controlled by oxygen fugacity, plays an essential role.

In previous investigations activated melt infiltration has been used for the production of MMCs using SiC ceramics in combination with low melting point alloys like bronze and Ag [4]. Our present interest is in combining high-melting alloys (e.g. Ni-, Fe-base) with oxide ceramics [5]. The influence of dwell time and activator content on the obtained MMC microstructure was investigated, with the main focus on the ceramic/metal interfaces and their bonding.

EXPERIMENTAL

Al_2O_3 particles and elemental Ti powder (99.23 % purity) were combined with commercially available organic binders to form homogeneous mixtures. After addition of a small amount of deionised water and homogenization in a ball mill, preforms were prepared by uniaxial pressing. The activated preforms (5 to 20 wt% Ti) were dried in a drying cabinet and subsequently infiltrated: Cleaned Ni-base alloy pieces (Tab. I) were placed on top of the preforms, which were then placed in alumina crucibles. Infiltration was carried out in a high vacuum furnace (Super VII, Centorr Vacuum Industries) with dwell times between 0.25 and 24 h at 1500 °C.

After removing the infiltrated samples from the crucibles, cross-sections were ground in preparation for characterization via optical microscopy (Zeiss Axiovert 100 A) and scanning electron microscopy (SEM, JEOL JSM-6300F). Lamellae for investigation by transmission electron microscopy (TEM, Philips CM-30) were milled using a focused ion beam (FIB, FEI DB235). A detailed description of the TEM sample preparation is given by Gasser et al. [6].

Table I. Chemical composition of Ni-base alloys.

	Approximate chemical composition [wt %], main elements
Hastelloy C4 (2.4610)	Ni >60, Cr 14-18, Mo 14-17, Fe <3, Co <2
Inconel 625 (2.4856)	Ni >58, Cr 20-23, Mo 8-10, Nb 3.15-4.15, Fe 5

INFLUENCE OF DWELL TIME AND ACTIVATOR CONTENT
Activator content

Activated preforms were infiltrated with Hastelloy at 1500 °C. Optical microscopy investigations revealed an inhomogeneous structure along the longitudinal sections of all short-time infiltrated samples. The gradient as a function of infiltration depth is more prominent in samples with high Ti-activation. The macroscopic infiltration depth is greater at higher Ti-contents.

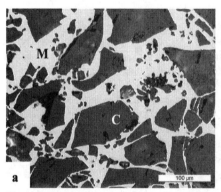 

Figure 1. Structural variation in an intermediately Ti-activated sample with infiltration depth a: 1 mm, b: 4 mm (0.5 h dwell time, M=Metal matrix, C=Ceramic). The arrow indicates the Ti-rich phase in the lower part of the sample.

In Fig. 1, an example of the upper and lower MMC structure of an intermediately Ti-activated sample is shown: In the upper part, coarse Al_2O_3 particles and Al_2O_3 fines were found in the metal matrix (Fig. 1a). The lower MMC structure exhibits a reddish looking phase in addition (Fig. 1b, medium gray) which was shown by EDS to be Ti-rich. Some porous areas were shown to persist. In general, the pore content increases with increasing infiltration depth up to the transition into the non-infiltrated ceramic structure.

Dwell time

Long-term experiments were carried out with preforms containing different levels of activation. Macroscopically, all samples were completely infiltrated. In contrast to the short-term samples, all long-term samples show no Ti-rich phase in their metal matrix and the fine Al_2O_3 fraction has vanished. Al_2O_3 fines were only found at the bottom of the samples with low Ti-activation. The intermediately and highly activated samples show a homogeneous structure over their whole longitudinal section. The sample with the highest activation shows only coarse Al_2O_3 grains surrounded by metal matrix in its structure.

These experiments prove that time is the most crucial parameter for the infiltration process. During short infiltration cycles, infiltration was cut short by the temperature being decreased. Infiltration proceeded faster at higher Ti contents, and consequently, complete infiltration was achieved only in the case of highly Ti-activated green bodies.

Furthermore, the rounding of Al_2O_3 grains in long term experiments shows that the melt has a slightly corrosive influence. This behavior seems to enhance bonding between the metal and the ceramic.

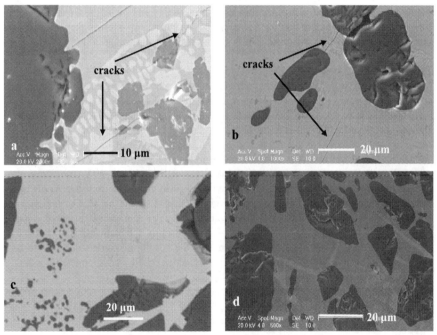

Figure 2. SEM images of the metal matrix (Hastelloy) in dependence on Ti activation and dwell time. a: 0.25 h, 20 wt% Ti (white globular phase: Mo, Cr-rich with Ni; medium grey: Ni with Cr and Mo), b: 24 h 20 wt% Ti, c: 0.25 h, 5 wt% Ti, d: 24 h, 5 wt% Ti.

Metal matrix

SEM investigations of polished Hastelloy-infiltrated samples showed distinct differences in

the metal matrices obtained. Comparison of low and high Ti-activation with dwell times of 0.
and 24 h, revealed that many cracks occurred in the highly-activated 0.25 h samples (Fig. 2a)
well as in the 24 h infiltrated samples (Fig. 2b). Samples with low Ti-activation showed little
no cracking in the metal matrix (Fig. 2c, 2d). When cracking occurs, it is mostly aligned with t
demixing lamellae. Sometimes cracks pass through the metal as well as the ceramic.

EDS measurements confirmed the homogeneity of the metal matrix of the low Ti-activat
samples. The short-time samples with high Ti-activation (Fig. 2a) showed white globular are
which were confirmed by EDS as Cr, Mo-rich with Ni, whereas the medium grey matrix consi
mostly of Ni with Cr and Mo.

CERAMIC/METAL INTERFACES AND BONDING

In order to obtain information about the interfaces and the nature of bonding between t
ceramic particles and the metal matrix, several cross-interface lamellae were milled by FIB
subsequent TEM investigation.

Since formation of a reaction layer should be more prominent with increasing dwell tim
most lamellae were prepared from ceramic/metal interfaces in long-term infiltrated Hastello
MMC samples. In some places, additional milling with FIB was carried out to investigate t
interface in a mechanically-unaffected state. The top surface of samples prepared for optical a
electron microscopy may be mechanically altered by polishing and grinding.

Long-term Al$_2$O$_3$/Hastelloy-MMC

The samples with high Ti-activation generally displayed unsatisfactory bonding with ma
cracks appearing at the interfaces. In some cases, the cracks were very large. "Secondary" crac
were also found in the metal matrix itself. Some metal surfaces at the boundary with the ceran
showed step-like features. It is assumed that at these locations no wetting of the ceramic by t
metal took place such that the metal developed its crystalline facets during cooling. The lamell
from long-term high and low Ti-activation samples that were extracted by FIB and examin
with TEM showed very smooth and sharp interfaces with no sign of a transition zone or diffusi
layer. Line-scans from TEM (Fig. 3) showed only little diffusion from the Al$_2$O$_3$ to the me
matrix. Neither a reaction zone nor a layer enriched in Ti could be seen at the interfaces. T
corresponding EDS line-scan showed no trace of a Ti-containing interlayer (Fig. 3) or
maximum in Ti-content. The decrease in counts for each of the elements is an artifact, which
due to the beam diameter of about 30 nm and the resulting overlap at the interface. TE
investigation of the low Ti-activated sample revealed good ceramic/metal bonding, but
reaction zone.

According to Espié et al. [7] and Eustathopoulos and Drevet [8], a reaction layer of Ti$_2$O$_3$ as
was found in sessile drop experiments on Al$_2$O$_3$ (Cu-Pd-Ti alloy, 1200 °C, 10' dwell time) sho
be in the order of at least 0.5 µm in thickness and therefore should be already recognizable
SEM.

No titanium oxides or other continuous reaction layers ≥ 50 nm were detected at
ceramic/metal interface. The interface shows a smooth transition from Hastelloy to Al$_2$O$_3$.

Short-term Al$_2$O$_3$/Inconel-MMC

The element distribution maps (Fig. 4) show inhomogeneities in the metallic part of
MMC. The edges of the Al$_2$O$_3$ grains exhibit lower Al and O concentrations than the rest of
grain, which is an effect of different sample thickness due to preparation. At the ceramic/me

interface, a narrow Ti-Ni-Nb fringe was detected, but it is not continuous along the interface. In the metal matrix, lamellae consisting mainly of Ti, Ni, and Nb were found.

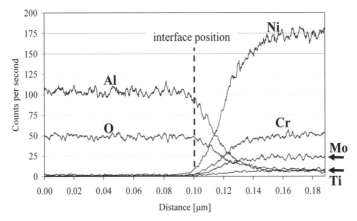

Figure 3. Line-scan at the ceramic/metal interface of an Al_2O_3/Hastelloy-MMC (24 h, 5 wt% Ti) lamella. No reaction layer was detected.

Cr- and Mo-rich areas were detected in places where no prominent Ti-Ni-Nb fringe exists. The Cr-Mo seems to bond directly to the ceramic interface. Mo and Cr were also detected in the gussets of the Ti-Ni-Nb lamellae.

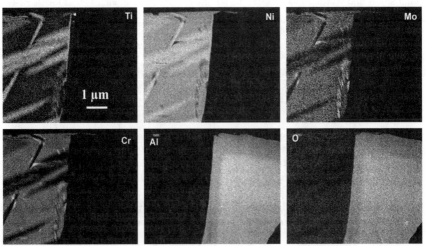

Figure 4. Element distribution images of Al_2O_3/Inconel-MMC-lamella (high Ti-activation, short dwell time).

79

Fe occurs in areas with medium Ni, Cr, Mo, and Nb concentrations. The Ti-Ni-Nb lamellae a well as the Cr-Mo gussets are to a large extent Fe-free. The O concentration in the metal part o the MMC shows an inhomogeneous distribution. O was found in areas with medium Ni, Cr, Fe Mo, and Nb concentrations, but not inside the lamellae containing high Ni-Ti-Nb concentrations.

The metal matrix of the MMC can be divided into three different areas: 1) The dark lamellae which are Ti-Ni-Nb-rich, in all likelihood consist of γ'-phase of the Ni_3Al-type, where Al i substituted by Ti and Nb. 2) The medium gray matrix (Cr-Fe-Ni-Mo-Nb- and O-containing could be a Ni solid solution. 3) The light gussets (Cr-Mo-rich phase) possibly represent th intermetallic σ phase. Further investigations are needed to verify these assumptions. The analysi is complicated by the fact that the areas of the different phases are too small (features on the orde of several 10 nm) for quantitative investigation.

SUMMARY

It is possible to fabricate MMCs from Al_2O_3 and Ni-base alloys by activated melt infiltratior Infiltration times of a few minutes up to several hours were possible due to the chemical stabilit of the ceramic component in the alloys. Microstructural analysis in general confirmed goo bonding at the ceramic/metal interface. Samples with low Ti-activation exhibit better interface than the highly Ti-activated samples, which feature some cracks between the ceramic and th metal. Independent of the appointed Ni-base alloy, the amount of Ti-activation and the infiltratio time, no reaction layer at the ceramic/metal interface could be verified.

REFERENCES

[1]C.-C. Lin, R.-B. Chen and R.-K. Shiue, "A wettability study of Cu/Sn/Ti active braze alloy on alumina," *Journal of Materials Science,* **36**, 2145-2150 (2001).

[2]P. Kritsalis, V. Merlin, L. Coudurier and N. Eustathopoulos, "Effect of Cr on Interfacia interaction and wetting mechanisms in Ni alloy / alumina systems," *Acta metall. mater.*, **40** [6 1167-1175 (1992).

[3]J.-G. Li, "Wetting of Ceramic Materials by Liquid Silicon, Aluminium and Metallic Melt Containing Titanium and Other Reactive Elements," *Ceramics International*, **20**, 391-412 (1994)

[4]Ch. Englisch and K. Berroth, "Keramik-Metall- oder Metall-Keramik-Komposite (Ceramic Metal- or Metal-Ceramic-Composites)," CH Pat. No. 692 296, April 30, 1997

[5]K. Lemster and J. Kübler, "Keramik-Metall- oder Metall-Keramik-Komposite (Ceramic Metal- or Metal-Ceramic-Composites)," PCT/CH 03/00702, patent pending.

[6]Ph. Gasser, U.E. Klotz, F.A. Khalid and O. Beffort, "Site Specific Preparation by Focuse Ion Beam (FIB) Milling for Transmission Electron Microscopy (TEM) of Metal Matri Composites," *Microscopy and Microanalysis*, accepted September 2003

[7]L. Espié, B. Drevet and N. Eustathopoulos, "Experimental Study of the Influence c Interfacial Energies and Reactivity on Wetting in Metal/Oxide Systems," *Metallurgical an Materials Transaction*, [25A] 599-605 (1994).

[8]N. Eustathopoulos and B. Drevet, "Interfacial bonding, wettability and reactivity in meta oxide systems," *Journal de Physique III*, **4** [10] 1865-1881 (1994).

MULTIFUNCTIONAL METAL-CERAMIC COMPOSITES BY SOLID FREE FORMING (SFF).

Rolf Janssen, Mark Leverkoehne[*], Jose J. Coronel
Advanced Ceramics Group
Technical University Hamburg-Harburg
Denickestrasse 15
21073 Hamburg
Germany

ABSTRACT

In recent years, the development of metal-ceramic composites with interpenetrating networks has found considerable interest due to their advantageous material properties. The idea of multifunctional metal-ceramic composites consequently arose from the combination of the 3A (alumina-aluminide alloys) and the LPIM (low-pressure injection molding) technology. Thereby, composites with locally controlled composition can be fabricated b y e xtrusion o f d ifferent m etal-ceramic p araffin w ax s uspension i n s olid free forming. For example, structures of a electrical conducting network ($\sim 10^{-3}$ Ωm) within an isolating matrix ($> 10^{6}$ Ωm) can be tailored as desired. Possible applications are structural parts with integrated heating or sensible functions, e.g. for indicating wear or crack propagation.

INTRODUCTION

For numerous applications, materials are needed which have functional (e.g. sensoric) properties in addition to mechanical strength. Multifunctional metal-ceramic composites with electrically conductive structures in an insulating matrix can fill these requirements. Local control of composition can be realized by solid free forming technology. The spatial composition can be varied by computer-controlled deposition of powder-loaded wax suspensions. Both components are metal-ceramic composites. The conductive component has a high metal content (above the percolation threshold of about 30 vol. %) so that a continuous network of metal phases is formed during reaction sintering. The insulating matrix component also contains metal phase to ensure compatibility during co-sintering, but the metal content is less than 10 vol. % so that the composite contains only isolated metal particles. Such composites have a number of potential applications, e.g. as damage sensor for detection of cracks (or wear) or as heating element[1].

A novel solid free forming technique was developed for the fabrication of such composites. The composite structure is built up layer-by-layer by co-extrusion of powder-loaded paraffin wax suspensions using two separate feed and nozzle systems[2]. Conductive and insulating component can be selected from different material systems. Zr_xAl_y-Al_2O_3

---

[*] Present address: Robert Bosch GmbH, Tuebinger Strasse 123, 72762 Reutlingen, Germany

and Cr-Al$_2$O$_3$ systems due to good high-temperature properties can be used for this purpose. Composition and sintering conditions can be varied in order to optimize mechanical and electrical properties. The effect of temperature and metal content on the electrical conductivity can also be determined. In the reactive Al/Al$_2$O$_3$/ZrO$_2$ system, regions with low Al content react to electrically insulating Al$_2$O$_3$/ZrO$_2$/ZrAl$_2$, while a high Al content leads to the formation of conductive ZrAl$_3$/ZrAl$_2$/Al$_2$O$_3$[3]. The reaction of Al and ZrO$_2$ to zirconium aluminides takes place during reaction sintering. In the Cr-Al$_2$O$_3$ system, a good densification in pressureless sintering was achieved up to a Cr content of 50 vol. % by addition of Al (1-2 wt%) to the starting powder mixture[4]. This novel pressureless reaction sintering process has been developed at the Advanced Ceramics Group for aluminothermic fabrication of metal-reinforced Al$_2$O$_3$ (alumina-aluminide alloys, 3A)[5,6,7]. The process was aiming to improve strength and fracture toughness compared to pure Al$_2$O$_3$ using low-cost starting powders. The aluminothermic reaction between Al and contained metal oxides takes place during sintering.

The suspension technology used was originally developed at the Advanced Ceramics Group for low-pressure injection molding (LPIM) of ceramic parts[8]. Main component is a low-melting paraffin wax, so that processing can take place at around 80 °C. It was successfully adapted to the special needs of milled metal-ceramic powders. Debinding was performed by a combination of capillary forces and thermal decomposition prior to sintering.

Model samples were produced to characterize and optimize the electrical and mechanical properties (model composites: conductive 1D-ligaments in an insulating matrix / rectangular samples with U-shaped ligaments / CT samples with multiple ligaments in one plane).

This work focuses on the fabrication and characterization of multifunctional metal-ceramic composites with electrically conductive structures in an insulating matrix. Local control of composition is realized by a novel solid free forming technique. The spatial composition can be varied by computer-controlled deposition of powder-loaded wax suspensions.

EXPERIMENTAL
Starting Powders and Investigated Compositions
Details of the starting powders and the compositions investigated are given in Tables I and II. The different compositions were always named after the volume content of metal in the starting powder mixture (see Table II). These names will often appear during presentation and discussion of results.

Table I. Details of the starting powders used.

powder	producer	trade name	median particle size, d_{50} [μm]
α-Al$_2$O$_3$	Condea Inc.	Ceralox HPA0.5	0.5
m-ZrO$_2$	Z-Tech Corp.	SF Extra	0.8
Al	Fa. Eckart	AS 081	20
Cr	Alfa Aesar	Cr 99.8%	5
Mo	Alfa Aesar	Mo 99.95%	16

Table II. Details of the compositions investigated[2].

	material system	Zr$_x$Al$_y$-Al$_2$O$_3$		Cr-Al$_2$O$_3$		Mo-Al$_2$O$_3$
	sample name	"Al10"	"Al44"	"Cr10"†	"Cr50"†	"Mo30"
starting	Al$_2$O$_3$ [vol%]	83,1	36,0	89,5	47,4	68,5
powder	ZrO$_2$ [vol%]	6,9	20,0	-	-	-
mixture	Al [vol%]	10,0	44,0	0,5	2,6	2,2
	Cr [vol%]	-	-	10,0	50,0	-
	Mo [vol%]	-	-	-	-	29,3
	powder density [g/cm³]	3,961	3,754	4,279	5,521	5,793
after	Al$_2$O$_3$ [vol%]	94,6	71,1	90,1	50,4	70,8
sintering	ZrO$_2$ [vol%]	-	-	-	-	-
	Zr$_x$Al$_y$ [vol%]	5,4	28,9	-	-	-
	Cr [vol%]	-	-	9,9	49,6	-
	Mo [vol%]	-	-	-	-	29,2
	t.D. [g/cm³]	4,036	4,156	4,285	5,543	5,810
	sintered density [%t.D.]	94,0	91,2	99,4	95,0	94,1

†equivalent to these samples: compositions "Cr15", "Cr20", "Cr25", "Cr30", "Cr35"

Particle Size Analysis

The particle size distribution of milled powder mixtures was measured by light diffraction. The Al starting powder used had a comparably high median particle size of around 20 µm (see Table I) for safety reasons. A fine particle size of around 1-2 µm was necessary to ensure a sufficient reactivity during sintering, so that the starting powder mixture was milled in an attrition mill using TZP balls (tetragonal zirconia polycrystals). A minimum in particle size seems to be reached at a point where milling effect and coarsening due to cold welding are in equilibrium[9].

Suspension Fabrication

Paraffin and different surfactants were melted at 100 °C in a glass beaker. The powder mixture was then added stepwise while mixing the suspension with a glass stirrer. Further homogenization was achieved by passing each suspension several times through a three-roller mill (Al$_2$O$_3$-roller mill; Exact GmbH, Germany). Each Al$_2$O$_3$ roll had a diameter of 80 mm and a length of 200 mm. The rolls were held at a constant temperature of 90 °C. High shear stresses in the small gap between the counter-rotating rolls lead to good homogenization of the suspensions, so that a screw-type mixer was not necessary. All suspensions were repeatedly taken off the roller mill by collector roll and blade and again homogenized until sufficient homogeneity had been achieved. Subjective criteria for sufficient homogeneity were the flow behavior and the optical impression of the suspension (the smaller the still contained particle agglomerates, the smoother and shinier the suspension surface). Powder mixtures with a high Al content were more difficult to disperse than pure alumina. In that case, as much powder as possible was added to the binder in the glass beaker and the rest of the powder was sprinkled in the suspension stepwise during homogenization on the three-roller mill. Other metal powders like Cr, Fe or NiAl were less problematic. Details on binder compositions optimized are given elsewhere[2,10].

The homogenized suspensions were further processed in suspension containers at the solid freeforming apparatus. Before deposition of ligaments in solid freeforming, each suspension was heated to around 90 °C, mixed for 2 h and subsequently degassed for 2 h (without mixing). Electromagnetic pressure valves were used to control the external pressure and, b y t hat, t he s uspension f low t hrough t he p ipe a nd t he h eated c opper t ube towards extrusion head and nozzle.

Solid Freeforming with Local Control of Composition

Fig. 1 shows schematically the solid freeforming technique for local control of composition which has been developed and assembled in the course of this work. XYZ-table and extrusion head were contained in a dust cover.

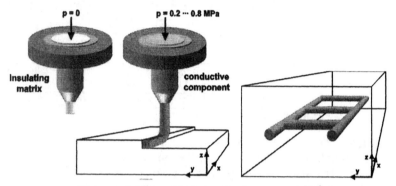

Figure 1. Solid freeforming technology (overview)[2].

Continuous ligaments could be extruded at low diameter variation using conical sapphire nozzles (outlet diameter: 250 µm). Processing conditions were 80-90 °C and 0.6-0.8 MPa. The optimal table speed of the XYZ-Table was determined empirically for each suspension and processing conditions by stepwise speed variation. When the table speed was too low, the ligaments were not deposited in straight lines. Too much material was extruded out of the nozzle at that constant processing pressure, so that rather loops and tangles formed. By reducing the processing pressure or increasing the table speed, the process could be adjusted in such a way that straight ligaments were deposited. When the table speed was too high or the pressure too low, the ligament tore off the nozzle so that only discontinuous segments or droplet-like structures were deposited. Different concepts are possible for layerwise build-up of a part by sequential deposition of ligaments. Two concepts were investigated during this work. In the first, ligaments are placed side by side in direct sequence. Alternatively, each second ligament place can stay empty at first and be filled later. The second procedure proved to be advantageous as gaps were filled better and less distortion occurred during solidification. The next layer was always placed in gaps of the previous layer, i.e. shifted half a ligament diameter. The ambient temperature was adjusted to around 35 °C by a series of infrared light bulbs. By that, the solidification speed of extruded ligaments was far lower and therefore a good bonding between single ligaments ensured.

Debinding, Sintering and Model Samples

Before sintering, the binder was taken out by thermal debinding in a powder bed (330°C for 40 hours). The samples were then sintered for two hours at a temperature of 1500 °C in argon atmosphere. After sintering, the samples were ground using a 107 μm diamond grinding disk to remove any sintering skin and to obtain parallel surfaces. Model samples like CT ones (24 x 25 x 2.4 mm^3) for measuring of R-curve were cut out of bigger plates with a high precision saw (Exact MCP, Exact GmbH, Germany) and also notched. Two holes of 5 mm diameter were drilled with a hollow diamond-coated drill. A sharp precrack was initiated by applying a load in a piezoceramic load cell. The notch was then cut out again so that only the sharp crack remained (crack length around 200μm). During the R-curve measurement, the crack tip was observed my optical microscopy. The load in K_i-direction was increased until the crack propagated, then released so that the crack arrested and the new crack length was measured. This was done repeatedly while measuring the ohmic resistance of the contacted ligaments. The crack surface was investigated by optical and also electron microscopy.

RESULTS AND DISCUSSION
Preparation of the paraffin-wax suspensions

The amount of dispersing agent has to be enough to fill the interparticle voids separating the particles so that the particles can move with respect to each other in shear or extensional flow, and to avoid agglomeration[11,12,13,14]. An excess in binder during molding can separate from the powder, leading to inhomogeneities in the molded compact and dimensional control problems[15,16]. The ideal case corresponds to the particles being in point contact, with no voids in the binder. Thus, there is an optimal binder concentration for a given powder[17]. Fig. 2 compares viscosity versus shear rate for different concentrations of Al$_2$O$_3$ powders. In all samples, the viscosity decreases with the shear rate. Further modifications in the binder composition (see Fig. 2) led to improving in the viscosity.

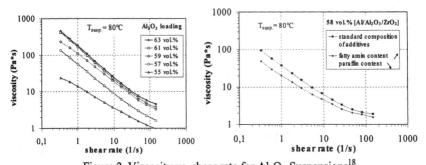

Figure 2. Viscosity vs. shear rate for Al$_2$O$_3$-Suspensions[18].

Solid Freeforming with Local Control of Composition and Multifunctional Porperties

By varying the Al content of the starting compositions, the 3A technology (alumina-aluminide alloys) can be used to obtain electrically insulating as well as conducting composites (see Fig. 3). The smallest possible ligament size is an important process parameter. At present, a minimum continuous ligament diameter of 240 μm has been achieved, with only s mall diameter variation (see Fig. 4). An e xample of the fabricated model samples is shown in Fig. 5. As the programming code necessary for the direct processing of computer-aided design (CAD) data was not yet fully established for the two components, the model samples were fabricated by a combined extrusion and pressing technique. The combined technique also proved to be time-saving for such planar structures.

A strong percolation behavior was observed in the electrical properties, i.e. close to the percolation treshold (at around 28 vol. % metal phase) small changes in composition lead to very large changes in resistivity (± 4 orders of magnitude). AC resistivity measurement proved to be a good tool for characterization of metal-ceramic composites with interpenetrating networks. The values obtained differed from DC values and were also dependant on the frequency. Further knowledge of AC resistivities and their dependence on the frequency is necessary, e.g. for the design of AC resistance heaters.

The produced model samples with ligaments of the second component in one plane showed a good bonding between the two components. In the Cr-Al$_2$O$_3$ and Zr$_x$Al$_y$-Al$_2$O$_3$ systems high interfacial strength was observed and no pores or debonding cracks were detected at the interface. A reaction layer with a thickness of 15 to 20 μm occurred in the Zr$_x$Al$_y$-Al$_2$O$_3$ system (see Fig. 4).

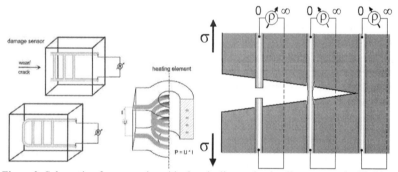

Figure 3. Schematic of a composite with electrically conducting ligaments embedded in an insulating matrix[18].

Figure 4. Left: Extrusion of single ligaments. Right: Samples green, brown and sintered.

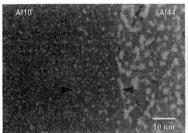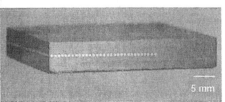

Figure 5. Left: Micrograph of a formed sample after sintering. Right: CT sample for detection of crack propagation[18].

CONCLUSIONS

The solid freeforming process presented in this work is well suited for the fabrication of two-component metal-ceramic composites with full 3D variability by extrusion of powder-loaded paraffin wax suspensions. Structures with high metal content and therefore good electrical conductivity can be incorporated in an electrically insulating matrix. A number of possible applications could be realized with such multifunctional composites, e.g. damage sensors and heating elements.

Currently, a minimum resolution of 240 to 280 µm (220 to 240 µm after reaction sintering) was realized by deposition of ligaments. Enhancement of the suspension homogeneity and better control of the processing pressure may further improve this resolution. The programming code has to be extended to allow the direct processing of CAD data for the two components. In addition, it may be possible to improve the generation of droplets in order to enable full variability at high resolution.

For planar structures of the electrically conductive component, the processing time can be significantly shortened by a combined pressing and solid freeforming approach. Model samples with planar structures of the second component were fabricated in the $Cr-Al_2O_3$ and $Zr_xAl_y-Al_2O_3$ systems. No cracks or pores were observed at the interface of such composites. Likewise, other 3A systems are suitable for the fabrication of multifunctional composites, provided that the sintering shrinkage is similar or adapted by different solid contents of the suspensions.

ACKNOWLEDGEMENTS

The authors are grateful for DFG (German Research Association) funding under E.5-08.032, DAAD (German Academic Exchange Office) funding under A9700435 and CONACyT funding (Mr. J. Coronel).

LITERATURE

[1]Crumm A.T., Halloran J.W. "Fabrication of Microconfigured Multicomponent Ceramics" J. Am. Cer. Soc., 1998, 81, 4, 1053-1057.

[2]Leverkoehne M. "Fabrication and Characterization of Multifunctional Metal-Ceramic Composites." FortschrBer. Düsseldorf: VDI-Verlag. VDI Reihe 5 Nr. 637 (2001).

[3]Zhe X., Hendry A. "In situ Synthesis of Hard and Conductive Ceramic Composites from Al and ZrO2 mixtures by Reaction Hot-Pressing". J. Mat. Sci. Let. [17] 687-689 (1998).

[4]Garcia D.E., Schicker S., Janssen R., Claussen N. "Nb- an Cr-Al2O3 Composites with Interpenetrating Networks." J. Eur. Ceram. Soc. 18 [6] 601-605 (1998).

[5]Claussen N., Garcia D.E., Janssen R. "Reaction Sintering of Alumina-Aluminide Alloys (3A)." J. Mat. Res 11 [11] 2884-2888 (1996).

[6]Schicker S., Garcia D.E., Bruhn J., Janssen R., Claussen N. "Reaction Processing of Al2O3 Composites Containing Iron and Iron Aluminides." J. Am. Cer. Soc., 80 [9] 2294-2300 (1997).

[7]Schicker S., Garcia D.E., Bruhn J., Janssen R., Claussen N. "Reaction Synthesized Al2O3-Based Intermetallic Composites." Acta Mater., 46 [7] 2485-2492 (1998).

[8]Sajko M., Kosmak T., Dirschel R. and Janssen R. "Microstructure and Mechanical Properties of Low Pressure Injection Moulded RBAO Ceramics" . J. Mat. Sci. [32] 2647-2654 (1997).

[9]Roeger M. "Aufbereitung und Verarbeitung von RBAO-Precursormischungen zur Herstellung hochfester Bauteile." Dissertation. AB Technische Keramik, TU Hamburg-Harburg (1997).

[10]Coronel J.J., Leverkoehne M., Janssen R. and Claussen N. "Design of Paraffin Suspensions Highly Loaded with Ceramic or Metal-Ceramic Powders" 25th Annual Conference on Composites, Advanced Ceramics, Materials, and Extructures: B. Ceramics Engineering and Science Proceedings. 22 [4] 67-74 (2001).

[11]Wildman R.D., Blackburn S. "Breakdown of agglomerates in ideal pastes during extrusion" Journal of Materials Science, 1998, 33, 5119-5124.

[12]Benbow J.J., Blackburn S., Mills H. "The Effects of Liquid-Phase Rheology on the Extrusion Behaviour of Paste" Journal of Materials Science, 1998, 33, 5827-5833.

[13]Evans J.R. "Plastics Technology Applied to Ceramic Suspensions" Dt. Keram. Ges., 1993, 8, 81-106.

[14]Novak S., Vidovic K., Sajko M., Kosmac T. "Surface Modification of Alumina Powder for LPIM" Journal of the European Ceramic Society, 1997, 17, 217-223.

[15]Bergström L., Schinozaki K., Tomiyana H., Mizuta N. "Colloidal Processing of a very fine BaTiO3 Powder. Effect of Particle Interactions on the Suspension Properties, Consolidation and Sintering Behaviour" J. Am. Cer. Soc., 1996, 80, 2, 291-300.

[16]Novak S., Dakskobler A., Ribitsch V. „The Effect of Water on the Behaviour of Alumina-Paraffin Suspensions for Low-Pressure Injection Moulding (LPIM)" Journal of the European Ceramic Society, 2000, 20, 2175-2181.

[17]German R. "Powder Injection Molding" Metal Powder Industries Federation, Princeton, 1990.

[18]Leverkoehne M., Coronel J., Dirschel R., Gorlov I., Janssen R., Claussen N. "Novel Binder System Based on Paraffin-Wax for Low-Pressure Injection Molding of Metal-Ceramic Powder Mixtures." Adv. Eng. Mat. 3 [12] 995-998 (2001).

SOLID FREEFORM FABRICATION OF A PIEZOELECTRIC CERAMIC TORSIONAL ACTUATOR MOTOR

Barry A. Bender, Chulho Kim and Carl Cm. Wu
Code 6351
Naval Research Laboratory
Washington, DC 20375

ABSTRACT

Through modifications in binder burnout schedule, firing schedule, and build design, robust torsional ceramic actuator segments were fabricated using laminated object manufacturing (LOM). The lead zirconate titanate (PZT) trapezoidal segments were incorporated into two polygonal torsional actuator tubes. The tubes were used to build a full cycle piezoelectric motor that achieved a speed of 400 rpm at resonant peak frequency.

INTRODUCTION

The continuing evolvement of high authority piezoelectric actuators (high force and high displacement output) employing progressive actuator designs is an important enabling step for future Naval systems. These high authority actuators can be used in piezoelectric ultrasonic motors and as vibration isolation machinery mounts in submarines. At the Naval Research Laboratory, Kim[1] and Wu[2] have created innovative actuators using PZT-based materials that can be used in the above applications. To fully optimize these designs, advanced processing and manufacturing techniques have to be developed such as solid freeform fabrication (SFF).

Solid freeform fabrication techniques generate dimensionally accurate prototypes from CAD files without the use of expensive tooling, dies, or molds. As a result, components with unique geometries can be manufactured in one operation. Since SFF is an additive layered process, components can be fabricated with designed material compositions in the z direction. Fabrication of production components can be created for immediate 'form, fit, and function' performance evaluation. Rapid iterative design optimization is possible since new tooling is not required. All of this leads to shorter lead times and reductions in the cost of development of specialized and/or novel functional ceramics such as high authority piezoelectric actuators.

One such novel high authority actuator being developed for the Navy is the torsional tube actuator.[1] It takes advantage of the high piezoelectric shear coefficient, d_{15}, of PZT. This shear effect can be directly transformed to generate angular displacement and torque. The tubular actuator then can be directly coupled to a rotor to create an efficient ultrasonic piezoelectric motor with a projected power and torque density ten fold that of electromagnetic motors.[3] The innovativeness of the torsional actuator is in using alternately poled PZT segments machined to a certain geometry. The machining of these segments is a long, tedious, expensive process. An alternative way to make these segments is to use the SFF technique of laminated object manufacturing. This paper describes in further detail the design of the piezoelectric motor and the optimization of the LOM process to make robust staves for use in the torsional tube actuator.

PIEZOELECTRIC TORSIONAL ACTUATOR MOTOR

The heart of the piezoelectric motor is the torsional tube actuator stator. The torsional actuator tube consists of an even number of piezoelectric ceramic segments, poled along their length such that the polarity alternates between adjacent segments (see Fig. 1). PZT-5A was chosen as the electroactive material due to its combination of high shear response, high d_{15} and high Curie temperature.[1] The ceramic segments are trapezoidal in cross-section. They are machined from blanks of PZT with specific geometries such that when the segments are joined together they form a polygonal tube that approaches being circular in cross section. The segments are poled using a continuous poling fixture[4] where the segments are placed in peanut oil at 100°C and moved slowly through a 10kV/cm field. The poled segments are assembled into a polygonal actuator tube using silver-filled epoxy which serves both as a joint and an electrode.[1]

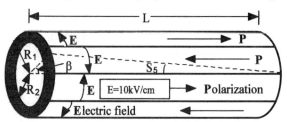

Fig. 1 Schematic of the segmented PZT torsional tube (L=6cm, R_1=1.6cm, R_2=1cm).

Application of an electric field, E, perpendicular to the polarization direction of the segments results in a twisting deformation of the tube due to the coherent piezoelectric shear response of the individual segments. The shear strain, S_5, induced in each segment is directly transformed into the angular displacement, β, of one end of the tube with respect to the other:

$$\beta = S_5 L/R \qquad (1)$$

where $S_5 = d_{15}E$ and R= $(R_1+R_2)/2$ (R_1 and R_2 are outer and inner radii of the tube- see Fig. 1). Therefore, the L/R ratio serves as a built-in amplification of the piezoelectric strain.

Additional amplification of twisting deformation occurs when the actuator is excited at the electromechanical resonance of the tube[5] (see Fig. 2):

$$\beta = S_5 Q_m L/R \qquad (2)$$

where Q_m is the mechanical quality factor of the actuator at resonance frequency. Ultrasonic piezoelectric motors take advantage of the high-power density of PZT by operating at resonance frequency where the amplification factor, Q_m, has been measured to be as high as 22.[1] To avoid frictional losses the innovative piezoelectric motor designed by Glazounov et al.[5] directly couples the torsional tube stator to the rotor. This is done by cementing a pair of one-way clutches to the actuator tube ends so that they could oscillate together with the tube ends (see Fig. 2). The one-way roller clutches accumulate the minute angular displacements over many cycles of the applied ac field at resonant frequency.[3,5] This allows for the motor to operate in a continuous mode of operation or in a precise stepwise manner.

Dielectric heating does occur when operating at resonant frequency. One reason for this is the lossiness of PZT-5A. The other reason is mechanical loss due to the stress/strain differences between the outer layer and inner layer of the ceramic staves. Even though the applied voltage across the electrodes is the same, E will be higher at the inner layer of the stave because the short length of the trapezoidal segment is 25% smaller than that outside larger length

90

of the segment. Then according to equation (1) there will be strain differences which will induce stresses. These strain differences could be eliminated by the use of compositionally graded PZT staves with varying d_{15} properties so as compensate for the 25% difference. One way to do this is solid freeform fabrication of the ceramic staves using LOM, which would also be a desirable alternative to the expensive tedious process of machining the trapezoidal ceramic PZT segments.

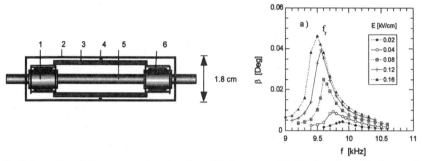

Fig. 2 Schematic of the piezomotor showing (1) the rotor, (2) the housing, (3) the torsional actuator stator, (4) the joint between the stator and the housing, (5) the rotor shaft, and (6) the one-way clutch. The graph shows how the angular displacement is amplified when operating at resonant frequency and increasing electric field E.

LOM OF TORSIONAL ACTUATOR SEGMENTS
The Original LOM Process
The LOM process is a commercially available SFF technology developed by Helisys (now Cubic Technologies, Carson, CA). Their LOMS system was developed for the fabrication of prototypes from CAD design files using laminates of paper. The system works by feeding LOM paper (paper with a heat-sensitive adhesive backing) to the build and laminating it to the built-up stack using thermocompression via a heated roller. The thickness of the build is then measured. This information is sent to the LOM computer which calculates the 2-D slice for that particular thickness from the prototype's CAD model. The computer drives a 50 watt CO_2 laser to cut the perimeter of that cross section one layer deep. The unwanted material is laser-diced. This process is repeated layer by layer until the part is finished. The build is then separated from the LOMS and the waste material is removed to reveal the fabricated 3-D part.

Modified Ceramic LOM Process
Javelin 3D (Salt Lake City, UT) has modified the LOM process to use sheets of ceramic tape in order to fabricate ceramic prototypes. Lamination is done via solvent welding where a coating of solvent is applied to the tape. The solvent attacks the tape surface making it tacky. It is then laminated to the build via mechanical compression. Recently, Javelin 3D has totally automated the ceramic-based LOM process through the development of the SteamRoller-2 (SR2, Javelin 3D). The SR2 works by picking up a sheet of ceramic tape via a vacuum onto a cylindrical drum. The tape is then transported by an arm to the lamination assist solvent (LAS) station where it is rolled against a paint roller and coated with the LAS. The tape is transported into the LOM and placed onto the build where it undergoes compression via automated lamination pressure strokes from the cylindrical drum (similar in motion to a steamroller). The LOM system then takes over and completes the rest of the LOM process cycle.

LOM Processing of Torsional Actuator Segments

The ceramic LOM process is a complex process[6] involving seven different processing steps. To make robust reproducible parts each step must be understood and optimized. This section describes the fabrication of trapezoidal torsional actuator segments and the modifications that were needed to produce functional segments.

Actuator segments were fabricated using PZT-5A (APC Products, Mackeyille, PA) green ceramic tapes made with a proprietary binder system (Richard E. Mistler, Inc., Morrisville, PA) that were 0.2 to 0.25 mm thick. The tapes were laser cut into sheetstock (20 cm x 10 cm) and placed onto the loading table of the SR2. The initial lamination parameter was three coating applications of the LAS (propanol + 10% polyethylene glycol (PEG-200)). Laser cutting was done at 100% power to prevent the formation of a brittle heat affect zone.[6] Laser cutting speed was adjusted to a rate of 11.5 cm/s which allowed the laser just to cut through one thickness of tape. The initial builds were 7.5 x 7.5 cm from which individual trapezoidal segments 0.5 cm in thickness and maximum width of 0.82 cm were decubed. The segments were fired slowly to 485°C to burn out the tape organics and bisqued at 840°C for handling strength. The staves were placed on setter sand and fired to 1245° for two hours in an enclosed environment.

Initial results were discouraging. The segments always came out warped which would make it impossible to put them together as staves in the torsional actuator tube. One way to overcome this problem was to place a frame around the segments in the build. Both a two-sided and a four-sided frame were tested. A base layer 2.5 mm in thickness was added to the build to relieve stresses that can occur during the LOM operation. Both frames worked well but the segments still showed a small amount of warping. Since the two-sided frame uses less material it was chosen as the standard build design. The final amount of warping was eliminated by using a four-step sintering process. Work on improving the strength of LOM PZT test bars had confirmed that PZT does not densify at a constant rate.[7,8] Since warping is often a result of non-uniform densification it could be reduced ideally by firing at a constant densification rate. Because the firing furnace had no dilatometry capability a four-step firing schedule was adapted. The four-step firing schedule used was 0.2°C/s up to 1100°C with a 30 minute hold, 0.2°C/s up to 1150°C with a 30 minute hold, 0.2°C/s up to 1200°C with a 30 minute hold, and 0.2°C/s up to 1245°C with a hold of one hour.

Upon machining the trapezoidal segments to their final dimensions it was discovered that the segments were quite weak (83% failed). Fractography (see Fig.3A) showed the presence of delaminations between layers of tape. These delaminations were associated with large interlaminar defects indicating that too much LAS was used. In the next build the number of LAS coatings was reduced from three to two. Fractography also showed that the bars failed at large flaws near their centers. This indicated that the builds were experiencing large overpressure of trapped gases during the binder burnout process. The same binder burnout schedule was used that was successful in earlier work.[6] However, it was realized that the torsional actuator builds were much larger in size (55 cm^2 vs. 12 cm^2). Researchers[9,10] modeling the binder burnout process showed that gas overpressure is very sensitive to part size. A four fold increase in part size leads to a three fold increase in overpressure.[9] As a result, the binder burnout schedule was modified by reducing the heating rates by 5 fold to 0.1°C/min.

The modifications showed improvements as upon machine finishing the bars only 33% failed. Fractography (see Fig. 3B) showed that the number and size of the delaminations were smaller but still too many. Therefore, the number of LAS coatings was dropped to one for the next build. For iteration two, failure again was occurring at a large central binder burnout defect (see Fig. 3B). In an effort to further reduce the overpressure the binder burnout schedule was modified. This time the initial temperature for temperature holds was reduced from 120 to 90°C. This is because decomposition kinetics change when heating rates change.[9] This was confirmed by TGA studies on the PZT-5A tape which showed a drop of 40 degrees in the temperature of 20% decomposition with a change of order of one magnitude in heating rates.

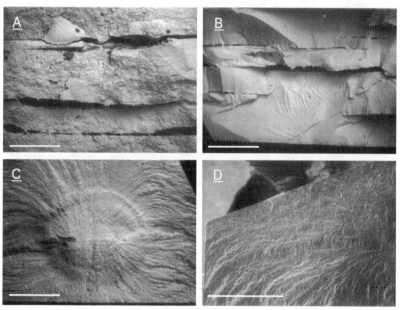

Fig.3 SEM fractographs of the LOM bars showing the reduction in size of the binder burnout flaws and lamination defects with each iteration (A- first iteration, B- second iteration, C- third iteration, and D- final iteration showing a robust stave). Note the line markers represent 1mm.

The modifications showed further improvements as the failure rate dropped to 17%. Though the interlaminar flaws were much smaller, 100% lamination efficiency had not yet been achieved. To obtain better efficiency the LAS solution was changed from propanol to butanol.[8] Failure (iteration 3) was at the center of the specimens but the flaws were much smaller (see Fig. 3C). It appeared that interlaminar defects were acting in concert with the pressure buildup during binder burnout to make these LOM builds even more sensitive to overpressure during the binder decomposition process. Since overpressures as low as 1.05 can lead to failure[10] steps had to be taken to lower the overpressure below the failure threshold. This was done by extending the hold times from 5 hours to 15 hours and adding two more hold times at 150 and 180°C.

The modifications led to the SFF of robust torsional actuator segments (see Fig. 3D). No bars failed during finishing machining. Only 8% of the bars failed during poling. The piezoelectric properties of the poled LOM bars were the same as their commercial counterpart. The staves were successfully incorporated into two torsional actuator tubes. The two tubes were then incorporated into a full cycle piezoelectric motor as shown in Fig. 4. The resonant peak frequency rotor speed was measured to be 400 rpm at 4.4 kHz and E= 240 V/cm.

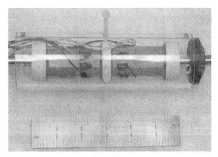 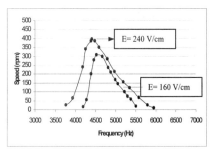

Fig. 4 Photograph of the full cycle piezoelectric motor and graph showing that the motor can obtain a speed of 400 rpm when operating at resonant frequency (4.4k Hz) and E=240 V/cm.

CONCLUSIONS

Through modifications in binder burnout schedule, firing schedule, and build design, torsional actuator segments were successfully fabricated using LOM. The segments were incorporated into two polygonal torsional actuator tubes. The tubes were used to build a full cycle piezoelectric motor that achieved a speed of 400 rpm at resonant peak frequency. In the future we plan to develop graded LOM segments, which will reduce dielectric heating of the piezoelectric motor and help obtain its projected efficiency of having a power and torque density ten times greater than that of electromagnetic motors.

REFERENCES

[1]C. Kim, A. Glazounov, F.D. Flippen, A. Pattnaik, Q. Zhang, and D. Lewis, "Piezoelectric Ceramic Assembly Tubes for Torsional Actuators," *Proc. SPIE*, **3675** 53-62 (1999).

[2]C. Cm. Wu, D. Lewis, M. Kahn, and M. Chase, "High Authority, Telescoping Actuators," *Proc. SPIE*, **3674** 212-19 (1999).

[3]C. Kim, "Piezoelectric Torsional Vibration Driven Motor," US Patent No. 6,417,601 (2002).

[4]C. Cm. Wu, "Continuous Poling of PZT Bars," *Ceram. Trans.*, **90** 21-31 (1998).

[5]A. Glazounov, S. Wang, Q. Zhang, and C. Kim, "Piezoelectric Stepper Motor with Direct Coupling Mechanism to Achieve High Efficiency and Precise Control of Motion," *IEEE Trans. On Ultrasonics, Ferroelectrics, and Frequency Control*, **47** [4] 1059-1067 (2000).

[6]B. A. Bender, R. J. Rayne, and C. Cm. Wu, "Solid Freeform Fabrication of a Telescoping Actuator Via Laminated Object Manufacturing," *Ceram. Eng. Soc. Proc.*, **21** [4] 143-50 (2000).

[7]W. Wersing, H. Wahl, and M. Schnoller, "PZT-Based Multilayer Piezoelectric Ceramics with AgPd-Internal Electrodes, *Ferroelectrics*, **87** 271-94 (1988).

[8]B.A. Bender, R. Rayne, C. Wu, C. Kim, and R.W. Bruce, "Solid Freeform Fabrication of PZT Ceramics Via an Automated Tape Laminated Object Manufacturing System," *Ceram. Trans.*, **154** 193-204 (2003).

[9]R. Shende and S. Lombardo, "Determination of Binder Decomposition Kinetics for Specifying Heating Parameters in Binder Burnout Cycles," *J. Am. Ceram. Soc.*, **85** [4] 780-86 (2002).

[10]L. Liau, B. Peters, D. Krueger, A. Gordon, D. Viswanath, and S. Lombardo, "Role of Length Scale on Pressure Increase and Yield of Polyvinyl Butryal-Barium Titanate-Platinum Multilayer Ceramic Capacitors During Binder Burnout," *J. Am. Ceram. Soc.*, **83** [11] 2645-53 (2000).

CENTRIFUGAL SINTERING

Y. Kinemuchi and K. Watari
National Institute of Advanced Industrial
Science and Technology
2266-98 Anagahora, Shimoshidami,
Moriyama, Nagoya 463-8560, Japan

S. Uchimura
Shinto V-Cerax Ltd.
3-1 Honohara,
Toyokawa 442-8505, Japan

ABSTRACT
Centrifugal sintering is an advanced technology that is specifically designed to sinter structures under constrained conditions, such as film on substrate and multi-layered ceramics. This technology consists of loading high centrifugal acceleration more than 100 km/s^2 onto specimens and heating. Owing to the distinctive pressing measure, pressing without molds, and anisotropic shrinkage during sintering are achieved. Using these advantages, self-pressurized diffusion bonding and crack-free constrained sintering were demonstrated.

INTRODUCTION
Ceramic parts tend to be downsized towards further improvement of properties, which is especially apparent in the electric device application. Among the application, ceramics components are sintered on substrates or with other components such as electrodes. Because of the other components, sintering of ceramics progresses under complicated stress conditions. Normally, sintering is accompanied by a linear shrinkage ratio larger than 10 %. Hence, the shrinkage mismatch between ceramics and other materials is the main reason for the occurrence of residual stress. In most cases, these stresses can not be released during sintering, leading to crack formation or de-lamination. [1] One approach to avoid the formation of these macro defects consists in controlling the viscous nature of sintering. [2] The common procedure for the control in industry is to mix a glass phase with ceramic powders. Other method is to apply external pressure, which was demonstrated by applying centrifugal force in the present study. The later is advantageous to obtain pure thick films without glass phase. Conventionally, the pressing during sintering has been carried out using molds that frequently cause surface reaction. On the contrary, centrifugal pressing utilizes the mass of the film, which is effective to minimize solid contact with other materials.

The centrifugal force can be applied to diffusion bonding of small ceramics onto substrates, like chip mounting. This technique is also important to integration of ceramics devices.

In this paper, trial results of centrifugal sintering and self-pressurized diffusion bonding are reported.

EXPERIMENTAL
Equipment
Figure 1 shows the schematic of the equipment used. Specimens are placed in a ceramic rotor which rotates up to 20,000 rpm. The radius of rotation is 80 mm, which results in a maximum acceleration of ~350 km/s^2. The maximum rotation is limited by the resonance of the rotor which is a function of its Young's modulus, mass and configuration. High Young's modulus and low density are desirable material properties to achieve high rotational speed. Therefore, ceramics are good candidates for the rotor material. Furthermore, high tensile strength at high temperature is

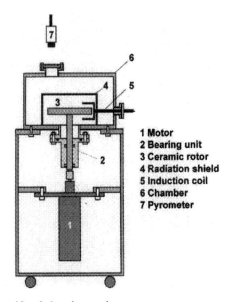

Figure 1. Schematic of centrifugal sintering equipment.

1 Motor
2 Bearing unit
3 Ceramic rotor
4 Radiation shield
5 Induction coil
6 Chamber
7 Pyrometer

required. Based on the above requirements, Si_3N_4 and SiC are adopted for the rotor material. In our design, the rotational number of resonance is settled at ~30,000 rpm. Heating is carried out by an induction heating. The maximum temperature is 1,000 °C. Temperature is monitored by a pyrometer. A chamber is kept airtight, so that an atmosphere of N_2, Ar or vacuum can be available during sintering.

Procedures

In order to clarify the effect of centrifugal force on constrained sintering, sintering of laminated ceramics was performed. Here, $BaTiO_3$ (Koujyundo Chemical Laboratory Co. Ltd., 99.9 % purity, 1 μm) and Ni (Koujyundo Chemical Laboratory Co. Ltd., 99.9 % purity, 2-3 μm) powders were sheet-casted and stacked alternately. In addition, LiF (Koujyundo Chemical Laboratory Co. Ltd., >99.9 % purity) and $BaCO_3$ (Koujyundo Chemical Laboratory Co. Ltd., 99.95 % purity) were added to $BaTiO_3$ to achieve low temperature sintering [3]. The amount of additives was weighed to be $BaCO_3$ of 2 mol% to that of $BaTiO_3$, and LiF of 0.5 wt% to that of a mixture of $BaCO_3$ and $BaTiO_3$. The layers were fabricated by sheet casting in which slurry was prepared mixing with poly vinyl butyral (PVB) and dibutyl phthalate (DBP) in an ethanol medium. The solid loading of the slurry was adjusted to be 20 vol%, and the content of the binder was 40 vol% to the volume of the powder. Here, the volume ratio of PVB and DBP was chosen to be 2 : 1. After drying, the thickness of the sheet was found to be 60 μm. These sheets were layered and then warm-pressed under 10 MPa at 100 °C. Finally, the thickness of green compact became 1.3 mm. Those compacts were subjected to centrifugal sintering where sintering temperature of 1,000 °C was maintained for 20 min in vacuum. Applied centrifugal acceleration was 87 km/s^2 generated by rotational number of 10,000 rpm. Meanwhile, the pellets of each

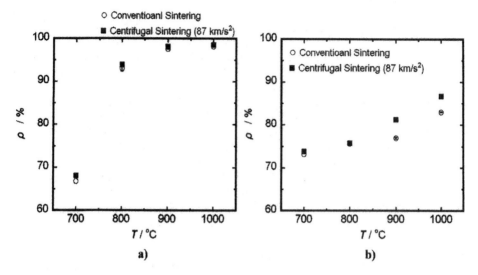

Figure 2. Sintering curves of a) BaTiO₃ and b) Ni as a function of firing temperature.

powder were sintered with or without centrifugal acceleration, in which the sintering temperature was varied from 700 to 1000 °C and held for 20 min. Here, the centrifugal acceleration was set also to be 87 km/s². Archimedean method and dimensional measurement were adopted to analyze sintering behavior, and SEM observation was performed for microstructural analysis.

Self-pressurized diffusion bonding was demonstrated for a joining of alumina. Here, the alumina (density: 99%, purity: 99.9%) was joined to stainless steel (JIS SUS304) using an interlayer of aluminum (99.9% purity). The thickness of the interlayer was 0.5 mm. The selection of these materials was based on published work by Nicholas et al.. [4] The surface of Al₂O₃ and stainless steel had a roughness of several micrometers, i.e., a ground surface without polishing. During diffusion bonding, the chamber was evacuated by means of a rotary pump, and a pressure of less than 100 Pa was maintained. The temperature was raised to 650 ˚C at a heating rate of 5 K/min, and held for 2 h. Furnace cooling was then carried out, during which a cooling rate of 7 K/min was measured above 500 ˚C.

RESULTS AND DISCUSSION

Sintering curves of BaTiO₃ and Ni are shown in Fig. 2. In the case of BaTiO₃, full densification was almost achieved above 900°C, which was a result of liquid formation at low temperature due to the addition of LiF which forms a liquid phase above 610 °C. This densification behavior was in good agreement with a previous report [3]. There was no clear difference in sintering curves between centrifugal sintering and conventional sintering. In the case of Ni sintering, densification improvement by several percents was observed for centrifugal sintering compared to conventional sintering. The difference in the effect of centrifugal acceleration was possibly caused by the difference in their densities: BaTiO₃ of 6.0 g/cm³ and Ni of 8.9 g/cm³. Based on the density difference, the maximum pressure for these pellets was

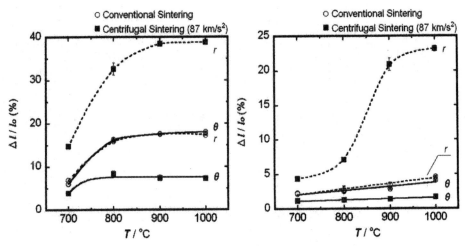

Figure 3. Linear shrinkage, $\Delta l/lo$, of a) BaTiO$_3$ and b) Ni as a function of firing temperature, T. Shrinkage direction is denoted as r and θ indicating radius and tangential directions, respectively.

estimated to be 0.9 and 1.3 MPa for BaTiO$_3$ and Ni pellets, respectively. Although the densification curves were almost similar between centrifugal sintering and conventional sintering, the liner shrinkage behavior was quite different between them as shown in Fig. 3. For conventional sintering, shrinkage occurred homogeneously, while anisotropic shrinkage was observed for centrifugal sintering. The anisotropy in shrinkage is caused by uni-axial pressing promoted by the centrifugal force. These results indicate that the applied centrifugal force was not strong enough to enhance sintering, yet it was enough to promote the rearrangement of particles. Importantly, shrinkage along the tangential direction of rotation was maintained below 10 %, and shrinkage mismatch between BaTiO$_3$ and Ni was kept below 5 %. In contrast, conventional sintering showed the mismatch of ~10 %. Therefore, co-firing of both materials is difficult for conventional sintering. To overcome such large mismatch in conventional sintering, shrinkage adjustment is carried out by adding a second phase. [5] However, additional phases which might result from unwanted reactants leading to deterioration of properties should be into consideration.

Sintering of thick films has similar problems. These include crack formation and separation of film from the substrate. In this case, shrinkage adjustments are not very effective. The way to achieve one dimensional shrinkage, i.e., no shrinkage along the parallel direction to the surface of substrate, is substantial solution. Application of centrifugal sintering might be one options So far, it was confirmed that the acceleration of 87 km/s^2 was beneficial to suppress macro defects of copper film with a thickness of 20 μm.[6]

The advantage of centrifugal force as mentioned above was confirmed by the sintering of layered ceramic components. Here, BaTiO$_3$ and Ni were co-fired. The microstructure of the cross-sectional area is shown in Fig. 4. In comparison with a SEM photograph of conventional sintered specimen (Fig. 4a), all of the layers sintered under centrifugal force showed homogeneous microstructure. On the contrary, conventional sintered specimen included large pores with a crack-like shape, which was a result of constraint due to a shrinkage mismatch

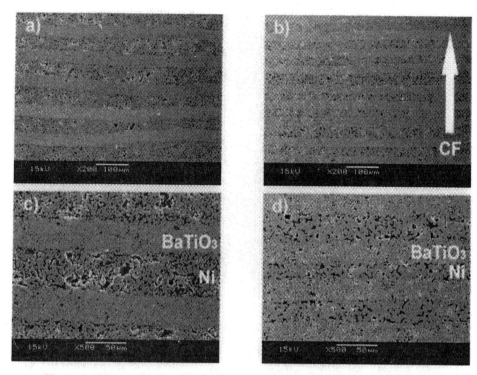

Figure 4. Microstructure of multi-layered ceramics. Figs. a) and b) show the cross-section of ceramics by conventional sintering and centrifugal sintering, respectively. Figs. c) and d) are magnified views of Figs. a) and b), respectively. The arrow in Fig. b) indicates the direction of centrifugal acceleration.

between both layers.

For the fabrication of three dimensional structural devices, bonding technology of solid parts to the structure is advantageous, since complex structure would be divided into several primitive parts. Thus the integration of these primitive parts enables to be shaped into a desirable structure. To join ceramics, diffusion bonding is the most feasible process to obtain high adhesion strength. Diffusion bonding is a solid-state form of bonding by applying external pressure which promotes interfacial matching of faced materials via plastic deformation, diffusion and creep. Typically, this procedure is carried out in the temperature range of 0.5-0.8 of the absolute temperature of the melting point of the material. Diffusion bonding is conventionally processed using hot pressing (HP), where specimens are clamped or inserted in dies, and subsequently pressurized mechanically while applying heat. Therefore, this conventional process is not suitable for the fabrication of three dimensional structural devices. On the contrary, centrifugal pressing is advantageous for this purpose because of its simple procedure and available for parts with complex shape. In order to confirm the possibility of diffusion bonding assisted by centrifugal force, joining of Al_2O_3 on stainless steel was performed. The resultant structure is shown in Fig. 5, showing the potential of centrifugal process for this field of application.

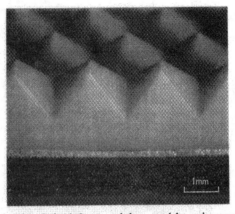

Figure 5. Diffusion bonded Al$_2$O$_3$ to stainless steel by using centrifugal pressing. The upper part with pyramidal structure corresponds to Al$_2$O$_3$ and The bottom part is stainless steel. Aluminum was used as an inter-layer.

SUMMARY

The viability of centrifugal process under heating in a controlled atmosphere has been demonstrated. It was found that the centrifugal process is advantageous to eliminate macro defects in constrained sintering, and to achieve joining of ceramic parts with complex shape. Since the present process does not require pressing media, contamination during processing would be minimized. Owing to these advantages, the present process would be helpful for the ceramics integration.

ACKNOWLEDGMENT

This work was financially supported by the New Energy and Industrial Technology Development Organization (NEDO).

REFERENCES

[1]R. K. Bordia and R. Raj, "Sintering Behavior of Ceramics Films Constrained by a Rigid Substrate," *Journal of the American Ceramics Society*, **68**[6] 287-92 (1985).

[2]R. K. Bordia and G. W. Scherrer, "On Constrained Sintering -I. Constitutive Model For a Sintering Body," *Acta Metallurgica*, **36**[9] 2393-97 (1988).

[3]S. Wang, T. C. K. Yang, W. Huebner and J. Chu, "Liquid-phase Sintering and Chemical Inhomogeneity in the BaTiO3-BaCO3-LiF system," *Journal of Materials Research*, **15**[2] 407-16 (2000).

[4]M. G. Nicholas and R. M. Crispin, "Diffusion Bonding Stainless Steel to Alumina using Aluminum Interlayers," *Journal of Materials Science*, **17** 3347-60 (1982).

[5]P. Z. Cai, D. J. Green and G. L. Messing, "Constrained Densification on Alumina/Zirconia Hybrid Laminates, I: Experimental Observation of Processing Defects," *Journal of the American Ceramic Society*, **80**[8] 1929-39 (1997).

[6]Y. Kinemuchi, K. Watari, H. Morimitsu, H. Ishiguro and S. Uchimura, "Ceramics Integration by Hot Centrifugal Pressing," *Journal of the Ceramic Society of Japan*, (In Press).

ALUMINA-BASED FUNCTIONALLY GRADIENT MATERIALS BY CENTRIFUGAL MOLDING TECHNOLOGY

Chun-Hong Chen, Tadahiro Nishikawa, Sawao Honda, and Hideo Awaji
Nagoya Institute of Technology, Nagoya, Japan, 466 8555

ABSTRACT

Ceramic-based functionally gradient materials (FGMs) were fabricated by centrifugal molding technique (CMT). Experiments were performed with colloidally processed submicron alumina powders and two kinds of tungsten powders (10vol.%). The powder mix was made into slurry by adding alcohol and hollow cylinders were cast using a stainless steel mold. A drying procedure was carried out after casting in a partial vacuum atmosphere to avoid cracks. Tungsten profiles were measured by microstructure observations and EDX analysis. Vickers hardness distributions also proved the compositional variation from inner to outer surface of the gradient cylinders. The result showed that the control of the composite formability has great influence on the composition gradient along the radial direction, which can be attained by varying the ratio of the fine to coarse powders. The slurry viscosity ranging from 100mPa·s to 300mPa·s was found to be suitable for the continuous gradient of hollow cylinders. Residual thermal stresses generated during the cooling process were also analyzed. The hoop residual stresses distributed from tensile on the inner surface to compressive on the outer surface due to the gradual material properties of the alumina-tungsten FGM cylinder. The results obtained indicate that stress concentrations at the interface can be relaxed by the continuous variation of the material properties.

INTRODUCTION

The initial concept for functionally gradient materials (FGMs) was proposed by Niino and his colleagues [1] in Sendai, Japan, as a kind of thermal barrier materials for space planes. The interest in FGMs research has grown rapidly in recent years because FGMs have many advantages, such as high temperature and corrosion resistance as well as reduced residual stress and thermal stress under severe environmental conditions. To attain these superior performances, FGMs can be designed by combining desirable properties of the constituent phase with a continuously gradual variation of other materials, such as ceramics and metals.

There are several approaches for the fabrication of FGMs, which can be divided into gas phase, liquid phase and solid phase method by starting materials. Some of the prevailing techniques used in practice to fabricate hollow cylinder FGMs include centrifugal casting [2-4] and centrifugal molding [5], based on a similar principle. Centrifugal molding is more economical and convenient method than centrifugal casting because the molten metals are not used in the former [6,7]. It is clear that particle separation might occur due to the differential settling during the consolidation [8,9]. However, the relationship between the slurry characters of starting materials and the gradients obtained needs to be studied. A common drawback in all the centrifugal methods is the inaccuracy of controlling the particle distribution.

Studies in this work are directed to the basic investigation of fabrication of hollow cylinder FGMs composed of ceramics and metals. The commercially available powders of alumina and tungsten were chosen because they don't react with each other at high temperature and there is a large difference in density and the similar thermal expansion coefficients at room temperature [10]. The emphasis of the investigation is to study the influence of slurry character, which is adjusted with different particle sizes, on the microstructure of FGMs formed by centrifugal molding technique. Calculations are also made on the thermally induced residual stresses arising from the fabrication process.

EXPERIMENTAL PROCEDURE

Starting Materials and Slurry Preparation

α-Al$_2$O$_3$ (TAIMEI Chemicals Co., Ltd.) with an average particle size of 0.25μm was used for the matrix. Tungsten powders (High purity Chemicals Co., Ltd) with particle sizes averaging 0.6μm and 3μm were also used. MgO with 0.25 wt% alumina was added as a sintering aid and grain growth inhibitor. Tungsten and alumina were used to prepare mixtures, where the ratio of fine to coarse tungsten varied from 0% to 100%. The mixtures were dispersed in ethanol solvent by a magnetic stirrer for 1h. The solid load of the slurry was maintained at 25% by volume through the experiment. Dolapix ED 85 (alkali free polyelectrolyte, Zschimmer & Schwarz GmbH & Co.) and poly vinyl butyral were used as a dispersant and binder, respectively. These dispersant and binder were completely soluble in alcohol and were added along with a minor dopant. While the solid load was used differently, a much larger deviation in density with different slurries could be measured, even in the presence of an optimal dosage of the dispersant and binder. This is because a competitive adsorption phenomenon exists in nonaqueous suspensions [11] and a phenomenon that should be studied further.

Polyacryl tubes of 20mm inner diameter were used for the sedimentation progress and the observation on dispersion of the suspension. The initial slurry heights were 100mm, and the depths of the clarification zones were measured with passage of time. The sediment packing volume fraction was calculated from the sediment height and the particle density. The apparent viscosity of slurries was studied using a rotational viscometer of B type. All the measurements were performed under isothermal condition at 25°C.

Processing and Characterization

The slurries were consolidated by centrifugal molding in a stainless mold with an inner diameter of 30mm and a height of 70mm. The centrifugal acceleration was 150g for compaction with centrifugation velocity set at 3000rpm. The centrifugation time to compact the given slurry was dependent on acceleration and was maintained for 30 minutes. This is in good agreement with the experimental results of Kim et al [12]. Centrifuged green bodies still contain about 10wt% water, which will evaporate during drying. It is well known that fast evaporation of the liquid during drying, especially at the end of the constant rate period can be lead to the formation of cracks in the green body, resulting in a low yield after sintering [13]. The supernatant liquid was poured out after centrifugation, and the mold, together with the sample, was dried in a vertical position. Drying was performed in a temperature-controlled chamber with partially vacuum, where the temperature was increased to 100°C with a slow heating rate of 5°C /h. Since the inner surface of mold was

102

precoated with paraffin wax, the tubes shrunk and released easily from the mold after drying. The dried and shrunk tubes had an approximate dimension of 3mm in thickness. The tubes were then heated gradually up to 673 for 2hs to vaporize the organic particles of dispersant and binder. The sintering was carried out at 1873K for 3hs. The heating and cooling rate was set as 200□/h and an atmosphere of H_2/N_2 with H_2 of 4% and rest N_2 was used through sintering process.

Crystalline phases of the fired bodies were characterized by XRD (Philips 1710). Relative density and porosity of sintered tubes were determined by the Archimedes method in distilled water as per JIS-R 1634. The microstructures were observed on the cross-sectioned surfaces along the radial direction using a scanning electron microscope (ABT-55) equipped with energy dispersion X-ray (EDX) analysis system. The compositional change along the radial direction was examined by EDX area analysis, which was performed on the minute area (300μm×225μm) at equal distances from the inner to outer side of specimens. Vickers hardness was also measured on the polished surface along the radial direction to confirm the compositional change.

RESULTS AND DISCUSSION

Suspension Stability and Viscosity

It is obvious that composition distributions along the centrifugal force differ by the different settling rate of powders under colloidal system. The difference in those structures is mainly dependent on not only rheology of suspensions, but also on the centrifugal parameters such as speed, time etc. A designed gradient and high green strength for handling were obtained when the alumina/W ratio of the mixture is set to 90/10 in volume and the concentration of binder and dispersant is 0.5wt%. All slurries were stabilized to about pH 8. The slurries and the related specimens were denoted as S-0, S-20, S-50 and S-100 in order of the mixing ratio of the fine to coarse W powders.

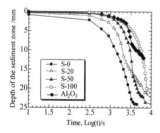 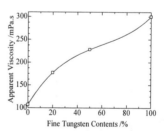

Fig.1 Sediment progress in different slurries   Fig. 2 Variation of the slurry viscosity with fine W contents

Figure 1 shows the effect of the fine W concentration on the sediment progress under gravity. As is obvious from the figure, there is seldom any initial settling rate and the interface settles faster and deeper with the decease of the fine W content. The preferential segregation under gravity while the slurry is poured into the mold is negligible for all slurries. There was no clear interface between slurry and liquid when the mixture of fine and coarse W was used. This was not true in the case of only coarse W since the settlement is fostered by weak agglomerate and gel construction formed by interactions among particles.

Figure 2 shows the apparent viscosity for slurries with different fine W contents. With rising fine powder content, the minimum of apparent viscosity increase from 100mPa•s to 300mPa•s. These results show the fine/coarse powder ratios are essential factor for the slurry character and bi-sized powders are desirable for centrifugal molding technique.

Microstructure

The X-ray diffraction patterns of the sintered specimens were procured and no other diffraction peak expect W and alumina was observed. Fig. 5 shows the sintered specimens. Bi-layered structure appears in S-0, where the lighter phase is alumina, the darker phase is W. No interface in S-20 is observed and there is a continuous gradient along the radial direction as mentioned later. Fig. 6 shows the difference in microstructures along the radius. The darker continuous matrix is alumina whereas the lighter phase is W. Fig. 6 illustrates that W powder increased along direction of the centrifugal force, Since there is very little W powder in the inner side as shown in S-0 and S-20. The composition gradient is slightly visible in S-50 and there are only homogenous microstructures regardless of the centrifugal speed in S-100. It turns out that the composition gradient is increased with the decrease of fine W content. This is because an increase in the fine W content raises the viscosity of the slurry and hinders its fluidity.

Fig.5 Sintered specimens of S-0 and S-20

Composition Gradient under Centrifugal Field

Figure 7 shows the variation of the relative W content measured by EDX. It is clear that the gradient along the radius is formed by using bi-sized W powders. A bi-layer structure is visible in S-0, whereas there is homogenous structure in S-100. The result shows the gradient can be controlled by the ratio of the fine to coarse W powders. The reason is that the viscosity of the slurry is adjusted by changing the ratio of the bi-sized W powders shown in Fig.2 regardless of the large difference in density of W and alumina.

The variation of Vickers hardness along the radius is shown as Fig. 8. Because the hardness of alumina is larger than that of W [10], there is lower value in the alumina-W composite with increasing W. The figure reveals that the Vickers hardness from the inner side to the outer side decreased due to the gradient in S-0, S-20 and S-50. It turns out that pure alumina exists on the inner side in S-20, and the unexpected lower value on the outer side is visible in S-0 since larger pores formed due to the W aggregate formation.

Figure 9 shows the residual stress distributions of the inhomogeous structures generated from the sintering, which are calculated based on the multi-layered model [10,14,15]. The radial stress is less than the hoop stress. Tensile hoop stresses appeared on the inner surface of the hollow cylinder and compressive stresses on the outer surface due to the difference in the thermal expansion coefficient of alumina and W. The radial stress is released by continuous gradient and the hoop stress jumped sharply by the formation of the interface in S-0. The analytical result shows the residual stress is released by selecting appropriate gradients, which can be controlled by the ratio of fine to coarse metal powder.

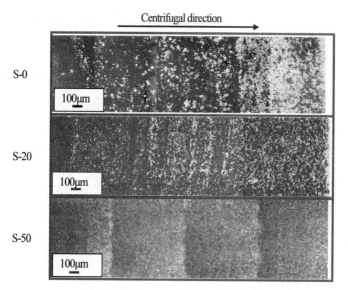

Fig.6 SEM photographs of aluimna-W hollow cylinders

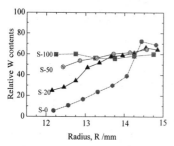

Fig.7 Relative W contents along the radius

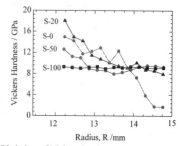

Fig. 8 Variation of Vickers hardness along the radius

CONCLUSIONS

Hollow cylinder of FGMs composed of alumina and W with two kinds of particle sizes was fabricated by centrifugal molding technique. A drying procedure was employed for hollow cylinders in partial vacuum to avoid drying and sintering cracks. SEM photographs and EDX analysis performed showed the different gradient of W content along the radial direction of hollow cylinders. It was in good agreement with measured Vickers hardness values and was obtained with the

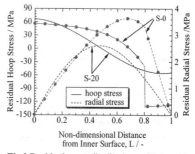

Fig.9 Residual stress distributions of S-0 and S-20

range of the slurry viscosity from 100mPa•s to 300mPa•s by the mixing ratio of the fine to coarse W powders. Residual thermal stresses generated after sintering were also analyzed, and the results show the hoop residual stresses distributed from tensile on the inner surface to compressive on the outer surface due to the composition gradient of the alumina-tungsten FGM cylinder. The stress concentration on the interface can be relaxed by the continuous variation of the material properties.

REFERENCES

[1] M. koizumi, "FGM activities in Japan", *Composites Part B*, **28**, 1-4 (1997).

[2] Y. Fukui, "Fundamental investigation of Functionally Gradient Material Manufacturing System using Centrifugal Force," *JSME Int. Series* □, **34**[1] 144-148 (1991).

[3] J. Zhang, Y-Q. Wang, B-L. Zhou, and X-Q Wu, "Functionally graded Al/Mg$_2$Si in-situ Composites prepared by Centrifugal Casting," *Journal of Materials Science Letters*, **17**, 677-79 (1998).

[4] J. W. Gao and C. Y. Wang, "Modeling the Solidification of Functionally Graded Materials by Centrifugal Casting," *Materials Science Engineering A*, **292**, 207-15 (2000).

[5] T. Nishikawa, A. Masuda, S. Honda and H. Awaji, "Thermal Shock Resistance of Alumina/Nickel Functionally Gradient Materials prepared by Centrifugal Casting," *Functionally Graded Materials*, **13**, 107-113 (1999).

[6] H. Nagae, M. Toriyama, K. Sobata and S. Kiriyama, "Centrifugal Casting of Alumina-Zirconia Mixed Powder," *Journal Ceramic Society of Japan*, **102** [4] 346-49 (1994).

[7] R. Sivakumar, T. Nishikawa, S. Honda and H. Awaji, "Processing of Mullite-Molybdenum Graded Hollow Cylinders by Centrifugal Molding Technique," *Journal Ceramic Society of Japan*, **110**, 472-475 (2002).

[8] F. F. Lange, "Powder processing science and technology for increased reliability," *Journal of the American Ceramic Society*, **72** [1] 3-15 (1989).

[9] J. C. Change, B. V. Velamakanni, F. F. Lange and D. S. Pearson, "Centrifugal Consolidation of Al$_2$O$_3$ and Al$_2$O$_3$/ZrO$_2$ Composite Slurries vs Interparticle Potentials: Particle Packing and Mass Segregation," *Journal of the American Ceramic Society*, **74** [9] 2201-04(1991).

[10] C. H. Chen, S. Honda, T. Nishikawa and H. Awaji, "A Hollow Cylinder of Functionally Gradient Materials Fabricated by Centrifugal Molding Technique," *Journal Ceramic Society of Japan*, **111**[7], 479-484(2003).

[11] K. Blackman, R. M. Slilaty and J. A. Lewis, "Competitive Adsorption Phenomena in Nonaqueous Tape Casting Suspensions," *Journal of the American Ceramic Society*, **84** [11] 2501-06 (2001).

[12] K-H Kim, S-J Cho, D-H Kim and K-J Yoon, "Centrifugal Casting of Large Alumina Tubes," *Journal of the American Ceramic Society*, **19**[11] 54-56(2000)

[13] G. W. Scherer, "Theory of drying," *Journal of the American Ceramic Society*, **73** [1] 3-14(1990).

[14] C. H. Chen and H. Awaji, "Transient and Residual Stresses in a Hollow Cylinder of Functionally Graded Materials," *Materials Science Forum*, **423-425**, 665-670(2003).

[15] H. Awaji and R. Sivakumar, "Temperature and Stress Distributions in a Hollow Cylinder of Functionally Graded Material: The case of Temperature-Independent Materials Properties," *Journal of the American Ceramic Society*, **84** [5] 1059-65 (2001).

INVESTIGATION OF A NOVEL AIR BRAZING COMPOSITION FOR HIGH-TEMPERATURE, OXIDATION-RESISTANT CERAMIC JOINING

K. Scott Weil and John S. Hardy
Pacific Northwest National Laboratory
P.O. Box 999
Richland, WA 99352

Jens Darsell
School of Mechanical and Materials Engineering
Washington State University
Pullman, WA 99164-2920

ABSTRACT

One of the challenges in developing a useful ceramic joining technique is in producing a joint that offers good strength under high temperature and highly oxidizing operating conditions. Unfortunately many of the commercially available active metal ceramic brazing alloys exhibit oxidation behaviors which are unacceptable for use in a high temperature application. We have developed a new approach to ceramic brazing, referred to as air brazing, that employs an oxide wetting agent dissolved in a molten noble metal solvent, in this case CuO in Ag, such that acceptable wetting behavior occurs on a number of ceramic substrates. In an effort to explore how to increase the operating temperature of this type of braze, we have investigated the effect of ternary palladium additions on the wetting characteristics of our standard Ag-CuO air braze composition.

INTRODUCTION

As the operating temperatures of advanced power generation equipment continue to be pushed upward by thermal efficiency considerations, there is an ever increasing need to develop materials suitable for these applications, particularly for use under oxidizing conditions. Ceramics are attractive because of their excellent high-temperature mechanical properties and their high level of wear and corrosion resistance. What limits their usefulness, however, is our current ability to economically manufacture large or complex-shaped ceramic components that exhibit reliable performance. One alternative is to fabricate small-sized, simple-shaped parts that can be assembled and joined to form a larger, more complex structure.[1] A number of joining technologies currently exist, including glass joining, diffusion bonding, reaction bonding, and active metal brazing. However, intrinsic to each is some form of trade-off in terms joint properties, ease of processing, and/or cost. For example active metal brazing requires a stringent firing atmosphere, either high vacuum or reducing-gas conditions, to prevent the active species, typically titanium, from pre-oxidizing. This represents higher capital expenses and operating costs than typical air-fired processes such as glass joining. In addition, recent studies on the oxidation behavior of active metal brazes have shown that they are unreliable at temperatures beyond 500°C, at which point they eventually oxidize completely, causing complete degradation of the joint.[2, 3]

A new brazing technique, referred to as air brazing, has recently been developed[4-6] and shown to offer both excellent joint strength and high temperature oxidation resistance.[7] Employing silver as the filler material and CuO as a wetting agent, joining can be readily conducted in air and acceptable wetting obtained on a number of ceramic surfaces with as little as 1mol% CuO added to the filler.[8] We are currently investigating how the Ag-CuO system can be compositionally modified to achieve greater wettability, better gap filling characteristics, and higher use temperature. In this paper, we will discuss our initial wetting results obtained on alumina substrates when palladium was added to our standard Ag-CuO braze formulations.

EXPERIMENTAL

Materials

Polycrystalline alumina (Al-23, Alfa Aesar, Inc.) discs measuring 50mm in diameter were employed as the wetting substrates in these studies. The alumina was approximately 98% dense and 99.7% pure, containing a small amount of silicate material. Prior to testing, one face on each of the discs was polished to a 10 μm finish, cleaned with acetone and methanol, air dried, and heated in air for four hours at 600°C to burn off any residual organic contamination. The wetting characteristics of the following three different series of braze compositions were determined by sessile drop experiments: the Ag-CuO binary and Pd-Ag-CuO brazes containing a molar ratio of palladium-to-silver of 1:3 or 1:1. In this paper, the different compositions within each braze family are identified by the molar ratio of metal (palladium and silver) to copper oxide. For example, a braze containing 80 mol% of Pd and Ag alloyed in a ratio of 1:3 and 20mol% CuO is labelled 80(1:3)20PAC, whereas a braze formulated using 90mol% of silver, no palladium, and 10mol% CuO is denoted by 9010AC. The brazes were prepared by mixing palladium powder (99.95% purity, 0.5 – 1.7μm average particle size, Alfa Aesar), silver powder (99.9% purity, 0.5 – 1 μm average particle size, Alfa Aesar), and copper powder (99% purity, 1 – 1.5 μm average particle size, Alfa Aesar) in the appropriate ratios to yield the target compositions of our study and cold pressing these mixtures into pellets measuring 0.6mm in diameter by ~0.6mm thick. Once pressed, the braze pellets were heat treated in hydrogen at 700°C for 24hrs to promote compositional homogeneity. In this condition, the pellets still retained between ~20 – 30% porosity, most of it suspected to be open. This is beneficial because during brazing in-situ oxidation of the copper to CuO (the wetting agent) is desired.

Testing and Characterization

Sessile drop experiments were carried out by placing a braze pellet atop the center of the polished face of an alumina disc. The melting and wetting behaviors of the braze pellet on the alumina substrate were observed in a static air box furnace equipped with a transparent quartz window through which the heated sample could be imaged and recorded via video. The samples were heated at 30°C/min to 900°C at which point the heating rate was decreased to 10°C/min. Each experiment was conducted by holding the furnace temperature at least 50°C above the liquidus of each braze for a period of 15 min to allow the bead of molten braze to reach its equilibrium shape. Frames from the end of the fifteen minute dwell at each targeted temperature were captured from the videotape and converted to computer images using VideoStudio6™ (Ulead Systems, Inc.) video editing software. These images were subsequently imported into Canvas™ (version 8.0.5, build 619; Deneba Systems, Inc.) graphics software for contact angle determination. Because the molten braze theoretically forms a spherical cap, a best-fit circle can be circumscribed around the surface of the molten braze. The resulting contact angle is consequently equal to one half the angle of arc formed by the rounded molten braze surface.

Microstructural analysis was conducted on samples from each of the three braze families to investigate the effects of varying amounts of ternary palladium addition on the room temperature microstructure of the bulk braze and the braze/alumina interfacial region. The samples were cross-sectioned, polished, and carbon coated, then examined using a scanning electron microscope (JEOL JSM-5900LV) equipped with an Oxford energy dispersive X-ray analysis

(EDX) system that employs a windowless detector for quantitative detection of both light and heavy elements.

RESULTS AND DISCUSSION

Contact angle measurements of the molten Ag-CuO and Pd-Ag-CuO brazes on alumina are shown as a function of temperature in Figure 1. Two series of data were collected on the binary brazes at 1000°C and 1100°C, whereas the 1:3 and 1:1 palladium to silver based brazes were held isothermally at 1250°C and 1350°C, respectively. In each case, complete melting was observed and the fifteen minute hold time used during the sessile drop experiments appeared to be long enough for a stable contact angle measurement to be obtained. Although not obvious in Figure 1 due to overlap of the data points, none of the pure metal brazes, including silver, displayed wetting on the alumina. In fact, the addition of palladium not only raises the liquidus temperature of the overall braze, as observed in preliminary differential scanning calorimetry measurements, but also causes a general increase in the contact angle of the braze for a given metal-to-copper oxide compositional ratio; in some cases to the point of a non-wetting condition, i.e. >90° contact angle. Increasing the concentration of CuO in the braze has the opposite effect by causing a reduction in the contact angle. From the macroscopic contact angle data in Figure 1 it appears that as little as 1mol% CuO in pure silver or in the 1:3 Pd:Ag alloy is enough to cause wetting, whereas ~10mol% of the oxide is needed in the case of the 1:1 alloy. Molar amounts of up to ~20 – 30% CuO appear to cause significant decreases in contact angle for all three families of brazes, while greater concentrations of the oxide have a much lesser effect. Back scattered electron images of comparable examples from the binary braze family and one of the ternary families, shown in Figures 2 - 4, suggest possible reasons for these trends.

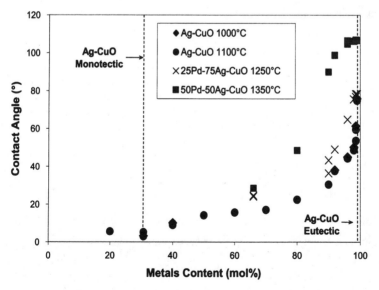

Figure 1:   Contact angle of Ag-CuO and Pd-Ag-CuO brazes on alumina in air as a function of total metals content.

The two Ag-CuO wetting specimens displayed in Figure 2 were both heat treated to a braze temperature of 1100°C, under the conditions described above, then furnace cooled to room temperature. As expected from their target compositions, the majority phase in each braze is pure silver (CuO is not soluble in silver at room temperature). Fine precipitates of CuO on the order of ~1μm in size are typically found in the bulk silver matrix away from the interface with the Al_2O_3. In the braze containing the lowest copper oxide content, 9901AC in Figure 2(a), discrete micron-size CuO particles are found along the braze/alumina interface. Found in the wide regions between these particles and in nearly perfect contact with the alumina interface is pure silver. In the case of the high CuO content braze, 4060AC in Figure 2(b), a nearly continuous interfacial layer of CuO approximately 1μm in thickness was observed covering the oxide substrate. This layer was occasionally disrupted by small patches or islands of interfacial silver.

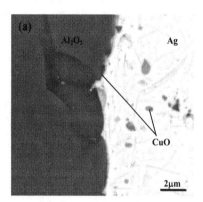 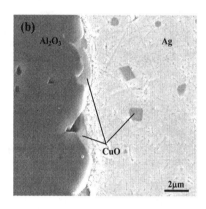

Figure 2 Cross-sectional SEM micrographs of Ag-CuO/Al_2O_3 interfaces: (a) 9901AC and (b) 6040AC. Each wetting specimen was heated in air at a final soak temperature of 1100°C before cooling to room temperature.

In the case of the ternary braze family 25Pd-75Ag-CuO, several interesting observations were noted at the micrscopic level. For example, in the 1mol% CuO sample that displayed a contact angle with the alumina of ~80°, SEM analysis reveals that very little wetting actually took place, as seen in Figure 3(a). Instead, small protuberances appear to grow on the surface of the braze near the faying surface of the alumina. At higher magnification, Figure 3(b), a small amount of adhesion between the braze and substate is apparent, but no CuO is observed in this interfacial region. The CuO is found as a patchy submicron skin covering the outside of the Pd-Ag filler. Apparently a more extreme form of segregation has occurred in this braze composition than was seen in the binary brazes. Unlike the Ag-CuO brazes in which CuO and Ag exhibit a mutual solubility in the liquid state, we suspect that when Pd is dissolved into solution the degree of CuO solubility decreases substantially causing segregation to initiate in the liquid state.

When 4mol% CuO is added to the system, the microstructure of the braze/substrate interface changes dramatically as seen in Figure 4(a). EDX analysis shows that along this interface a thin, 1μm thick CuO-Al_2O_3 diffusion zone forms. We have observed this phenomenon previously between the binary brazes and alumina substrates at $T_{braze} > 1100°C$[7] and found that this result is consistent with the CuO-Al_2O_3 phase diagram, which displays complete solubility between CuO and Al_2O_3. Unlike the binary brazes however, nearly all of the copper from the 96(1:3)04PAC

110

braze appears to be tied up in this interdiffusion zone, with very little being found within the filler matrix.

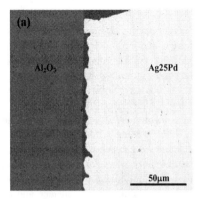 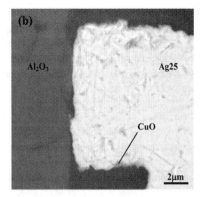

Figure 3   Cross-sectional SEM micrographs of 25Pd-75-Ag-CuO/Al₂O₃ interfaces: (a) 99(1:3)01PAC (low mag) and (b) 99(1:3)01PAC (high mag). The specimen was heated in air at a final soak temperature of 1250°C before cooling to room temperature.

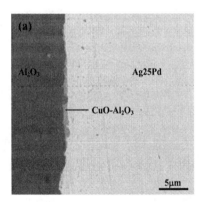 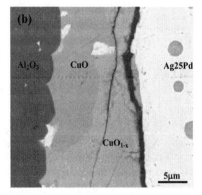

Figure 4   Cross-sectional SEM micrographs of 25Pd-75-Ag-CuO/Al₂O₃ interfaces: (a) 96(1:3)04PAC and (b) 66(1:3)34PAC. Each wetting specimen was heated in air at a final soak temperature of 1250°C before cooling to room temperature.

Interestingly, this interdiffusion zone was not observed in the braze containing a much higher concentration of CuO, 66(1:3)34PAC in Figure 4(b). Instead, the CuO forms a thick continuous layer along the braze/alumina interface. Also, the copper oxide exhibits two distinct stoichiometries in this interfacial layer: a darker phase determined to be CuO (Cu:O of ~1:1) and a lighter phase which appears to be a stoichiometric copper oxide with a Cu:O ratio of 3:1. We have labelled this phase CuO_{1-x} because Cu_3O is not known to be a thermodynamically stable form of copper oxide. Multiple spot analysis of this phase along a 5mm section of this sample via EDX indicates that the oxide exhibits a definite 3:1 metal to oxygen stoichiometry. We are currently conducting additional tests to confirm the formation of this phase. Numerous cracks were observed running through this phase, as seen in Figure 4(b), suspected to be caused by thermal mismatch with the alumina substrate.

CONCLUSIONS

The addition of palladium to our standard Ag-CuO air brazing compositions was investigated as a potential means of increasing the use temperature of the braze. Reported here are our intial wetting and microstructure results on these ternary brazes. The findings from macroscopic contact angle experiments conducted on alumina substrates indicate that palladium does cause a measurable increase in the wetting angle, but that as little as 1mol% CuO would provide an acceptable level of wetting in the Ag-CuO and 25Pd-75Ag-CuO braze families and ~10mol%CuO in the 50Pd-50Ag-CuO brazes. However, microstructural analysis revealed that while this is true for the binary braze system, little wetting actually took place in the 99(1:3)01PAC braze. A more promising microstructure was observed in the 96(1:3)04PAC sample, which displayed a thin, non-continuous $CuO-Al_2O_3$ interdiffusion zone between the braze and the substrate. While increases in the CuO concentration show improved macroscopic wetting, a thick interfacial layer of CuO segregates to the alumina faying surface and tends to form microsopic cracks, suspected to initiate upon cooling.

ACKNOWLEDGMENTS

The authors would like to thank Nat Saenz, Shelly Carlson, and Jim Coleman for their assistance. This work was supported by the U. S. Department of Energy, Office of Fossil Energy Advanced Research and Technology Development Program. The Pacific Northwest National Laboratory is operated by Battelle Memorial Institute for the United States Department of Energy (U.S. DOE) under Contract DE-AC06-76RLO 1830.

REFERENCES

[1] S.D. Peteves, M. Paulasto, G. Ceccone, and V. Stamos, The Reactive Route to Ceramic Joining: Fabrication, Interfacial Chemistry, and Joint Properties," Acta Materialia, **46** [7] 2407-14 (1998).

[2] J-H Kim and Y-C Yoo, "Bonding of Alumina to Metals with Ag-Cu-Zr Brazing Alloy," *Journal of Materials Science Letters*, **16** [14] 1212-15 (1997).

[3] J.P. Rice, D.M. Paxton, and K.S. Weil, "Oxidation Behavior of a Commercial Gold-Based Braze Alloy for Ceramic-to-Metal Joining," *Proceedings of the 26th Annual Conference on Composites, Advanced Ceramics, Materials, and Structures* (2002).

[4] C.C. Shüler, A. Stuck, N. Beck, H. Keser, and U. Täck, "Direct Silver Bonding - An Alternative for Substrates in Power Semiconductor Packaging," *Journal of Materials Science: Materials in Electronics*, **11** [3] 389-96 (2000).

[5] K.M. Erskine, A.M. Meier, and S.M. Pilgrim, "Brazing Perovskite Ceramics with Silver/Copper Oxide Braze Alloys," *Journal of Materials Science*, **37** [8] 1705-9 (2002).

[6] J.S. Hardy, J.Y. Kim, and K.S. Weil, "Joining Mixed Conducting Oxides Using an Air-Fired Electrically Conductive Braze," *Journal of the Electrochemical Society*, in press.

[7] J.Y. Kim, J.S. Hardy, and K.S.Weil, "Effects of CuO Content on the Wetting Behavior and Mechanical Properties of a Ag-CuO Braze for Ceramic Joining," *Journal of the American Ceramic Society*, in review.

[8] K.S. Weil, J.S. Hardy, and J.Y. Kim, "A New Technique for Joining Ceramic and Metal Components in High Temperature Electrochemical Devices," *Journal of Advanced Materials*, in press.

JOINING OF ADVANCED STRUCTURAL MATERIALS BY PLASTIC DEFORMATION

D. Singh, N. Chen, K. C. Goretta, and J. L. Routbort
Energy Technology Division
Argonne National Laboratory, Argonne, IL 60439, USA

F. Gutierrez-Mora
Departmento de Fisica de la Materia Condensada
Universidad de Sevilla, Sevilla 41080, Spain

ABSTRACT

Superplastic deformation has been employed to join nickel aluminide (Ni_3Al) at temperatures of 1000-1200°C and strain rates ranging from 1×10^{-5}/s to 5×10^{-3}/s. Optimum joining was achieved at 1100°C for strain rates ranging from 1×10^{-4}/s to 9×10^{-4}/s. Corresponding flow stresses were 47-123 MPa with permanent strains $\approx 10\%$. The room-temperature stress-strain of the joined flexure bars made at 1100°C was similar to that of monolithic samples. The role of surface oxidation conducted at 1100-1200°C on the joint quality was also investigated.

INTRODUCTION

Recently, joining of various alumina- and zirconia-based structural ceramics using high-temperature superplastic deformation has been demonstrated [1-3]. Joining by superplastic deformation entails heating the parts to elevated temperatures and deforming them under a constant displacement rate. Total strains in the joined parts are typically <10%. Dense joints are formed by grain sliding and interpenetration of grains at the interface [1-3]. Joining by superplastic deformation requires no surface preparation, results in minimal deformation, and occurs at lower temperatures compared to conventional diffusional bonding [4].

Intermetallic compounds, such as aluminides of nickel and titanium, have excellent elevated-temperature mechanical properties and are finding applications as a structural material in aerospace, automobile, and energy industries. Since many of these applications require joining of simpler shapes to fabricate a complex part, it is important to develop techniques for joining of intermetallics. In the past, diffusional bonding has been applied for joining intermetallics and metal alloys [5,6]. However, the microstructural and mechanical properties of these materials are adversely affected by exposure to extremely high temperatures (>75% of the solidus temperature) [7,8]. In this regard, a technique is needed to achieve lower temperature joining of metallic systems, especially intermetallics.

This paper addresses the applicability of the high-temperature plastic joining of intermetallic compounds. Joining of intermetallics such as nickel aluminide (Ni_3Al) is demonstrated by using superplastic deformation technique. Effects of strain rates and joining temperatures are investigated. Joint quality is evaluated by metallographical analysis and mechanical tests. In addition, the role of Ni_3Al oxidation and its effect on joining are briefly discussed.

EXPERIMENTAL DETAILS

Material

To demonstrate plastic self-joining of intermetallics, an alloy with reasonable plasticity at moderate temperatures was selected. Ni_3Al has been shown to exhibit superplastic behavior at temperatures >1000°C [9]. Ni_3Al alloy used in this study was same as studied by Mukhopadhayay et al. [9]. This alloy contained approximately 0.02 wt.% B to provide grain refinement and enhanced ductility. The grains are equiaxed with average grain size ranging from 2-6 μm as shown in Figure 1.

Joining

Right parallelepiped specimens (≈3 x 3 x 5 mm) were cut from the as–received Ni_3Al billet. No surface preparation of the specimens was done. All specimens were joined by uniaxial compression, at a constant crosshead displacement rate, on an Instron Model 1125 (Canton, MA) equipped with an atmosphere-controlled high-temperature furnace. To ensure superplastic deformation of Ni_3Al during joining, strain rates and joining temperatures were determined from Ref. 9. Joining temperatures varied from 1000°C to 1200°C with strain rates ranging from 1 x 10^{-5}/s to 9 x 10^{-4}/s. Since Ni_3Al is prone to oxidation, all tests were conducted in a reducing atmosphere of He-26% H_2. Moreover, prior to start of the test, the furnace chamber was repeatedly flushed with He-26% H_2. Test chamber temperature was allowed to equilibrate for 10 min before loading the sample. All joinings (including heating and cooling of furnace) were completed within 5 hours. Total permanent strains on the specimens were ≈0.1.

Joint Evaluations

Joined samples were sectioned and polished to investigate the joint interface by scanning electron microscopy (SEM) using a Hitachi S-4700-II (Tokyo, Japan) field emission microscope.

To measure the mechanical strength of the Ni_3Al joint and compare it with strengths of monolithic Ni_3Al, large sized specimens (3 x 3 x 10 mm) were also joined. Specimens for room-temperature mechanical tests were joined at 1100°C and a strain rate of 5 x 10^{-4}/s. This joining resulted in flexure bar samples with final dimensions of ≈3 x 3 x 18 mm. Mechanical behavior of the joined and monolithic Ni_3Al samples was evaluated using 3-point bend tests (span=15mm) on Instron Model 4505 (Canton, MA) machine at room temperature and a constant crosshead speed of 0.5 mm/min.

Oxidation Study

A limited study was conducted to investigate the extent of oxidation that the Ni_3Al underwent at the joining atmosphere and temperatures used in this study. In this regard, parallelepiped Ni_3Al samples were metallographically polished on two mutually perpendicular surfaces. Samples were exposed to elevated temperatures (1100-1200°C) under identical atmospheres and times as those used for joining. Subsequently, one of the surfaces was slightly polished and examined by SEM. Furthermore, energy dispersive X-ray analysis (EDAX) was conducted to identify the elemental composition of various oxidation-related phases that formed.

Fig. 1.   Transmission Electron Micrograph of Ni₃Al used in this study.
(Courtesy of A. Sergueeva, Univ. of California, Davis)

RESULTS AND DISCUSSION

Figure 2 shows the typical engineering stress-strain plot obtained in a Ni$_3$Al joining experiment conducted at 1100°C and a strain rate of 5 x 10^{-4}/s. The stress increased monotonically up to ≈43 MPa; thereafter stress remained constant with increasing strain, indicative of steady-state plastic flow of the material. Typically, cumulative strains in the sample were ≈0.1 before samples were unloaded, as shown by the sharp drop in the stress in Fig. 2.

The various temperatures (1000-1200°C) and strain rates ($\dot{\varepsilon}$) used for Ni$_3$Al joining are summarized in Table 1 along with the steady state flow stress and joint quality assessment made using SEM. As expected, at lower temperatures, flow stresses were higher, and they increased with increasing strain rates at a specific joining temperature.

Ni$_3$Al joint quality appears to be temperature dependent. Optimal joining in Ni$_3$Al was achieved at 1100°C for all the strain rates investigated. Fig. 3a shows the high-magnification SEM micrograph of an excellent interface formed at 1100°C and a relatively high strain rate of 9 x 10^{-4}/s. None of the samples attempted at 1050°C or 1200°C showed pore-free joints. At these temperatures, the joint interface typically had trapped porosity, as shown in Fig. 3b.

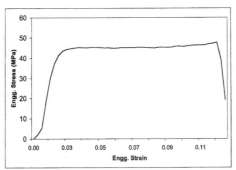

Fig. 2.   Typical engineering stress-strain response during superplastic
joining of Ni$_3$Al at 1100°C and strain rate of 1 x 10^{-5}/s.

Table I. Summary of tests on Ni$_3$Al joining

T (°C)	$\dot{\varepsilon}$ (s^{-1})	Flow Stress (MPa)	Comments
1000	1 x 10^{-4}	122.5	porous joint
	1 x 10^{-5}	41.1	porous joint at one edge
1050	5 x 10^{-4}	89.4	cracked through
	9 x 10^{-4}	216.1	cracked through
	1 x 10^{-4}	47.0	excellent joint
1100	5 x 10^{-4}	74.7	excellent joint
	9 x 10^{-4}	111.2	excellent joint
	9 x 10^{-4}	122.5	excellent joint
	1 x 10^{-4}	6.3	porous joint
1200	9 x 10^{-4}	41.8	porous joint
	5 x 10^{-3}	48.4	porous joint

At 1200°C, inadequate joining is probably due to oxidation-related effects. To confirm this, a study was undertaken that compared surface oxidation for Ni$_3$Al samples exposed to 1100°C and 1200°C for 30 minutes (equivalent to time spent at the joining temperature in an actual experiment). As shown in Fig. 4, SEM micrographs for the two oxidized samples clearly show the differences in the extent of oxidation. Oxidation products (light spots) are quite sparse for the 1100°C sample (Fig. 4a). However, for the 1200°C sample (Fig. 4b), oxidation is quite extensive, with continuous light regions of the oxidation product extending from the sample surface to depths of ≈120 μm. Moreover, for the 1200°C samples, oxidation products cover the sample surface (as shown by arrows in Fig. 4b). We believe that this surface coverage precludes joining at elevated temperatures.

Oxidation products on the 1200°C samples were analyzed using EDAX. The dark region (Fig. 4a) showed strong peaks of nickel and aluminum, which is consistent with the base material. However, light regions showed the strong presence of aluminum and oxygen and relatively lower amounts of nickel. This observation indicates that aluminum is being oxidized to aluminum oxide at the surface. These results are consistent with recent work by Gao et al. [10], where surface formation of aluminum oxide occurred when Ni$_3$Al was exposed to 900°C at oxygen partial pressures as low as 10^{-22} atm.

Since superplastic joining is based on grain-boundary sliding, it will be strongly dependent on joining temperature. At low temperatures (<1050°C), the lack of good bonding observed in Ni$_3$Al is probably due to limited grain boundary sliding and, hence, limited interpenetrations of grains to form a good joint.

Figure 5 compares the mechanical response of monolithic and joined Ni$_3$Al bars tested in the three-point-bend mode at room temperature. Based on the optimal joining conditions determined from Table 1, joined Ni₃Al bars were prepared at 1100°C at a strain rate of 5 x 10⁻⁴/s with total strain of ≈0.07. Both types of Ni₃Al showed a linear increase in stress, up to 700-800 MPa, with increasing crosshead displacement. Beyond that point, the material begins to plastically deform. The yield stress is consistent with literature data for Ni₃Al alloys [11]. It is interesting to

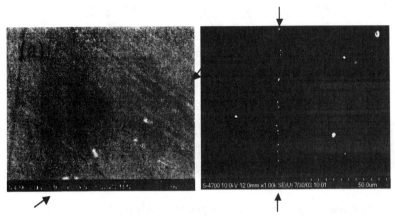

Fig. 3. (a) Excellent joint in Ni₃Al bonded at 1100°C and a strain rate of 9 x 10⁻⁴/s and (b) porous interface formed in Ni₃Al joined at 1200°C and a strain rate of 9 x 10⁻⁴/s. Arrows indicate the location of interface.

Fig. 4. Extent of surface oxidation in Ni₃Al exposed at (a) 1100°C and (b) 1200°C.

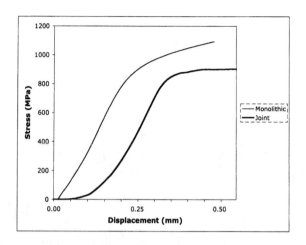

Fig.5. Stress-vs.-displacement behavior at room temperature for monolithic and superplastically joined Ni₃Al.

find that the mechanical behavior of the two specimens is quite similar. Moreover, the data show that strength of the Ni₃Al joint is >800 MPa, or as high as the yield stress of the monolithic Ni₃Al. Mechanical property evaluation further confirms the efficacy of the superplastic joining technique.

SUMMARY

Superplastic deformation has been shown to be applicable for joining intermetallic alloys. Nickel aluminide (Ni₃Al) samples were joined by uniaxial compression at elevated temperatures. Optimum joining was achieved at 1100°C and strain rates ranging from 1×10^{-4}/s to 9×10^{-4}/s. Corresponding flow stresses ranged from 47-123 MPa with permanent strains of ≈10%. Comparison of mechanical response of the joined samples with monolithic samples showed that the joining process does not have any adverse effects on the material. At high temperatures (>1200°C), oxidation phases seem to inhibit joining in Ni₃Al.

ACKNOWLEDGMENTS

This work was supported by the Office of FreedomCar and Vehicle Technlogies of the U.S. Department of Energy, under Contract W-31-109-Eng-38 at Argonne National Laboratory. The authors are grateful to the program manager, Dr. S. Diamond for his support, Prof. A. K. Mukherjee (Univ. of California, Davis) for providing the nickel aluminide sample, and Dr. J. P. Singh (Argonne National Laboratory) for flexure testing. The electron microscopy was performed in the Electron Microscopy Collaborative Research Center at Argonne National Laboratory.

REFERENCES

1. J. Ye and A. Domininquez-Rodriguez, "Joining of Y-TZP Parts," Scripta Metall. Mater. **33** 441-6 (1995).
2. F. Gutierrez-Mora, A. Dominguez-Rodriguez, J. L. Routbort, R. Chaim, and F. Guiberteau, "Joining of Yttria-tetragonal Stabilized Zirconia Polycrystals using Nanocrystals," Scripta Mater. **41** 455-60 (1999).

3. K. C. Goretta, F. Gutierrez-Mora, J. J. Picciolo, and J. L. Routbort, "Joining Alumina/Zirconia Ceramics," Mater. Sci. & Eng. **A341** 158-162 (2003).
4. C. Scott and V. B. Tran, "Diffusion Bonding of Ceramics," Am. Ceram. Bull. **64** 1129 (1985).
5. N. Ridley, M. F. Islam, and J. Pilling, "Isostatic Diffusion Bonding of Superplastic Ti-6Al-4V and Super Alpha-2 Intermetallic Alloy," in Proc. 7th Int. Titanium Conf., San Diego, July 1992, (Ed.) F. H. Froes and I. Caplan, TMS, Warrendale, PA, pp. 1625-1632 (1993)
6. H. Y. Wu, S. Lee, and J. Y. Wang, "Solid-State Bonding of Iron-Based Alloys, Steel-Brass, and Aluminum Alloys," J. Mater. Process Technol. **75** 284 (1998).
7. K. Iizumi, A. Ogawa, and N. Suzuki, "Diffusion Bonding Properties of Ti-4.5%Al-3%V-2%Fe—2%Mo Alloy," J. Jpn. Inst. Light Met. **49** 368 (1999).
8. H. Somekawa, H. Watanabe, T. Mukai, and K. Higashi, "Low Temperature Diffusion Bonding in a Superplastic AZ31 Magnesium Alloy," Scripta Mater. **48** 1249-1254 (2003).
9. J. Mukhopadhyay, G. Kaschner, and A. K. Mukherjee, "Superplasticity in Boron Doped Ni_3Al Alloy," Scripta Metall. **24** 857-862 (1990).
10. W. Gao, Z. Li, Z. Wu, S. Li, and Y. He, "Oxidation Behavior of Ni_3Al and FeAl Intermetallics under Low Oxygen Partial Pressures," Intermetallics **10** 263-270 (2002).
11. R. N. Wright and V. K. Sikka, "Elevated Temperature Tensile Properties of Powder Metallurgy Ni_3Al Alloyed with Chromium and Zirconium," J. Mat. Sci. **23** 4315-4318 (1988).

MECHANICAL PERFORMANCE OF ADVANCED CERAMIC PARTS JOINED BY PLASTIC FLOW

F. Gutierrez-Mora and A. Dominguez-Rodriguez
Departmento de Fisica de la Materia Condensada
Universidad de Sevilla, Sevilla 41080, Spain

D. Singh, N. Chen, K. C. Goretta, and J. L. Routbort
Energy Technology Division
Argonne National Laboratory, Argonne, IL 60439, USA

ABSTRACT

Four-point-bend tests have been performed on samples of yttria-stabilized zirconia that contained 0-80% alumina joined by plastic deformation to the same or a different composition. The fracture strength of joints made by the same composition was equal to the strength of the monolithic material. Fracture of joints made between different compositions occurred at the position of maximum tensile residual stress, not at the interface. Measured strengths were in accord with the calculated residual stresses.

INTRODUCTION

Joining by plastic deformation has been applied to various advanced ceramics (yttria-stabilized zirconia (YSZ)/alumina composites [1-3], mullite [4], and SiC and TiC whiskers in a zirconia-toughened alumina (ZTA) matrix [5]. Joining by plastic deformation has many advantages over other joining techniques, such as brazing or diffusional bonding in terms of applicability or mechanical performance, especially at high temperatures at which a metallic interlayer, e.g., one used in brazing, will exhibit diminishment of its properties [6,7].

Joining by plastic deformation has few, if any, serious deficiencies when similar materials are being joined [1]. However, some issues remain to be addressed when dissimilar materials are to be joined. The presence of residual stresses that are generated during cooling after joining materials with different thermal expansion coefficients (CTEs) poses additional challenges. Recently, thermal residual-stress distributions have been simulated by finite-element analysis (FEA) of two YSZ/alumina composites with differing compositions that were joined by plastic deformation. Subsequently, FEA results were compared with residual stresses that were experimentally determined in joined samples by an indentation technique [1].

In the present work, the flexural strengths of joined YSZ/alumina composites with varying compositions (and residual stresses) have been evaluated using 4-point bend tests. Failure location with respect to the joint interface has been established by scanning electron microscopy (SEM). Fracture mechanics principles, in conjunction with fractographic analysis, were used to explain the strengths of joined ceramics in the presence of residual stresses.

EXPERIMENTAL DETAILS

Dense YSZ/alumina samples of various compositions (alumina volume fractions that ranged from 0 to 80%) were prepared by typical ceramic processing routes described in Ref. 1. The resultant as-sintered monolithic YSZ and YSZ/alumina composites were cylinders approximately of 10 mm in diameter and 10 mm high. As-sintered samples without any surface preparation were used for joining.

All specimens of similar and dissimilar compositions were joined by uniaxial compression, at a constant crosshead displacement rate, on an Instron Model 1125 (Canton, MA) equipped with an atmosphere-controlled high-temperature furnace. All joinings were conducted at strain rates of $\approx 10^{-5}$ s^{-1}, at temperatures that ranged from 1250 to 1350°C to ensure plastic flow of the YSZ/alumina composites. All joining experiments were conducted in an Ar atmosphere. Test chamber temperature was allowed to equilibrate for 10 min before the samples were loaded. Total permanent strains on the specimens were ≈ 0.1.

As-prepared and joined samples were cut into $\approx 2 \times 2 \times 15$ mm bars for flexure testing. Samples were carefully polished to 1-μm diamond paste and the edges were beveled to avoid edge effects during testing. Room-temperature four-point bending tests were carried out at constant crosshead speed with a stainless steel four-point bending fixture in an Instron Model 4505 (Canton, MA). The inner load span was 9.5 mm and the outer load span was 14 mm; the loading rate was 1.3 mm/min. Strength was calculated from the maximum load to failure. Typically, 3-6 specimens were tested per sample type.

To identify the failure causing flaws in the joined samples, fractographic analysis with a scanning electron microscope (Model S-4700-II, Hitachi, Japan) was conducted on the fracture surface of the flexure-tested samples.

RESULTS AND DISCUSSION

Table 1 shows the strengths of monolithic YSZ, alumina, and their composites. The strength of monolithic samples vs. alumina content follows a well-established trend [8]. The strengths varied from 300 MPa for monolithic alumina to 1030 MPa for fully sintered YSZ. The strength of ZTA composites used in this investigation peaks at 20 vol.% alumina and subsequently decreases with further alumina additions.

Table 2 summarizes the four-point bending tests carried out on joined samples of similar and dissimilar composition. Self-joined samples of ZTA (50% volume fraction of each, ZT50A) exhibit a strength of 620 ± 100 MPa. Joined and monolithic ZT50A

Table I. Summary of flexure test data for base materials

Material	Flexural Strength (MPa)
Al$_2$O$_3$ sintered	300
ZT80A	560 ± 70
ZT60A	580 ± 80
ZT50A	650 ± 100
ZT20A	1020 ± 150
YSZ sintered	1030

samples did not exhibit a significant difference in strength. The strength of the joined sample was within the experimental value for the monolithic material (Table 1) of 650 ± 100 MPa. Furthermore, of six samples tested, at least four fractured at locations away from the interface as shown in Fig. 1. This finding is further evidence of the efficacy of the plastic deformation technique for joining similar materials.

As expected, because residual stresses were generated, flexure test results for dissimilar joint materials differed significantly. The distribution of residual stresses generated during cooling from the joining temperature has been simulated by FEA for joined pieces of ZTA with two compositions and is shown in Fig. 2 [1]. A high tensile stress perpendicular to the joint interface develops in the material with lower CTE. In our case, this material corresponds to the composite with the lower YSZ volume fraction, because the CTE of alumina is considerably lower than that of YSZ [1]. This superposition of high tensile residual stress affects the mechanical performance of the joined sample, as discussed below.

Flexure tests were carried out on joints between zirconia toughened with 60 vol.% (ZT60A) and 40 vol.% (ZT40A) alumina. A large scatter (200-530 MPa) was observed in the strength values of three joined samples that were tested. The fracture sites of all samples were carefully inspected. It was observed that fracture took place in the ZT60A portion at distances from the interface indicated in Table 2. All interfaces between materials were intact, as shown in Fig. 3. Fracture always occurred in the section with tensile residual stress. If the failure locations are represented with the relevant residual-stress distribution (Fig. 2), it is clear that fracture takes place where the tensile stresses in the direction affected by the flexure test (i.e., perpendicular to the interface) are maximum.

The lower strengths of 360 and 201 Mpa, observed for the two ZT60A/ZT40A joined samples, were believed to be due to larger flaws. To confirm this, fractography was conducted on the two low-strength samples. SEM photomicrographs, one of which is in Fig. 4, revealed surface damage on the tensile surfaces in two samples.

Table 2. Summary of flexure test data on joined samples

Joint	Flexural Strength (MPa)	Fracture-to-Interface Distance (μm)
ZT50A/ZT50A	620 ± 100	695
ZT60A/ZT40A	530	195
	360	270
	201	540
ZT60A/ZT20A	500 ± 50	692 ± 108
ZT60A/ZT0A	440 ± 80	381 ± 99

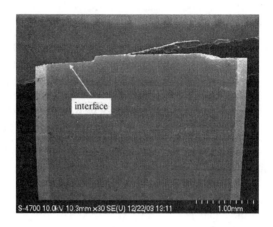

Figure 1. SEM photomicrograph showing fracture location in ZT50A/ZT50A
sample tested in flexure.

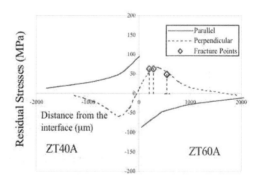

Figure 2. Distribution of residual stresses simulated by FEA for ZT40A/ZT60A joint.
Experimental data correspond to distance from interface at which fracture occurred
in flexure testing.

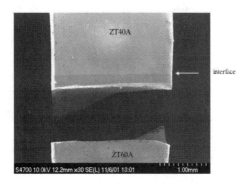

Figure 3.   SEM photomicrograph showing fracture location in ZT60A/ZT40A
sample tested in flexure.

It is conceivable that damage was introduced during fabrication or handling of these specific
ZT60A/ZT40A samples.

To study factors that influence the strength of the joined parts with the residual stress
distribution, we performed a second series of joining experiments. These experiments consisted
of bonding zirconia toughened with 60 vol.% alumina (ZT60A) to zirconia toughened with 20
vol.% alumina (ZT20A) and monolithic YSZ. It is expected that tensile residual stress in ZT60A
section will increase as it is joined to compositions with lower alumina compositions because of
higher CTE mismatch. For the ZT60A/ZT20A samples, flexural strengths were 500 ± 50 Mpa;
for ZT60A/YSZ, were 440 ± 80 MPa; a 25% reduction in strength from the monolithic strength.
Thus, qualitatively, as the tensile residual stresses increased in ZT60A, a monotonic decrease in
the ZT60A strength occurred in its joint configuration. Once again, the failure locations were
away from the interface in both sets of joints, as shown in Fig. 5, indicating strong joints. These
results once again demonstrate the feasibility of the plastic joining technique for dissimilar
materials with differing CTEs.

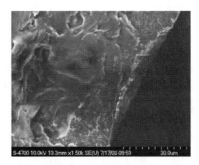

Figure 4.   SEM photomicrograph showing large flaw on tensile surface of ZT60A section
from fractured ZT60A/ZT40A joined sample.

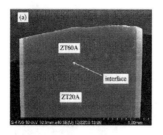 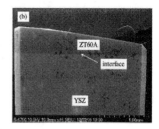

Figure 5. SEM photomicrographs showing failure location in (a) ZT60A/ZT20A and (b) ZT60A/YSZ joints.

SUMMARY

The strength of yttria-stabilized zirconia (YSZ)/alumina parts joined by high-temperature plastic deformation was measured by flexure tests. Samples with varying amounts of alumina (0-80%) were used in this work. Strength of joints fabricated with the same composition showed strength similar to that of monolithic material. This observation demonstrates the efficacy of the joint and validates the joining technique. For joints made from dissimilar compositions, fracture occurred at locations away from the joint interface. These failure locations have been shown to be consistent with the peak residual stresses calculated by finite-element analysis. Variations in the fracture behaviour produced by joining dissimilar materials are predominantly controlled by the residual stresses that arose from the joining process.

ACKNOWLEDGMENTS

This work was supported by the Office of FreedomCar and Vehicle Technlogies of the U.S. Department of Energy, under Contract W-31-109-Eng-38 at Argonne National Laboratory. The authors are grateful to the program manager, Dr. S. Diamond for his support and Dr. S. Majumdar (Argonne National Laboratory) for FEA calculations. The electron microscopy was performed in the Electron Microscopy Collaborative Research Center at Argonne National Laboratory.

REFERENCES

1. F. Gutiérrez-Mora, K. C. Goretta, S. Majumdar, J. L. Routbort, M. Grimdisch, and A. Domínguez-Rodríguez, "Influence of Internal Stresses in Superplastic Joining of Zirconia Toughened Alumina," Acta Materialia, 50 3475-3486 (2002).
2. K. C. Goretta, F. Gutiérrez-Mora, J. J. Picciolo, and J. L. Routbort, "Joining Alumina/Zirconia Ceramics", Mater. Sci. Eng., A341 158-162 (2003).
3. F. Gutierrez-Mora, K. C. Goretta, J. L. Routbort, and A. Dominguez-Rodriguez, "Joining of Ceramics by Superplastic Flow," in Advances in Ceramic Matrix Composites VII, Ceramic Transaction #128, edited by N. P. Bansal, J. P. Singh and H. T. Lin, p. 251. The American Ceramic Society, Westerville, Ohio (2001).
4. A. Sin, J. J. Picciolo, R. H. Lee, F. Gutiérrez-Mora, and K. C. Goretta, "Synthesis of Mullite Powders by Acrylamide Polymerization," J. Mater. Sci. Lett., 20 [17] 1639-1641(2001).
5. N. Chen, F. Gutierrez-Mora, R. E. Koritala, K. C. Goretta, J. L. Routbort, and J. Pan, "Joining Particulate and Whisker Ceramic Composites by Plastic Flow", Comp. Struct., 57 135-139 (2002).

5.   N. Chen, F. Gutierrez-Mora, R. E. Koritala, K. C. Goretta, J. L. Routbort, and J. Pan, "Joining Particulate and Whisker Ceramic Composites by Plastic Flow", Comp. Struct., **57** 135-139 (2002).

6.   M. R. Locatelli, B. J. Dalgleish, K. Nakashima, A. P. Tomsia and A. M. Glaeser, "New approaches to Joining Ceramics for High-Temperature Applications", Ceram. Int. **23** 313-322 (1997).

7.   M. M. Schwartz, Ceramic Joining. ASM International, Materials Park, OH, (1990).

8.   P. B. Becher, "Transient Thermal Stress Behavior in ZrO_2-Toughened Al_2O_3," J. Am. Ceram. Soc., **64** [1] 37-39 (1981).

PHYSICAL CHARACTERISATION OF TRANSPARENT PLZT CERAMICS PREPARED BY ELECTROPHORETIC DEPOSITION

Thomas Nicolay
Department of RF and Microwave Engineering
Saarland University, Bld. 22
66123 Saarbrücken, Germany

Erik Bartscherer
Department of Powder Technology
Saarland University
66123 Saarbrücken, Germany

ABSTRACT

In order to archive transparent PLZT Ceramics, green bodies have been prepared by electrophoretic deposition (EPD) process. The submicron PLZT powder was produced by an improved mixed oxide route. Compared to the strongly varying optical quality of transparent ceramics by using cold pressed green bodies the number and size of pores could significantly be reduced. The optical and electrical properties of the samples are presented and discussed. Additionally dielectric constant and polarisation properties of the samples are shown in detail and compared with other compositions and properties of typical ferroelectric ceramics. The results show that EPD is a very suitable method for the production of green bodies with any size, complex shapes and highest optical properties.

INTRODUCTION

The optical and physical properties of PLZT ceramics open a wide spectrum in technical applications for these materials e.g. photonic crystals, electrooptical shutters, color filters and linear light gate arrays for printers. In the 1970s Haertling and Land prepared the first transparent PLZT ceramics [1],[2]. Up to now the main handicap for commercial use are very high production costs. The preparation of PLZT powders by coprecipitation [3] or hydrothermal synthesis [4] delivers high quality powders compared to the simple mixed oxide process, but these improved preparation strategies come along with higher expenses.

In order to decrease the production costs and as shown in [5], we produced a sub micron PLZT powder by an improved mixed oxide route. By using nanosized Zirconia- and Titania powders we have been able to synthesise a low cost PLZT powder suitable for the preparation of transparent specimen by sintering. The abandonment of the hot pressing process denotes further cost reduction of the preparation process. By the use of electrophoretic deposition (EPD) green bodies of high quality could be formed. Compared to the strongly varying optical quality of transparent ceramics by using cold pressed green bodies the number and size of pores could significantly be reduced [6].

This paper presents and discusses the properties of the samples in details. The results show that the improved mixed oxide route together with EPD is a very suitable method for the production of transparent PLZT ceramics with any size, complex shapes and good optical properties at moderate production costs.

EXPERIMENTAL

The used PLZT powders have been prepared and characterized as described in [7]. An excess of lead oxide was added before the calcination step. For the preparation of green bodies by electrophoretic deposition the membrane method [8] was used. The EPD was done with a PLZT suspension with a 40 % solid content and additional 0.5 % based on the mass of PLZT powder used Dispex A (Ciba). The pH value of the suspension was equilibrated at 10 using TMAH. For the deposition 30 V was applied for 6 minutes. According to [9], densities of ~96% of theoretical can be achieved with air atmosphere only. To improve this value, two step sintering was performed under oxygen atmosphere at a temperature of 1170 °C for 72h. Compared to [10], the use of nanosized powders allows us to reduce the sintering temperature by 50 °C. The PLZT samples were embedded in PLZT powder in a closed crucible in order to minimize the loss of lead oxide. The obtained ceramics were characterized by SEM (Jeol), microscopy (microscope: Leica, CCD and software: Softimage, Germany)and by X-ray diffraction (Siemens D500 with a position sensitive detector, λ=CuKa$_{\alpha a}$). The UV/VIS spectra were measured with a Perkin Elmer Lambda 35 UV/VIS spectrometer. The dielectric properties of the PLZT 9/65/35 samples were studied using a HP impedance/gain analyzer (Model: HP4194A).

RESULTS AND DISCUSSION

X-ray Diffraction Spectrum

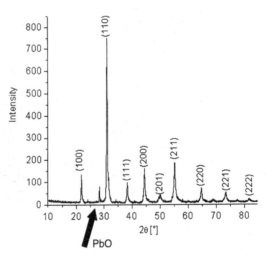

Figure 1: XRD trace of PLZT 9/65/35 with PLZT phase formation

Figure 1 shows the X-ray diffraction (XRD) spectrum of PLZT 9/65/35 powder calcinated 900 °C for one hour. The XRD pattern indicates single phase formation and the sharp peaks the XRD indicate good crystallinity. All lines could be indexed on the basis of a tetragonal cryst structure [11]. Figure 1 still shows some impurities (arrow), which could be identified as lea oxide. In order to obtain fully dense sintered specimen, the excess of lead oxide is necessary compensate the volatilization of lead oxide during sintering. Additionally the excess of lead oxide provides higher densification rates via liquid-phase sintering.

Electrical Characterization

Figure 2 shows the temperature variation of the dielectric constant at a frequency of 1 kHz. Starting at room temperature the dielectric constant rises up to a value of 12200. At the Curie point a temperature of T_C= 86 °C was measured. This result perfectly matches with the in [12] derived linear decrease of T_C as the La content in x/65/35 compositions increases.
Extending temperature beyond this value results in a decreasing dielectric constant. Other dielectric values of the samples are shown in Table 1. The values are in close agreement with those reported in literature [1],[2],[13].

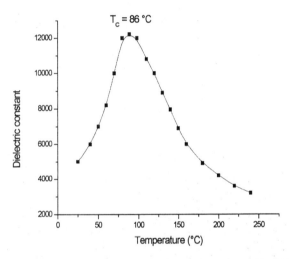

Figure 2: Temperature variation of dielectric constant for PLZT 9/65/35 at 1 kHz

Table 1: Curie temperature and dielectric values for PLZT 9/65/35 at 1 kHz

PLZT	T_C (°C)	k at 300K	tan θ	k at T_C
9/65/35	86	5100	0,009	12200

The hysteresis loop (polarization versus electric field) is one of the most important measurements for the characterization of the electrical behavior. Measurements of the polarization have be done with a modified Sawyer-Tower bridge at a frequency of 0.1 Hz. Figure 3 illustrates the hysteresis loop character of the PLZT system at room temperature. The graph shows a narrow, nonlinear loop behavior which is typical for quadratic electrooptic ceramics. The ceramic PLZT 9/65/35 is an electrically induced ferroelectric and becomes birefringent if an electric field is applied but relaxes to an isotropic (nonbirefringent) state when the field is removed. At an electric field of 5.5 kV/cm^2 the rotation of the vibration direction of the polarized

light is equal to 90°. Using monochromatic light and extending the electric field beyond thi
value results in a progression of repeating light and dark bands, as in an interferometer.

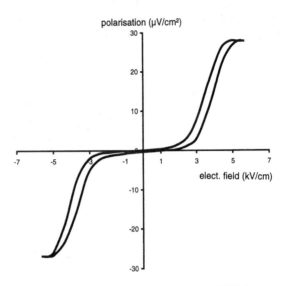

Figure 3: Hysteresis loop for PLZT (9/65/35)

Optical Properties

Figure 4 illustrates the transmission curves for two PLZT 9/65/35 samples with differe
thickness (d_1=1,05 mm, d_2=3,45 mm). The material shows a high absorption below 400 n
which is typical for PLZT 9/65/35. Due to the surface refection loss of 31% (calculated with
index of refraction n=2.5 and assuming orthogonal incidence), the theoretical maximu
transmission is 69%. According to the results in Figure 4 the optical transmission of the sampl
nearly reaches this value and shows the excellent quality of the samples. The transmission resu
are comparable with those prepared by hot pressing [8]. The sample on the picture inserted
Figure 4 has a diameter of 12 mm, a thickness of 1.05 mm and the distance to the background
10 mm. The ceramic doesn't show any optical distortion or impurities.

Using EPD for green body preparation it is possible to obtain transparent ceramics sintered
full density (Figure 5, left). Compared to the grain size of hot pressed samples (~20 – 25 μ
[1],[2]) the optimized preparation route used in this paper allows us to achieve a mean grain s
of 5 μm (Figure 5, right). This results in a high degree of optical uniformity on a macrosco
scale. The good results show that hot isostatic pressing as used in [6] can be abandoned whi
enormously reduces production costs and eliminates limits in shaping.

CONCLUSIONS

PLZT powders prepared by the improved mixed oxide route, using nano sized ox
powders, are suitable for the production of high quality transparent ceramics. The electrical a
optical properties match those known from literature. The paper shows that expensive product
steps as hot pressing, vacuum sintering and hot isostatic pressing can be abandoned. The gr

body preparation by EPD instead of cold pressing or casting results in higher optical quality of the samples. Since the evaporation of lead oxide takes place during the sintering process the achieved lower sintering temperature is an ecological benefit. With the combination of simple powder synthesis and EPD it is possible to produce transparent ceramics of any size and with complex shapes, enabling near net shaping processes, impossible for hot pressing. Compared to state of the art industrial methods it was possible to reduce production costs and the technical expense.

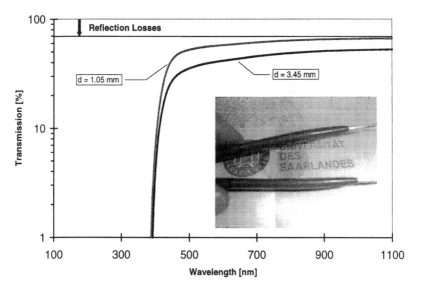

Figure 4: Optical transmission spectrum of electrooptic PLZT (9/65/35)

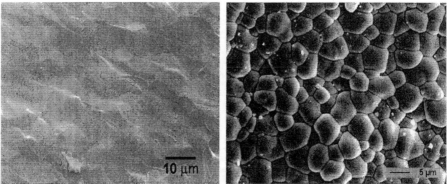

Figure 5: SEM pictures of transparent samples PLZT 9/65/35 (left side: fracture surface, right side: surface)

133

REFERENCES

[1]   G. H. Haertling, "Improved Hot-Pressed Electrooptic Ceramics in the (Pb, La)(Zr, Ti)O System," *J. Am. Ceram. Soc.*, 303-309 (1971).

[2]   G. H. Haertling and C. E. Land, "Hot-Pressed (Pb, La)(Zr, Ti)O$_3$ Ferroelectric Ceramics fc Electrooptic Applications," *J. Am. Ceram. Soc.*, 1-11 (1971).

[3]   Y. Abe, T. Tanaka, K. Kakegawa and Y. Sasaki, "Synthesis of compositionall homogeneous Pb(Zr$_x$Ti$_{1-x}$)$_{1-y}$(Mg$_{1/3}$Ta$_{2/3}$)$_y$O$_3$ by wet-dry combination method and i dielectric properties," *Mater. Lett.*, 308-313 (2001).

[4]   N. Texier, C. Courtois, M. Traianidis and A. Leriche, "Powder process influence on th characteristics of Mn, W, Sb, Ni-doped PZT," *J. Eur. Ceram. Soc.*, 1499-1502 (2001).

[5]   E. Bartscherer, K. Sahner and R. Clasen, "Preparation of PLZT powders from nano size oxides," *26th Annual Conference on Composites, Advanced Ceramics, Materials an Structures*, 601-607 (2002).

[6]   E. Bartscherer, A. Braun, M. Wolff and R. Clasen, "Improved Preparation of Transpare PLZT Ceramics by Electrophoretic Deposition and Hot Isostatic Pressing," *27th Annu Conference on Composites, Advanced Ceramics, Materials and Structures*, 593-600 (2003

[7]   R. Clasen, "Forming of compacts of submicron silica particles by electrophoret deposition," *2nd Int. Conf. on Powder Processing Science*, 633-640 (1988).

[8]   G. S. Snow, "Fabrication of Transparent PLZT Ceramics by Atmosphere Sintering," *J. Ar Ceram. Soc.*, **56** [2] 91-96 (1973).

[9]   T.F.Murray, R.H. Dungham, "Oxygen Firing Can Replace Hot Pressing for PZT," *Cerar Ind.*, **82** 74-77 (1964).

[10] K. Nagata, H. Schmitt, K. Stathakis and H.E. Müser, „Vacuum Sintering of Transpare Piezo-Ceramics," *Ceramurgia Int.*, **3**, 53-56 (1977).

[11] M. Cerqueira, R.S. Nasar, E.R. Leite, E. Longo, J.A. Valera, "Sintering ar characterization of PLZT /9/56/35)," *Ceramics International*, 231-236 (2000).

[12] S. Shah, M.S. Ramachandra Rao, Preparation and dielectric study of high-quality PLZ x/65/35 x=6,7,8) ferroelectric ceramics," *Appl. Physics A, Materials Science & Processi* (June 2000).

[13] H.R. Rukimini, R.N.P. Choudhary, D.L. Prabhakara, "Sintering dependent physical a dielectric properties of Pb$_{0.91}$La$_{0.09}$(Zr$_{0.65}$Ti$_{0.35}$)$_{0.9775}$O$_3$," *J. of Materials Science Letters.*, 1705-1708 (2000).

FABRICATION OF MICROSTRUCTURED CERAMICS BY ELECTROPHORETIC DEPOSITION OF OPTIMIZED SUSPENSIONS

H. von Both
Institute of Microsystem Technology / Materials Process Technology, University of Freiburg
Georges-Köhler-Allee 102
D-79110 Freiburg, Germany

M. Dauscher and J. Haußelt
Institute of Materials Research III, Forschungszentrum Karlruhe
Hermann-von-Helmholtz-Platz 1
D-76344 Eggenstein-Leopoldshafen, Germany
and
Institute of Microsystem Technology / Materials Process Technology, University of Freiburg
Georges-Köhler-Allee 102
D-79110 Freiburg, Germany

ABSTRACT

Ceramics with microstructured surfaces open new possibilities in applied microsystem and micro chemical engineering technology [1]. For excellent surface roughness and dimensional accuracy the use of submicron and nanosized ceramics is necessary. The processing of such powders by conventional processing techniques like micro injection moulding, slip casting or axial pressing is difficult. However, dispersing, deagglomeration, and stabilization of such particles in alcohol results in suspensions which can be used for electrophoretic deposition (EPD) [2-5]. EPD produces homogeneous green bodies with high particle packing density, low surface roughness and high mechanical strength [5]. In case of fabrication of microstructures, the deposition electrodes simultaneously serve as so-called "lost" substrates. The thermal removal of the substrates is followed by sintering the green parts to full density [6]. The present paper focuses on the fabrication of microstructured alumina ceramics by electrophoretic deposition of optimized suspensions. Emphasis is put on the increase of particle packing densities of the green bodies and therefore on the minimization of sintering shrinkage for high dimensional accuracy. Relating to particle packing theory [7-8], the packing density is directly influenced by the particle size distribution of the starting powders. A precise knowledge of the influence of particle size distribution on packing densities allows the fabrication of microstructured ceramic green bodies by EPD with high green densities. Therefore, optimized suspension and deposition conditions of the sintered ceramic microstructured components show low surface roughness and high dimensional accuracy.

INTRODUCTION

The availability of microstructured ceramic components with excellent chemical, mechanical, and physical properties offers advantages in a lot of new application fields related to microsystem and micro chemical engineering technology. The use of submicron and nanosized ceramics is necessary for structural details in the micrometer range, excellent surface roughness and dimensional accuracy. Suspension based technologies offer the advantages that submicron and nanosized particles can be totally dispersed and that the green density of the component is much higher than the solids loading of the suspension. A special benefit of the potential assisted EPD is, that the deposition rate of the particles is nearly independent of the particle size and therefore the processing of powders with nanosized and multimodal particle size distributions for enhanced packing density is possible. A maximum density is achieved when the large particles form a hexagonally packed lattice with the interstices completely filled by smaller particles. For bimodal size distributions the ideal ratio of large to small particles is greater than 6.5. Thus, an optimized particle size distribution should provide a homogeneous green body with a packing density as high as possible.

To verify and use this particle packing theory, analyses were carried out with ethanolic suspensions containing alumina powder mixtures with defined bimodal particle size distributions. The packing densities of the green bodies, achieved by EPD, were studied as a function of the fraction of small powder. Suspensions with optimized particle size distribution were used for the fabrication of microstructured ceramic components.

EXPERIMENTAL PROCEDURE
The experiments were carried out with an organic, electrosterically stabilized suspension [9-1] containing two different alumina powders with a median particle size of 250 nm and 1320 nm, surface area of 12.8 m²/g and 3.2 m²/g, and a theoretical density of 3.95 g/cm³. The powders were suspended with a magnetic stirrer in ethyl alcohol, which contained an organic aliphatic acid a dispersing agent. Homogenization and dispersing of the particles was carried out for 5 min in a ultrasonic bath and for one hour at 200 min⁻¹ in a planetary ball mill (PM 400/2, Retsch GmbH Hanau, Germany).

The fabrication of the ceramic green bodies was done by electrophoretic deposition of the positively charged particles onto flat platinum cathodes for density measurements and onto lo substrates for microstructures, respectively. To ensure a uniform deposition rate of the particles, constant current of 100 µA (13 mm distance of the electrodes) was chosen for 30 to 60 min (240 SourceMeter, Keithley Instruments, Cleveland, USA). Debindering and sintering was carried o in a chamber furnace (RHF 17/3E, Carbolite, Ubstadt-Weiher, Germany) for 1 hour at 1550° under flowing air (5 dm³/min).

Fig. 1. Micro milled master mold (left), replicated silicon rubber mold, and microstructured electrode of a triple mirror array (right).

The manufacturing of microstructured ceramics needs a master mold, produced for instant by stereolithography or mechanical micro engineering [12]. Inverting the master mold in silic rubber (fig. 1) followed by a second replication into electrically conductive paraffin produces microstructured electrode. Figure 2 shows different types of such microstructured lost substrate

Fig. 2. Microstructured electrically conductive paraffin electrodes.

The particle size distributions of the starting powders and the binary mixtures were measured on diluted suspensions by laser scanning diffraction (LS 230 Beckman Coulter, Krefeld, Germany).
The green densities of the deposited ceramics were characterized using mercury porosimetry [13] and Archimedes' principle [14-15], by infiltrating and covering the green bodies with paraffin. The sintered densities were analyzed using Archimedes' principle, by infiltrating the sintered bodies with water after evacuation.

RESULTS

Particle size distributions

Figure 3 shows the particle size distributions of the small and large starting powders within suspension and the bimodal curve of the powder mixtures as well. The existence and level of the plateaus depends on the fraction of small particles.

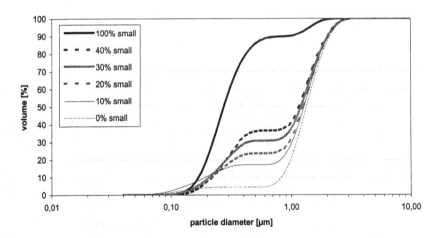

Fig. 3. Particle size distribution of the starting powders and the binary mixtures.

Green densities

The relative densities of the green bodies after drying is shown in fig. 4 as a function of small particles. Suspensions with 30 v% and 45 v% of solids loadings exhibit a density maximum at 10 to 25 % of small particles. The suspension with a powder loading of 30 v% yields a density maximum of 70 ± 1 % t.d. at a small particle volume fraction of 20 %.With increasing the solids loading to 45 v%, the maximum enhances to 73 % and the fraction of small particles reduces to 10 %. Selected green density analyses by mercury porosimetry shows 2 % higher densities with the same dependence of fraction of small particles.

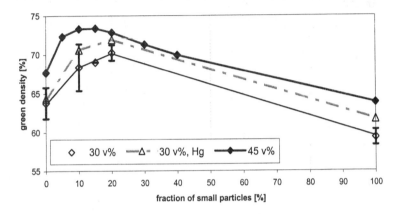

Fig. 4. Relative green densities after drying as function of solids loading and fraction of small particles.

Deposition on microstructured electrodes

The electrophoretic deposition of optimized suspensions with a solids loading of 30 v% was carried out on microstructured lost substrates. Figure 5 shows a sintered triple mirror array without visible cracks, warpage, and a good molding accuracy.

Fig. 5. SEM images of sintered triple mirror array of alumina ceramic.

The three-dimensional structure analysis of the sintered triple mirror array ceramic in fig. 6 by confocal laser scanning microscopy (LSM 5 Pascal, Carl Zeiss, Stuttgart, Germany) demonstrates isogonal surfaces with an area angle of 90° and an edge length of 850 μm.

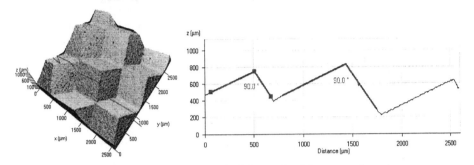

Fig. 6. Three-dimensional structure analysis by confocal laser scanning microscopy.

DISCUSSION

The presented results of the particle size distributions demonstrate, that the used alumina powders do not tend to agglomerate and therefore well dispersed suspensions with defined bimodal size distribution can be produced. The solids loading of the suspensions and the particle size distribution directly influence the green density of the samples. A relative green density up to 73 % resulting in a low sintering shrinkage if can be attained by using highly concentrated suspensions together with a certain fraction of small particles (10 to 25 %). The location of the maximum and the deviation to the theoretical amount at 27 % of small particles [7-8] is caused by the size distribution of the starting materials which shows a characteristic width and fraction of small and large particles.

The formation of homogeneous, densely packed green bodies by electrophoretic deposition of submicron and nanosized ceramic particles on a microstructured lost electrode is possible by the use of optimized suspensions and deposition conditions. This allows the subsequent processing steps drying, debinding of the substrate, and sintering to microstructured ceramic component of high density. The good molding accuracy with isogonal structures and without visible cracks and warpage is demonstrated by the analysis by scanning and three-dimensional laser scanning microscopy.

CONCLUSIONS

The presented work shows the electrophoretic deposition of nanosized and submicron alumina particles on a microstructured electrically conductive substrate, which was removed in a subsequent processing step. Optimized deposition conditions with a defined electric field are necessary to achieve a dense particle packing and to prevent the formation of bubbles by electrolysis of the solvent. The influence of binary powder mixtures is discussed on the basis of particle size distributions and green density measurements. A precise knowledge of the influence of particle size distribution on packing densities allows the fabrication of microstructured ceramic green bodies by EPD with high green densities and low sintering shrinkage. In conclusion, this work shows the possibility to fabricate microstructured sintered ceramic components with high dimensional accuracy via EPD.

ACKNOWLEDGEMENTS

The authors thank the colleagues of the Forschungszentrum Karlsruhe (Germany) for the microstructured master molds and mercury porosimetry measurements. Furthermore, the authors thank the colleagues at the Institute of Microsystem Technology (University of Freiburg, Germany) for the confocal laser scanning analyses and F. Umbrecht (Diploma thesis, University of Freiburg, Germany) for the manufacturing of the microstructured electrodes.

REFERENCES

[1] H.-J. Ritzhaupt-Kleissl and J. Haußelt, Ceramic Forum International, 9 (1999).

[2] Hamaker H.C., „Formation of a deposit by electrophoresis", Trans. Faraday Soc., 36, 279-287 (1940)

[3] R. Moreno and B. Ferrari, „Effect of the slurry properties on the homogeneity of alumina deposits obtained by electrophoretic deposition", Mat. Res. Bul., 35, 887-897 (2000)

[4] P.M. Biesheuvel and H. Verweij, „Theory of Cast Formation in Electrophoretic Deposition", J. Am. Ceram. Soc., 82 (6), 1451-1455 (1999)

[5] P. Sarkar and P.S. Nicholson, "Electrophoretic Deposition (EPD): Mechanisms, Kinetics and Application" J. Am. Ceram. Soc., 79 (1996).

[6] H. von Both and J. Haußelt, "Ceramic Microstructures by Electrophoretic Deposition of Colloidal Suspensions", in Electrophoretic Deposition: Fundamentals and Applications, The Electrochemical Society, Proceedings Volume 2002-21, pp. 78-85 (2003)

[7] D.J. Cumberland and R.J. Crawford, The Packing of Particles, Elsevier Science, (1987).

[8] J.S. Reed, Introduction to the Principles of Ceramic Processing, Wiley (1988).

[9] R.G. Horn, J. Am. Ceram. Soc., 73, pp. 1117-1135 (1990).

[10] B. Russel, D.A. Saville, and W.R. Schowalter, Colloidal Dispersions, Cambridge University Press (1989).

[11] N. Israelachvili, Intermolecular and surface forces, Academic Press, London, (1997).

[12] W. Menz, J. Mohr, and O. Paul, Microsystem technology, Wiley-VCH, New York (2001

[13] DIN 66 133, „Bestimmung der Porenvolumenverteilung und der spezifischen Oberfläche von Feststoffen durch Quecksilberintrusion" (1993)

[14] DIN 51 056, „Prüfung keramischer Roh- und Werkstoffe: Bestimmung der Wasseraufnahme und offenen Porosität" (1985)

[15] DIN EN 623-2, „Prüfverfahren für Hochleistungskeramiken; Allgemeine und strukturelle Eigenschaften; Bestimmung von Dichte und Porosität" (1991)

LOW COST PROCESS FOR MULLITE UTILIZING INDUSTRIAL WASTES AS STARTING RAW MATERIAL

Kachin Saiintawong, Shigetaka Wada

Graduate School
Department of Materials Science
Faculty of Science, Chulalongkorn University
Pathumwan, Bangkok, 10400 Thailand.

Angkhana Jaroenworaluck

National Metal and Materials Technology Center
114,Thailand Science Park, Paholyothin Rd.,
Klong 1, Klong Luang,
Pathumthani, 12120 Thailand.

ABSTRACT

Mullite synthesised for low cost processing can be produced by using waste powders from two Thai industrial sources; (i) aluminium oxide from a surface coating process, and (ii) silica from rice husks. Aluminium oxide obtained as a waste product powder was treated and calcined to give a specific BET surface area, 105 m^2/g. SiO_2 was obtained from the treatment of rice husks. The silica from the treated rice husks is present as the amorphous phase and has a specific BET surface area of 291 m^2/g. The resulting silica has a high purity, >99.0%. The treated aluminium oxide and the silica from the rice husk were mixed together in a ball mill, dried and pressed into pellet shaped compacts at various pressures, using conventional methods. The pellets were heated to maximum sintering temperatures in the range 1400-1700°C with different holding times; these varied for each of the maximum sintering temperatures. The density of the resulting sintered mullites was measured by Archimedes' method. The XRD profiles and SEM micrographs obtained showed only the pure mullite phase to be present when sintered to 1500°C for 1 hr. No other phases were apparent. The maximum relative density of the sintered mullite was found to be ~98 %, a value obtained from the pellets sintered at 1700°C for 3 hr. The mullite ceramics were prepared for testing in 3 point bending strength and methods compared to commercial mullite materials. Mullite ceramics prepared from Sumitomo, A21, were mixed with silica from rice husk treated with HCl, and proprietary method have shown higher bending strength than materials prepared using powder from the aluminium surface coating mixed with rice husk silica treated with HCl and proprietary method. The mechanical properties have been correlated with the formation of a glassy phase formed around the mullite grains.

INTRODUCTION

Mullite ($3Al_2O_3.2SiO_2$) is a ceramic material that is a suitable candidate for many high temperature applications because of its high melting point (~1830°C)[1], high thermal shock resistance, low coefficient of thermal expansion, high creep resistance, high chemical stability and low thermal conductivity.[2] Starting raw materials for synthesis of mullite are usually mixtures of various kinds of Al_2O_3 and SiO_2 in suitable proportions to give a stoichiometric composition. The method of synthesis used will depend on the type of precursor raw materials employed in the processing route. Conventional methods will employ kaolinite, alumina, bauxite, quartz, and various kinds of alumino-silicate materials; and this process route will generally require a higher sintering temperature. Admixtures of sols and salts that combine with silica and alumina sol, silica and alumina alkoxides, the precursor for sol-gel method,[3,4] can be processed at a significantly lower temperature, but the material costs are considerably higher and often the process is unsuited to commercial production routes. The temperature for formation of mullite by the sol-gel method is lower than for the conventional methods, because the starting materials have a very fine particle size and consequently a high specific surface area. However the disadvantage of this method is the difficulty of moving to production on an industrial scale.[5]

This present paper deals with the synthesis of mullite by a conventional method using raw materials which have a high specific surface area and low cost, such as silica from rice husk and alumina from a surface coating process. The objective of the work has been to produce in Thailand mullite powder which is lower in cost, using local raw materials and to improve the processing route in order to have potential for manufacture. Moreover, a secondary objective will be to prepare mullite ceramic by the reaction sintering of this mullite powder.

EXPERIMENTAL PROCEDURES

Materials: Two kinds of silica powder prepared from rice husks were used. The first was processed by washing in HCl acid solution[6] followed by burning at 650°C. The purity and specific surface area of the silica were 99.06 % and 182 m^2/g, respectively. The second powder was processed by proprietary method followed by burning at same temperature. The purity and specific surface area of the silica were 98.30 % and 291 m^2/g, respectively.

Two different alumina powders were used as raw materials. The first was A-21 (α-Al$_2$O$_3$ Sumitomo, Japan), having a purity 99.60 % and a specific surface area of 19 m^2/g. The second was derived from aluminium hydroxide waste (a mixture of boemite and gibbsite, purity 98.78 % and specific surface area 105 m^2/g. The chemical composition and specific surface area of starting materials are shown in table 1.

Methods: Al$_2$O$_3$ and SiO$_2$ powders, in a stochiometric composition, i.e. a weight ratio of 72:28, were wet mixed in a ball mill with alumina balls for 2 hr. The mixture of starting materials is shown in table 2. After milling the slurry was dried on a hot plate with magnetic stirring to prevent segregation. The dried mixture was pressed into pellets 2.0 cm in diameter and 0.5 cm in thickness by a biaxial hydraulic press at 40 MPa. The specimens were then fired in an electric kiln in the temperature range 1300-1700°C with a heating rate of 5°C/min; soaking times of 1, 3 and 5 hr were selected to determine the effect of soaking time on the degree of mullitization.

Bulk density and water absorption were measured by Archimedes's method. Phase analysis was measured by X-Ray diffraction (XRD). Microstructures were observed on polished surfaces by scanning electron microscopy (SEM JEOL 6301 F). 3 point bend strength of mullite ceramics were measured from the polished surface of samples (specimen size: rectangular bars, 8.0 mm×8.0 (W) and length 50.0 mm) using a universal testing machine, Instron 4502, with a span length of 20.0 mm and crosshead speed of 0.3 mm/min

Table 1. Composition of Starting Raw Materials

	SiO_2 treated by HCl Acid Method	SiO_2 treated by Proprietary Method	Al_2O_3 A-21	Al_2O_3 -X
SiO_2 (%)	99.06	98.30	0.01	0.36
Al_2O_3 (%)	0.80	0.71	99.60	98.78
Fe_2O_3 (%)	0.01	0.00	0.01	0.04
CaO (%)	0.06	0.83	0.00	-
MgO (%)	0.00	0.03	0.00	0.39
Na_2O (%)	0.00	0.04	0.25	0.44
K_2O (%)	0.03	0.07	0.00	-
MnO_2 (%)	0.02	0.02	0.00	-
Surface Area (m^2/g)	182	291	19	105

Table 2. Composition of Raw Materials for Each Formulation

Formulation	Alumina Source	Silica Source
1	Alumina A-21	Silica from rice husk treated by HCl method
1-A	Alumina-X	Silica from rice husk treated by HCl method
1-S	Alumina A-21	Silica from rice husk treated by proprietary method
1-S-A	Alumina-X	Silica from rice husk treated by proprietary method

RESULTS AND DISCUSSION

1. Synthesis of Mullite Powder

Generally, the mullitization temperature for synthesis mullite of conventionally processed powders is above 1600°C depending on factors such as starting raw materials, particle size of the raw materials, impurity content and the milling process used for preparation of precursor mix. The XRD results obtained from each sample after firing at various temperatures are shown in figure 1. Cristobalite, alumina and mullite peak, are seen to be present after firing at 1300°C. At 1400°C, the intensity of the cristobalite and alumina peaks have decreased and the peak of mullite increased dramatically. Above 1450°C the powder has become fully mullitised while cristobalite and alumina peaks have totally disappeared. When compared with commercial mullite powder, the synthesised mullite shows a XRD pattern identical to commercial mullite. For the other formulations used, the transformation to the mullite phase was also complete at temperatures above 1450°C.

2. Sintering Mullite Ceramic

Reaction bond sintering was used to produce a mullite ceramic from the mixture of alumina and silica using each composition. The relative density of the powder compact is shown in figure 2. The relative density of the compact which used alumina-x from waste was higher than 90 % on firing at 1600°C whereas the sample which used alumina A-21 had a relative density lower than 80 % at same firing temperature. When the sintering temperature was increased up to 1700°C, the relative density of sample 1-S-A and 1-A was increased slightly. On the other hand the samples 1 and 1-S significantly increased in density at 1700°C. The relative density of sample 1 is almost 98 % theoretical. The reason why the relative density of sample that used alumina-x was higher than the sample which used alumina A-21 at 1600°C may well be due to the impurities in alumina-x such as alkali oxides. Such oxides can cause the formation of a glassy phase between grain boundaries of mullite, and also reside at the triple points in the sintered mullite. However, the presence of this glassy phase has not noticeably affected the density of samples derived from alumina A-21 sintered at temperatures up to 1700°C.

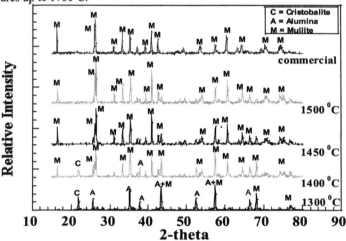

Figure 1. XRD pattern of the synthesis mullite at various sintering temperatures Formulation 1 is compared with commercial mullite powder.

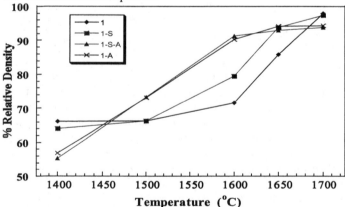

Figure 2. The relationship between relative density and sintering temperature for each formulation.

The results for the bending strength are shown in figure 3. It is apparent that the bending strength is related to the bulk density attained in the processing. At 1700°C samples 1 and 1-S, which had been combined with alumina A-21, have a higher bulk density than formula 1-A and 1-S-A that used alumina from the waste product of the surface coating process. The mechanical strength of mullite ceramics which used the Sumitomo alumina from Japan was higher than other formulations that also contained commercial mullite. It is suggested that this is because alumina A-21 has low alkali oxide present, whereas the impurities are at a lower level than in alumina-x. It can be implied that glassy phase from alumina A-21 is lower than in the other alumina powders. These impurities were able to generate a glassy bonding phase between the grains of mullite and consequently the bend strength of mullite that used alumina-x have a lower value than mullite made from alumina A-21.

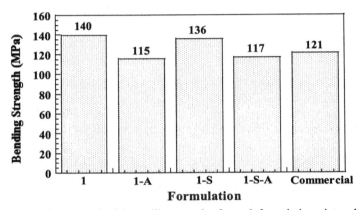

Figure 3. The bending strength of the mullite ceramics for each formulation, sintered at 1700°C.

3. Microstructural Differences

The microstructure of the mullite ceramic is shown in figure 4. For formulation 1-S, the specimen presents a novel microstructure consisting of fused mullite (2/1-mullite) which rod-like grains, whereas formulation 1-A has an acicular grain shape and a higher aspect ratio than formulation 1-S. For formulation 1 and 1-S-A porosity appears to be present between the mullite grains which derives from the glassy phase that had been removed by etching with HF acid.

The characteristic rounded shape of the mullite in the micrographs is often due to the presence of a glassy phase at the grain boundaries which changes the surface energy and therefore the shape of the grain. The effect is most apparent in figure 4 (c). The affect can also be enhanced by the composition and viscosity of the liquid phase, which affects the degree of wetting of the interfaces, where there is little wetting or little liquid phase present than the crystal would be expected to form their equilibrium shape which is often characteristic and represented by sharp planes and boundaries. These planes often have low crystallographic indices (close packing, low surface energies) and give rise to the characteristic morphology.

Figure 4. Microstructure of mullite ceramics for each formulation sintered at 1700°C.

CONCLUSIONS
1. Mullite has been prepared by using waste products from other industries such as sili⋅ derived from rice husks and alumina from a surface coating process. The mullite formation reacti⋅ was almost fully completed at a temperature 1450°C after 5 hr.

2. The sintered mullite ceramic can be fabricated by a reaction bonding processing rou⋅ which gave relative density ~ 98 %. This was obtained from a sample sintered to 1700°C for 3 ⋅ The microstructure of each formulation differed and depended on starting materials utilised.

3. The microstructure of the mullite showed differences in the grain shape and grain size f⋅ each formulation, due to the nature of the glassy phase at the grain boundaries.

ACKNOWLEDGEMENTS
K Saiintawong acknowledges for support from a scholarship of Thailand Graduate Institu⋅ of Science and Technology (TGIST) and the authors wish to thank National Metal and Materia⋅ Technology Center (MTEC) for the use of facilities and further support.

REFERENCES
[1]K.S. Mazdiyasni and L.M. Brown, "Synthesis and Mechanical Properties of Stoichiomet⋅ Aluminum Silicate(Mullite)," *Journal of the American Ceramic Society*, **55** [11] 548-552 (1972⋅
[2]S.Somiya and Y. Hirata , "Mullite Powder Technology and Application in Japan ," *Ceram⋅ Bulletin*, **70** [10] 1624-1632 (1991).
[3]H.Schneider, K.Okada and J.A. Pask, *Mullite and Mullite Ceramics*, John Wiley & Sons, W⋅ Sussex, 1994.
[4]A.M.L. Marques Fonseca, J.M.F.Ferreira, I.M.M. Salvado and J.L. Baptista, "Mullite bas⋅ compositions prepared by sol-gel techniques," *Journal of Sol-gel Science and Technology*, ⋅ 403-407 (1997).
[5]K.Okada, N. Otsuka and S. Somiya, "Review of Mullite Synthesis Route in Japan," *Ceram⋅ Bulletin*, **70**[10] 1633-1640 (1991).
[6]C.Real, M.D.Alcala and J.M.Criado, Preparation of Silica from Rice Husks, *Journal of ⋅ American Ceramic Society*, **79**[8] 2012-16 (1996).

LOW-COST PROCESSING OF FINE GRAINED TRANSPARENT YTTRIUM ALUMINUM GARNET[*]

HeeDong Lee, Tai-Il Mah, and Triplicane A. Parthasarathy
UES, Inc.
4401 Dayton-Xenia Rd.
Dayton, Ohio 45432, U.S.A.

ABSTRACT

Transparent polycrystalline yttrium aluminum garnet (YAG) with a fine grain size (1~2 μm) was produced via a sinter-HIP process using highly reactive nano-sized powders that were synthesized by partial and full combustion processes at 200° - 240°C. The primary particle size was estimated to be between 30 - 50 nm, and the powders appeared to be weakly agglomerated. The densities after sintering at 1550° - 1650°C were 95.0 - 99.0 % of the theoretical density and the samples appeared translucent. Further HIPing at temperatures between 1500° and 1550°C resulted in completely dense polycrystalline YAG that showed a visibly higher transparency that was comparable to a single crystal YAG. The details of microstructural characterization (SEM), phase identification (XRD), and visible and IR transmittance (Spectrophotometer) are presented.

INTRODUCTION

Yttrium aluminum garnet (YAG:$Y_3Al_5O_{12}$) belongs to a cubic crystal system with a unique crystal structure of a non-close packed oxygen sublattice and distinct physical properties. YAG is considered to be the most creep resistant oxide with a high melting point of 1970°C, and a high Young's modulus (E = ~340 GPa) and hardness (Hv = ~19 GPa) that are comparable to sapphire (E = ~380 GPa, Hv = ~20 GPa) [1,2]. Thus, YAG is a very attractive oxide compound for high temperature structural applications.

Another attractive feature of YAG is its high efficiency of energy transfer and radiation damage resistance, which has made it a popular laser host material. The transmittance of YAG, in its single crystal form, extends from the UV (0.2 μm) range to the mid-IR (5.5 μm) range, which indicates that YAG is a potential material for IR and laser windows, along with missile domes, where mechanical integrity is also important. For such applications, high optical transparency is demanded, and single crystal YAG is preferred. However, the optical quality of transparent polycrystalline YAG is reportedly comparable to that of single crystal YAG. Polycrystalline YAG offers better strength, if fine-grained, and lower cost. Recently, transparent polycrystalline YAG has been successfully fabricated through vacuum sintering processes with highly reactive solid precursors or YAG powders [3-10].

The high quality of transparent YAG relies heavily on the complete elimination of pores because significant optical scattering can result from residual pores. The total volume of the residual pores must be restricted below a few ppm in order to achieve the desirable optical quality. Because of such extreme requirements, the current processes that produce transparent polycrystalline YAG have primarily been conducted under high vacuum (< 10^{-6} torr) and high

---

[*] U.S. Patent pending

temperatures (> 1750°C) with lengthy soaking periods (> tens of hours), both of which add considerably to the cost.

To date, the powders that have been used to produce the transparent polycrystalline YAG have been prepared by mixing fine powders of alumina (Al_2O_3) and yttria (Y_2O_3) [6] or by co-precipitating the water-based precursors (nitrates, sulfates, chlorides) with different precipitants (urea, ammonium hydroxide, ammonium hydrogen carbonate, etc.) [11 - 13]. The primary particle size of the final powders appeared in the nano-size range, and the powders were claimed to be well dispersed because of soft agglomeration. The resultant transparent polycrystalline YAG materials consisted of grains ranging in size from a few to tens of microns and exhibited outstanding optical transmittance comparable to single crystal YAG. It has also been reported that YAG powders could be successfully prepared by combustion [14], spray (or flame) pyrolysis [15], sol-gel process [16,17], and citrate method [18]. A transparent YAG, however, has not been produced with such powders.

In the present study, we introduce a novel combustion powder synthesis as well as a two-step densification process, namely a sinter-HIP process. By combining these two technologies, we were able to produce transparent polycrystalline YAG that exhibited an excellent optical quality comparable to that of single crystal YAG. The process can be carried out at much lower temperatures with shorter processing times compared to the current vacuum sintering, and can easily be scaled up. This process is therefore believed to be highly economical. In the present work, the combustion powder synthesis and a two-step densification process was used to fabricate undoped and 2 at. % Nd-doped transparent polycrystalline YAG for IR and laser windows and high power lasing applications, respectively.

EXPERIMENTAL PROCEDURE

The powders were synthesized through two different combustion processes. First, partial combustion (PC) was attempted by mixing commercial α-Al_2O_3 (AKP53, d_{50}: 0.2 µm, Sumitomo Chemical) with an aqueous yttrium nitrate solution at a mole ratio of 5:3 of Al:Y. Appropriate amounts of reductants (amino acids) were then added. Second, full combustion (FC) was initiated by dissolving yttrium nitrate and aluminum nitrate in water with a mole ratio of 5:3 of aluminum and yttrium, and proper amounts of reductants (amino acids) and oxidant (ammonium nitrate) were then added. Both precursors (PC and FC) were dried and self-combusted at temperatures around 200° - 240°C. The as-combusted PC-YAG and FC-YAG were further calcined at 1000° and 1100°C with a 2 hour hold. The as-calcined powders were mildly ball milled for 24 hours with high purity alumina grinding media (Tosoh) using ethyl alcohol as a liquid medium. The alumina balls were weighed before and after ball milling to monitor potential contamination. The as-milled calcined powders were further processed with a binder (PEG), and finally granulated by passing through a 200 mesh nylon sieve. The same process was used to fabricate Nd-doped powders of YAG. The as-processed powders were cold pressed into disc shapes with dimensions of 12.5 mm diameter and 3 mm thickness. The cold pressed discs were then cold isostatically pressed at 200 MPa. The cold isostatically-pressed discs were heat treated at 700°C in air for binder removal and then sintered at temperatures between 1550° and 1650°C for 5 hours in an alumina tube muffle furnace. Finally, the as-sintered YAG was hot isostatically pressed (HIPed) at temperatures of 1450° - 1550°C for 5 hours under 200 MPa argon pressure. The HIPed samples were surface ground and polished with a 1 µm diamond slurry for optical transmittance measurements.

X-ray powder diffractometry (Rigaku) was used to identify the formation of crystalline phases during various heat treatments. The morphology and particle size of powders, and microstructure of the sintered and HIPed samples were characterized by high resolution scanning electron microscopy (Sirion). The in-line transmittance in the wavelength from the visible (400 nm) to the near IR (2500 nm) range was measured using a Cary 5E spectrophotometer. The sample dimensions for transmittance measurement were a 12 mm diameter with a 2 mm thickness.

RESULTS

The details of the morphologies of the as-combusted powders (PC-YAG and FC-YAG) are seen in Fig.1 (a) and (b). The snow-like flaky structure and porous large agglomerates were typical for both cases. The agglomerates were very soft, and could be readily de-agglomerated by ball milling. After calcining at 1000°C for 2 hours, the primary particles are clearly seen in Fig.1 (c) and (d). The primary particle size of the as-calcined FC-YAG was estimated to ~ 30 nm, whereas the PC-YAG consisted of 50 - 100 nm sized particles. It is worthy to note that YAM ($Y_4Al_2O_9$) as well as YAG forms at this temperature, so the particle size of the as-calcined powders should not represent the initial particle size of the as-combusted powders.

(scale bar: 5 μm)

(scale bar: 200 nm)

Fig.1. Powder morphology of the as-combusted (a) PC and (b) FC powders. The primary particle size of the as-calcined powders (1000°C) are shown for (c) PC and (d) FC powder.

As noticed in Fig.1(c), it is also difficult to distinguish the particle size of any unreacted yttrium oxide and aluminum oxide. It is likely that the unreacted aluminum oxide could retain its initial average particle size of 200 nm, but this is not observed in Fig. 1(c). The as-isostatically-pressed green discs showed densities of ~45 - 50 % of the theoretical density ($\rho_{YAG} = 4.55$ g/cm^3). The as-sintered bodies were single phase YAG, identified through x-ray diffraction analysis, and the densities were ~95.0 - 99.0 % of the theoretical density (measured by Archimedes method). The samples appeared translucent.

The results of x-ray analyses of PC and FC-YAG are summarized in Fig. 2. The as-combusted PC-YAG was virtually a mixture of alumina and nano-sized yttrium oxide. During calcination at

1000°C, orthorhombic YAM (ICDD file No. 14-0475) began to form, but peaks of unreacted alumina and yttrium oxide were also seen. After exposure at 1600°C for 2 hours, the as-combusted PC-YAG powder was completely converted to a single phase cubic YAG (ICDD file No.33-0040). The as-combusted FC-YAG powder consisted of a small amount of hexagonal YAP (YAlO$_3$: ICDD file No. 16-0219) with an amorphous phase. After a 2 hour calcination at 1000°C, only single phase cubic YAG was detected. Further heating at 1600°C confirmed the single phase cubic YAG through an increase in peak intensity.

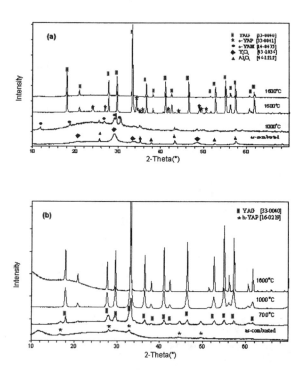

Fig. 2. The results of x-ray powder analyses for (a) PC and (b) FC-YAG powders.

Figure 3 shows the microstructures of the HIPed YAG samples (PC-YAG and FC-YAG). In both samples, the pores look completely eliminated during HIPing, and the grains are uniformly distributed, with sizes ranging from ~1 - 2 μm. Both HIPed YAG specimens showed a similar grain size distribution even at slightly different HIPing temperatures. Based on the scale of the grain size, and from previous work, very good mechanical strength is expected.

The transmittance results are plotted in Figure 4 along with a single crystal Nd doped YAG with a thickness of 1.2 mm. The transmittance of the HIPed FC-YAG looked nearly identical to the single crystal. The small deviation can be attributed to a slightly different polishing quality. The HIPed PC-YAG exhibited a lower transmittance than the HIPed FC-YAG, but it still looks promising. The primary advantage of the PC-YAG is its simple powder production and highly economic aspect; therefore, further refinement is needed. The 2 at.% Nd-doped polycrystal YAG also exhibited the same level of transparency.

(a)   **(b)**

Fig. 3. Microstructures of HIPed (a) PC-YAG at 1550°C and (b) FC-YAG at 1500°C. Both YAG samples were sintered at 1600°C for 5 hours. (Scale bar: 2 μm)

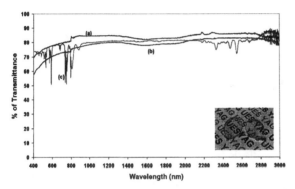

Fig. 4. Transmittance of HIPed (a) FC-YAG at 1500°C and (b) PC-YAG at 1550°C, and (c) single crystal YAG. (Inset picture shows HIPed FC-YAG.).

CONCLUSION

Through combustion processes, highly reactive nano-sized undoped and 2 at. % Nd-doped YAG powders (FC-YAG) were produced. The calcined powders were softly agglomerated and readily de-agglomerated by a conventional ball milling process. Dense compacts from this powder were sintered at 1550° - 1650°C and further HIPed at ~1500°C, resulting in a completely dense body that exhibited a high quality of optical transmittance comparable to that of single crystal YAG. The same process was applied to produce a highly reactive oxide mixture (PC-YAG) that was converted to a single phase YAG (doped and undoped), which showed a visibly high transparency after HIPing. The transparent YAG produced at the present conditions consisted of uniformly distributed fine grains ranging from 1 to 2 μm in size. Thus, high mechanical strength can be anticipated. Finally, it can be expected that the newly developed combustion powder syntheses, along with a two-step densification process, namely sinter-HIP, can provide a great economic advantage for producing large-sized transparent polycrystalline YAG compared to the previous processes.

*This program was supported by UES IR&D, and partly through MDA SBIR Phase I (MDA 03-19) contract number : F33615-03-M-5439.

REFERENCES

[1]T.A. Parthasarathy, T. Mah, and K. Keller, "High Temperature Deformation Behavior of Polycrystalline YAG," Ceram. Eng. Sci. Proc., 12 [9-10] 1767-73 (1991).

[2]T.A. Parthasarathy, T.-I Mah, and K. Keller, "Creep Mechanism of Polycrystalline Yttrium Aluminum Garnet," J. Am. Ceram. Soc., 75 [7] 1756-59 (1992).

[3]G. de With and H. J. A. van Dijk, "Translucent $Y_3Al_5O_{12}$ Ceramics," Mater. Res. Bull., 19 1669-74 (1984).

[4]M. Sekita, H, Haneda, T. Yanagitani, and S, Shirasaki, "Induced Emission Cross Section of Nd: $Y_3Al_5O_{12}$ Ceramics, J. Appl. Phys., 67 [1] 453-58 (1990).

[5]M. Sekita, H, Haneda, and S, Shirasaki, "Optical Spectra of Undoped and Rare-Earth Doped Transparent Ceramic $Y_3Al_5O_{12}$, J. Appl. Phys., 69 [6] 3709-18 (1991).

[6]Akio Ikesue, Isao Furusato, and Kiichiro Kamata, "Fabrication of Polycrystalline, Transparent YAG Ceramics by a Solid-State Reaction Method," J. Am. Ceram. Soc., 78 [1] 225-28 (1995).

[7]Akio Ikesue, Toshiyuki Kinoshita, Kiichiro Kamata, and Kunio Yoshida, "Fabrication and Optical Properties of High-Performance Polycrystalline Nd:YAG Ceramics for Solid-State Laser," J. Am. Ceram. Soc., 78 [4] 1033-40 (1995).

[8]Ji-Guang Li,Takayasu Ikegami, Jong-Heun Lee, and Toshiyuke Mori, "Low-Temperature Fabrication of Transparent Yttrium Aluminum Garnet (YAG) Ceramics without Additives," J. Am. Ceram. Soc., 83 [4] 961-63 (2000).

[9]Jianren Lu, Ken-ichi Ueda, Hideki Yagi, Takagimi Yanagitani, Yasuhiro Akiyama, and Alexander A. Kaminski, "Neodymium doped yttrium aluminate garnet $(Y_3Al_5O_{12})$ nanocrystalline ceramics-a new generation of solid state laser and optical materials," J. Alloys and Compounds, 342 220-25 (2002).

[10]Y. Rabinovitch, D. Tetard, M.D. Faucher, and M. Pham-Thi, "Transparent polycrystalline neodymium doped YAG: synthesis parameters, laser efficiency," Optical Materials, 24 345-51 (2003).

[11]N. Matsushita, N. Tsuchiya, K. Nakatsuka, and T. Yanagitani, "Precipitation and Calcination Processes for Yttrium Aluminum garnet precursors Synthesized by Urea Method," J. Am Ceram Soc., 83 [4] 961-63 (2000).

[12]Ji-Guang Li, Takayasu Ikegami, Jong-Hun Lee, and Toshiyuki Mori, "Characterization of yttrium aluminum garnet precursors synthesized via precipitation using ammonium bicarbonate as the precipitant," J. Mater. Res., 15 [11] 2375-86 (2000).

[13]Ji-Guang Li, Takayasu Ikegami, Jong-Hun Lee, and Toshiyuki Mori, "Well-sinterable $Y_3Al_5O_{12}$ powder from carbonate precursor," J. Mater. Res., 15 [7] 1514-1523 (2000).

[14]J. McKittrick, L.E. Shea, C.F. Bacalski and E.J. Bosze, "The influence of processing parameters on luminescent oxides produced by combustion synthesis," Displays 19 169–172 (1999).

[15]Y. C. Kang, Y. S. Chung and S. B. Park, "Preparation of YAG : Europium red phosphors by spray pyrolysis using a filter-expansion aerosol generator," J Am Ceram Soc., 82 [8] 2056-60 (1999).

[16]G. Gowda, " Synthesis of yttrium aluminates by the sol-gel process," J. Mater. Sci. Lett., 5 1029-32 (1986).

[17]S. M. Sim, K.A. Keller, and T. Mah, "Phase formation in yttrium aluminate garnet powders synthesized by chemical methods," J. Mater. Sci., 35 713-17 (2000).

[18]M. K. Cinibulk, "Synthesis of yttrium aluminum garnet from a mixed-metal citrate precursor," J. Am Ceram Soc., 83 [5] 1276-78 (2000).

GAS-PRESSURE SINTERING OF SILICON NITRIDE WITH LUTETIA ADDITIVE

Naoki Kondo,* Masato Ishizaki,** and Tatsuki Ohji*

*Synergy Materials Research Center, National Institute of Advanced Industrial Science and Technology (AIST),
Shimo-shidami 2268-1, Moriyama-ku, Nagoya, 463-8687 Japan
**Synergy Ceramics Laboratory, Fine Ceramics Research Association (FCRA),
Shimo-shidami 2268-1, Moriyama-ku, Nagoya, 463-8687 Japan

ABSTRACT

Gas-pressure sintering conditions of silicon nitrides containing lutetia (Lu_2O_3) and silica (SiO_2) as sintering additives were investigated. The amounts of Lu_2O_3 and SiO_2 were fixed to 8 and 2 wt.%, respectively, and sintering temperature and settlement condition inside a graphite crucible were varied. Densification behavior, grain size, chemical composition and formation of secondary crystalline phases were investigated. Specimens were successfully densified in every condition, and detected secondary crystalline phase was only $Lu_2Si_2O_7$. However, grain size and chemical composition were affected by sintering temperature and settlement condition. Mechanical properties of the sintered specimens were also strongly affected by their sintering condition. These results were discussed in relation to the phase diagram.

INTRODUCTION

Silicon nitride is a candidate material for ceramic components in gas turbine engine systems, such as combustor liners and blades. To improve operating temperature, highly heat resistant silicon nitride must be required. One conventional way to achieve high temperature strength is the control of sintering additives. Degradation of silicon nitrides at high temperatures has been considered in relation to softening of the glassy grain boundary phase and melting of secondary crystalline phases. By choosing sintering additives and making refractory grain boundary phases, high temperature strength can be improved. Presently, the best additive for high temperature use is believed to be lutetia (Lu_2O_3). Some products with lutetia additive were already developed,[1-4] however, suitable gas-pressure sintering condition has not been reported.

The authors has been investigated the suitable sintering conditions of silicon nitrides containing lutetia (Lu_2O_3) and silica (SiO_2) as sintering additives.[5] Microstructures including grain size and formation of secondary crystalline phases were quite sensitive to their composition namely, the ratio of Lu_2O_3 and SiO_2. The specimen with 8wt.% Lu_2O_3 and 2wt.% SiO_2 exhibited the highest strength at elevated temperatures. However, considerable amount of sintering additives went out during sintering. This might affect their properties. Thus, in this work, the amounts of Lu_2O_3 and SiO_2 were fixed to 8 and 2 wt.%, respectively, and sintering conditions, i.e., temperature and settlement inside a graphite crucible, were varied to investigate the composition after sintering. Strengths and fracture toughness of the fabricated specimens were measured, and the results were discussed in relation to the conditions.

EXPERIMENTAL

Silicon nitride (SN-E10, Ube Industries. Ltd., Tokyo, Japan), Lu_2O_3 (Shin-Etsu Chemical Co. Ltd., Tokyo, Japan) and SiO_2 (99.9%, 0.1 µm, Hokko Chemical Co. Ltd., Tokyo, Japan) powders were used. The amounts of Lu_2O_3 and SiO_2 were fixed to 8 and 2 wt.%, respectively. Phase diagram of Si_3N_4 - Lu_2O_3 - SiO_2 ternary system is shown in Fig. 1.[6] The nominal composition is located inside the triangle Si_3N_4 - Lu_2SiO_5 - $Lu_2Si_2O_7$ as mentioned in the phase diagram. However, silicon nitride raw powder usually contains a small amount of oxygen as SiO_2 and H_2O. The actual composition should move to the SiO_2-rich region.

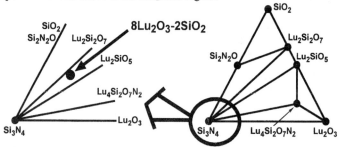

Fig. 1   Phase diagram of Si_3N_4-Lu_2O_3-SiO_2 ternary system.

The powders were ball milled for 72h in ethanol, then dried and sieved. The mixed powder was compacted using a steel die, followed by cold isostatic pressing (CIP) with a pressure of 400 MPa. The dimensions and relative density of the CIPed green body were 53 x 38 x 7 mm and ~53 %, respectively. Six CIPed green bodies were settled in a carbon crucible. Inside the crucible, three conditions were selected as follows. i) CIPed green bodies were covered by CIPed SN-E10 plates and SN-E10 powder. ii) CIPed green bodies were covered by SN-E03 powder. iii) CIPed green bodies were covered by dense silicon nitride plates, whose sintering additive was Lu_2O_3. SN-E03 powder is also provided by Ube Industries. Ltd.; this powder has larger grain size than SN-E10. Amount of oxygen contained in SN-E10 and SN-E03 powders were 1.3 and 0.8 wt.%, respectively. CIPed green bodies were soaked for 1h at 1700°C, 1h at 1800°C and 6h at 1950°C in 0.9 MPa N_2 atmosphere. Additionally, the bodies covered by SN-E10 were sintered at 1900°C instead of 1950°C. Heating and cooling rates were both 10°C·min^{-1}. The sintered bodies were heat-treated for 24h at 1500°C to crystallize grain boundary glassy phase. Hereafter, owing to their sintering conditions, the specimens were designated as E10-1900, E10-1950, E03-1950 and DENSE-1950, respectively.

All the specimens provided for investigation were taken from inner four sintered bodies. Chemical analysis was conducted to examine the amount of Lu and O. X-ray diffraction analysis was conducted to identify crystalline phases. Microstructures of the specimens were examined by scanning electron microscopy (SEM). Specimens (3 x 4 x 40 mm) for measuring bending strength were cut from the sintered bodies. Three-point bending strength was measured following ISO 14704-2000 or JIS R1601 with a span of 30 mm and a displacement rate of 0.5 mm·min^{-1} in air at room temperature (R.T.) and 1500°C. Fracture toughness was measured by Vickers Indentation-Fracture (IF) method (JIS R1607). Indentation load and time were 196 N and 15 sec, respectively.

RESULTS AND DISCUSSION
All the specimens were successfully densified. Densities of E10-1900, E10-1950, E03-1950 and DENSE-1950 were 3.32, 3.33, 3.34 and 3.36, respectively. E10-1900, E10-1950 and

DENSE-1950 shrunk almost uniformly, while outmost two plates of E03-1950 were distorted to inside. As E03-1950 was put into E03 powder, CIPed bodies and E03 powder contacted directly. This might affect the shrinkage behavior.

Amount of Lu and O after sintering are shown in Fig. 2. Considerable amount of sintering additives went out during sintering. The amount of Lu_2O_3 gone out was from 0.01 to 0.3 wt.%, and they were large in order of E10-1950, E10-1900, E03-1950 and DENSE-1950. The amount of O in DENSE-1950 was the smallest; it was about 0.3% smaller than the other sintered bodies. Oxygen in SN-E10 and SN-E03 powders seemed to go out during sintering. This suppressed the discharge of O from CIPed bodies. In above calculations, vaporization of Si and N was not taken into account, though they also should have vaporized out during sintering. Mol ratio of SiO_2/Lu_2O_3 in E10-1900, E10-1950 and E03-1950 were 2.6, 2.6 and 2.5, respectively, therefore, formation of $Lu_2Si_2O_7$ and Si_2N_2O were expected as secondary crystalline phases. The ratio in DENSE-1950 was 2.0. Thus only $Lu_2Si_2O_7$ was expected.

X-ray diffraction analysis, shown in Fig.3, revealed that the secondary crystalline phase actually formed in every specimen was only $Lu_2Si_2O_7$. This result seemed to be inconsistent to the presumption from the chemical analysis.

Chemical analysis can reveal the total amount of Lu and O included in a specimen. Condition of the element is unclear. In this case, the condition of O is probably a problem. Oxygen must exist as grain boundary glassy phase without crystallization. Additionally, small amount of O can be solved into silicon nitride grains. Therefore, presumption of secondary crystalline phases based on the chemical analysis has an error in calculation.

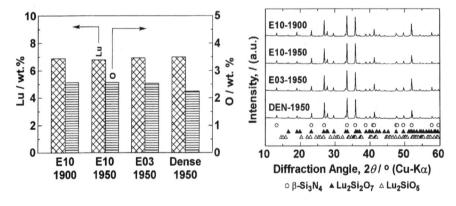

Fig. 2　Amount of Lu and O after sintering.　Fig. 3　Result of X-ray diffraction analysis.

Microstructures of the specimens are shown in Fig. 4. Densities of the specimens were somewhat different, however, the amount of residual pores were not large in any specimen. All the specimens consisted of rod-like silicon nitride grains. Grain size was fine in order of E10-1900, E10-1950, E03-1950 and DENSE-1950. The difference between E10-1900 and E10-1950 is due to the sintering temperature. The difference among E10-1950, E03-1950 and DENSE-1950 is very likely due to their composition.

All the three specimens had Si_3N_4 and $Lu_2Si_2O_7$ as crystalline phases. Thus, excepting the uncrystallized glassy phase, composition of the specimens is near the line Si_3N_4 - $Lu_2Si_2O_7$. There is an eutectic region inside the triangle Si_3N_4 - Re_2SiO_5 - $Re_2Si_2O_7$ (Re = Y, Yb).[7]

Existence of an eutectic point is also presumed in the Lu system, though it has not been reported
As the SiO_2/Lu_2O_3 ratio becomes smaller, the composition moves toward the eutectic region
The ratio was large in order of E10-1950, E03-1950 and DENSE-1950, though the difference is
not so large.   DENSE-1950 is closer to the eutectic region, therefore, slightly larger amount of
glassy phase should form, resulted in enhanced grain growth.   In this way, grain growth seemed
quite sensitive to the ratio.

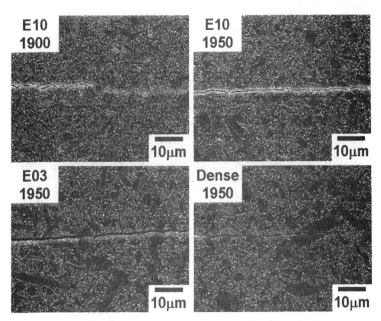

Fig. 4   Microstructures of the fabricated specimens.
Cracks were introduced by Vickers Indentation.
They run from left to right.

Measured strength and fracture toughness are shown in Fig. 5.   E10-1900 showed the
highest strength at R.T., but its strength largely decreased and became the lowest at 1500°C.   On
the contrary, E03-1950 showed the highest strength of about 480MPa at 1500°C, and it was
almost the same as the strength at R.T.   Fracture toughness of E10-1900 was high of about
$MPa \cdot m^{1/2}$, while those of other specimens were very low of less than 4 $MPa \cdot m^{1/2}$.   Crack
introduced by Vickers indentation were shown in Fig. 4.   The crack in E10-1900 was
substantially deflected.   Many pulled-out grains were observed in the micrographs.   Therefore
toughening mechanisms such as deflection, bridging and pull-out acted effectively, resulted :
high fracture toughness.   On the contrary,   the cracks in other specimens ran almost strai
resulted in low fracture toughness.

The raw powder composition was the same, however, mechanical properties of the sintere
specimens were quite different.   As mentioned above, composition of the specimens is near the
line Si_3N_4 - $Lu_2Si_2O_7$.   The smallest amount of grain boundary glassy phase exists when the
composition is located on the line Si_3N_4 - $Lu_2Si_2O_7$.   Existence of glassy phase at elevate

temperatures leads to strength degradation. Smaller amount of glassy phase is advantageous to achieve high temperature strength. If E03-1950 has the best composition to reduce grain boundary glassy phase at 1500°C, it must exhibit the highest strength. For the present, this seems a plausible reason why E03-1950 showed the highest strength at 1500°C.

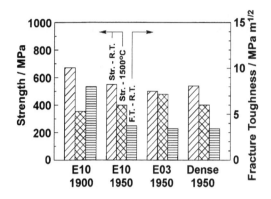

Fig. 5
Strength (Str.) and fracture toughness (F.T.) of the fabricated specimens.

However, the above discussions are not enough to explain all the differences among the four specimens. Fracture toughness of E10-1900 and E10-1950 is an example. These two specimens had the same SiO_2/Lu_2O_3 ratio, and E10-1950 consisted of larger grains. This seemed advantageous for E10-1950 to achieve higher fracture toughness. But the result is opposite. For further discussion, structure, composition and crystallization of grain boundary glassy phase in atomic level must be investigated.

ACKNOWLEDGEMENT
    This work has been supported by AIST, METI, JAPAN, as part of the Synergy Ceramics Project. Part of the work has been supported by NEDO. The authors are members of the Joint Research Consortium of Synergy Ceramics. The authors wish to thank Dr. A. Tsuge, Ceramic Research Institute, AIST, JAPAN, for his kind advice on chemical analysis.

REFERENCES
[1] K. Tanaka, S. Tsuruzono, and H. Terazono, "Characteristics and Application of New Silicon Nitride Materials, SN281 and SN282, for Ceramic Gas Turbine Components"; pp 248-52 in *Proc. of 6th International Symposium on Ceramic Materials and Components for Engines* (Oct. 19-23, 1997, Arita, Japan), eds by K. Niihara, S. Hirano, S. Kanzaki, K. Komeya, and K. Morinaga, 1997.
[2] T. Nakashima, T. Tatsumi, I. Takehara, Y. Ichikawa, and H. Kobayashi, "Research & Development of CGT30"; pp 233-36 in *Proc. of 6th International Symposium on Ceramic Materials and Components for Engines* (Oct. 19-23, 1997, Arita, Japan), eds by K. Niihara, S. Hirano, S. Kanzaki, K. Komeya, and K. Morinaga, 1997.
[3] S.-Q. Guo, N. Hirosaki, Y. Yamamoto, T. Nishimura, and M. Mitomo, "Improvement of High Temperature Strength of Hot-Pressed Sintering Silicon Nitride with Lu_2O_3 Addition," *Scripta Mater.*, **45** [7] 867-874 (2001).

[4] N. Kondo, M. Asayama, Y. Suzuki and T. Ohji, "High Temperature Strength of Sinter Forged Silicon Nitride with Lutetia Additive", *J. Am. Ceram. Soc.*, **86** [8] 1430-32 (2003).

[5] N. Kondo, M. Ishizaki, and T. Ohji, "Fabrication of Silicon Nitride with Lutetia Additive," Proc. 4th Int. Symp. on Nitrides (Nov.17-19, 2003, Mons, Belgium), in press.

[6] N. Hirosaki, Y. Yamamoto, T. Nishimura and M. Mitomo, "Phase Relationships in the Si_3N_4 - SiO_2 - Lu_2O_3 System," *J. Am. Ceram. Soc.*, **85** [11] 2861-63 (2002).

[7] J. S. Vetrano, H.-J. Kleebe, E. Hampp, M. J. Hoffmann, M. Rühle and R. M. Cannon, "Yb_2O_3 - fluxed Sintered Silicon Nitride," *J. Mater. Sci.*, **28** 3529-3538 (1993).

USE OF COMBUSTION SYNTHESIS IN PREPARING CERAMIC-MATRIX AND METAL-MATRIX COMPOSITE POWDERS

K. Scott Weil and John S. Hardy
Pacific Northwest National Laboratory
Richland, WA 99352

ABSTRACT

A standard combustion-based approach typically used to synthesize nanosize oxide powders has been modified to prepare composite oxide-metal powders for subsequent densification via sintering or hot-pressing into ceramic- or metal-matrix composites. Copper and cerium nitrate salts were dissolved in the appropriate ratio in water and combined with glycine, then heated to cause autoignition. The ratio of glycine-to-total nitrate concentration was found to have the largest effect on the composition, agglomerate size, crystallite size, and dispersivity of phases in the powder product. After consolidation and sintering under reducing conditions, the resulting composite compact consists of a well-dispersed mixture of sub-micron size reinforcement particles in a fine-grained matrix.

INTRODUCTION

Improvements in lightweighting the materials used in freight truck and automotive engine and body components translate directly into increased fuel utilization and reduced emissions. One of the most promising families of materials in this respect are the discontinuously reinforced composites (DRC), which include discontinuous metal matrix and ceramic matrix composite materials (DMMCs and DCMCs, respectively). Using computational mechanics, it is possible to quantitatively describe the effects of the DRC microstructure (i.e. reinforcement composition, size, shape, and volume fraction, matrix composition and grain size, and residual stress at the reinforcement/matrix interface) on its overall thermomechanical behavior. For example from the standpoint of particle size, an optimized DMMC should consist of a homogeneous dispersion of sub-micron size reinforcement particles approximately $0.1 - 0.5\mu m$ in diameter, with $1 - 2vol\%$ of these in the $10 - 100nm$ range to enhance creep resistance.[1] Typically the type of analysis used in deriving these results assumes that the sub-micron reinforcement phase is uniformly dispersed throughout the matrix, with no particle clustering; a microstructural state that is very difficult to achieve in practice. In fact, it is experimentally well known that the presence of particulate clusters can severely degrade the mechanical properties of both DMMCs and DCMCs by acting as a source of crack initiation.[2,3] The objective of our study is to develop a DRC processing technique in which sub-micron size and uniform dispersion are simultaneously attained with the reinforcement phase so that in subsequent investigations we can examine the combined effect of these two variables on the mechanical behavior of the composite.

EXPERIMENTAL
Processing

There are a number of processes by which DRCs can be fabricated, including plasma spray deposition, exothermic dispersion, and melt oxidation. One of the most straight-forward and economical approaches is by powder processing in which the constituent matrix and reinforcement powders are mechanically blended and milled, then consolidated in a pressing operation (such as isostatic pressing, coining, roll compaction, or forging), and finally densified via presureless sintering or hot pressing. However, intimate mechanical mixing of sub-micron powders can be extremely difficult to accomplish, with phase segregation typically plaguing the end results. In an effort to avoid this problem, we have employed a chemical synthesis technique, developed at PNNL[4] in the early 1990's, to prepare a multi-phase powder that when consolidated would form the composite material of interest. Referred to as the glycine-nitrate combustion synthesis process, this technique consists of two basic steps: the formation of an aqueous metal nitrate-glycine solution and the subsequent heating of this solution to dryness and eventual autoignition, at which point a self-sustaining combustion reaction takes place that produces the final powder product. In this process, the glycine serves two purposes: (1) it prevents precipitation of the metal salts as the water is evaporated, thereby ensuring that the metal ions remain molecularly mixed in solution up to the point of combustion or chemical conversion, and (2) it acts as the fuel for the combustion reaction by undergoing oxidation with the nitrate ions during heating as shown in the example reaction [1] below:

$$12 \, Fe(NO_3)_3 \; + \; 8 \, H_2NCH_2CO_2H \; \Rightarrow \; 6 \, Fe_2O_3 \; + \; 20 \, H_2O \; + \; 16 \, CO_2 \; + \; 13 \, N_2 \qquad [1]$$

| metal nitrate | amino acid, glycine | metal oxide | waste gases |
| (oxidizer) | (fuel) | (product) | |

In the following study, the glycine-nitrate process (GNP) was explored as an in-situ method of preparing a series of as-mixed composite powders, which could be consolidated and densified by the traditional methods of cold pressing and sintering.

Materials

Cerium oxide and copper were chosen as the model system on which we conducted our investigation. Copper metal and its oxides do not display any reactivity with CeO_2, making it an ideal inert matrix in which to construct a series of DMMCs or conversely a model ductile reinforcement phase for a family of DCMCs. In addition, because copper can be readily reduced from the oxide state back to metallic form without a similar reduction reaction occuring in the ceria, the final densification heat treatment step is relatively straight-forward. Stock copper and cerium nitrate solutions were prepared in 1M concentration by dissolving the respective reagent-grade salts [$Cu(NO_3)_2 \cdot 6H_2O$ and $Ce(NO_3)_3 \cdot 6H_2O$, Alfa Aesar] in de-ionized water. The cation concentration of each solution was adjusted to match the target within ±0.5%, as determined by EDTA titration. The two nitrate solutions were then mixed by weight in the ratios necessary to meet the desired compositions of the final as-prepared composites. In this way, the following range of target powder compositions was prepared: 10vol% Cu/90vol% CeO_2 to 90vol% Cu/10 vol% CeO_2 in 20vol% increments. Glycine powder (99.5%, Alfa Aesar) was then dissolved into each of the mixed aqueous solutions, after which a 40ml aliquot was drawn and heated in a 4L stainless steel beaker on a hot plate. The solution was allowed to boil to dryness, at which point it autoignited or spontaneously "burned". A fine stainless steel mesh was used to cover the beaker and retain the exothermically synthesized product within. In addition to varying the ratio of nitrate salts in each solution, the proportion of glycine-to-total nitrate content was also examined as a potential processing variable. It has been previously established that "fuel-rich", stoichiometric, and "fuel-lean" conditions can have a significant effect on the stoichiometry and particle size of the GNP product.[5] For each copper-cerium nitrate composition, solutions with three different glycine-to-nitrate ratios were prepared: the stoichiometric ratio (as given in Reaction [1] above), twice the stoichiometric amount (fuel-rich), and half the stoichiometric amount (fuel-lean).

Characterization

The as-synthesized composite powders were characterized by particle size analysis (PSA), X-ray diffraction (XRD), scanning and transmission electron microscopy (SEM and TEM), and energy dispersive X-ray analysis (EDX). PSA was performed to determine powder agglomerate size using a Microtrac Model S3000 Particle Size Analyzer. X-ray diffraction (XRD) analysis was carried out with a Philips Wide-Range Vertical Goniometer and a Philips XRG3100 X-ray Generator over a scan range of 20–80° 2Θ with a 0.04° step size and 2s hold time. XRD pattern analysis was conducted using Jade 6+ (EasyQuant) software. All of the XRD specimens were prepared using the same specimen holder and preparation technique to ensure sample consistency. Initial SEM analysis of the powders and of cross-sectioned compacts was conducted using a JEOL JSM-5900LV equipped with an Oxford Energy Dispersive X-ray Spectrometer (EDS) system that employs a windowless detector for quantitative detection of both light and heavy elements, while more detailed microstructural analysis was completed in a JEOL 2010F Field-emission gun Transmission Electron Microscope (FEG-TEM). On the TEM, an Oxford EDS was also used for compositional measurements. A 0.7 nm diameter probe allowed compositional information to be obtained on single particles within an agglomerate.

RESULTS AND DISCUSSION

The matrix of fifteen synthesis runs were characterized with respect to the composition and quantity of phases formed, average agglomerate size of the composite powder, and average size of the ceria and various copper/copper oxide crystallites within the agglomerates. Initial analysis indicates that the fuel-to-oxidizer ratio affects the type and quantity of phases formed in the multicomponent product. An example of this is shown in Figure 1 for powders synthesized at a target composition of 50vol% CeO_2 in copper (in the as-reduced state) under the three different glycine-to-nitrate solution compositions. As expected, a higher fuel content in the precursor affords a more reducing condition during combustion, i.e. measurable amounts of metallic copper and Cu_2O are formed. Conversely, the fuel-lean precursors yielded completely oxidized

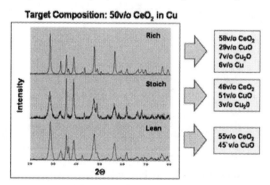

Figure 1:   XRD phase analysis of three post-combusted powders with an as-reduced target composition of 50vol% CeO_2 in copper. With the exception of glycine-to-nitrate ratio, all three powders were synthesized under the same conditions.

cerium and copper species upon reaction. This trend was found to be dependent on the ratio of copper-to-cerium in the original solution. As shown in Figure 2, the combustion conditions are significantly more reducing in the copper-rich (cerium-poor) precursors, as determined by the average oxidation state of the copper in the as-combusted powder. We presume that this is due to the lower combustion temperature exhibited by these compositions; i.e. the $Ce(NO_3)_3 \Rightarrow CeO_2$ reaction is significantly more exothermic than the $Cu(NO_3)_2 \Rightarrow CuO$ reaction.

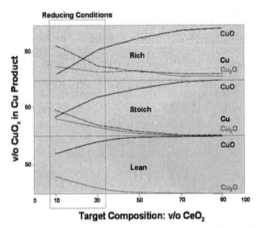

Figure 2   The effect of fuel-to-oxidizer ratio and cerium-to-coper nitrate ratio on the composition of phases formed in the as-reacted powders.

The glycine-to-nitrate ratio also has a strong effect on the average size of the cerium and copper product phases. As shown in Figure 3, the fuel-lean burns generally yielded a product with an average crystallite size 2 – 3 times finer than the fuel-rich precursors. However, the correlation between copper-to-cerium nitrate ratio and crystallite size is almost nil. The effect of these two process variables on the degree of phase dispersion is more difficult to quantify and

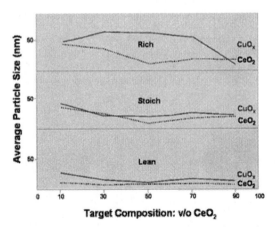

Figure 3   The effect of fuel-to-oxidizer ratio and cerium-to-coper nitrate ratio on average crystallite size in the as-reacted powders.

162

our attempts to determine this are only qualitative to date. Shown in Figure 4 are TEM micrographs of the typical agglomerate structures observed in two as-combusted powder products. Note that some phase segregation is seen in the agglomerate on the left, synthesized

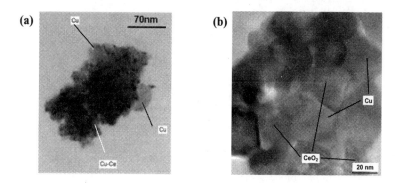

Figure 4   TEM micrographs of typical agglomerates in two 50vol% CeO₂/Cu product powders (as-reduced target composition) from: (a) a fuel-lean precursor and (b) a fuel-rich precursor.

from a fuel-lean mixture, whereas the fuel-rich sample appears to afford better dispersion of the cerium and copper phases. A summary of our findings is given in Table 1. In general, combustion of a stoichiometric to fuel-rich mixture is the preferred means of obtaining well-dispersed, sub-micron CeO₂ and CuOₓ crystallites in the powder product. Results from an initial SEM study of compacts made from these powders indicate that this microstructural state, i.e. sub-micron reinforcement particles well dispersed in the matrix, can be retained in the final densified product, as seen in Figure 4.

Table 1. Effects of glycine-to-nitrate and Ce:Cu nitrate ratios on phase composition and dispersion and agglomerate and constituent crystallite size in the as-combusted composite powders.

Glycine-to-Nitrate Ratio	Ratio of Metal Nitrates	
	High Ce(NO₃)₃/Cu(NO₃)₂	Low Ce(NO₃)₃/Cu(NO₃)₂
Fuel-Rich	Most reducing condition Good phase dispersion Agglomerates are large (~480nm) Large crystallite size	Moderate reducing condition Good phase dispersion Agglomerates are large (~300nm) Large crystallite size
Stoichiometric	Moderate reducing condition Good phase dispersion Agglomerates are moderate (~120nm) Large crystallite size	Oxidizing condition Good phase dispersion Agglomerates are moderate (~100nm) Moderate crystallite size
Fuel-Lean	Very slightly reducing condition Phase segregation Agglomerates are small Fine crystallite size	Oxidizing condition Phase segregation Agglomerates are small Fine crystallite size

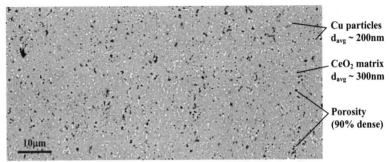

Figure 5   Cross-sectional SEM micrograph of a 10vol% Cu in CeO₂ composite. The composite was fabricated from fuel-rich powder by cold pressing the powder into a ~45% dense compact and sintering in Ar/2.75mol% H₂ at 900°C for 4hrs.

CONCLUSIONS

The glycine nitrate process was investigated as a in-situ means of preparing DMMC and DCMC composite powders with a well-dispersed mixture of sub-micron phases. It was observed that the fuel-to-oxidizer ratio in the powder precursor can significantly affect the crystallite size, degree of dispersion, and composition of the phases formed in the as-combusted powder. Fortunately, the ratio of copper-to-cerium nitrate has much less effect, making it possible to synthesize both metal and ceramic matrix powders by this approach. In general, fuel-rich and stoichiometric GNP precursors yield post-combusted powders with excellent phase dispersion, but somewhat larger agglomerate and individual crystallite sizes than the fuel-lean precursors. Initial consolidation experiments conducted on these powders suggest that the degree of dispersion and the sub-micron size of the reinforcement phase are retained in the final densified compact.

ACKNOWLEDGMENTS

The authors would like to thank Nat Saenz, Shelly Carlson, and Jim Coleman for their assistance. This work was supported by the U. S. Department of Energy, Office of Energy Efficiency and Renewable Energy, FreedomCAR Program. The Pacific Northwest National Laboratory is operated by Battelle Memorial Institute for the United States Department of Energy (U.S. DOE) under Contract DE-AC06-76RLO 1830.

REFERENCES

[1]K.S. Weil, "The Scaling Effect of Particle Size in Discontinuously Reinforced Metal Matrix Composites," *Materials and Design*, in press.

[2]O. Sbaizero and G. Pezzotti, "Tailoring the Microstructure of a Metal-Reinforced Ceramic Matrix Composite," *Transactions of the ASME: Journal of Engineering Materials and Technology*, **122** [3] 363-7 (2000).

[3]F. Petit, P. Descamps, M. Poorteman, F. Cambier, and A. Leriche, "Contribution of Crack-Bridging to the Reinforcement of Ceramic-Metal Composites/Definition of an Optimum Particle Size," *Key Engineering Materials*, **206-213** [2] 1189-92 (2002).

[4]L.A. Chick, L.R. Pederson, G.D. Maupin, J.L. Bates, and G.J. Exarhos, "Glycine-Nitrate Combustion Synthesis of Oxide Ceramic Powders," *Materials Letters*, **10** [1,2] 6-12 (1990).

[5]L.A. Chick, G.D. Maupin, and L.R. Pederson, "Glycine-Nitrate Synthesis of a Ceramic-Metal Composite," *NanoStructured Materials*, **4** [5] 603-15 (1994).

MECHANICAL RELIABILITY OF Si_3N_4

Kedar Sharma, P. S. Shankar, and J. P. Singh
Energy Technology Division
Argonne National Laboratory
Argonne, IL 60439

M. K. Ferber
Oak Ridge National Laboratory
Oak Ridge, TN 37831

ABSTRACT

Silicon nitride ceramics are leading candidates for advanced gas turbines. Testing of miniature disk specimens in a biaxial flexure mode has the potential to provide strength data needed for reliable performance prediction of the ceramic turbine components. To this end, the biaxial flexure strength of silicon nitride disks (with and without surface machining) was measured by a ball-on-ring fixture. The biaxial strength was compared with the measured strength of conventional flexure bars in order to assess the validity of the biaxial testing. The biaxial strength was higher than the strength of the flexure bars. This discrepancy may be related to the larger stressed volume of the flexure bars. Strength of both the biaxial disks and flexure bars was dependent on the severity of surface machining. As expected, coarser machining resulted in a lower strength. Since the surface of the flexure bars needs to be machined before testing to meet the standard geometrical requirement, conventional flexure bars may not provide reliable strength data for performance prediction of turbine components with as-processed surfaces. Unlike the flexure bars, the biaxial disks do not require surface machining because of their small size. Furthermore, fractography indicated that the flexure bars are more prone to fail from the specimen corners, making stress analysis and strength prediction difficult and complex. On the other hand, the biaxial disks always fail from the center of the disk, as predicted by analysis. Therefore, biaxial testing of miniature disks is the preferred mode of strength evaluation for performance prediction.

INTRODUCTION

Silicon nitride (Si_3N_4) ceramics with reinforcing agents and refined microstructures are leading candidates for high-temperature structural applications in advanced gas turbines[1]. The high performance of Si_3N_4 ceramics comes from their low density, refractoriness, and good corrosion and oxidation resistance. Recent studies [2] have concluded that in spite of the potential of Si_3N_4 ceramics, lack of expected mechanical reliability has been a key problem in actual service conditions. This problem arises from two factors. First, because the test specimens are often fabricated differently from the turbine components, the properties measured in the laboratory may not reflect those of the actual turbine components. In addition, the specimen surface for laboratory testing is generally machined and ground to meet the desired test standards, whereas the actual turbine components may exhibit as-processed surfaces, which may result in significantly different strength. Second, reproducing environmental effects on the properties of gas turbine components is difficult in laboratory testing.

An approach to avoid the above problems is to evaluate the mechanical properties of the components by testing miniature disk specimens with as–processed surfaces, directly machined from actual components. To validate this approach, we evaluated the mechanical strength of Si_3N_4 by biaxial testing of miniature disk specimens with the tensile surface in as-processed and ground conditions. The biaxial strength of Si_3N_4 miniature disks was compared with the strength of conventional flexure bars measured in a four-point bending mode. In addition, fractography was performed to identify and characterize critical flaws in the biaxial disks and flexure bars. The results of the fractographic evaluation were correlated with strength behavior to assess the applicability of biaxial testing of miniature disks for performance prediction.

EXPERIMENTAL PROCEDURES

Flexure bars (45 mm × 7 mm × 3 mm) were machined from Si_3N_4 (SN282 and AS800) plates, and the tensile edges were ground smooth to avoid edge failure. These bars were fabricated with their tensile surface in both as-processed and machined conditions. The bars were loaded in a mechanical testing machine in a four-point bend mode with a loading span of 15 mm and a support span of 30 mm. The cross-head speed was maintained at 1.27 mm/min. Using the load at fracture, the strength of the flexure bars was calculated by substituting the load at fracture into the standard bending equation [3].

Miniature biaxial disks (6.5-mm diameter and 0.5-mm thick) were prepared from the broken halves of four-point flexure bars of SN282 Si_3N_4. The disks were prepared by machining one face of the broken flexure bars to desired thickness and then core-drilling them to make the biaxial disks of desired diameter. Three sets of disks were prepared. One set consisted of disks with an as-processed surface, while the other two sets consisted of disks whose surface were machined to a 9- and 23- μm diamond finish. The biaxial test was performed in a ball-on-ring arrangement. The diameter of the support ring was 5 mm, and the cross-head speed was maintained at 1.27 mm/min. The biaxial flexure strength was calculated using Equation 1 [4],

$$\sigma_f = \left\{ \frac{3P(1+\upsilon)}{4\pi h^2} \right\} \left\{ 1 + 2\ln\frac{A}{B} + \frac{(1-\upsilon)}{(1+\upsilon)}(1 - \frac{B^2}{2A^2})(\frac{A}{R})^2 \right\} \tag{1}$$

where σ_f is the flexure strength, P is the measured fracture load, A is the radius of the lower support ring, B is the effective radius of contact of the loading ball on the specimen, R is the specimen radius, h is the specimen thickness, and υ is Poisson's ratio.

RESULTS AND DISCUSSIONS

Figure 1 shows the strength measured for biaxial flexure disks and four-point flexure bars of SN282 and AS800 Si_3N_4 specimens with as-processed surface in tension. The four-point strength of the as-processed SN282 and AS800 is 442±46 MPa and 459±102 MPa, respectively. Clearly, the four-point flexure strength is lower than the biaxial flexure strength, which is 673±39 MPa and 858±68 MPa for the SN282 and AS800, respectively. However, note that the four-point bend bars and biaxial disks have different effective volumes. Effective volume, $V_{effective}$, is the volume of a specimen that has the same probability of failure under different loading conditions (such as biaxial and four-point bend). Thus, for a proper comparison, strength values of the four-point bend bars and biaxial disks must be normalized on the basis of effective volume. In this study, the statistical Weibull approach was used to account for the effects of the effective volume of the specimens on strength. Thus, assuming the same probability of failure for flexure bars and biaxial disks of different volume, the two-parameter form of the Weibull

166

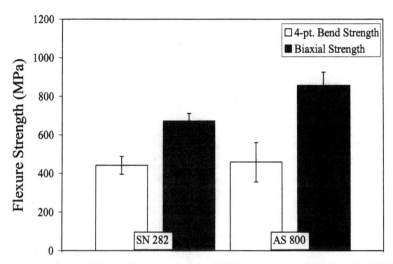

Figure 1. Four-point and biaxial flexure strength of Si_3N_4 specimens (SN282 and AS800) with as-processed surface in tension.

equation [5, 6] can be simplified in the following manner,

$$\sigma_{biaxial} / \sigma_{4\text{-pt.}} = \left(V_{effective, 4\text{-pt.}} / V_{effective, biaxial} \right)^{1/m} \tag{2}$$

where $\sigma_{biaxial}$ and $\sigma_{4\text{-pt.}}$ are the biaxial and four-point flexure strength, respectively; $V_{effective,biaxial}$ and $V_{effective, 4\text{-pt.}}$ are the effective volume of the biaxial and four-point specimens, respectively; and m is the Weibull modulus.

From Equation 2, the fracture strength ratio of the biaxial disk to flexure bar was estimated to be 1.79 and 2.34 for SN282 and AS800, respectively, which exceed the measured strength ratios of 1.51 and 1.75. For this calculation, m was taken as 11 and 21 for SN282 and AS800, respectively, based on experimental data. This overestimation suggests the biaxial strength prediction based on the Weibull equation may not take into account the complete stress-state effects on strength [7].

The effect of surface finish on flexure strength was determined with rectangular bars and biaxial disks. The results, shown in Figure 2, indicate that the strength of the specimens with the as-processed surface is different from that of the specimens with the machined surface. Furthermore, the strength depends on the severity of surface machining: coarser surface finish results in lower strength as compared to the finer surface finish. Thus, for turbine components with the as-processed surface, strength prediction based on flexure bar testing may not be reliable because the surfaces of these bars need to be machined before testing to meet the standard geometrical requirement. Unlike the flexure bars, the miniature biaxial disks do not require surface machining, because of their small size.

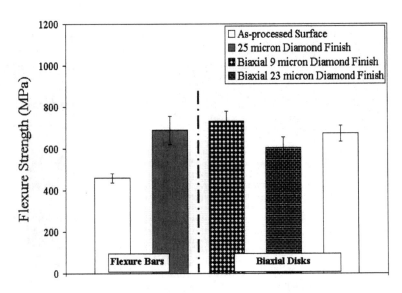

Figure 2. Measured strength of AS800 Si$_3$N$_4$ flexure bars and SN282 Si$_3$N$_4$ miniature biaxial disk with different surface finish.

The broken halves of the four-point bend and biaxial flexure specimens (with as-processed and machined surfaces) were examined by fractography to identify and characterize failure initiating critical flaws. Flexure bars and biaxial disks with machined surface failed primarily from machining flaws, whereas specimens with the as-processed surfaces failed from processing flaws (pores and agglomerates). Figure 3 shows typical crack initiation sites (critical flaws) in flexure bars and biaxial disks. Figure 3a shows a typical edge flaw in a flexure bar representing a processing-induced microstructural discontinuity. Figure 3b shows a processing-induced pore as a critical flaw in a biaxial disk specimen. Failure in flexure bars initiated primarily from edge flaws, making stress analysis and strength prediction difficult and complicated. On the other hand, the miniature biaxial specimens consistently failed from the center of the disks, as predicted by the analysis. Furthermore, as discussed before, the strength of flexure bars depends on the severity of surface machining, and hence may not be reliable in the predicting performance of turbine components. This problem is eliminated for miniature biaxial disks because they do not need surface machining due to their small size. Thus, biaxial testing of the miniature disks is a preferred mode for strength evaluation and reliability prediction, and has great potential for turbine applications.

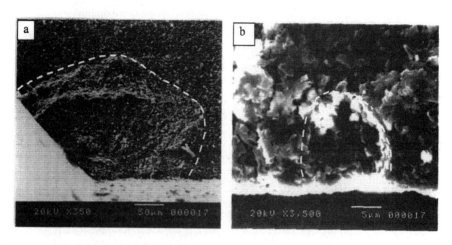

Figure 3. Critical flaws in fractured bar and disk specimens: (a) edge flaw in a flexure bar and (b) pores at the center of a biaxial disk.

ACKNOWLEDGMENT

This work was supported by the U.S. Department of Energy, Office of Fossil Energy, National Energy Technology Laboratory, under Contract W-31-109-Eng-38.

REFERENCES

[1]Tennery, V. J., "Ceramic Materials and Components for Engines" in *Proceedings of the Third International Symposium*, Las Vegas, Nevada, Nov. 1988.

[2]Lin, H. T., Ferber, M. K. and Roode, M. van, "Evaluation of Mechanical Reliability of Si_3N_4 Nozzles after Exposure in an Industrial Gas Turbine," Invited Paper for 7th International *Symposium on Ceramic Materials and Components for Engines*, Germany, June 19-21, 2000.

[3]Cranmer, D. C., and Richerson D. W., "Mechanical Testing Methodology for Ceramic Design and Reliability," Marcel Dekker Inc., New York, 1998.

[4]Thiruvengadaswamy, R., and Scattergood, R. O., "Biaxial Flexure Testing of Brittle Materials," *Scripta Metallurgica*, **25**, 2529-2532, (1991).

[5]Shetty, D. K., Rosenfield, A. R., Duckworth, W. H., and Held, P. R., "Biaxial Flexure Studies of a Glass-Ceramic," *Journal of The American Ceramic Society*, **64** [1] 1-4 (1981).

[6]Shetty, D. K., Petrovic, J. J., Rosenfield, A. R., and Duckworth, W. H., "Correlation of Uniaxial and Biaxial fracture Strengths of Ceramics" in *Proceedings of the 6th International Conference on Fracture*, Pergamon Press, Oxford, 1994.

[7]Duckworth, W. H. and Rosenfield, A. R., "Effects of Stress State on Ceramic Strength" in *Energy and Ceramics*, Edited by F. Vincenzini, Elsevier Scientific Pub., Amsterdam, 1980.

CORRELATION OF FINITE ELEMENT WITH EXPERIMENTAL RESULTS OF THE SMALL-SCALE VIBRATION RESPONSE OF A DAMAGED CERAMIC BEAM

Scott R. Short, Ph.D., P.E.
Department of Mechanical Engineering
Northern Illinois University
DeKalb, IL 60115

Shihong Huo
Department of Mechanical Engineering
Northern Illinois University
DeKalb, IL 60115

ABSTRACT

Finite element analysis results are compared to experimental results obtained during impact-excitation testing of a small ceramic beam. The experimental data were obtained by monitoring the transient vibration response of the beam using a non-contact pressure transducer (a microphone). Impact excitation was achieved by dropping a small metal sphere on the test beam. Making small cuts of various sizes using a metallurgical-sectioning saw simulated defects in the beam. The finite-element analysis data were reduced for comparison with the experimental data using a state-space model written in the software program Matlab®. Conclusions are made concerning the use of finite element models in precise modal analysis studies and the viability of the described experimental technique for the nondestructive evaluation of small ceramic components.

INTRODUCTION

Use of "spectral fingerprinting" to characterize materials and components is one of the most discussed areas of vibration-based engineering, including that involving nondestructive evaluation (NDE). Interest in the technique throughout history [1] along with numerous experimental and theoretical studies [2][3][4][5] have resulted in a detailed specification for its application, ASTM C1259 [6], in addition to the recent development of a powerful production NDE method referred to as resonant ultrasound inspection (RUI) [7][8]. From perusal of the related literature, it appears that the RUI method is capable of the type of production-level NDE sought by the manufacturing industry for years. The method can discriminate between "good" and "bad" components with a high level of confidence. A few limitations exist, however, with the production-environment RUI method. For example, the components being tested must vibrate in multiple resonant frequencies, damping cannot be too great, and a significant number of good and bad components must first be tested in order to adequately train the "brain" of the process, the pattern-recognition software. On the other hand, the RUI-technique literature makes an accurate observation about the superiority of using multiple resonant-frequency overtones to characterize the structural integrity of component, rather than merely relying on the fundamental (first harmonic) resonant frequency. Nonetheless, general research continues into vibratory NDE methods based on the simple and single impulse excitation of a test component; the goal being to determine if the transient response can be processed in some way to wrest useful information from it. Therefore, the desire for an accurate, inexpensive, transient vibration NDE test capable of being applied to a small number of components (even just one) continues.

Chen & Short [9] previously used digital signal processing (DSP) along with impact excitation testing to identify shifts in the fundamental natural frequency of a small ceramic beam containing small cuts made using a precision metallurgical sectioning saw. They discussed the role that the relative size of the component under test plays as to whether a shift in the natural frequency is the more accurate indicator of the structural integrity versus some measure of vibration damping. In a follow-up to that work, the current research uses finite element analysis to predict the theoretical frequency shifts and corresponding mode shapes to be expected when a small ceramic beam containing saw cuts is set into motion by impact excitation. Upon coupling the finite element analysis (FEA) data with a state-space model written in the software program Matlab®, a more accurate representation of the transient vibration response of the ceramic beam can be made and compared with the experimental data [10].

EXPERIMENTAL METHOD

Chen and Short [9] studied a 3 mm x 4 mm x 45 mm ceramic beam made of chemical-vapor-deposited (CVD) silicon carbide (SiC). The beam was excited into transient vibration by dropping a 4-mm-diameter copper shot on it from a height of 10 cm. The beam was supported using cotton thread at locations equal to 0.224L from each end. The impact-excitation test system included a signal-pickup system (microphone, preamplifier, amplifier, and power supply), a data acquisition system (hardware and software), and a suspension system (specimen, wires, ball, and guide tube). The frequency response of the microphone was ±2 dB flat below 100 kHz. A digital filter was used to isolate the lowest natural frequency (fundamental). Data was sampled at 1.25 MHz utilizing an algorithm written in the software program Labview®. Defects of various sizes and location were represented by partial cuts along the edge of the beam made using a precision metallurgical-sectioning saw as shown in Fig. 1.

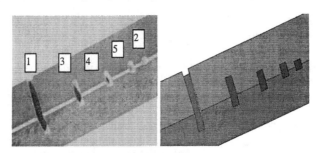

Figure 1. Left: ceramic beam showing order of saw cuts; Right: FEA model of beam.

FINITE ELEMENT AND STATE-SPACE ANALYSIS

The following material properties were provided by the silicon carbide beam's manufacturer: density = 3.214 g/cm^3; Poisson's ratio = 0.21. The manufacturer could not provide an accurate value for the dynamic tensile modulus of elasticity requiring that it be measured, giving a value of 4.48e8 Kg/s^2 mm via ASTM C1259. The non-contact pressure transducer (microphone) was used to measure this property.

The finite element software program, Ansys,® was then used to determine the first six modes of natural vibration for the beam containing no cuts. Of the first six modes of natural vibration that occur for a free-free beam, the second and fourth modes are out-of-plane vibration modes and hence are not considered further. The fifth vibration mode is rotational and is therefore also not considered further, assuming that the impact is centered with respect to the width of the beam. The first three modes of transverse vibration are shown in Fig. 2. However, the third mode (the second mode of transverse vibration) is physically impossible due to the fact that the beam was impacted at mid-length. Therefore, only the first (fundamental) and the sixth (third transverse vibration) modes affect the transient flexural vibration response of the beam. It is noted that the work by Sansalone and Street [11] shows that short-duration impact from steel spheres on concrete slabs are capable of exciting frequencies up to 80 kHz. The range of frequencies produced in the current research is actually much greater.

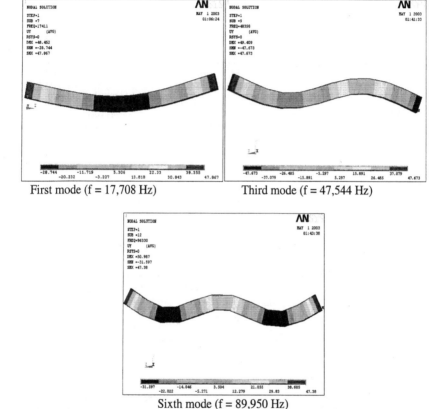

First mode (f = 17,708 Hz)   Third mode (f = 47,544 Hz)

Sixth mode (f = 89,950 Hz)

Figure 2. First three transverse vibration modes of the free-free beam.

The contribution of each mode on the transient vibration response of the beam was determined by performing a state-space analysis in Matlab®. For each mode, the Ansys® output file, providing information for all the nodes of the FEA model, was written out to the Matlab® state-space algorithm depicted in Figure 3. The state-space results shows that the sixth mode (third mode of transverse vibration) has far less effect on the transient vibration behavior of the beam due to the low amount of energy associated with this mode.

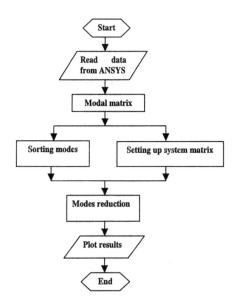

Figure 3. Matlab state-space algorithm.

In order to develop an accurate solid model for use in the finite element analysis of the beam with the saw cuts, a traveling microscope was used to measure the geometry of the saw cuts in the actual test beam. Due to the difficulty in map meshing the solid model of the beam containing the saw cuts, tetrahedral finite elements were used in the FEA model, along with a free-meshing technique (see Figure 4). The results of a convergence analysis of the finite element modeled beam indicate that a relatively fine mesh (>40,000 elements) is required to accurately model the beam's dynamic behavior.

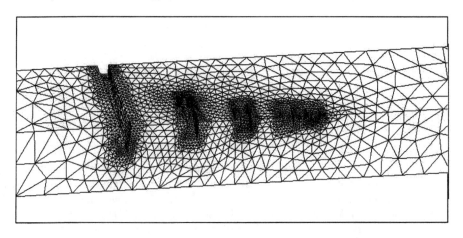

Figure 4. Meshed FEA model of beam containing saw cuts.

A comparison of the finite element and experimental results is shown in Table 1 and a typical shift in fundamental frequency is shown in Figure 5 for the beam with the 4th and 5th cuts.

Table I. Relative Shift in Fundamental Frequency

Data Source	No cuts	Cut#1	Cut#2	Cut#3	Cut#4	Cut#5
FEA	-	1490	3	88	36	9
Experiment	-	1211	2	66	16	4

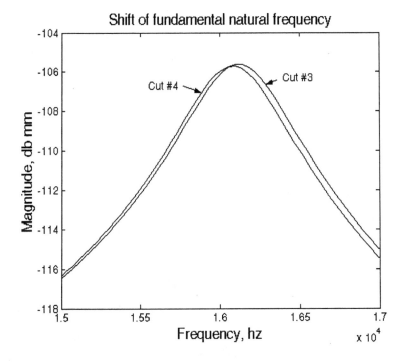

Figure 5. Fundamental frequency shift between beam with 4th cut and with 5th cut.

CONCLUSIONS

An NDE method of characterizing the structural integrity of small ceramic components involving impact excitation using a dropped mass and a non-contact pressure transducer (microphone) was investigated. Saw cuts made to the beam along its length simulated defects. If enough elements are used, finite element analysis is capable of characterizing the dynamic behavior of the beam with enough resolution to predict relatively small changes in frequency shift with changes in structural integrity. Finite element analysis indicates that if the beam is supported at the nodes of the fundamental resonant frequency, since other harmonic frequency

175

nodes occur elsewhere, the supports will be located in anti-node areas (or at least partially) and will most likely act to suppress these higher modes of vibration. Higher harmonics could be excited if the support points and point of impact excitation are changed. The relative strength of the harmonics depends on where the component is struck and on the parameters related to the dropped mass. To emphasize a particular harmonic, the component should be struck at a point to maximize the amplitude of the mode. Such a conclusion applies not only to a simple bar, but also to a component of general shape, such as a turbine blade.

The strong correlation of the results of finite element modeling with previous experimental results indicate that impact-excitation testing using a non-contact pressure transducer (microphone) is capable of measuring shifts in the fundamental frequency of small ceramic components, as long as the size of the anomalies is not too small.

REFERENCES
[1] G.F. Curtis and P.A. Lloyd, "Schizeophonics, then and now," Chartered Mechanical Engineer, London, England, October, pp. 55-60 1980.
[2] P. Cawley and R.D. Adams, "The mechanics of the coin-tap method of non-destructive testing," *J. Sound and Vibration*, 122 (2), pp. 299-316, 1988.
[3] J.A. Coppola and R.C. Bradt, "Thermal-Shock Damage in SiC," J. Amer. Cer. Soc., Vol. 56, No. 4, pp. 214-218, 1973
[4] S. Spinner and W.E. Tefft, "A Method for Determining Mechanical Resonance Frequencies and for Calculating Elastic Moduli from these Frequencies," ASTM Proceedings, Vol. 61, pp. 1221-1238, 1961.
[5] H. Demarest, Cube resonance method to determine the elastic constants of solids, J. Acoust. Soc. Am. 49, pp. 768-775, 1971.
[6] ASTM C1259-94 Dynamic Young's Modulus, Shear Modulus, and Poisson's Ratio for Advanced Ceramics by Impulse Excitation of Vibration, Annual Book of ASTM Standards, Vol. 15.02, American Society for Testing Materials.
[7] J. Maynard, Resonant Ultrasound Spectroscopy," Physics Today, 49(1), pp. 26-31, Jan. 1996.
[8] Quasar International WEB site; http://www.quasarintl.com.
[9] S.R. Short, S. Chen, "DSP Algorithms Applied to Acoustic Impact of Advanced Ceramics," *Proceedings of the 25th Annual International Conference on Advanced Ceramics & Composites*, Jan. 21-26, 2001, pp. 65-73.
[10] M.R. Hatch, *Vibration Simulation Using MATLAB and ANSYS*, New York, CRC, 2001.
[11] M.J. Sansalone, W.B. Street, *Impact-Echo, Nondestructive Evaluation of Concrete and Masonry*, Bullbrier Press, 1997.

MACRO-MICRO STRESS ANALYSIS OF POROUS CERAMICS BY HOMOGENIZATION METHOD

Yasushi Ikeda[1], Yasuo Nagano[1] and Hiroshi Kawamoto[1]
1) Synergy Ceramics Laboratory, Fine Ceramics Research Association
2-4-1, Mutsuno, Atsuta-ku, Nagoya, 456-8587, Japan

Naoki Takano[2]
2) Dept. of Manufacturing, Faculty of Engineering, Osaka University
2-1, Yamadaoka, Suita-city, 565-0871, Osaka, Japan

ABSTRACT
The 3-D image modeling of the porous alumina was obtained by X-ray CT, and microscopic volume elements were obtained from it. Porous alumina including spherical pores 60μm and 20μm in diameter with porosities 7.05% and 1.76% were studied by homogenization method, which calculated homogenized elastic constants for the microscopic volumes. These constants were applied to simulate 4-point bending tests and to analyze macroscopic stress distributions coupled with FEM calculation. Microscopic stress distributions were analyzed using the FEM result. The maximum microscopic stress concentration near a spherical pore were about 2 times higher than the average stress one.

1. INTRODUCTION

Porous ceramic materials are in widespread use and continue to be developed as the key materials for many industrial fields, e.g., filters for environmental purification systems, food manufacturing, and high temperature fuel cells. The mechanical properties of these porous materials are very sensitive to their 3-dimensional structures. Therefore, it is very important to analyze such materials taking into account their 3-dimensional morphology of the pore structure. The purpose of this work is to develop an image-based modeling method using high resolution 3-dimensional computed tomography (CT), which makes it possible to reconstruct the complicated porous structures of porous ceramics nondestructively with fairly rapid method. From the 3-dimensional CT data, binary-processed images can be obtained and a voxel mesh computational calculation is carried out by homogenization method[1]. After calculating homogenized elastic matrixes and homogenized elastic constants, macroscopic finite element method (FEM) analyses and microscopic voxel mesh analyses of stress concentration are carried out.

2. METHOD AND PROCEDURES

2.1 Principle of The Image-Based Modeling and Simulation Analysis of Stress Concentration By Homogenization Method

The outline of this study is shown in Fig.1 in the form of a flow chart. Our final purpose is to analyze macroscopic and microscopic stress concentrations in porous ceramic materials, which will be an origin of fracture of the materials. In this work, we prepared solid porous alumina samples by normal pressure sintering for imaging by microscopy and high resolution X-ray CT. The three-dimensional structure of the porous alumina was reconstructed by a commercially available computer program.

The homogenizing method was carried out using the voxel mesh method to obtain the homogenized elasticity of the specimen[2), 3)]. The homogenized elasticity obtained was used for VOXELCON HG (Quint Corp., in JApan) calculation, which calculated the macroscopic stress distribution for the global simulation model of the specimen under stress-loading, such as 4-point bending test. Using the calculated stress and displacement data for the global macroscopic model, the microscopic stress distribution in the small selected volume, which structure is the same as the RVE of the specimen, was calculated[4)]. The obtained voxel data were expressed as the histogram of stresses by newly a developed V-SEM software.

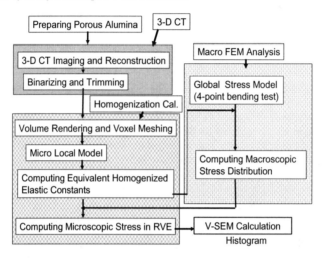

Fig. 1 Outline of the image-based modeling and the calculation of the homogenization method

2.2 Image-Based Modeling Of Porous Alumina Specimens By X-Ray CT

We prepared two types of porous alumina specimens. The pore sizes were 60 μm in the specimen A and 20 μm in the specimen B, and the porosities were 7.69 and 1.76 %, respectively. These specimens were imaged by 3-dimensional CT. The method of the 3-D CT system for image-based modeling is shown in Fig. 2. The high-resolution X-ray CT was using a micro-focus X-ray generator and a 2-D image detector. In this setup, a small specimen is put on a slowly rotating table. Continuous exposing by a corn beam of X-rays produced 2-D projection images of the specimen. The projection image was magnified by taking the large distance between the specimen and the image detector, and the small distance between the specimen and the focal point of the X-rays. We used the CT system by Uni-hite Corp. in Japan and the computing software of ACTIS+3 (Bio Imaging Research Co.). The pixel number of the 2-D image was set as 1024×1024 with a pixel size 3μm×3μm. The image size of cross section of specimens was about 1.5 mm×1.5 mm.

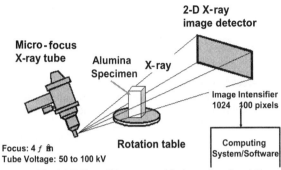

Fig. 2 3-D X-ray CT system used for image-based modeling

Figure 3(a) shows a typical 3-D CT image of the specimen A (60μm). The spherical pores are indicated with sufficiently high contrast in the images. The slice thickness of the CT was 3μm, so that the voxel size was about 3μm×3μm×3μm. We see many pores with realistic manner. The voxel number of the 3D image was about 500×500×240, so that the imaged region of the specimens was about 1.5mm×1.5mm×0.72mm. The total voxel number, 60 million, was too large for following computer simulation. Therefore, we need to extract a smaller volume from the 3D image, such as 100×100×100, i.e., 1 million voxels. The extracted small image should have a representative structure of the specimen in a extent, such as isotropic and with similar porosity to the bulk specimen. After we selected such microscopic volume, this was transferred to the voxel mesh computing software named as VOXELCON HG. In this software, the microscopic volume was prepared as the image-based model of the specimen. The volume of the model was 300μm×300μm×300μm and was more 100 times larger than the average pore volume. Figure 3(b) shows the image-based models of the RVE for the specimens A.

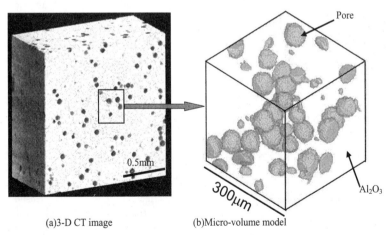

(a)3-D CT image   (b)Micro-volume model

Fig. 3 Image-based model of micro-volume of the specimen

In the models, pores are shown by gray color and alumina base is given by transparency (reversed images) to see the aspect of the inside. Homogenization method had been applied to the microscopic volume element.

3. RESULTS OF HOMOGENIZATION CALCULATION

3.1 Calculation of Homogenized Elasticity

Calculation of homogenized elasticity was carried out on these image-based models to input 400 GPa as Young's Modulus and 0.24 as Poisson's ratio of pure alumina matrix. Homogenized elastic constants of the specimen A and B are calculated taking account of the morphology of pore structures. The specimen A showed a little smaller average Young' modulus, 340 GPa than that of the specimen B, 390 GPa, which result is due to the large sizes of pores and larger porosity in the specimen A. On the other hand, Poisson' ratios are almost same for both specimens.

3.2 Macro-Micro Stress Analyses of 4-Point Bending Test

Using the obtained equivalent elastic constants, a simulation stress analysis of 4-point bending test, in which a load was given as obtaining the maximum tensile stress of 100 MPa at the bottom surface of the specimen. Under the macroscopic stress condition, the homogenization calculation was carried out to analyze the microscopic stress distribution in the RVE microscopic volume under the combining the macro-micro stress analyses equations.

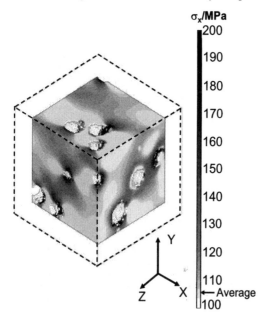

Fig. 4 Local stress distribution in the micro-volume of Specimen A

Using the macroscopic result of the FEM global analysis, we considered a microscopic volume, which had the pore structure of the RVE, in the region of maximum tensile stress of the specimen, that is, at the center of the bottom surface, and calculated the local stress distribution in the microscopic volume. Again we used homogenized elastic constants obtained above and calculated the local stress distribution in the microscopic volume. In fact, the calculated local stresses for the x-axis direction are distributed widely around the average values, 104 MPa for the specimen A. However, in the present study, the higher tensile stressed regions (0-200 MPa) are only interested. Fig. 4 shows the local stress distribution with higher tensile values in the microscopic volume of the specimen A. The average tensile stress, 104 MPa, in this case coincides well with the global external stress of 100 MPa.

4. DISCUSSIONS

In order to confirm the homogenized elasticity obtained for the RVE, we calculate the one for another larger volume cell (4 times larger, about 4 million voxels, which may be the largest voxel number allowable to conventional personal computers.). In this model, the porosity is 6.68% and the calculated modulus is 348.2, 343.8 and 343.0MPa for the x, y and z-axis, respectively. The obtained homogenized Young's modulus is plotted in Fig.5 with a square mark, which is very fitting to the line between the porosity and the young's modulus. Therefore, small difference can be seen between the data of some models including the large one and our RVE, we consider that our RVE model are the average model of the specimen A.

Next we consider the meaning of the obtained distribution of local stress in the RVE. Figure 6 show the histograms of the local stress with logarithm scale for the specimens A. Since the local stress is given for each voxel, the histogram gives the frequency of the voxels which have the corresponding values of the stress distribution. In order to distinguish the voxels with the same stress values, but being at the different distances from the pore surfaces, different patterns are used, that is, the voxels are classified with the distance from pores. The voxels indicated with d=1 are those locating at the surface of pores, and those with d=n are locating at the distance of the n-th

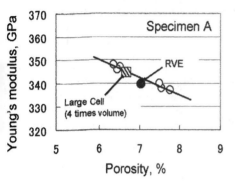

Fig.5 Variation of homogenized Young's modulus with porosity

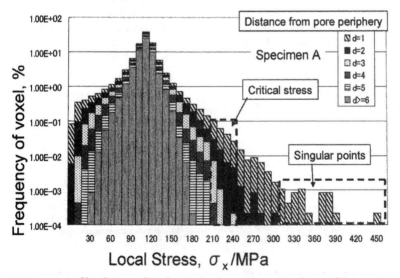

Fig. 6 Histogram of local stress distribution in the microscopic volume of the specimenA

voxels from pores. The calculated local stresses distributed in the range higher than 300 MPa, which are mostly at d=1, have very small frequencies, such as 1 to 10 in total 1 million voxels. There are possibilities that these voxels are isolated singular points due to the sharp boundary conditions of the alumina and vacancy (air) at the surfaces of pores used in the voxel-mesh calculation. Therefore, to evaluate and to discuss the stress concentration at the voxels, d=1, has the possibility to lead to erroneous results, that is, over-estimation. We think that the distributions at the voxels of d=2 and d=3 are more important for the discussion, and they give more realistic evaluation for the stress concentration. We consider that the stress values in the histogram where the voxels at d=2 and d=3 are appearing with d=1, and occupying more than 0.01% of frequency are very important for the contribution to fracture. From the Fig. 6, such critical stresses are seemed to be the values between 210 to 240 MPa in the specimen A. Comparing these values with the macroscopic value of the stress, 100 MPa, which is expected to appear in the case of non-porous dense alumina, we obtain the values of 2.1 to 2.4 as the critical local stress concentration factor in the case of the specimen A. Using the critical stress concentration factor, we estimate the flexural strength of the specimen A. The typical 4-point bending strength of non porous alumina specimen was measured as 440 MPa, so that considering with the above stress concentration factor, we can expect the strength of the specimen A decreases to the value in the range of 180 to 210 MPa for the specimen A. In our late experiments of 4-point bending test of similar specimens to the specimen A, we obtained 183 MPa as the average flexural strength, which is the comparable value to the low limit of the above expected stress region. The experimental values of the strength are scattered between 169 and 195 MPa, so that higher values are in better agreement with those of our estimation.

CONCLUSION
Image-based modeling and stress analysis by homogenization method were studied for porous alumina ceramics. 3-D CT with high resolution was successful for making the image-based model of microscopic volume element of porous materials. Homogenized elastic analysis was successfully carried out to give equivalent elastic constants and macro-micro stress distribution analyses of the porous alumina ceramics. The histogram of stress distribution was obtained and discussed from the point of stress concentration and fracture strength of the material.

ACKNOWLEDGEMENTS
This work has been supported by NEDO, as part of the Synergy Ceramics Project promoted by METI, Japan. The authors are members of the Joint Research Consortium of Synergy Ceramics.

REFERENCES
1) Ikeda, Y., Nagano, Y. and Kawamoto, H., *Proc. 27th Int. Cocoa Beach Conf. Advanced Ceramics and Composites, A*, pp.177-182 (2003).
2) Takano, N., Zako, M. and Ishizono, M., *J. Comput.-adid Mater. Design*, Vol.7, pp.111-132(2000)
3) Kimura, K., Takano, N., Kubo, F., Ogawa, S., Kawamoto, H. and Zako, M., *J. Ceram. Soc. Japan*, Vol.110, pp.567-575 (2002) (in Japanese)
4) Ikeda, Y., Nagano, Y., Kawamoto, H. and Takano, N., Proc. Atem'03, OS08W0101 (2003)

X-RAY AND NEUTRON DIFFRACTION STUDIES ON A FUNCTIONALLY-GRADED Ti₃SiC₂–TiC SYSTEM

X-RAY AND NEUTRON DIFFRACTION STUDIES ON A FUNCTIONALLY-GRADED Ti_3SiC_2–TiC SYSTEM

I.M. Low and Z. Oo

Materials Research Group, Department of Applied Physics, Curtin University of Technology, GPO Box U1987, WA. 6845 Australia

ABSTRACT

Ti_3SiC_2–TiC composites with graded interfaces have been processed through a high-temperature vacuum heat-treatment. X-ray diffraction (XRD), synchrotron radiation diffraction (SRD) and neutron diffraction (ND) have been used to characterise the phase evolution and the graded composition in this system. Results of SRD and ND in the temperature range 1000-1500°C show that the TiC layer commenced to form near the surface at 1200°C and grew rapidly in thickness with rising temperature. Depth-profiling of the TiC layer by XRD and SRD has revealed a distinct gradation in phase composition.

INTRODUCTION

Ti_3SiC_2 is a ternary carbide with a micro-layered microstructure and it displays a unique combination of mechanical, electrical, thermal and physical properties [1-7]. For instance, its electrical and thermal conductivities are higher than those of pure Ti, its thermal shock resistance is comparable to those of metals, and its machinability is similar to that of graphite. It is also relatively light (4.5 g/cm³), oxidation resistant, exhibits a ultra-low friction coefficient ($\mu = 3 \times 10^{-3}$), and is elastically rigid. However, with a Vickers hardness (H_v) of only 4 GPa and a Young's modulus (E) of 320 GPa, the low value of H_v/E suggests a mechanical behaviour of Ti_3SiC_2 somewhat similar to that of ductile metals.

However, unlike traditional binary carbides (eg. WC, SiC and TiC), which are among some of the hardest (> 25 GPa), stiffest, and most refractory (~2000°C) materials known, the ternary carbide Ti_3SiC_2 is relatively soft, not wear resistant, and has lower thermal stability (~1700°C). To counteract this, Barsoum and co-workers improved the surface hardness and oxidation resistance of Ti_3SiC_2 by using both carburization and silicidation to form surface layers of TiC and SiC [8].

Here, an alternative method for the formation of TiC layer on Ti_3SiC_2 is proposed. This approach involves the controlled heat-treatment of Ti_3SiC_2 in vacuum at high temperature. It has been proposed that the very low partial pressure of oxygen in the vacuum treatment may facilitate the surface formation of TiC via the following thermal dissociation reaction [9]. By interfacing Ti_3SiC_2 and TiC it is hoped that the combination of the hard-wearing surface layer of TiC and the tough under-layer of Ti_3SiC_2 will produce a composite that is stronger and more resistant to fatigue, wear and damage.

In this paper, results on the evolution of phase composition of vacuum heat-treated Ti_3SiC_2 at elevated temperatures are described. The development of TiC in the temperature range 1000-1500°C was monitored dynamically using neutron diffraction. Depth-profiling of the near-surface composition of TiC layer has been conducted by x-ray diffraction and grazing incidence synchrotron radiation diffraction.

EXPERIMENTAL METHOD

Sample Preparation

Ti$_3$SiC$_2$ samples were fabricated by reaction-sintering and hot-isostatic-pressing of Ti, SiC and C powders. The powder compacts were initially prepared by mixing in the proper molar ratio, cold pressed, followed by reaction-sintered in a vacuum furnace (~10^{-5} torr) at 1500°C for 1 h, and finally hot-isostactically-pressed (HIPed) in argon at 1650°C for 2 h with a pressure of 150 MPa. In order to study the temperature range of thermal dissociation of Ti$_3$SiC$_2$ in vacuum to form TiC by x-ray and synchrotron radiation diffraction, thin slices (~2 mm) of HIPed samples were heated-treated in an Elatec^{TM} vacuum furnace (~2 ×10^{-5} torr) at 900 – 1500 C for 1 - 8 h. Figures 1 and 2 show the respective microstructure and composition of as-HIPed Ti$_3$SiC$_2$ before and after the vacuum heat-treatment. Samples were accurately weighed before and after the vacuum-treatment to monitor mass change due to thermal dissociation of Ti$_3$SiC$_2$ to form TiC at various temperatures.

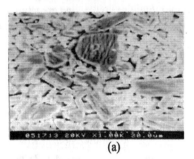
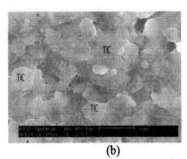

(a)   (b)

Figure 1. Typical scanning electron micrographs of HIPed Ti$_3$SiC$_2$ (a) before and (b) after vacuum heat-treatment at 1500°C for 8h to form TiC.

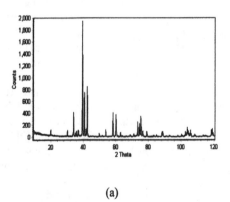
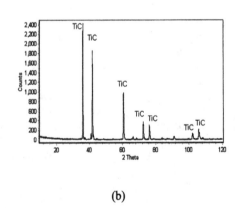

(a)   (b)

Figure 2. Typical XRD plots of HIPed Ti$_3$SiC$_2$ (a) before and (b) after vacuum heat-treatment at 1500°C for 8h to form TiC.

X-ray Diffraction (XRD)

Laboratory XRD patterns of vacuum-treated Ti_3SiC_2 samples were obtained with an automated Siemens D500 Bragg-Brentano instrument using $CuK\alpha$ radiation ($\lambda = 0.15418$ nm), produced at 40 kV and 30 mA over the 2θ range $5° - 130°$, step size 0.04 and counting time 2.4 s/step. Samples were mounted onto aluminium sample holders using a viscous adhesive and adjusted to the correct height with a glass slide. To obtain the graded composition profiles, the vacuum-treated samples were lightly polished with an emery paper to reach the desired depth. The relative phase abundance of Ti_3SiC_2 and TiC formed was computed using the Rietveld refinement method.

Synchrotron Radiation Diffraction (SRD)

The purpose of this experiment was to ascertain the near-surface phase composition of Ti_3SiC_2 heat-treated in vacuum at various temperatures. Synchrotron radiation diffraction (SRD) patterns of vacuum-treated samples were collected using the BIGDIFF diffractometer at the Australian National Beamline Facility in Tsukuba, Japan. Imaging plates were used to record the patterns over 2θ of 5-90°. The diffractometer was operated in Debye-Scherrer mode under vacuum with wavelength of 0.8 Å and an incidence angle of 1.0°. An incident beam of height 10 mm and width 150 μm and an exposure time of 20 min were used for each run. The relative phase abundance of TiC formed was estimated using the relative peak intensity ratio of (111) for TiC, and (102) for Ti_3SiC_2.

Neutron Diffraction (ND)

High temperature time-of-flight neutron diffraction was used to monitor the real time structural evolution of phase development of HIPed Ti_3SiC_2 in vacuum from 20 to 1500°C. Neutron diffraction data were collected using the Polaris medium resolution, high intensity powder diffractometer at the UK pulsed spallation neutron source ISIS, Rutherford Appleton Laboratory. Two diffraction patterns were collected at 17°C and at every 100°C between 1000 and 1500°C with a rate of 20°C/min. A total of 6 patterns were collected at 1500°C. The data acquisition times were 1 h (~200μAh) for the room temperature diffraction pattern, and 15 min (~50μAh) for each of the diffraction patterns collected at elevated temperatures. The relative phase abundance of TiC formed was estimated using the relative peak intensity ratio of (111) for TiC, and (102) for Ti_3SiC_2.

RESULTS AND DISCUSSION

Evolution of Phase Development

Figure 3 shows the relative abundance of TiC formed at various temperatures in vacuum. There is clearly no apparent dissociation of Ti_3SiC_2 to form TiC at temperatures below 1100°C. It dissociates initially slowly at 1100-1200°C but the process becomes quite rapid from 1250 to 1500°C. However, in addition to TiC, a transient phase of Ti_5Si_3C is also observed [10]. This phase is believed to form at the initial dissociation stage of Ti_3SiC_2 and it eventually converts to the stable TiC at elevated temperature. Figure 4 shows the in-situ neutron diffraction of Ti_3SiC_2 heat-treated in vacuum from 17-1500°C with a dwelling time 3 h at 1500°C. Below 1500°C, Ti_3SiC_2 is relatively stable and it commences to dissociate rapidly to form TiC only when the temperature approaches 1500°C. The relative abundance of TiC formed increases very rapidly with dwelling time of 1 to 3 h.

The apparent discrepancy in the commencement temperature for rapid Ti_3SiC_2 dissociation to form TiC as revealed by SRD and ND can be attributed to the difference in depths of information provided by both techniques. The former provides the near-surface information while the latter gives the average or bulk information which results in a much lower averaged phase content. This discrepancy also implies that the mechanism of Ti_3SiC_2 dissociation is *surface-initiated* rather than a *bulk* process. This suggests that a *diffusion-controlled* process may be responsible for the observed dissociation which starts at the surface and progresses slowly into the bulk. This further implies that the initially formed TiC surface layer may be of only a few nanometres at ~1100°C which can only be detected by SRD due to its high brightness and excellent sensitivity. The initial TiC surface layer formed will increase rapidly in thickness with a rising temperature. With a dwelling time of 3 h at 1500°C, it may reach a thickness of ~100 µm which can be readily detected by ND. This hypothesis of surface-initiated dissociation in vacuum-treated Ti_3SiC_2 is corroborated by the depth-profiling results to be presented in the next section.

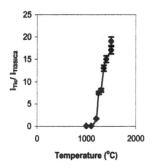
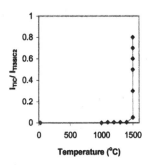

Fig. 3: The relative abundance of TiC formed at various temperatures in vacuum as revealed by synchrotron radiation diffraction.

Fig. 4: The relative abundance of TiC formed at various temperatures and dwelling time (0-3 h) at 1500°C in vacuum as revealed by in-situ neutron diffraction.

The effect of prolonged dwelling time during vacuum heat-treatment at 1500°C has been observed to increase the abundance and thickness of the TiC layer formed, indicating a time-dependent dissociation process [9]. A dwelling time of more than 10 h would be necessary to form a sufficiently thick layer of pure-TiC on Ti_3SiC_2.

The exact surface chemistry of this TiC formation in vacuum remains unclear. It is proposed here that in the presence of a very low oxygen partial pressure as in vacuum, the surface of Ti_3SiC_2 may undergo a high temperature thermal dissociation process to form TiC and Ti_5Si_3C in two stages as follows:

$$3Ti_3SiC_2 + O \rightarrow 4TiC_{(s)} + Ti_5Si_3C_{(s)} + CO_{(g)}$$

$$Ti_5Si_3C + 7O \rightarrow TiC_{(s)} + 3SiO_{(g)} + 4TiO_{(g)}$$

According to the above equations, the mass of samples following the heat-treatment should decrease as the temperature increases due to an increased volatility of CO, SiO and TiO. This hypothesis is collaborated by the observation of an increase in mass loss in vacuum as the heat-treatment temperature increases (Fig. 5)

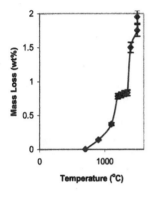 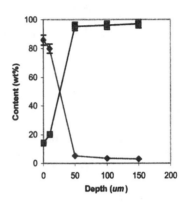

Fig. 5: Variation of mass loss as a function of temperature during vacuum heat-treatment.

Fig. 6: Depth-profiling of relative abundance of TiC ($\bullet$) and Ti_3SiC_2 ($\blacksquare$) in a vacuum-treated sample at 1500°C for 1 h.

Depth-Profiling of Phase Composition

The presence of graded composition in a vacuum-treated Ti_3SiC_2 sample at 1500°C for 1 h is clearly revealed by XRD as shown in Fig. 6. Before polishing with an emery paper, the as-treated surface contained ~85wt% TiC. The content of TiC decreased rapidly from the surface to the bulk after a depth of ~50 μm where it levelled off at ~3-5 wt%.

The graded nature of the near-surface composition of vacuum-treated Ti_3SiC_2 has also been verified by the variation of phase composition at different grazing angles [11,12]. The abundance of TiC was maximum on the surface and decreased rapidly as the angle or depth increased from 0.5 to 5.0°. These results support the above-mentioned hypothesis of a surface-initiated process for the formation of TiC. It is further proposed here that the dissociation to form TiC occurs initially on the surface by a nucleation and growth mechanism around the residual TiC grains. As the dissociation progresses, this mechanism propagates into the bulk of the sample through a diffusion process.

CONCLUSIONS

The feasibility of using the high-temperature vacuum heat-treatment to deposit a graded layer of TiC on Ti_3SiC_2 through a controlled dissociation process has been verified in this study. The graded nature of this system has been confirmed by x-ray diffraction and synchrotron radiation diffraction. Much work is still needed to examine the microstructure and to measure the microhardness and wear-resistance of the TiC layer. However, the nature of the nanostructured TiC layer may pose considerable challenges both in microstructural and mechanical properties evaluation.

ACKNOWLEDGMENTS

This work was supported by fundings from AINSE (Project 02/075), ISIS (Proposal RB13248) and ASRP (ANBF proposal 01/02-AB-36). Prof. M.W. Barsoum of Drexel University provided some Ti_3SiC_2 samples for initial study.

REFERENCES

[1] M.W. Barsoum and T. El-Raghy, "Synthesis and Characterisation of a Remarkable Ceramic: Ti_3SiC_2," *Journal of the American Ceramic Society,* **79**[10] 1953-58 (1996).

[2] M.W. Barsoum, D. Brodkin and T. El-Raghy, "Machineable Layered Ceramics for High Temperature Applications," *Scr. Met. et. Mater.,* **36**[3] 535-40 (1997).

[3] I.M. Low, S.K. Lee, M.W. Barsoum and B.R. Lawn, "Contact Hertzian Response of Ti_3SiC_2 Ceramics." *Journal of the American Ceramic Society,* **81**[2] 225-30 (1998).

[4] I.M. Low, "Vickers Contact Damage of Micro-layered Ti_3SiC_2," *Journal of the European Ceramic Society,* **18**[4] 709-15 (1998).

[5] M.W. Barsoum and T. El-Raghy, "Room Temperature Ductile Carbides," *Metall. Mater. Trans.,* **30A**[2] 363-70 (1999).

[6] T. El-Raghy, A. Zavaliangos, M.W. Barsoum and S.R. Kalidindi, "Damage Mechanisms Around Hardness Indentations in Ti_3SiC_2," *Journal of the American Ceramic Society,* **80**[3] 513-18 (1997).

[7] M.W. Barsoum, T. Zhen, S.R. Kalidindi, M. Radovic and A. Murugaiah, "Fully Reversible Dislocation-based Compressive Deformation of to 1 GPa," *Nature Materials.* **2**[1] 107-10 (2003).

[8] T. El-Raghy and M.W. Barsoum, "Diffusion Kinetics of the Carburisation and Silicidation of Ti_3SiC_2," *J. Appl. Phys.,* **83**[1] 112-18 (1998).

[9] I.M. Low, P. Manurung, R.I. Smith and D. Lawrence, "A Novel Processing Method for the Microstructural Design of Functionally Graded Ceramic Composites," *Key Engineering Materials,* **224-226** 465-70 (2002).

[10] I.M. Low, "Characterisation of Vacuum and Argon Heat-Treated Ti_3SiC_2," pp.191-192 in *Proc. of AUSTCERAM 2002* (Eds. I.M. Low & D.N. Phillips), 30 Sept – 4 Oct. 2002, Perth, WA.

[11] I.M. Low, M. Singh, P. Manurung, E. Wren, D.P. Sheppard and M.W. Barsoum, "Depth Profiling of Phase Composition and Texture in Layered-Graded Al_2O_3- & Ti_3SiC_2-Based Systems using X-ray and Synchrotron Radiation Diffraction," *Key Engineering Materials,* **224-226** 505-510 (2002).

[12] I.M. Low, "Depth-profiling of phase composition in a novel Ti_3SiC_2–TiC system with graded interfaces," *Materials Letters,* in press.

MODELING OF TRANSIENT THERMAL DAMAGE IN CERAMICS FOR CANNON BORE APPLICATIONS

J.H. Underwood, M.E. Todaro, G.N. Vigilante
Army Armament RD&E Center, Benet Laboratories
Building 115, Watervliet, NY 12189

ABSTRACT

Laser heating tests and thermomechanical modeling were performed for various ceramics to determine their suitability as a thermal barrier material at a cannon bore. Included were ZrO_2, Al_2O_3, SiAlON, Si_3N_4, and three types of SiC. The heating was a 4 millisecond pulse of a Nd:YAG laser resulting in about 0.9 J/mm^2 total heat input, typical of cannon bore heating during firing. Finite difference and solid mechanics modeling of laser and cannon-gas heating were used to characterize the near-surface thermal damage. Peak model temperature was used to deduce the peak *compressive stress* from thermal expansion, which was then compared to high-temperature *compressive strength* of a ceramic as determined from hot-hardness tests. Transient thermal stress leads to failure when it exceeds the reduced high-temperature compressive strength, causing permanent compressive strain and subsequent tensile residual stress and cracking upon cooling. The results indicate that ZrO_2 and Al_2O_3 are unsuitable for large caliber cannon use, showing considerable damage in tests and modeling, while Si_3N_4, SiAlON and SiC showed only limited damage for cannon firing conditions.

INTRODUCTION

Higher and more sustained cannon combustion-gas temperatures have led to interest in ceramics as a thermal barrier material at a cannon bore. Some initial work with SiC [1] showed cracks at a surface heated by a laser to similar temperature and duration as severe cannon firing. Finite difference and solid mechanics analysis of the thermal damage indicated that failure near the surface occurred when the transient thermal stress exceeded the reduced high-temperature compressive strength, leading to permanent compressive strain and subsequent tensile residual stress and cracking upon cooling. The purpose here is to perform laser-heating tests with seven additional ceramics and to compare damage from laser heating with that predicted from modeling hot-gas heating typical of cannon firing. In this case, the finite difference calculation of transient temperature includes detailed time-varying values of cannon gas temperature and convection coefficient, allowing a more realistic characterization of cannon thermal damage.

CERAMIC MATERIALS AND PROPERTIES

Table I lists the seven ceramics studied, along with properties E, v and α used in the analysis. The thermal conductivity, k, and diffusivity, δ, were determined for temperatures of 300-1270 K using a laser flash method [2]. Compressive strengths over 300-1070 K were determined from elevated temperature hardness tests, using the known

Table I – Ceramics investigated and selected properties

ceramic	Elastic Modulus E; GPa	Poisson's Ratio ν --	Thermal Expansion α; K^{-1}	Thermal Conduction k; W/m K	Thermal Diffusivity δ; cm^2/s	Compressive Strength S_C; GPa
ZrO_2	210	0.23	12.E-6	$12.7\,T^{-0.247}$	$0.126\,T^{-0.434}$	$531\,T^{-0.882}$
Al_2O_3	370	0.22	8.5E-6	$9610\,T^{-1.011}$	$175\,T^{-1.324}$	$780\,T^{-0.845}$
SiAlON	320	0.25	3.3E-6	$28.3\,T^{-0.170}$	$0.714\,T^{-0.503}$	$35.0\,T^{-0.283}$
Si_3N_4	310	0.27	3.2E-6	$339\,T^{-0.442}$	$14.1\,T^{-0.844}$	$33.3\,T^{-0.324}$
SiC-2	410	0.14	4.8E-6	$2000\,T^{-0.567}$	$64.8\,T^{-0.927}$	$182\,T^{-0.569}$
SiC-1	430	0.17	4.9E-6	$5240\,T^{-0.687}$	$218\,T^{-1.086}$	$226\,T^{-0.590}$
SiC-3	410	0.14	4.5E-6	$33600\,T^{-0.922}$	$1110\,T^{-1.287}$	$445\,T^{-0.665}$

relationship [3] that compressive strength, S_C, is well approximated as one third of hardness. Vickers hardness, HV, was used in the tests here in units of kg/mm^2. The resulting expression for S_C in units of GPa is:

$$S_C = (1/3) \text{ HV} \bullet 0.0098 \text{ GPa/(kg/mm}^2) \tag{1}$$

Power-law expressions were fitted to the k, δ and S_C data at room and elevated temperature, as shown in Table I for each of the seven ceramics and in the example plots of Figure 1 for Si$_3$N$_4$. The power law fits well and remains positive at high temperature, a requirement for thermal-mechanical modeling.

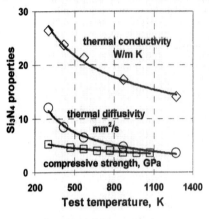

Figure 1. Example model input data for Si$_3$N$_4$ Figure 2. Hot gas input data for cannon firing

THERMO-MECHANICAL MODELING

Modeling of the transient temperatures and associated thermal damage at a cannon bore is similar to that in prior work [1,4], but with two improvements, summarized in Figures 1 and 2. First, the current modeling incorporates compressive strength vs temperature (Figure 1), gleaned from hot hardness measurements over a range of temperature for each of the seven ceramics. Prior work used literature values of hardness for generally similar materials. Second, the current model also incorporates detailed time-varying cannon combustion gas temperature and convection coefficient data (Figure 2) in the finite difference calculation of the near-bore temperature distribution. The hot gas data in Figure 2 were obtained from interior ballistic calculations at the axial location of most severe erosion damage in tank cannon firing. Note the much higher convection coefficient during the first millisecond, a significant change from the constant value over several milliseconds used in prior modeling work.

PULSED LASER HEATING RESULTS

A Nd:YAG laser described in prior work [1,5] was used here to apply a single, uniform, circular heating pulse to the surface of each 2 mm thick, 8 mm square ceramic sample. The heating pulse diameter was 1.8 mm for SiAlON, 2.6 mm for ZrO$_2$, and 3.4 mm for the other materials. Two or more samples of each of the seven ceramics were heated, and the results here are from the sample whose total heat input was closest to 0.9 J/mm^2, measured by calorimetry [5]. An analysis of the laser pulse profile showed a rapid increase and slow decrease in heating during the pulse, similar to the convection coefficient plot for cannon heating in Figure 2 but longer in duration. Based on this analysis, a constant heat-input pulse of 4 ms was used to approximate laser heating discussed later.

The 4 ms, 0.9 J/mm^2 laser heating tests showed very limited damage for the three types of SiC. Only one of two SiC-2 samples cracked as in prior work [1], and none of the SiC-1 and SiC-3 samples cracked. However, all ZrO$_2$, Al$_2$O$_3$, Si$_3$N$_4$ and SiAlON samples cracked. Metallographic cross-sections were prepared (unetched), from which optical and scanning electron micrographs were made, as shown in Figure 3. Two general features of the cracking seem to be [i] cracks normal to the surface with consistent 0.1-0.3 mm spacing and some opening; and [ii] cracks roughly parallel to the surface and about 0.1 mm below the surface with less opening. It appears that the normal cracks are formed by the mechanism of *thermal expansion – permanent compressive deformation – tensile residual stress* discussed in prior work and here. Further, it is suggested that the parallel cracks occur after the normal cracks have formed and opened, thereby allowing tension or shear stresses to develop near the tip of an opened normal crack and to cause cracking in directions other than normal to the surface, as discussed by Evans & Hutchinson [6].

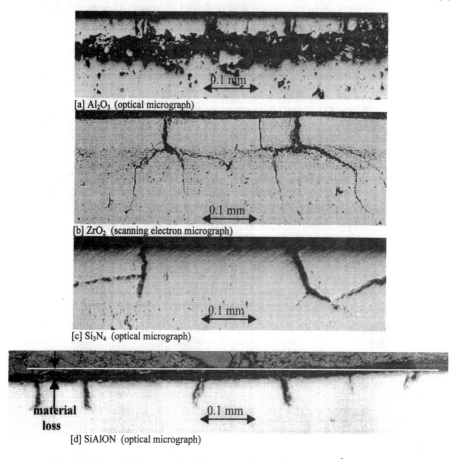

[a] Al_2O_3 (optical micrograph)

[b] ZrO_2 (scanning electron micrograph)

[c] Si_3N_4 (optical micrograph)

[d] SiAlON (optical micrograph)

Figure 3. Damage in four ceramics after one 4 ms laser pulse at 0.9 J.mm^2 total heat input

Some specific comments on the Figure 3 micrographs follow. The Al_2O_3 sample appears to have undergone significant fragmentation in the cracked areas, so much so that the presence of parallel cracks is much obscured. The ZrO_2 sample shows the clearest indication of normal cracks forming first and opening, followed by parallel cracks that later revert back to normal cracks. Note also the segments of surface material outlined by cracks, indicating impending fragmentation, as well as the distorted surface near the opened cracks, indicating that bending rotation may have occurred as a result of crack-face contact. The Si_3N_4 sample shows parallel cracks leading from near the tip of normal cracks, another indication that the normal cracks occurred first. The SiAlON sample shows small cracks (not easily seen in this photo) with changed direction at the tip of the normal cracks. Of more interest is the apparent loss of material in the laser-heated area of the SiAlON sample. The left edge of the photo is near the center of the 1.8 mm diameter laser "spot", and the right edge is near the outside of the spot. The line and arrows indicate a 0.02 mm loss of material, believed to be due to melting or decomposition of some constituent of the SiAlON ceramic. A general array of spherical globules was noted in the laser-heated area using a laser-scanning microscope before this sample was sectioned. None of the other six ceramics showed material loss or globules.

191

THERMO-MECHANICAL MODELING RESULTS

Temperature Distributions

Modeling of cannon gas heating and 0.9 J/mm² laser heating was performed to obtain the near-surface peak temperature distributions (and the resulting thermal stresses discussed later) for the seven ceramics and to compare the relative severity of the two heating modes. Figure 4 compares the laser and cannon heating temperatures for four of the ceramics, and Table II summarizes some of the key model results for all seven ceramics.

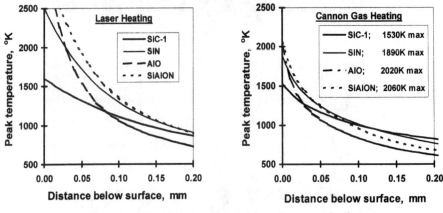

Figure 4 – Peak model temperatures for laser and cannon gas heating

Table II – Summary of model results for laser and cannon gas heating

	Conductivity, k, at 1270K W/m °K	Laser Heating; heat input, Q = 0.9 J/mm²				Cannon Gas Heating; Fig.2		
		T_{SURF} °K	a_{MEAS} mm	T at a_{MEAS} °K	a_{CALC} mm	T_{SURF} °K	a_{CALC} mm	Q J/mm²
ZrO₂	2.21	--	0.16	790	0.16	2430	0.09	0.34
Al₂O₃	7.20	3390	0.04	1710	0.13	2020	0.08	0.62
SiAlON	8.50	3070	0.04	2080	0.01	2060	0	0.53
Si₃N₄	14.1	2510	0.11	1230	0.01	1890	0	0.67
SiC-2	33.2	1710	0.04	1440	0.03	1580	0.01	0.81
SiC-1	37.3	1600	0	---	0.02	1530	0.01	0.84
SiC-3	45.2	1400	0	---	0	1430	0	0.90

As expected, the temperature distributions in Figure 4 show a rapid drop with depth. Also, some of the materials show significantly higher peak temperatures in laser heating than in cannon heating, particularly the materials with low thermal conductivity and diffusivity at high temperatures. In the laser heating model, 0.9 J/mm² is injected at the surface regardless of the material's thermal properties, whereas in cannon heating, materials with lower thermal conductivity and diffusivity will tend to reject heat transfer from the gas as the surface temperature goes up. It is interesting to review the surface temperatures for laser and cannon heating, shown in Figure 4 and Table II. Note that the order of surface temperatures, from high to low, is well predicted by the inverse of elevated temperature conductivity, shown in Table II for 1270 K. At the risk of over-simplicity, this indicates a "rule of thumb" that transient surface temperatures follow the inverse of elevated temperature conductivity.

Results for ZrO₂ and the other two types of SiC were not shown in Figure 4, to maintain clarity of the plots. The other SiC results were little different from those shown. The model temperatures for ZrO₂ were very high (due to low conductivity) and so far above the temperatures of available properties that the results were suspect.

Compressive Failure Predictions

The temperature distribution results, discussed above, are used to calculate the biaxial transient compressive stresses, S_T, using the following expression:

$$S_T = E\, \alpha\ (T - 300\ ^\circ K) / (1 - v) \tag{2}$$

where E, α and v are from Table I and T is the peak model temperature at a given location in the heated ceramic (plots such as Figure 4). When the transient compressive stress from Eq. 2 exceeds the elevated temperature compressive strength from Eq. 1, that is, when $S_T \geq S_C$, a permanent compressive displacement takes place, resulting in tensile residual stress and crack formation upon cooling.

Plots that show this model comparison of transient compressive stress with compressive strength for six of the seven ceramics are shown in Figures 5 and 6, for laser heating and cannon heating, respectively. The laser heating

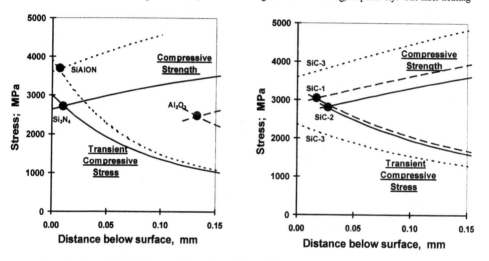

Figure 5 - Compressive failure predictions for model results from six ceramics with transient laser heating

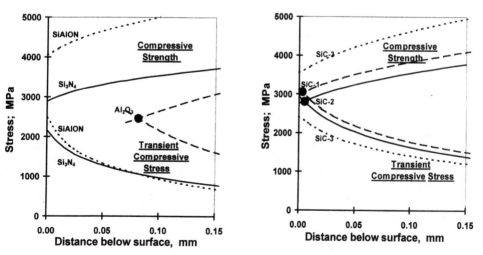

Figure 6 - Compressive failure predictions for model results from six ceramics with transient cannon heating

193

model results in Figure 5 predict that all ceramics except SiC-3 would crack, with crack depths approximated by the data points shown on the plots at the intersection of the compressive strength and transient compressive stress curves. These intersection points are listed in Table II as a_{CALC}. Cracking was predicted for ZrO_2, but these results were not shown, as discussed earlier in reference to Figure 4.

The cannon heating model results in Figure 6 show a marked improvement over the laser heating results, in that the predictions of crack depth (at the curve intersection points) are always at a smaller crack depth than those for laser heating. This is primarily a result of the rejection of cannon heat at high surface temperatures vs the injection of laser heat regardless of temperature, but it is also a result of the shorter heat pulse duration of cannon firing. As Table II shows, the cannon firing heat input (Q) is generally lower for materials with lower thermal conductivity at high temperature (1270K). These lower Q values indicate that the cannon heating was somewhat less severe than laser heating with Q=0.9 J/mm^2.

Finally, a ranking of suitability of the ceramics as a cannon bore thermal barrier material can be made using data from Table II. The preferred ceramics would be those with a_{CALC} = 0 and the lowest surface temperature for the cannon heating results. This rationale would rank SiC-3, Si_3N_4 and SiAlON as the most suitable; SiC-1 and SiC-2 as intermediate; and Al_2O_3 and ZrO_2 as the least suitable for a thermal barrier material under cannon bore firing conditions. This analysis and ranking of ceramics for cannon applications could be improved by additional thermal property and hot hardness data at temperatures up to those at which the damage occurs, indicated by the T at a_{MEAS} data in Table II.

SUMMARY

(i) Thermal damage modeling in ceramics was improved: by using detailed temperature-varying cannon gas temperature and convection coefficient data as input to finite difference transient temperature calculations; and by comparing compressive strength from hot hardness with near-surface compressive stress from transient heating.

(ii) Thermal damage cracks following a cannon-firing level of laser heating were observed in ZrO_2, Al_2O_3, SiAlON, Si_3N_4, and one of three types of SiC. The cracking mechanism is believed to be thermal expansion resulting in permanent compressive deformation followed by tensile residual stress and cracking upon cooling. Less damage is indicated for cannon heating because of the rejection of heat at high surface temperatures compared with the surface injection of laser heat, but also because of the shorter heat pulse of cannon firing.

(iii) Model prediction of thermal damage from typical tank cannon combustion-gas heating ranks one type of SiC as #1, Si_3N_4 as #2 and SiAlON as #3, in respect to the best resistance to thermal cracks. Ranking of resistance to thermal cracks is based primarily on the ratio of transient compressive stress to compressive strength being below unity and, secondarily, on a low value of transient surface temperature.

ACKNOWLEDGEMENTS

The authors are pleased to acknowledge: the help of R. Carter and J. Swab of the Army Research Laboratory for determining thermal and mechanical properties of the ceramics; the work of L. Bell and R. Yeckley of Kennametal, Inc. for determining the elevated temperature hardness of the ceramics; and the work of S. Smith and C. Rickard of Benet Laboratories for metallographic characterization of thermal damage in the ceramics.

REFERENCES

[1] J.H.Underwood, P.J.Cote and G.N.Vigilante, "Thermo-Mechanical and Fracture Analysis of SiC in Cannon Bore Applications", *Ceramic Engineering and Science Proceedings*, **24**, 3 503-508 (2003).

[2] Laser Flash Thermal Conductivity Data supplied by Netzsch Instruments, Inc., Bedford, MA under Contract W813LT-2077-6009 with the Army Research Laboratory.

[3] N.E.Dowling, "Hardness Tests"; pp. 139-147 in *Mechanical Behavior of Materials*, 2nd ed., Prentice Hall, Upper Saddle River, NJ, pp.139-147, 1999.

[4] J.H.Underwood, A.P.Parker, G.N.Vigilante and P.J.Cote, "Thermal Damage, Cracking and Rapid Erosion of Cannon Bore Coatings", *Journal of Pressure Vessel Technology*, **125** 299-304 (2003).

[5] P.J.Cote, G.Kendall and M.E.Todaro, "Laser Pulse Heating of Gun Bore Coatings", *Surface and Coatings Technology*, **146-147** 65-69 (2001).

[6] A.G. Evans and J.W. Hutchinson, "The Thermomechanical Integrity of Thin Films and Multilayers," *Acta. Metal. mater.*, **43**, 7 2507-2530 (1995).

STRENGTHENING OF CERAMICS BY SHOT PEENING

Wulf Pfeiffer and Tobias Frey
Fraunhofer Institut für Werkstoffmechanik (IWM)
Wöhlerstr. 11
79108 Freiburg, Germany

ABSTRACT

Shot peening is a common finishing procedure to improve the static and cyclic strength of metal components. Recent investigations showed that, under specific shot peening conditions, also in brittle ceramics high compressive stresses up to more than 1 GPa can be introduced near the surface which increase the near-surface strength. The presentation compiles the actual potential of the shot peening process for alumina and silicon nitride ceramics and gives detailed information about the influence of shot peening parameters on the residual stress state, dislocation density and topography. The effect of the residual stresses on the static and cyclic near-surface strength is quantified. In addition, the potential of shot peening for a recovery of machining induced damage is shown.

INTRODUCTION

Shot peening is a common procedure to improve the static and cyclic strength of metal components. It is based on a multiple localized plastic deformation of near-surface regions. This results in a surface layer which is improved by strain-hardening (increase of dislocation density) and macroscopic compressive residual stresses.

Non-transformation toughened ceramics show the typical brittle material behavior of failure before deformation at room temperature. Thus, strengthening of ceramics due to deformation induced compressive residual stresses has been thought to be not possible.

Recent investigations show however [1, 2] that, under specific shot peening conditions, also in brittle materials like ceramics high compressive stresses up to more than 1 GPa can be introduced near the surface. Exposing these strengthened surfaces to loading situations, which are characterized by a steep near-surface stress gradient, a boost of load capacity could be obtained. Such loading situations exist in, e.g., roller and sliding bearings or metal forming tools.

The aim of this paper is to compile the present potential of the shot peening process for 3 different types of ceramics.

EXPERIMENTAL DETAILS

Materials investigated

Commercially available alumina and silicon nitride ceramics were investigated. Table I summarizes the materials and the most important material characteristics.

Table I. Materials investigated and most important material characteristics

material	type / manufacturer	4-P bending strength	fracture toughness
alumina	A61 / Kennametal Hertel	400 MPa [5]	4.0 MPa m$^{1/2}$ [3]
silicon nitride	N3208 / H.C. Starck Ceramics	877 MPa [4]	4.2 MPa m$^{1/2}$ [4]
silicon nitride	XN8110/ H.C. Starck Ceramics	860 MPa [3]	6.2 MPa m$^{1/2}$ [5]

Shot Peening

The shot penning tests were carried out with a modified injection system. The pressurized air and the shot were applied to the jet nozzle in two different tubes. The shot is accelerated in the nozzle to a high velocity and hits the sample surface. The shot used was tungsten carbide beads with a diameter of 610 – 690 μm. The peening pressure ranged from 0.2 MPa up to 0.4 MPa. Using bead flux of 700 g/min typical peening times were 280 up to 840 seconds for flat samples and – due to the larger surface - 22 minutes for rolling samples. The resulting Almen-intensity was in the range of 0.22 up to 0.28 mm A. The most important shot peening parameters are indicated in the figures.

Determination of residual stresses and dislocation density

The residual stresses and dislocation densities were determined by X-ray diffraction (XRD). The full width at half maximum (FWHM) was calculated to characterize the dislocation density. The macroscopic residual stresses were derived from the peak shift using the well known $\sin^2\psi$-method [6]. The most important measurements parameters are summarized in Table II. The penetration depth from which 67% of the diffracted X-rays arise was in the range of 8-10 μm. For depth probing of residual stresses, it was necessary to remove stepwise thin surface layers by polishing using 3 μm diamond abrasives. This mechanical material removal resulted in additional small residual stresses.

Table II. Parameters of residual stress evaluations by XRD

material	radiation	lattice plane	X-ray elastic constant ½ s_2
alumina	CrKα	{220}	3.15 GPa^{-1}
silicon nitride	CrKα	{411}	3.89 GPa^{-1}

Determination of static and cyclic load capacity

The static and cyclic load capacity was determined using the ball-on-plate test. The advantage of this test is the high surface sensitivity due to the rapidly decreasing tensile stresses with depth. The tests were performed using an electro-mechanical and a servo-hydraulic testing machine. The load capacity was determined by stepwise increasing the load until cracks could be observed by optical microscopy. In case of the cyclic experiments 1,000,000 cycles with a frequency of 60 Hz were applied before inspection of the surface. At least five tests were performed per load step. A silicon nitride ball with a diameter of 10 mm was used for flat specimens, in case of bearing rollers made of XN8110 a 4.76 mm ball had to be used to anticipate a failure of the indenter-ball.

RESULTS
Residual Stress and static load capacity

The relationship between near-surface residual stresses and static load capacity was evaluated in detail for silicon nitride (N3208) and alumina. Figure 1 correlates the near-surface stress states and the load capacities of polished and differently shot peened alumina and silicon nitride samples.

Shot peening allowed creating up to 1.3 GPa compressive residual stresses near the surface. These compressive stresses shifted the load needed to introduce a Hertzian cone-crack into the surface by a factor of 3 (silicon nitride) and 9 (alumina), respectively. The increase of near-surface strength was so high, that the load limit of the ball-on-plate testing device (11.7 kN) was reached in some cases and the samples passed the test without any crack. Thus, for the most effective shot peening process no error bar could be calculated for the fracture load.

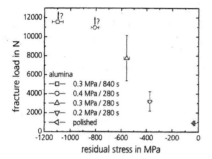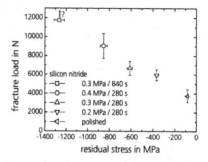

Fig. 1: Fracture load versus residual stress of (left) alumina and (right) silicon nitride samples (N3208) in polished and different shot peened condition.

The depth effect of the shot peening process was evaluated through depth probing of residual stresses and width of the diffraction lines (FWHM, a measure for the dislocation density) for silicon nitride N3208 and alumina. Figure 2, left, shows that a lower peening pressure (0.2 MPa) results in the maximum compressive stress at the immediate surface whereas a higher peening pressure (0.3 MPa) results in maximum compressive residual stresses up to 2.0 GPa 25 – 30 μm below the surface. The peening process may affect an up to 80 μm thick near-surface. The depth distributions of the dislocation densities (Fig. 2, right) and macroscopic residual stresses are similar in shape. This indicates that the main driving force for the creation of compressive residual stress is micro-plastic deformation. This applies in particular to alumina, where a higher dislocation density leads to macroscopic stresses being comparable to those in silicon nitride.

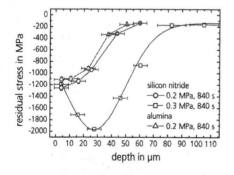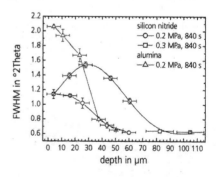

Fig. 2: Depth distribution of residual stresses (left) and width of the diffraction lines (right) of differently shot peened silicon nitride (N3208) and alumina samples.

197

Cyclic load capacity

The influence of shot peening on the cyclic load capacity was determined in detail for silicon nitride N3208 using cyclic ball-on-plate test. Figure 3 shows the fracture probability as a function of the load for the shot peened and the polished reference surfaces.

For a fracture probability of 50 %, the polished reference samples achieved a static load capacity of about 4 kN and a cyclic load capacity of 2.5 kN. The cyclic load capacity of the shot peened samples reached 10.5 kN which is a gain of a factor of 4. The shot peening treatment increased the static load capacity to more than 16 kN. The capacity of the testing device was not high enough to introduce any cracks into the shot peened surface. From the result of no failure up to 16 kN a gain of the static load capacity of a factor of at least 5 can be concluded.

Fig. 3: Fracture probability of polished and shot peened samples (N3208), respectively, determined in static and cyclic ball-on-plate tests. Note that in case of the static tests of the shot peened samples the capacity of the test equipment was exceeded.

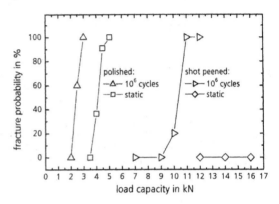

Topography

During shot peening each hit by a bead will produce a localized deformation accompanied by the creation of dislocations in the near surface crystallites. The superposition of many localized deformations will result in an overall roughness of the surface. Figure 4 shows a typical topography resulting from a shot peening process and compiles the average roughness values of the alumina and silicon nitride surfaces in polished and different shot peened conditions.

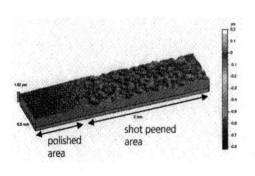
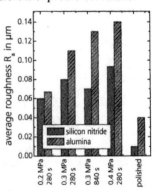

Fig. 4: (Left) Topographical map comparing a polished and a shot peened (0.2 MPa) Si_3N_4 surface area. (Right) Average roughness R_a of polished and shot peened Al_2O_3 and Si_3N_4 samples.

The shot peening of the polished surface leads to a significant increase of the roughness. Nevertheless, the roughness of shot peened surfaces is comparable to surfaces in ground or lapped condition.

Recovery of machining damage

The static load capacity of rollers (Ø 8 mm x 8 mm, XN8110) was determined before and after shot peening in rough-ground and finished surface condition, respectively. In contrast to all other ball-on-plate tests described in this paper, 4.76 mm diameter indenter-balls had to be used to anticipate a failure of the indenter-ball.

Comparing the load capacity of the finished and rough-ground rollers (see Fig. 5), a significant drop of strength is obtained due to machining damage intoduced into the near-surface of the rough-ground rollers. Shot peening the rough-ground rollers not only compensates for the machining damage, but increases the load capacity above the level of the well-finished rollers. From the result that the load capacity is comparable to finished + shot peened rollers, it can be concluded, that the shot peening process completely eliminated the influence of machining damage on the near-surface strength of the rollers.

Fig. 5: Fracture probability of rollers (XN8110) in polished, ground and additional shot peened condition, respectively (static ball-on-plate tests).

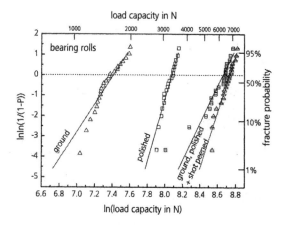

Thermal stability

The high temperature stability of the shot peening induced residual stresses was evaluated for silicon nitride (N3208) through XRD. Figure 6 shows, that tempering samples at 800 °C leads to a 50 % reduction of compressive residual surface stresses within the first 5 hours. A further reduction could not be obtained within the next 15 hours.

CONCLUSIONS

The presented investigations show that, using a novel shot peening process for brittle materials, micro-plastic deformation and high compressive residual stresses up to 2 GPa can be introduced into the near-surface of ceramics, declining within the first 100 µm. Tempering experiments, performed at 800 °C on silicon nitride revealed a satisfying stability of the shot peening effect. The shot peening process can dramatically increase the near-surface strength of ceramics. The static and cyclic load capacity tests show an increase of the load capacity by a factor 4, at least. The investigations performed on roller bearing components in different surface conditions show that also roughly machined surfaces, which suffer from machining damage, can be improved dramatically by the shot peening process. Further investigations will concentrate on increasing the depth of the compressive residual stress field, on the effect of the residual stresses on the bending strength of ceramics and on the application of this promising technique to other types of ceramics.

Fig. 6: Decline of shot peening induced residual stresses in silicon nitride N3208 at 800°C.

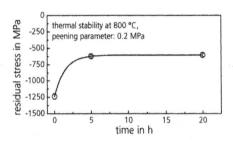

ACKNOWLEDMENT
Part of the investigation was sponsored by the Deutsche Forschungsgemeinschaft (DFG).

REFERENCES
[1] W. Pfeiffer and T. Frey "Strengthening of Ceramics by Shot Peening", Materials Science Forum, Vols. 404-407, p. 101-108 (2002).
[2] W. Pfeiffer and T. Frey, "Shot Peening of Ceramics: Damage or Benefit?", ceramic forum international Cfi/Ber. DkG 79 No.4, E25 (2002).
[3] M. Op de Hipt "Randzonencharakterisierung hartbearbeiteter und tribologisch belasteter Al_2O_3 und SiC-Keramiken", Fortschritt-Berichte VDI Reihe 5 Nr. 521, VDI-Verlag (1998).
[4] M. Rombach, "Experimentelle Untersuchungen und bruchmechanische Modellierung zum Versagensverhalten einer Siliciumnitridkeramik unter Kontaktbeanspruchungen", PhD Thesis University Karlsruhe (1995).
[5] Data sheet H.C. Starck Ceramics.
[6] P. Müller, E Macherauch "Das $\sin^2\psi$-Verfahren der röntgenographischen Spannungsmessung", Z. ang. Phys. 13, p. 305-312 (1961).

Solid Oxide Fuel Cells

DOE FE DISTRIBUTED GENERATION PROGRAM

Dr. Mark C. Williams
US Department of Energy, National Energy Technology Laboratory,
PO Box 880, 3610 Collins Ferry Road, Morgantown, WV 26507-0880

ABSTRACT

The U.S. Department of Energy's (DOE) Office of Fossil Energy's (FE) National Energy Technology Laboratory (NETL), in partnership with private industries, is leading the development and demonstration of high efficiency solid oxide fuel cells (SOFC) and fuel cell turbine hybrid power generation systems for near term distributed generation (DG) market with emphasis on premium power and high reliability. NETL is partnering with Pacific Northwest National Laboratory (PNNL) in developing new directions in research under the Solid-State Energy Conversion Alliance (SECA) initiative for the development and commercialization of modular, low cost, and fuel flexible SOFC systems. The SECA initiative, through advanced materials, processing and system integration research and development will bring the fuel cell cost to \$400/kilowatt (kW) for stationary and auxiliary power unit (APU) markets. The President of the U.S. has launched us into a new hydrogen economy. The logic of a hydrogen economy is compelling. The movement to a hydrogen economy will accomplish several strategic goals. The U.S. can use its own domestic resources – solar, wind, hydro, and coal. The U.S. uses 20 percent oil but has only 3 percent of resources. Also, the U.S. can reduce green house gas emissions. Clear Skies and Climate Change aim to reduce CO_2, NOx, and SO_2 emissions. SOFC's have no emissions so they figure significantly in these DOE strategies. In addition, DG – SOFC's, reforming, energy storage – has significant benefit for enhanced security and reliability. The use of fuel cells is expected to bring about the hydrogen economy. However, commercialization of fuel cells is expected to proceed first through portable and stationary applications. This logic says to develop SOFC's for a wide range of stationary and APU applications, initially for conventional fuels, then switch to hydrogen. Like all fuel cells the SOFC will operate even better on hydrogen than conventional fuels. The SOFC hybrid is a key part of the FutureGen plants. FutureGen is a major new Presidential initiative to produce hydrogen from coal. The highly efficient SOFC hybrid plant will produce electric power and other parts of the plant could produce hydrogen and sequester CO_2. The hydrogen produced can be used in fuel cell cars and for SOFC DG applications.

This U.S. DOE fuel cell funding shown in Figure 1 is only for SOFC. If other U.S. agencies, Japanese, European, and industry cost share, SOFC funding is around \$100 MM/year. The DOE FE SECA budget has shown strong growth since a small amount of funds were allocated for initial work in 2000. The SECA program has grown as earlier programs ended. On the right, Industry Team funding is in blue and Core Technology Program funding is in green. The Industry Team/Core Technology funding has been 60/40.

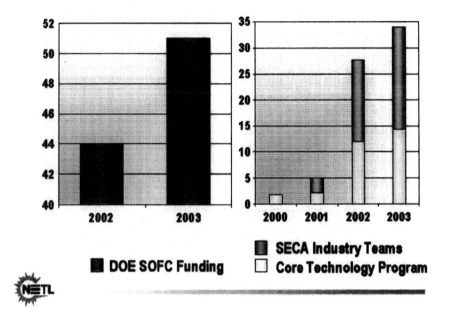

U.S. DOE SOFC and SECA Funding ($M)

DOE SOFC Funding

SECA Industry Teams
Core Technology Program

Figure 1. SOFC Funding

DOE FE SOFC Programs and Technology

Tubular SOFC

The Tubular SOFC program in DOE FE is scheduled to wind down in FY2004. It is, in a sense, the precursor program to the SECA program. It was through the tubular program that many of the attributes of SOFC were first demonstrated. The 200 kW pressurized fuel cell turbine hybrid demonstration at the National Fuel Cell Research Center in Irvine was a great learning experience and operated successfully for 3,000 hours. The unit achieved a projected 52 percent electrical efficiency. The first-of-a-kind demonstration did establish excellent emissions and efficiency standards.

A 250 kW atmospheric pressure tubular SOFC power system with heat extraction or combined heat and power (CHP) is now operating in a Canadian Facility at Kinetrics (Figure2). Partners in this project are Siemens Westinghouse Power Corporation (SWPC), OPT Consortium, and DOE/NETL. Project planned efficiency is 45 percent fuel to electricity net ac to grid (LHV) 75 percent thermal. The atmospheric CHP product will represent SWPC's first commercial offering and has already operated for 1000 hours.

250 kW CHP Fuel Cell System
@ Kipling Research Facility Toronto, Canada

Photo Courtesy of Siemens Westinghouse

Figure 2. Kinetrics SOFC

Solid State Energy Conversion Alliance (SECA)

The SECA Program is the main thrust of the DOE FE DG Fuel Cell Program (Figure 3). SECA SOFC supports Climate Change, FutureGen, Clear Skies, and Homeland Security. SECA is also recognized as part of the Hydrogen Program. Achieving the SECA goals should result in the wide deployment of the SOFC technology in large high volume markets. This means benefits to the nation are large and cost is low, which is the SECA goal. Less expensive materials, simple stack and system design, and high volume markets are the three criteria that must be met by a fuel cell system to compete in today's energy market. Near zero emissions, fuel flexibility, modularity, high efficiency, simple CO_2 capture will provide a national payoff that gets bigger as these markets get larger.

Overall, the SECA Program is progressing extremely well. In fact, there is early interest from auto manufacturers in SECA type fuel cells as evidenced by BMW's arrangement with Delphi, one of the SECA industry team developers, to put a compact fuel cell unit for auxiliary power in BMW trucks by 2007.

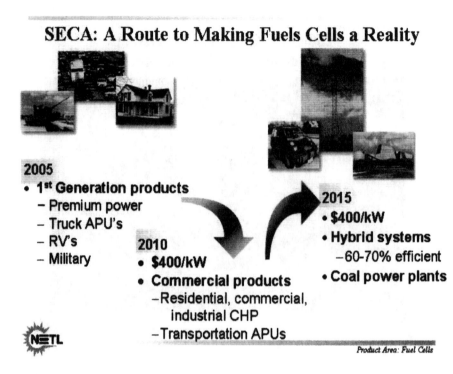

Figure 3. SECA

The SECA program is dedicated to developing innovative, effective, low-cost ways to commercialize SOFCs. The program is designed to move fuel cells out of limited niche markets into widespread market applications by making them available at a cost of $400 per kilowatt or less through the mass customization of common modules. SECA fuel cells will operate on today's conventional fuels such as natural gas, diesel, as well as coal, gas, and hydrogen, the fuel of tomorrow. The program will provide a bridge to the hydrogen economy beginning with the introduction of SECA fuel cells for stationary (both central generation and distributed energy) and auxiliary power applications.

The SECA program is currently structured to include competing industry teams supported by a crosscutting core technology program. SECA has six industry teams (Figure 4) working on designs (Figure5) that can be mass-produced at costs that are ten-fold less than current costs. The SECA core technology program is made up of researchers from industry suppliers and manufacturers as well as from universities and national laboratories all working towards addressing key science and technology gaps to provide breakthrough solutions to critical issues facing SECA.

SECA INDUSTRIAL TEAMS

Figure 4. Six SECA Industry Teams

The SECA industry teams are making very good progress. Delphi, in partnership with Battelle, is developing a 5 kW, planar, 700°C - 800°C, anode-supported SOFC compact unit for the DG and APU markets. Delphi is expert at system integration and high-volume manufacturing and cost reduction. They are focused on making a very compact and light-weight system suitable for auxiliary power in transportation applications.

General Electric is initially developing a natural gas 5 kW, planar, 700°C - 800°C, anode-supported SOFC compact unit for residential power markets. GE is evaluating several stack designs and is especially interested in extending planar SOFCs to large hybrid systems. They also have a radial design that can simplify packaging by minimizing the need for seals. GE has made good progress in achieving high fuel utilization with improved anode performance using standard materials by optimizing microstructure.

SWPC is developing 5-10 kW products to satisfy multiple markets. SWPC has developed a new tube design for their 5 kW units that use flat, high power density (HPD) tubes. This allows for a shorter tube length and twice the power output compared to their current cylindrical tube. It also results in more efficient manufacturing, assembly, and better volumetric power density.

207

Team	Design	Manufacturing
Cummins-SOFCo	• Electrolyte supported • 850 C • Thermally matched materials • Seal-less stack	• Tape casting • Screen printing • Co-sintering
Delphi-Battelle	• Anode supported • 750 C • Ultra compact • Rapid transient capability	• Tape casting • Screen printing • 2–stage sintering
General Electric	• Anode supported • 750 C • Hybrid compatible • Internal reforming	• Tape calendering • 2–stage sintering
Siemens Westinghouse	• Cathode supported • 800 -950 C • Redesigned tubular • Seal-less stack	• Stack extrusion • Plasma spray
Accumetrics	• Tubular • 800 C	• Stack Extrusion
Fuel Cell Energy/MSRI	• Anode Supported • Planar • 700-800 C	• Tape Casting • Spray Coating

NETL

Figure 5. SECA Industry Team Designs

Cummins and SOFCo (formerly McDermott) are developing a 10 kW product initially for recreational vehicles that would run on propane using a catalytic partial oxidation reformer. The team has produced a conceptual design for a multilayer SOFC stack assembled from low-cost "building blocks." The basic cell, a thin electrolyte layer (50-75 micron), is fabricated by tape casting. Anode ink is screen-printed onto one side of the electrolyte tape, and cathode ink onto the other. The printed cell is sandwiched between layers of dense ceramic that will accommodate reactant gas flow and electrical conduction. The assembly is then co-fired to form a single repeat unit.

In FY 2003, we began funding the two additional SECA industry teams – FuelCell Energy and Acumentrics. These industry teams represent additional industry design alternatives that will enhance the prospects of success of SECA fuel cells for a broader market.

Fuel cells are a universal power source. Multi-fuel, modular, clean, efficient, they can power just about anything. Commercial trucks, military vehicles, aircraft, and ships are all potential applications. To lower the cost of fuel cells as much as possible it is important to put fuel cells in all of these applications. A number of other government agencies and offices, federal and state are directly supporting work that helps SECA (Figure 6). DOE views this combined effort representing the spectrum of applications as essential to a successful SECA program.

Other Pathways to High Volume
With Help from our Friends

Figure 6. SECA Partners

SECA Core Research Issues

Developing the solid oxide fuel cell which meets both performance and cost targets is a matter of complex trade-offs. General attributes of SOFC systems are shown in Figure 7.
Within each sub-system and its components there exist research needs. Figure 8 presents a matrix of general research priorities. The needs may be different for the various system alternatives and their design – anode supported, cathode supported, and electrolyte supported; planar, radial tubular. While progress has been made in power density and utilization, the cathode remains an important area of research if low temperatures are to be achieved. To achieve low overpotentials at reduced temperature requires optimization over composition and structure. Mixed ionic and electronic conducting cathodes with sufficient catalytic activity are being considered. Seals are a long-standing issue in some SOFC designs. The requirements on the seal are demanding to ensure thermal cyclability and gas tightness. The use of low-cost metallic interconnects is highly desirable for some designs. Fundamental R&D needs were recently identified by a group of SOFC experts. These are shown in Figure 9.

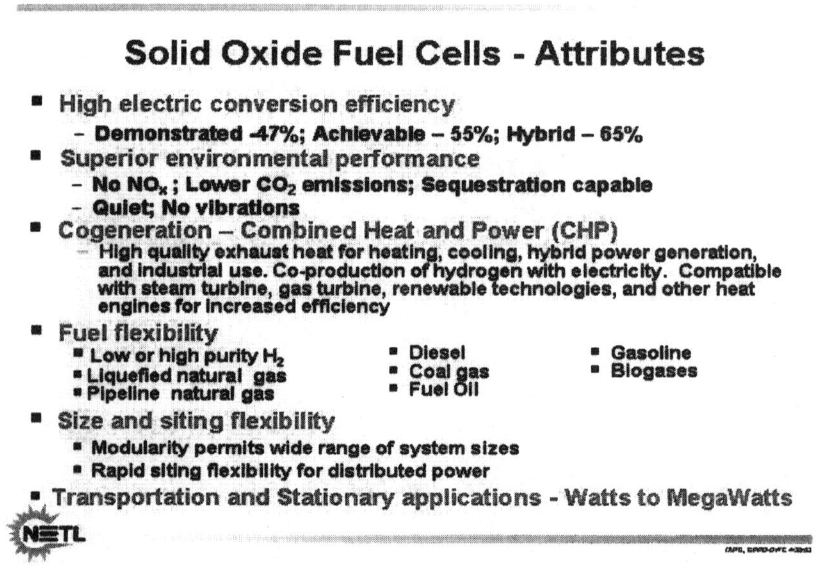

Solid Oxide Fuel Cells - Attributes

- High electric conversion efficiency
  - Demonstrated -47%; Achievable – 55%; Hybrid – 65%
- Superior environmental performance
  - No NO_x; Lower CO_2 emissions; Sequestration capable
  - Quiet; No vibrations
- Cogeneration – Combined Heat and Power (CHP)
  - High quality exhaust heat for heating, cooling, hybrid power generation, and industrial use. Co-production of hydrogen with electricity. Compatible with steam turbine, gas turbine, renewable technologies, and other heat engines for increased efficiency
- Fuel flexibility
  - Low or high purity H_2
  - Liquefied natural gas
  - Pipeline natural gas
  - Diesel
  - Coal gas
  - Fuel Oil
  - Gasoline
  - Biogases
- Size and siting flexibility
  - Modularity permits wide range of system sizes
  - Rapid siting flexibility for distributed power
- Transportation and Stationary applications - Watts to MegaWatts

NETL

Figure 7. SOFC Attributes

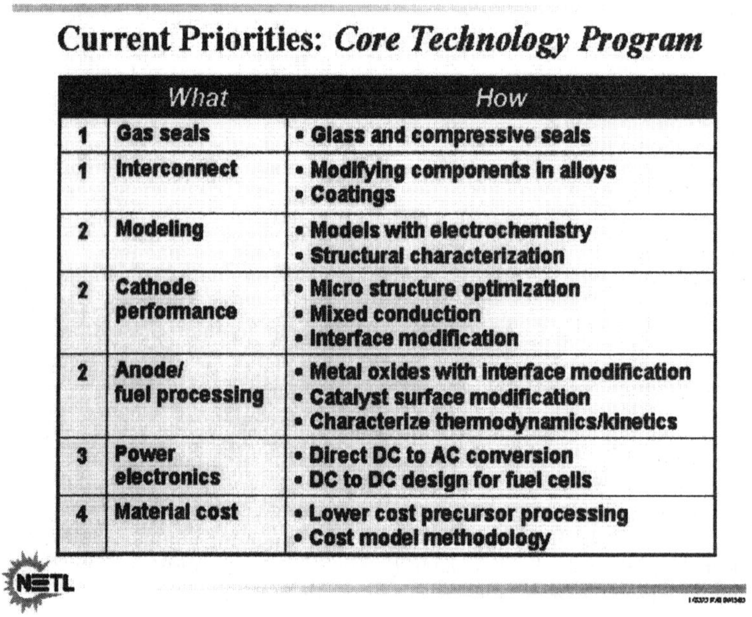

Current Priorities: *Core Technology Program*

	What	How
1	Gas seals	• Glass and compressive seals
1	Interconnect	• Modifying components in alloys • Coatings
2	Modeling	• Models with electrochemistry • Structural characterization
2	Cathode performance	• Micro structure optimization • Mixed conduction • Interface modification
2	Anode/ fuel processing	• Metal oxides with interface modification • Catalyst surface modification • Characterize thermodynamics/kinetics
3	Power electronics	• Direct DC to AC conversion • DC to DC design for fuel cells
4	Material cost	• Lower cost precursor processing • Cost model methodology

NETL

Figure 8. Core Research Priorities

Recommended SOFC Basic Research Topics

Electrolyte/Electrodes	Improved Performance • Understanding relationship between micro-, meso-, and nano-structures and electrocatalysis • Oxygen reduction mechanisms • Solid state and gas/solid interactions between dissimilar material interfaces • Understanding defect equilibria, electronic structure and transport in mixed ionic/electronic conductors • Understanding effect of material impurities on cell kinetics and stability • Surface characterization and modification to enhance electrocatalytic activity. • Extending percolation and other macro homogenous theories to include multiple phase interfacial, surface and nanostructure effects. • Using advanced in-situ characterization techniques (e.g., Advanced Photon Source, Neutron Spallation Source, etc.) to probe the local environment of low concentration dopants and defects.
Anode and Fuel Processing	Oxygen, Sulfur, and Carbon Tolerance • Re-oxidation when exposed to air • Sulfur tolerance • Carbon tolerance • Internal reforming • Cost
Seals	Fundamental understanding of hydrogen solubility and transport in SOFC interconnect and seal materials.

Figure 9. Recommended R&D Needs

SUMMARY

Commercialization of fuel cells will occur first in portable and stationary markets. Stationary fuel cells still need a viable DG market since residential and commercial markets are the largest initial markets. Transportation fuel cells could create the hydrogen economy. All fuel cells use hydrogen so the hydrogen economy is welcome. Moreover, SOFC's fuel flexibility provides a bridge to the hydrogen economy. SOFC's can also be used to generate hydrogen from natural gas and coal, concentrate CO2, and generate electric power in FutureGen plants.
Energy/hydrogen storage, CO2 separation, hydrogen generation (reform, electrolyze) and fuel cell electric power generation are DG technologies. SECA fuel cells may even ultimately be used in transportation, propulsion applications.

REFERENCES

1) Fuel Cells--A Handbook (Revision 4, November 1998), Report Number DOE/FETC-99/1076 (CD).
2) M. Williams, "Status of SOFC Development and Commercialization in the US", 6th International SOFC Symposium, in Proceedings, pp. 3-9 (1999).
3) M. C. Williams, "Fuel Cells - Realizing the Potential", Fuel Cell Seminar, Portland, November 1, 2000, in Proceedings Fuel Cell Seminar, pp. (2000).
4) M. Williams, "Energy Decision Magazine Roundtable on DG", during Electric Power 2000, Cincinnati, Ohio, in Energy Decision, June pp. 26-32, 2000.
5) M. Williams, "Energy Futures: Advanced Fuel cell Power Systems," Brookings Institute Seminar, "Science and Technology Policy: Current and Emerging Issues," June 14, 2000.
6) "SECA Workshop Proceedings," June 1-2, 2000, Baltimore, MD.
7) N. Minh, et. al., in Solid Oxide Fuel Cells VI, S. C. Singhal and M Dokiya, Editors, PV 99-19. p.68, The Electrochemical Society Proceedings Series, Pennington, NJ (1999).
8) J. Thijssen, et.al., Conceptual Design of POX SOFC 5 kW net System, Arthur D. Little, Inc. Report # 71316 (to be published), Cambridge MA (2000).
9) M. Williams and S. Singhal, "Mass-produced ceramic fuel cells for low-cost power," in Fuel Cell Bulletin, no. 24, pp. 8-11 (September 2000).
10) M. Williams, "Fuel Cells: Realizing the Potential for Natural Gas," Key Note Address in Proceedings Fuel Cells 2000, Philadelphia, PA, September 27, 2000.
11) M. C. Williams, "Fuel Cells - Realizing the Potential", Fuel Cell Seminar, Portland, Oregon, November 1-3, 2000, in Proceedings Fuel Cell Seminar, pp.
12) M. Williams, "Status of SOFC Development and Commercialization in the US", 7th International SOFC Symposium, Japan, in Proceedings, (2000).
13) M. C. Williams, "Distributed generation fuel cells and power reliability," #1605, Energy 2001, Baltimore, Md., 2001.
14) M. Williams, "Fuel Cells and the World Energy Future," in Proceedings PowerGen, Orlando, FL, 2001.
15) David A. Berry, Wayne A. Surdoval, and Mark C. Williams, "The solid state energy conversion alliance - program to produce mass manufactured ceramic fuel cells," in Proceedings ACS Fuel Cell Symposium, Chicago August 2001.
16) M. C. Williams, "New Direction in the US Fuel Cell Program," in Proceedings 7th Grove Fuel Cell Symposium, September 11, 2001.
17) Catherine Greenman, "Fuel Cells: Clean, Reliable (and Pricey) Electricity," The New York Times, May 10, 2001, p. D8.
18) "SECA Workshop Proceedings," March 29-30, 2001, Arlington, VA.
19) "SECA Workshop Proceedings," March 21-22, 2002, Washington, DC.
20) "SECA Workshop Proceedings," April 15-16, 2003, Seattle, WA.

LANTHANUM GALLATE ELECTROLYTE FOR INTERMEDIATE TEMPERATURE SOFC OPERATION

S. Elangovan, B. Heck, S. Balagopal, D. Larsen, M. Timper and J. Hartvigsen
Ceramatec, Inc.
2425 South 900 West
Salt Lake City, UT 84119-1517

ABSTRACT

Among the various oxygen conducting materials, Sr and Mg doped lanthanum gallate (LSGM) is attractive for use as an electrolyte in intermediate temperature Solid Oxide Fuel Cells (SOFC). The LSGM compositions show high oxygen conductivity as well as excellent stability over the range of oxygen partial pressures encountered in SOFC operation. Lowering the operating temperature is expected to provide a more expeditious route to commercial realization of Solid Oxide Fuel Cell (SOFC) based power systems. The intermediate operating temperature (600 to 800°C) allows the use of lower cost materials in both the stack and the balance of plant, and significantly slows down potential deleterious materials interaction, thus extending the useful long-term performance and thermal cycle capability. The primary challenge in long-term stability of LSGM-based fuel cell operation is the chemical reaction between the nickel anode and the LSGM electrolyte. A chemical modification to the anode was found to reduce the reaction. The modified anode composition is also shown to perform equivalent to standard NiO based anode. Single cells using thin (30 micron) electrolyte supported on a thicker anode were found to operate at a power density of greater than 500 mW/cm^2 at 700°C. Stack test verification using thick electrolyte supported cells show promising performance. Stack tests using thin electrolytes are planned.

INTRODUCTION

Lowering the operating temperature of the Solid Oxide Fuel Cell (SOFC) is considered essential to commercialize SOFC based power systems. The intermediate operating temperature (600 to 800°C) allows the use of lower cost materials in both the stack and the balance of plant. It also significantly slows down potential deleterious materials interaction, thus extending the useful stack life. When the operating temperature is in the range of 600 to 700°C, it is also possible to reform hydrocarbon fuels within the stack. The endothermic nature of this reaction provides a significant cooling effect, thereby reducing the parasitic losses associated with excess air requirements for stack cooling.

The operating temperature of the fuel cell system is largely governed by the electrolyte material. Electrolyte materials such as doped bismuth oxide or doped ceria exhibit superior ionic conductivity at lower temperatures. However, the material instability of bismuth oxide and the mixed conduction of ceria under fuel gas conditions are difficult to overcome. Sr and Mg doped LaGaO$_3$ (LSGM) has been shown to have an ionic conductivity at 800° C equivalent to that of yttria stabilized zirconia (YSZ) at 1000° C. Unlike other materials possessing a high oxygen ion conductivity lanthanum gallate is stable in both fuel and air. Ishihara et al., reported a power density of ~ 0.35 W/cm^2 at 800° C in single cells (1). Goodenough et al., (2,3) have reported extensive materials characterization information.

The performance degradation of LSGM based cells has limited the wide use of the material by SOFC developers. The primary degradation of LSGM single cells is electrode related (4), and in particular occurs at the anode-electrolyte interface. Investigations by Huang and Goodenough (3) suggested that the likely formation of an insulating phase, $LaNiO_3$, at the anode – electrolyte interface may be the source of degradation. In addition to long-term stability, the reaction between the anode and electrolyte limits the cell fabrication options. For example, anode supported thin electrolyte approach that has been successfully demonstrated in zirconia based cells cannot be readily employed because of the reactivity. The gallate electrolyte based cells presently have been fabricated only as electrolyte supported cells. A zirconia matrix supported thin gallate electrolyte has been reported (5). Again, the reactivity with zirconia to form lanthanum zirconate insulating phase at the interface is of concern. The focus of our current program is to modify the anode composition such that the long-term stability of gallate cells is not limited by the anode reactivity. Additionally, fabrication of thin anode supported cells becomes a viable option.

EXPERIMENTAL

Powder mixtures of LSGM and the modified anode material were calcined at 1250 and 1350°C for four to six hours. It was determined that not only the severity of the reaction but also the nature of the second phase depends on the reaction temperature. At 1250°C the reaction phase formed was La_2NiO_4, while the predominant phase at 1350°C was $LaNiO_3$. A chemical modification to the anode reduced the reaction considerably [6]. Initial single cell evaluation was done using 500 micron thick electrolyte to verify the performance of the modified anode composition. Sr doped $LaCoO_3$ was used as the cathode.

Combinations of tape casting (for the substrate) and screen printing (for the electolyte) techniques were employed to fabricate electrolyte layers that are 20 to 30 microns in thickness. Various amounts and types of pore formers were used in the slip formulation for tape casting to control the final porosity and shrinkage of the anode layer. Electrochemical characterization of thin electrolyte cells was performed at various temperatures. Long-term stability of the cell performance was also evaluated at various temperatures.

RESULTS

The modified anode composition is shown to have equivalent initial performance as the standard NiO based anode. The performance of two electrolyte supported LSGM cells, using the modified anode is shown in Figure 1. Both cells showed an area specific resistance of 0.6 ohm-cm^2 at 800°C.

The bi-layer sintering evaluations were done at 1350 – 1400°C to provide a 10 to 30-micron thick LSGM electrolyte supported on the anode. Nickel diffusion into the electrolyte region was observed in initial trials. Compositional analysis of the electrolyte region using the Energy Dispersion Spectrometry (EDS) showed that the nickel diffusion destabilized the perovskite phase resulting in a multiphase composition. Process modifications such as lower nickel content and a reduction in sintering temperature were introduced to lower the reactivity between the electrolyte and the anode. Figure 2a shows a thin (~ 30 micron) LSGM electrolyte supported on a thick anode structure. Initial bi-layers that were fabricated successfully resulted in an anode porosity of less than 10%. Attempts to increase the anode porosity through the use of pore formers (principally carbon black) have shown promising results. A micrograph of a bi-layer is shown in Figure 2b, with an estimated anode porosity of 20 – 25%. Additional process modifications are currently under way to increase the porosity while maintaining acceptable tape shrinkage during sintering.

The thin electrolyte button cells were tested at different temperatures. The cells showed an area specific resistance of 0.5 ohm-cm^2 at 700°C corresponding to a peak power density of 0.5 W/cm^2 at low fuel utilizations. The cell performance is shown in Figure 3. The performance decreased significantly when the fuel utilization was increased, consistent with the low anode porosity of about 10%. The cell performance as a function of fuel utilization is shown in Figure 4. Cells were also tested for their long-term stability at various temperatures. The performance of a cell tested at 700°C is shown in Figure 5. The performance was stable at 0.5 W/cm^2 for over 2,000 hours of test duration.

Tape casting and sintering processes for fabricating 10x10 cm cells are in progress. Cells using both thick electrolyte (300 microns shown in Figure 6) and supported thin electrolyte (30 micron) have been fabricated. A stack, shown in Figure 7, using electrolyte-supported cells was tested. The 8-cell stack used two variations of experimental metal interconnects. The overall stack performance as a function of time is shown in Figure 8. The stack power increased over the first 50 hours of operation and then declined at a rate of approximately 10% per 1,000 hours. Fuel utilization was maintained between 30 and 45%. Post-test analysis are planned to investigate the changes in electrode microstructure and to characterize the oxide scale of the metal interconnects in order to determine potential sources of performance degradation. Stack tests using thin electrolyte cells are planned to verify the promising single cell performance.

SUMMARY

LSGM electrolyte based cells have shown good performance characteristics at the intermediate operating temperature of 700°C. Button cells tested for long-term stability showed stable performance. A stack using electrolyte supported cells was tested. Post-test analysis indicated that chrome evaporation from the metal interconnects may be the primary source of performance degradation. No reaction product at the anode – electrolyte interface was observed.

REFERENCES

1. T. Ishihara, M. Higuchi, H. Furutani, T. Fukushima, H. Nishiguchi and Y. Takita, J. Electrochem. Soc., 144, 5, 1997.

2. K. Huang, M. Feng and J.B. Goodenough, J. Am. Cer. Soc., 79, 4, 1996.

3. K. Huang and J. Goodenough, Final Report to EPRI, Report No. TR-108742, October 1997.

4. K. Huang and J. Goodenough, *Superior Perovskite Oxide-ion Conductor: Strontium- and Magnesium- Doped LaGaO₃,* EPRI Final Report No. TR-108742 (1977).

5 J. W. Yan, Z. G. Lu, Y. Jiang, Y. L. Dong, C. Y. Yu, and W. Z. Li, J. Electrochem. Soc. 149(9) A1132 (2002).

6 S. Balagopal, I. Bay and S. Elangovan, Proc. Fifth European SOFC Forum, p. 233 (2002).

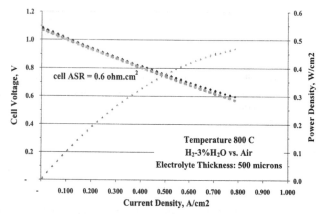

Figure 1: Performance of LSGM single cell
(Current-Voltage characteristics of two cells are shown)

Figure 2. Cross Section of Thin LSGM Supported on Anode Structure

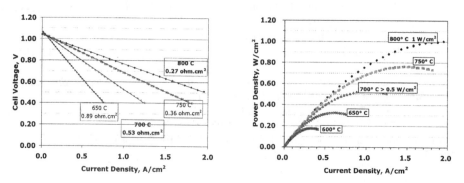

Figure 3. Thin Electrolyte Cell Performance

216

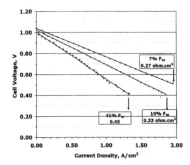

Figure 4. Effect of Fuel Utilization

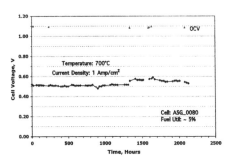

Figure 5. Long-term Single Cell Performance

Figure 6. 10-cm Electrolyte Supported Cells

Figure 7. A photograph of an 8-cell Stack

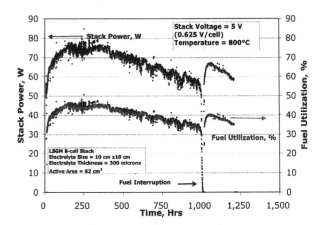

Figure 8. Performance of an 8-cell Electrolyte Supported Gallate Stack

ACKNOWLEDGEMENT

This work is supported by the U.S. Department of Energy under a Phase II SBIR Grant No. DE-F2-03-01ER83212.

217

SOLID OXIDE FUEL CELL DEVELOPMENT AT FORSCHUNGSZENTRUM JUELICH

Ludger Blum, Hans-Peter Buchkremer, L.G.J. (Bert) de Haart, Heinz Nabielek, Jo Willem Quadakkers, Uwe Reisgen, Robert Steinberger-Wilckens*, Rolf W Steinbrech, Frank Tietz, Ico Vinke
Forschungszentrum Juelich
D-52425 Jülich
Germany

ABSTRACT

Solid Oxide Fuel Cells (SOFCs) are a promising power generation technology due to their high electrical efficiency, multi-fuel capability, potential role in carbon sequestration and possibilities for coupling with a gas turbine. SOFC development is, however, fraught with various problems of high-temperature operations, cost-effective materials and manufacturing processes etc. To solve these problems, we have assembled and tested around 150 SOFC stacks in the last 8 years. Our present design consists of thin electrolyte, planar anode substrate cells in stacks with metallic interconnects featuring internal manifolding with counterflow.

The first in a series of large stacks was operated in 2002. All ferritic parts were made of commercial steel type X10CrAl 18 (Ferrotherm 4742). The 40-cell stack delivered 9.2 kW_{el} in hydrogen operation and 5.4 kW_{el} with methane as fuel. The average degradation rate of around 10% per 1 000h at 850°C is consistent with results published on characteristics of unprotected ferritic steel interconnects and our own laboratory experience.

A new series of short stacks was assembled with interconnects manufactured from the modified ferritic steel Crofer22 APU. The new series of stack tests was operated up to 4 000 hours with degradation rates between 2 and 3% per 1 000 hours of operation, a marked improvement over earlier stacks. However, the target of development is directed towards 0.75%/ 1 000h for commercial operations.

INTRODUCTION

Forschungszentrum Jülich has been active in the field of SOFC technology for over a decade. Research covers all aspects from materials development and characterisation, over cell and stack design, modelling and manufacturing, to testing of cells, stacks and systems.

Since the mid-nineties several generations of SOFC stacks have been designed and tested incorporating the anode substrate type cells developed in Jülich. The main topics addressed in the development of the stack technology are the electrical contact between interconnect and electrodes, the gas tightness of the cells and sealing areas and the corrosion of and/or evaporation from the metallic interconnect. These components all affect the durability and thermal cyclability of the stack. Simultaneously, modifications to the stack design were made in order to reduce the number of manufacturing steps.

*Presenting author

PBZ, Forschungszentrum Jülich, D-52425 Jülich, Germany
Tel. +49 2461 61 5124, Fax +49 2461 61 6695, e-mail r.steinberger@fz-juelich.de

STACK DESIGN

The 5th generation stack design, labelled 'E-design' (Fig. Ia), for planar anode substrate typ cells and metallic interconnects was the 'work horse' at Forschungszentrum Jülich used fc testing SOFC materials, cells and manufacturing processes in cell and stack development from i introduction in the year 2000 (ref. i) up to 2002. It has a relatively thick metallic interconnec with machined gas channels on both sides and a metallic 'picture frame' for the cell, brazed to th interconnect by a metal solder. Glass sealants are applied to the planar cell and frame surface using an automated dispenser. A fine Ni-mesh is spot-welded to the ribs separating the ga channels on the fuel side of the interconnect in order to improve the electrical contact betwee interconnect and anode substrate. On the air side a contact layer of lanthanum cobaltite (LC) usually sprayed onto the ribs of the interconnect.

The follow-up 6th generation stack design, the 'F-design' line, was developed towards reduction in manufacturing steps and a greatly simplified processing. This resulted in noticeab reduced costs due to less material expense and halved workshop time. The F-design, shown Fig. Ib, is hydraulically very similar to the E-design (counter-flow configuration). The ga channels have been omitted on the anode side, though, leaving the fuel distribution function to coarse Ni-mesh, which simultaneously provides the electrical contact between interconnect ar anode substrate. The selected Ni-mesh proved to have a pressure drop which was only twice th of the gas channels used in the E-design and therefore within tolerable limits.

In the new F-design the metal frame holding the cell is no longer brazed to the interconnec This saves the time consuming and costly high temperature vacuum metal brazing step. Instea the metal frame is 'glazed' to the interconnect with the same glass-ceramic which is being use already for sealing the repeating units in the stack. This 'glazing' and sealing occurs now in single process during the first heating-up of the assembled stack.

A 7th generation, G-design, is similarly aimed at providing a working basis for investigatic into materials properties but in this case also includes application derived criteria from th auxiliary power unit (APU) stack development. It has been developed as a first step towards light weight stack design and will also deliver information on mechanical properties and transie operation. As a consequence it is based on manufacturing technology used in automobile indust (sheet metal forming) and has the potential for a 90% reduction in weight (cf. Fig. Ic) compared to the F-design.

RESULTS

Testing in the 5-kW range

Mid-April 2002 saw the first Jülich multi-kilowatt, 40-layer stack (40 cells of 20 x 20 cm²) operation. It was a E-design stack with all ferritic parts made of the commercial steel type 1.474 (X10CrAl 18). The stack delivered 9.2 kW$_{el}$ in hydrogen operation and 5.4 kW$_{el}$ with methar (ref. ii). A small fraction of hydrogen was added to simulate pre-reformed natural ga Nevertheless, the stack operated at near to 100% internal reforming. It was run at 0.5 A/cm² wi a fuel utilisation around 59%. After a reduction of the furnace temperature the stack continued operate in thermal self-sustaining mode, still producing between 4.0 and 4.5 kW$_{el}$ with methar Current density was lowered to 0.3 A/cm² after 1000 h, until approx. 1250 h of total operatic were reached. Average degradation was in the range of 10%/1000 h, typical for the unprotecte ferritic steel 1.4742, at an average temperature below 850 °C.

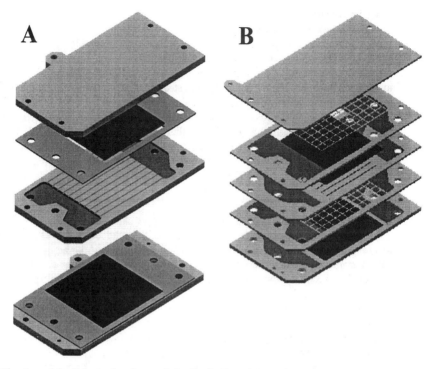

Fig. Ia and b: Schematic views of the E- (left) and F-Design (right) for stacks with planar anode type cells and metallic interconnects.

Figure Ic. G-design stamped sheet metal 'repeating unit'.

Long Term Durability Tests

FZJ has adopted the design with stainless steel interconnector plates due to its potential for low cost materials and high electrical conductivity. Still, the problems of chromium evaporation from the steel and the corrosion behaviour of steel in SOFC relevant atmospheres need to be continually addressed. One step was the development of the JS-3 steel (now commercially available as CroFer22APU) (ref. iii). This steel displays reduced volatile chromium formation and a high conductivity and stable oxide scale.

E-design stacks with JS-3 steel. A series of E-design short-stacks (2 cells 10 x 10 cm²) was produced in 2002 from JS-3 coated with the cathode contact layer LCC2. Long term testing indicated an improved durability compared to the stacks with the 1.4742 ferritic steel (ref. ii). Fig. II shows the time dependence of one of these short-stacks operated over 4000 h with hydrogen at 0.3 A/cm² and 800 °C. The degradation rate for both cells was between 2 and 3%/1000 h.

Thermal Cycling Tests

The satisfactory operation of short-stacks with JS-3 interconnects over some 2000 h gave rise to a first targeted test of thermal cycling capabilities (Fig. III). After an initial period of 1800 h of galvanostatic operation, a short-stack was subjected to regular heating up (2K/min, approx. 5 to ? h in total) and cooling down cycles (natural cooling down in the furnace, approx. 20 h to 220°C). Weekends were used to cool down to 75°C in a 48 hour cycle (see insert in Fig. III).

The stack was cycled 20 times, operated for 500 hours in order to determine the steady-state operation and again cycled 20 times. The initial degradation of the cells was also around 2-3%/1000 h. As can be seen from Fig. III, the ageing did not depend on the number of cycles and did not remarkably change with the number of cycles.

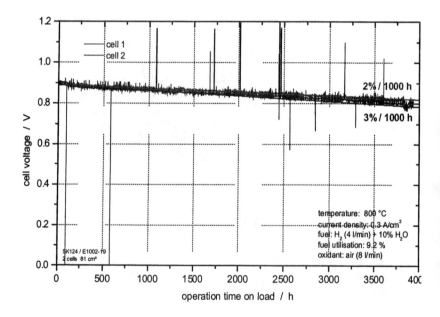

Figure II. Time dependence of the cell voltages in an E-design short-stack during galvanostatic operation at 0.3 A/cm² and 800 °C with H_2

222

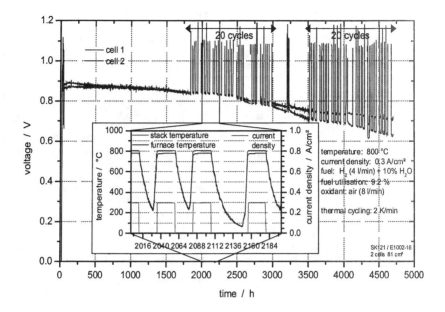

Figure III. Time dependence of the cell voltages in an E-design short-stack during galvanostatic operation at 0.3 A/cm² and 800 °C with H₂, followed by two periods of 20 thermal cycles each; insert shows the time-temperature (down to 220 °C and 75 °C, resp.) profile of the cycles.

Post-operational analysis showed that the glass sealant used was of high quality and had hardly suffered in the testing. This would explain that apparently no major leakages occurred and the stack remained in a stable operational mode.

Tests with F-design stacks

After a first series of encouraging tests with short-stacks manufactured according to the new F-design, a number of larger stacks with 20 x 20 cm² area cells were manufactured, assembled and operated. Fig. IV shows the current-voltage and current-power characteristics of a 10 cell F-design stack operated both on hydrogen and (simulated partially pre-reformed) methane. With hydrogen a maximum power of nearly 2.5 kW was reached, with methane operation 2.0 kW. Fuel utilisation was in the range 60 to 70%. With similar 10-cell E-design stacks operated under the same conditions the maximum power reached was substantially less; 1.6 kW with hydrogen and 1.0 kW with methane, respectively (ref. ii). Since both metallic parts of the F-design repeating unit, interconnect and cell frame, only have to be machined at one side, the flatness of both parts is improved, which in combination with the coarse Ni-mesh results in a far better and more reproducible contacting of cell and interconnect. The single layers in the stack all show power densities close to the best values obtained for single cells measured in ceramic housing with low fuel utilisation, i.e. 0.63 W/cm² (ref. i).

The flattening of the current-voltage curves at higher current densities (see Fig. IV) is caused by the temperature increase in the stack during operation under load. This is more pronounced in

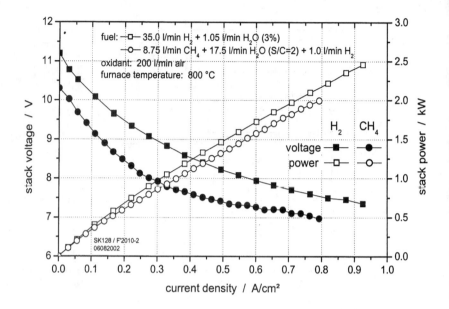

Figure IV. Stack voltage and power vs. current density for an F-design stack (10 cells 20 x 20cm²) operated with hydrogen and methane

the curve for methane operation, because at lower current densities the cooling due to the methane reforming reaction at the anode still dominates the temperature profile.

Lowering the Operating Temperature

An alternative route to reduce corrosion phenomena in stack operation is lowering the operating temperature. Since this also considerably reduces the current density more efficient cathode materials need to be employed. LSCF cathodes currently used at FZJ deliver similar results to the standard LSM cathodes but at temperatures approx. 100 K lower. This corresponds to results reported in literature in the past years (for instance refs. (v), (vi), (vii)).

The reduction in temperature offers access to extending the durability of stacks to lifetimes that are relevant for practical operation (around 5 000 hrs. in APU, above 20 000 hrs. for stationary applications). The combination of new steel materials, new cathodes and protective coatings could well extend the operation time of stacks towards 40 000 hrs. which would be necessary in order to address the market of stationary combined heat and power (CHP).

Fig. V shows the voltage plot of the short stack over a testing time of over 2500 hrs. The stack displayed a degradation around 3,5% on both cells. This is still relatively high in comparison to Fig. III considering the much lower temperature. We believe the effect is due to corrosion problems we have encountered in the interaction of our standard glass sealings with the CroFer22 APU. A more careful adaptation of both components to each others chemical properties is believed to alleviate the problem.

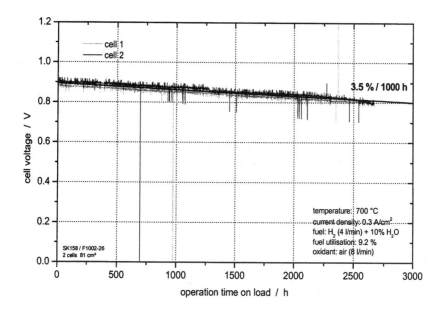

Fig. V: Operation of a short stack (2 layers) with LSCF cathodes at 700°C. The average degradation over 200 hrs. was 3,5% per 1000 hrs at 300 mA/cm².

Application Designed Stacks and Packaging

As stated earlier, the E- and F-design stacks are aimed at laboratory operation for generating results on materials properties and interactions under real-life SOFC conditions. Due to the overall weight of the stacks they can hardly be practical for actual applications. Also, they do not possess a thermal housing and are operated in a furnace environment.

APU cassette design. SOFC are being discussed as main elements in APU (auxiliary power units) for on-board electricity production on vehicles (road, marine, rail, aircraft etc.). They offer not only the benefit of a high electricity conversion efficiency but also multi use for instance in cold-start. On the other hand a mobile application puts forward very ambitious requirements that will be hard to meet.

Following the involvement of FZJ with developing SOFC components for APU applications a complete re-design of the F-design was performed. The resulting design was labelled ‚G-design' and is shown in Fig. Ic. It maintains the gas flow characteristics and geometry of the F-design whilst at the same time integrating 'picture frame', SOFC cell and interconnect into a single 'repeating unit'. The parts are welded together and then joined into a stack by ceramic materials. The approximately 90% weight savings are essential for SOFC stacks in mobile applications. At the time of writing the welding processes and joining have been understood and first stacks are awaiting their electrochemical characterisation.

Packaging of SOFC Stacks. In the context of the set-up of a 20 kW SOFC system at FZJ which includes all balance of plant (BoP) components, a new concept for packaging the 'hot' parts of the system (meaning parts operating above 400°C) has been developed. The resulting module design is shown in Fig. VI. Pre-reformer, recuperator, afterburner and stack are integrated into a common housing that is thermally insulated and can be operated in self-sustaining mode without the conventional furnace for balance of heat flow. The pressure for contacting and sealing of the stack is applied through spring loaded bolts on the cold side of the insulation.

This design reduces volume and weight by avoiding additional piping. Because all components are manufactured out of a ferritic chromium steel similar to that used for the interconnect plates of the stack thermal mismatch is avoided and no additional effort for compensation of expansion differences is needed. The high integration of components in the housing results in a high efficiency of heat transfer on one hand, but also in a close link of the processes in the system. The components are designed for combination with a stack with nominal power of 5 kW, based on a cell size of 20×20 cm^2. Operating results from the system expected until end of 2004, need to show whether this concept also functions satisfactorily in transitional modes.

OUTLOOK

The more recent designs for stacks with planar anode substrate type cells and metallic interconnects at FZJ transcend the 'work horses' for testing components, cells and manufacturing processes in cell and stack development and move towards 'real world' SOFC applications. Work now has to be completed on stacks suitable for mobile application and for stationary use. The aim is to design all components for optimal industrialised processing, low weight, high flexibility and low cost. The oncoming EU-funded project RealSOFC will concentrate on a full understanding of ageing processes and materials that show considerably less degradation. Integration of the results are expected to move stack lifetime above 20 000 hrs.

ACKNOWLEDGEMENTS

The authors gratefully thank all members of the Jülich anode substrate SOFC development team.

Most of the work was performed within the project 'Zellen und Stackentwicklung für planar SOFC (ZeuS)' funded by the German Federal Ministry of Economics and Technology (BMWi' Part of the work was funded by the EC within the projects 'Decentralised Power Generation Plants based on Planar SOFC Technology - Proof of Concept (ProCon)'.

REFERENCES

i.   L.G.J. de Haart, I.C. Vinke, A. Janke, H. Ringel and F. Tietz, in *Solid Oxide Fuel Cell VII*, H. Yokokawa and S.C. Singhal, Editors, PV 2001-16, p. 111, The Electrochemical Society Proceedings Series, Pennington, NJ, (2001)
ii. L. Blum, L.G.J. de Haart, I.C. Vinke, D. Stolten, H.-P. Buchkremer, F. Tietz, G. Blaß, D. Stöver, J. Remmel, A. Cramer and R. Sievering, in *5th European Solid Oxide Fuel Cell Forum*, J. Huijsmans, Editor, Vol. 2, p. 784, European Fuel Cell Forum, Oberrohrdorf (2002)

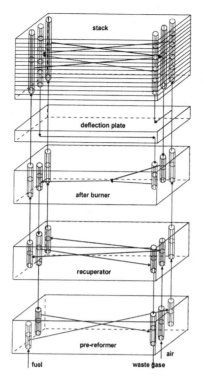

Fig. VI: Integrated Stack Concept – "Integrated Module"

iii. J.P. Abellán, V. Shemet, F. Tietz, L. Singheiser, W.J. Quadakkers and A. Gil, in *Solid Oxide Fuel Cells VII*, H. Yokokawa and S.C. Singhal, Editors, PV 2001-16, p. 811, The Electrochemical Society Proceedings Series, Pennington, NJ, (2001)

iv. O. Teller, W.A. Meulenberg, F. Tietz, E. Wessel and W.J. Quadakkers, in *Solid Oxide Fuel Cells VII*, H. Yokokawa and S.C. Singhal, Editors, PV 2001-16, p. 895, The Electrochemical Society Proceedings Series, Pennington, NJ, (2001)

v. Ahmed, Love, Ratnaraj, in *Solid Oxide Fuel Cells VII*, H. Yokokawa and S.C. Singhal, Editors, PV 2001-16, p. 904, The Electrochemical Society Proceedings Series, Pennington, NJ, (2001)

vi. Perry, Murray, Barnett, in *Solid Oxide Fuel Cells VI*, H. Yokokawa and S.C. Singhal, Editors, PV 1999, p. 369, The Electrochemical Society Proceedings Series, Pennington, NJ, (1999)

vii. Honegger et al. in *Solid Oxide Fuel Cells V*, H. Yokokawa and S.C. Singhal, Editors, PV 1995, p. 321, The Electrochemical Society Proceedings Series, Pennington, NJ, (1995)

DEVELOPMENT OF MOLB TYPE SOFC

Hitoshi Miyamoto and K. Mori
Takasago Research & Development Center
Mitsubishi Heavy Industries, LTD.,
2-1-1 Shinhama Arai,
Takasago, 676-8686 Japan

A. Nakanishi, M. Hattori and Y. Sakaki
Electric Power Research & Development Center
Chubu Electric Power Company, Inc.,
20-1 Kitasekiyama, Odaka, Midori-ku,
Nagoya 459-8522 Japan

T. Mizoguchi, S. Kanehira,
K. Takenobu and M. Nishiura
Kobe Shipyard and Machinery Works
Mitsubishi Heavy Industries, LTD.,
1-1-1, Wadasaki Hyogo-ku,
Kobe, 652-8585 Japan

ABSTRACT
Chubu Electric Power Company, Inc. (CEPCO) and Mitsubishi Heavy Industries, Ltd. (MHI) have jointly developed MOLB(Mono-block Layer Built) type SOFC since 1990. In 1994, an advanced type MOLB was developed. This cell consists of only two parts, an active layer and an interconnect plate. The electrolyte has unique embossed shape and both plates are manufactured by low cost wet process.

The joint projects are divided into three, 1 kW, 5 kW and several 10 kW projects. In our latest developments, several 10 kW class module was manufactured and tested from 2000 to 2001. The successful tests resulted in a maximum output of 15 kW in a total operating period of 7,500 hours, with the fuel of reformed pipe line natural gas, and also with the direct internal reforming process. Based on these operational results, an automatic operation control system was developed and the module components were improved.

A part of the study was supported by New Energy and Industrial Technology Development Organization (NEDO) of Japanese government.

INTRODUCTION
Fuel cell development has accelerated in recent years primarily in a viewpoint of its high efficiency and minimal environmental impact. Among several types of the fuel cells, SOFC, which has higher operating temperature, is expected for potential use from small scale distributed power supplies to large-scale power plants in the near future. SOFC promises high electric power efficiency and uses the high temperature gas effectively.

Since 1990, CEPCO and MHI have jointly developed and evaluated advanced planar type SOFC. The SOFC has unique structure and commonly called MOLB.

In 2001, CEPCO and MHI have jointly started the technical development on thermally self-supporting SOFC module, which is expected to serve as the basis for the future SOFC system.

The project is partially funded by NEDO. This paper summarizes the major progress in MOLB type SOFC developments.

PROGRESS OF MOLB TYPE SOFC
Cell Components

The MOLB stack consists of two parts, an active layer and an interconnect with a seal material as shown in Fig. 1.

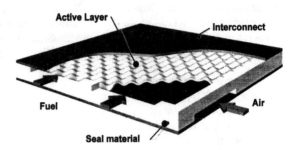

Fig. 1   Structure of MOLB type SOFC

The PEN active layer, air electrode/electrolyte/fuel electrode assembly, is responsible to the power generating reaction. The active layer is shaped into three dimensional dimples which have both convex and concave shape. The active layer also acts as gaseous paths for cathode and anode by itself. The large active surface gives high performance and capability for configuring compact SOFC system.

Fig. 2 shows typical V-I characteristics under hydrogen-air environment [1]. Even dilute fuel concentration, MOLB gives lower internal cell resistance and can produce 0.35W/cm^2 class power density. The power density values are very high even though MOLB adopt self-supporting electrolyte. The estimation of the cell performance is satisfactory for the practical use. The details of the simulation for the MOLB cell performance are described elsewhere [1].

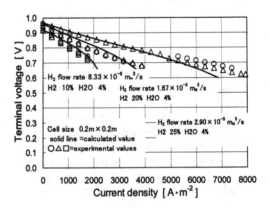

Fig. 2 Typical V-I curves of MOLB type SOFC

The MOLB adopts ceramic interconnect throughout its development stages. The interconnect works as an electrical connection of the active layers in series and works as gas separator. Only ceramic interconnect can promise the matching of the thermal expansion with the active layer at long term endurance for oxidizing environments. Also ceramic interconnect should resist the reduced environment as shown in Fig. 3[2], the MOLB interconnect had overcome the contradiction with the thermal expansion coefficient, the electric conductivity and the expansion under the reduced environment.

The unique ceramic seal material is applied for the gas sealing at the stack ends, and the manifolds are set to supply or discharge anode gas and cathode gas to or from the stack.

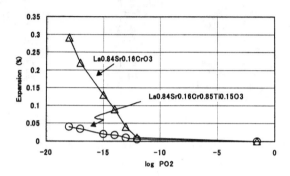

Fig. 3 Expansion of interconnect in reduced environment [2]

Stack Development

The stack development of MOLB is summarized in Table I. The joint research of MHI and CEPCO on the MOLB type SOFC had started in 1990. In 1992, 1 kW class test succeeded for the first time in the flat plate SOFC using 150 mm sized 3 stacks with 40 cells [3],[4]. Since then, the development of large stacks was started and in 1996, 200 mm sized 2 stacks with 40 cells produced maximum power 5.1 kW [5],[6].

Table I Development history of MOLB type SOFC

Year	1992	1993	1994	1995	1996	1997	1998	1999	2000
Power	1kW Class			5kW Class		Several 10 kW Class			
Size	150mm × 150mm			200mm × 200mm					
Type	Conventional MOLB			Advanced MOLB			T-MOLB		
Stack	1kW	1kW	1kW 2kW	5kW	1kW	2kW		15kW	
Module									

For several years since 1997, the reliability and durability of stack were improved. The vertically tandem connection of MOLB stacks, named T-MOLB, was developed as shown in Fig. 4. The units of SOFC stacks are assembled horizontally, and gases flow vertically, these are just the opposite of the conventional MOLB design. T-MOLB will solve the problems in attaining

231

larger capacities and a large number of stacks assembly, maintaining reliability and facilita
multi-layer piling up.

Since July 2000, power generation tests have been carried out by using 300 cell of T-MO
type SOFC. The module is shown in Fig. 5.

Fig. 4 T-MOLB type stack   Fig. 5   15kW SOFC module

The module was operated 7,500 hours. The results in short were as follows. Power out
was 15 kW, the maximum power density 0.24W/cm^2 and the direct internal reforming operat
was carried out for 2,437 hours [7][8][9].

Currently, a thermally self-supporting module, which is essential for the configuration
future systems, is under development.

DEVELOPMENT OF THERMALLY SELF-SUPPORTING MODULE
Features of Module

To put the SOFC in practical use, it is essential to develop the thermally self-suppor
system that can maintain the operating temperature at 1273K by the fuel cell reactions. H
thermally self-supporting means that SOFC module is operated without any additional exte
heat such as electrical heater. In order to solve the problems in thermally self-supporting mod
the following two measures were adopted. First, during the start-up, module temperature
raised by pipeline natural gas burning, and second, the heat recovery system was adopte
maintain SOFC at 1273K by utilizing the heat from power generation.

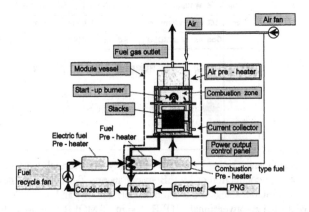

Fig. 6   Process diagram of T-MOLB type SOFC module

As shown in Fig. 6, the module consists of the fans to ventilate the fuel and supply air to the stack, heat exchangers to preheat the fuel and air, startup burner to raise the module temperature and the power output control panel to manage the SOFC output. The appropriate circulation of fuel gas keeps MOLB type SOFC from excessive thermal stress and enables the direct internal reforming for pipeline natural gas.

Verification of Integrated Heat Exchangers
The module developed in 2000 in Fig. 5 had individual air heat exchangers for each stack for raising air temperature by hot exhaust gas.
The performance of each stack could be identified preferentially in this system. It is necessary to integrate the auxiliary units such as heat exchangers in the future large scale SOFC power generation module.
As shown in Fig. 7, a large and integrated air preheated system was developed, where air pre heaters are installed for every five stacks as compared with being installed for each stack in 2000.

Fig. 7   Appearance of air pre heater

Verification of Automated Control of Module Operation
The SOFC power generation system consists of the stacks, the auxiliary units such as heat exchanger, and the control units. In the future SOFC power generation plant, operations of raising temperature and generating power should be automated. Therefore, operation control was automated to monitor the temperatures of the stacks and the auxiliary units including heat exchangers, and to obtain as well the stacks performance data.
In the conventional method of module operation, it was difficult to prospect the behavior of each unit in the module when some operating conditions were changed. The operation parameters of each unit were adjusted manually according to the individual situation by the aid of the operation panel.
Therefore, a digital control system (DCS), which was well suited to monitor the control and operation of the unit, was employed. The operational know-how, which was obtained through manual operation, was converted into DCS logics. This contributed to realize an integrated automatic control of the amount of fuel, airflow rate, temperature at stack inlet and the outlet. Fig. 8 shows the appearance of the module.
Fig. 9 shows the outline of the module operation procedure and Table II shows the automation of the operating steps.
Prior to the tests with actual stacks, preliminary test for the automated operation control

using dummy stacks was conducted, where the control logics and the parameters were modified and adjusted.

The startup burner to raise temperature in the cathode gas, above all, was mainly optimized. Adjusting the fuel flow to the start-up burner controls the temperature in the cathode gas. However, since the air temperature at stack inlet responded dully to the flow of burner fuel because of the heat capacity, the air temperature at stack inlet was observed to fluctuate due to delayed control. We confirmed that the fluctuation of cathode gas temperature control could be suppressed by changing the feed back control to the forward control.

Fig. 8 SOFC and Automation control panel

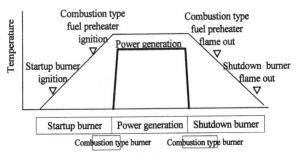

Fig. 9 Process of module operation

Table II Automatic operation control of SOFC

Process	Operation	Control
Temperature rise	by startup burner	
	by combustion type fuel pre heater	Automated control
Power generation	Load change	by DCS
Temperature fall	by startup burner	
	by combustion type fuel pre heater	

After the preliminary test with dummy stacks, the power generation test was performed by using the module loaded with 200 mm sized 10 cells stacks. As shown in Fig. 10, the automatic operation control was successfully demonstrated from startup to shut down processes without any abnormal output or temperature behavior of stack.

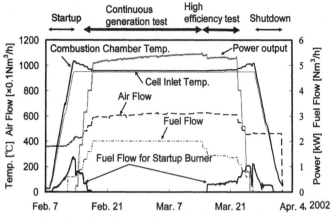

Fig. 10　5 kW class module automatic operating test

After the 5 kW class module startup and continuous generation test, the thermal cycle and load change tests were carried out following the schedule as shown in Fig. 11. The startup and shutdown rates, the cell temperature change rates, were set automatically at 10, 20 and 50 K/h as shown in Fig. 12. And the electric load were changed automatically at 10, 20, 50, 75, 100 and 150 A/h (3%/min) as shown in Fig. 13. It was confirmed that MOLB type SOFC module could endure rapid start up and stop operations, and can endure the rapid change of the electric power output as the same rates as the conventional thermal power plants.

In 2003, the latest 10 kW class module was completed as shown in Fig. 14[10]. The thermal efficiency of MOLB type SOFC module will be evaluated in addition to the electric output power as being estimated on the elemental module until 2002. The module will be operated in 2004.

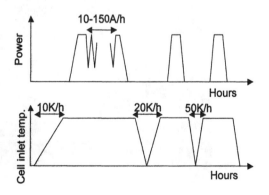

Fig. 11　Load change test schedule

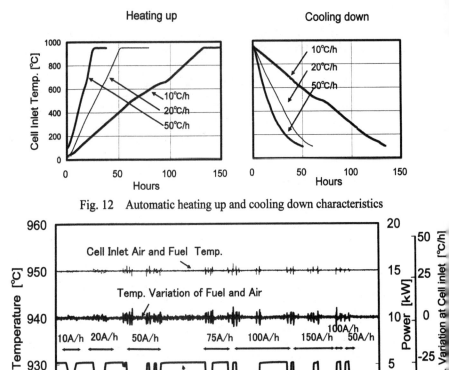

Fig. 12   Automatic heating up and cooling down characteristics

Fig. 13   Automatic load change test

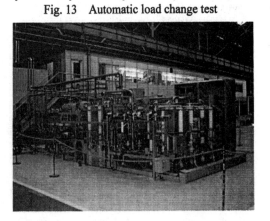

Fig. 14   10 kW class thermally self-supporting module

SUMMARY

The MOLB type SOFC has been developed jointly with Mitsubishi Heavy Industries. Ltd. and Chubu Electric Power Company Inc. since 1990. The MOLB is a unique planar type SOFC consisting of fully ceramics components. The simple structure showed superior durability and reliability. The wet manufacturing process will promise the reduced costs. The power generation tests for single and 10 cells stack showed high power density of 0.35W/cm^2 and automated startup and shut down tests using 5 kW class module has been carried out quite successfully .

In the studies conducted by NEDO project, the integration of air pre heaters and the automatic operation control were verified. These results will serve us not only as preparations for the development of a thermally self-supporting module but also as the basis for future SOFC power generation system. A part of the study was supported by the NEDO; authors greatly acknowledge the assistance of NEDO.

REFERENCES

[1]H. Miyamoto, M. Hattori and M. Minemoto, "Numerical Analysis of Planar Solid Oxide Fuel Cells by Two-Dimensional Approximation," *Kagakukogaku Ronbunshu (J. of Chemical Engineering Society Japan)* **28** [4] 443-450 (2002).

[2]M. Hattori, H. Yoshida, Y. Esaki, and Y. Sakaki, H. Miyamoto, K. Mori, F.Nanjo, K. Takenobu, and T.Matsudaira, "Development of MOLB-type SOFC," *Proc. of the 4th Int. Conf. on New Energy Systems & Conversions*, 283-286, Osaka, Japan (1999)

[3]Y. Yamauchi, Y. Yoshida, N. Hisatome and F. Nanjo"SOFC Development status of Mitsubishi Heavy Industries," *Proc. of 1994 Fuel Cell Seminar*, 639-642, San Diego (1994)

[4]Y. Esaki, M. Hattori, F. Nanjo, M. Funatsu, K. Takenobu and H. Miyamoto, "Development of 5kW Class MOLB-type SOFC," *Proc. of the 2nd IFCC*, 243-246, Kobe (1996)

[5]Y. Sakaki, Y. Esaki, M. Hattori, H. Miyamoto, T. Satake, F. Nanjo, T. Matsudaira and K. Takenobu, "Development of MOLB type SOFC," *Solid Oxide Fuel Cells VI*, PV97-40, 61-64, Pennington, NJ(1997)

[6]M. Hattori, H. Yoshida, Y. Esaki, Y. Sakaki, H. Miyamoto, F. Namjo, K. Takenobu and T. Matsudaira, "Development of MOLB-type SOFC," *Proc. of the 1998 Fuel Cell Seminar*, 511-514, Palm Springs (1998)

[7]Y. Sakaki, H. Yoshida, A. Nakanishi, Y. Esaki, M. Hattori, F. Nanjo, K. Takenobu and H. Miyamoto, "Development of MOLB type SOFC," *Proc. of the 3rd IFCC*, 345-348, Nagoya (1999)

[8]A. Nakanishi, H. Yoshida, M. Hattori, Y. Sakaki, H. Aiki, K. Takenobu and H. Miyamoto, " Development of Several 10kW Class MOLB type SOFC," *Proc. of the 2000 Fuel Cell Seminar*, 799-782, Portland (2000)

[9]Y. Sakaki, A. Nakanishi, M. Hattori, H. Miyamoto,　H. Aiki and K. Takenobu," Development of MOLB type SOFC," *Proc. of Solid Oxide Fuel Cells VII*,　PV01-16, 72-77, Tsukuba (2001)

[10]Y. Sakaki, A. Nakanishi, M. Hattori, H. Miyamoto,　H. Aiki and K. Takenobu, *Proc. of Solid Oxide Fuel Cells VIII*,　PV03-7, 53-56, Paris (2003)

DEVELOPMENT OF ADVANCED CO-FIRED PLANAR SOLID OXIDE FUEL CELLS WITH HIGH STRENGTH

Zhien Liu, Garry Roman, Jeff Kidwell,
Tom Cable, Rich Goettler
SOFCo-EFS Holdings, LLC
1562 Beeson Street, Alliance, OH 44601

Dennis Larsen, Jenna Pike and S. Elangovan
Ceramatec, Inc.
Salt Lake City, Utah 84119

ABSTRACT

Co-fired planar solid oxide fuel cells (SOFCs) were fabricated using tape-casting and screen printing technique. The mechanical strength of bare electrolyte and co-fired electrolyte supported cells (ESC) were measured. Testing results indicate that the co-fired cell strength is significantly lower than that of bare electrolyte. One of the major reasons for co-fired cell weakness is the thermal expansion coefficient mismatch between the anode and electrolyte, leading to the formation of microcracks within the anode. These microcracks can propagate through the anode-electrolyte interface and into the electrolyte substrate. Co-fired ESC performance and mechanical strength have been significantly improved through development of an alternative electrolyte material and improved electrodes.

INTRODUCTION

Solid oxide fuel cells have received increasing attention in recent years as a promising technology for a range of mobile power and distributed generation applications. However, to become commercially viable, substantial improvements must be made in SOFC stack performance and reliability while continuing to reduce cost. In 2000, SOFCo-EFS Holdings LLC (SOFCo) successfully demonstrated a co-fired electrolyte-supported cell. SOFCo is presently developing a second generation co-fired cell which has the potential of meeting commercial targets for cell performance and strength.

From a design standpoint, planar SOFCs are generally electrolyte supported or electrode supported. In both approaches the primary goal is to achieve high performance and long-term reliability. For electrode supported cells, most of the effort has been directed towards anode-supported cells (ASC), primarily because of the better match in sintering temperature between the anode material and the electrolyte. Besides, the relatively thick anode maintains the thermal expansion induced tensile stress in the anode at an acceptably low level while imposing a favorable compressive stress in the electrolyte. There has been limited activity in the pursuit of suitable cathode-supported cell designs for a planar SOFC. There are advantages and disadvantages for each of the two primary designs. The ESC design has potential higher strength and toughness than ASC (ASC strength linked to the porous anode strength of ~100MPa). ESC also have a lower tensile stress in the electrolyte than the ASC during an anode reoxidation event. The ASC design has lower internal resistance and lower tensile stress in the anode than ESC after firing.

The lower resistance is the primary reason for the substantial attention directed at the ASC design. However, the low strength presents challenges with stack assembly and sealing.

From a SOFC processing point of view, there are two basic categories: post-fired cells and co-fired cells. In a post-fired cell process, the electrolyte (or electrolyte and anode bilayer) is pre-sintered. Next, the anode and cathode are applied onto the fired electrolyte and the cell is sintered again. Alternatively, the anode and cathode can be applied separately onto the electrolyte, allowing each electrode to be fired at different individual temperatures. In the co-fired cell process, both the anode and cathode are applied to green electrolyte tape and the resulting tri-layer is fired in one step. The post-fired cell process allows tailoring of sintering temperatures for each component (electrolyte, anode and cathode) to obtain the desired microstructure for optimum cell performance. However, the cycle time is long and the cost is high due to the multiple firing operations. The primary benefit of a co-fired cell is that only one firing run is required, thereby shortening the process cycle time and lowering the fabrication cost. However, some material issues exist for co-fired cells, such as shrinkage mismatch, sintering temperature mismatch between the cathode and electrolyte, and potential deleterious interactions between the electrodes and electrolyte at the sintering temperature. Unless the material set and the process parameters are carefully selected, these issues can result in low mechanical strength and poorer operational performance for a co-fired SOFC.

FIRST GENERATION CO-FIRED ELECTROLYTE-SUPPORTED CELLS

SOFCo has been developing co-fired planar solid-oxide fuel cells for several years. By implementing multilayer ceramic processing technology (tape cast electrolyte and screen printed electrodes), the SOFCo team has successfully developed and fabricated co-fired electrolyte supported cells, including both button cells (3cm diameter) and standard 10cm x10cm production cells[1,2]. The most critical issue in co-firing the electrolyte and electrodes in one step is to obtain a dense electrolyte without detrimental interaction between the electrodes and the electrolyte. Shrinkage mismatch between the components has to be solved to obtain cells that meet flatness tolerance for subsequent stack construction. Co-fired cells inherently have good bonding between the electrodes and the electrolyte, and thus have the potential to provide low interfacial resistance. With the same materials set, SOFCo has successfully fabricated co-fired cells having ~25% lower area-specific resistance (ASR) compared to post-fired ESC. Figure 1 shows a typical V-I curve for the first generation co-fired ESC. The anode (Ni-YSZ cermet) and cathode (composite materials) were both screen-printed onto a 6.5YSZ green tape and then co-fired at a lower sintering temperature than is optimum for the electrolyte. After sintering, the electrode thickness was typically in the range of 15 to 40 microns and electrolyte thickness was about 180 microns. It can be seen that the maximum power density is ~270 mW/cm^2. Figure 2 shows the 1000 hour performance test data for a co-fired ESC. There is a slight degradation during the first 100 hours, and the voltage then becomes stable during the remainder of the test duration. The testing results for three different cells are plotted, indicating repeatability.

240

The two major issues that have been recognized for early co-fired ESC are lower than desired power density and the poor mechanical strength, which leads to cell cracking

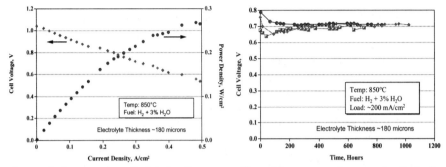

Figure 1: Performance of a co-fired single cell using a YSZ electrolyte

Figure 2 Long-term performance of co-fired cells using a YSZ electrolyte

during stack operation under the constrained load. SEM examination clearly shows the presence of hairline microcracks on the anode surface (see Figure 3). The formation of microcracks on the anode surface can be attributed to the thermal expansion coefficient mismatch between the anode material (CTE $\cong 13.2 \times 10^{-6}$ $^\circ$C^{-1} for Ni-YSZ cermet) and the electrolyte (CTE=10.6×10^{-6} $^\circ$C^{-1} for YSZ). For a co-fired cell, the electrodes have good bonding with the electrolyte, which contributes to good cell performance. However, cracks in the anode are more likely to propagate through a strong interface into the electrolyte, thereby lowering the mechanical strength of the cell. In fact, mechanical tests showed that co-fired cells had lower strength (~160 MPa) compared to the bare electrolyte (strength ~280 MPa).

SECOND GENERATION CO-FIRED ELECTROLYTE SUPPORTED CELLS
Strength Improvement
Fabrication of stronger cells with improved reliability is a very important issue for a co-fired ESC. The SOFCo team has focused on improving the strength of co-fired cells and has made significant progress. In addressing the materials issues, the anode composition and microstructure have been optimized to reduce surface microcracks. The strength of the bare electrolyte was significantly improved through surface strengthening. The principle of surface strengthening is to print a thin (5 to 20 microns) stronger and tougher layer (such as 3YSZ, 4ScSZ, or 6ScSZ) onto a thicker electrolyte with higher conductivity and lower strength (such as 6.5YSZ, 8YSZ, 10ScSZ, etc.) to form a multiple layered structure. In the work described below, 3YSZ was screen printed onto a green 6.5YSZ electrolyte and then co-fired. Other combinations are presently under study. Simultaneously, the fracture toughness was also improved to help prevent the propagation of microcracks from the anode into the electrolyte.

The bend strength of cells and bare electrolyte were measured with and without surface strengthening by using a bend test apparatus, developed in-house. The test method employed subjects strips of the ceramic material (either electrolyte or cells) to a gradually decreasing uniform bend radius. The bend radius at failure can be directly correlated with strength. Test specimens measuring approximately 50mm long and 6mm wide were cut from full-size, standard thickness production cells or electrolyte using either a diamond scribe and snap technique or a thin diamond saw blade. The former is significantly faster, although not usable on high fracture toughness materials. With either preparation method, it was very important to subsequently diamond hone or polish the edges of the specimen to eliminate small chips that could act as failure initiation sites. It was also essential that cells or electrolytes were

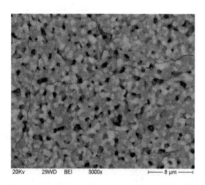

Figure 3 Anode surface of a co-fired ESC, YSZ as electrolyte

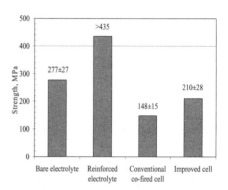

Figure 4 Electrolyte and co-fired cell strength before and after improvement using YSZ as electrolytes

made using standard production processes to assure that representative manufacturing flaws and material characteristics are tested. The pure bending technique primarily exercises surface flaws and is particularly good at investigating the effect of different electrode compositions or processes on overall cell strength. Also, factors relating to the electrode-to-electrolyte interface strength, such as cracking, delamination, and microstructure can be effectively investigated using this measurement technique. Once the specimen is installed in the fixture, the apparatus applies pure bending to the entire specimen with a gradual decrease in the radius of curvature until fracture is achieved. The radius of curvature at the point of fracture is measured directly by the apparatus and is used, in conjunction with the measured specimen thickness, to calculate the failure stress at the specimen's tensile stress surface.

Figure 4 summarizes the initial results for mechanical strength data of bare electrolyte and co-fired cells before and after strength improvement. For cells, the strength value in the bar chart was obtained with the cathode in tension. With the anode in tension, the strength was164±6MPa for standard cells and 211±64MPa for improved cells, respectively. The best strength of 230±35MPa was obtained for the improved cell with the cathode in tension. The electrolyte with strengthening layer did not break at 435MPa, which is the limit of the testing equipment. Generally, after improvement, bare electrolyte strength increased by more than 50%. Co-fired cell strength was typically increased by ~30%. From the standard deviation, it can be seen that co-fired cells with a strengthening layer have a higher variation in strength, which indicates that the screen printing process may introduce flaws in the strengthening layer, such as screen mesh marks. With the optimization of ink and fabrication processes, co-fired cell strength is expected to improve further. Based on stress analysis of a fuel cell stack, the current co-fired cell strength is above the minimum strength required to eliminate cell cracking during stack operation.

Based on short and long-term electrochemical testing of co-fired cells, the improved cells having improved strength exhibited a similar ASR to conventional co-fired cells. However, the long-term degradation has significantly improved with an extrapolated power degradation of nearly 0% per 1000 hours, based on 300 hours of operation from a number of different cells.

Performance Improvement

Since conventional co-fired cells generally have lower power density (less than 300 mW/cm^2), improvement of co-fired cell performance is, of course, very important for development of a commercially viable product. SOFCo's cost model analysis indicates that cell ASR, or power density, is the most sensitive factor affecting cell and stack cost. To insure a high strength, conventional co-fired cells require a relatively thick electrolyte, resulting in a high ASR contribution to the cell. In addition, the YSZ electrolyte material has higher resistance than other candidate electrolyte materials, such as scandia-stabilized zirconia, doped ceria, and doped LaGaO$_3$. However, in a commercial product, SOFCs will also need to have sufficient strength to withstand the thermal and mechanical stresses to which they are subjected to during stack assembly and operation.

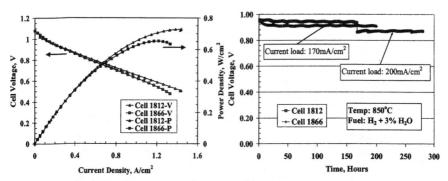

Figure 5 Electrochemical performance of second generation co-fired ESC using ScSZ electrolyte (at 850°C)

Figure 6 Long term performance of SOFCo second generation co-fired ESC using a ScSZ electrolyte

To reduce internal resistance and improve power density of a co-fired ESC, the SOFCo team selected scandia-doped zirconia as a new electrolyte material. Extensive evaluation of the new electrolyte material included testing for conductivity, mechanical strength, fracture toughness, and degradation. A series of finite element models were used to determine the minimum required strength, fracture toughness, and thickness of the new electrolyte material needed to survive the imposed mechanical and thermal stresses. A second generation ESC was developed by incorporating a thinner electrolyte using an optimized composition and using improved tape casting and thick film printing processes. The typical performance of SOFCo's second generation co-fired ESC is shown in Figure 5. The maximum power density is in the range of 600 to 700 mW/cm^2, more than 100% improvement compared to a conventional co-fired ESC. The most important observation is that the second generation co-fired ESC also shows very stable performance, as shown in Figure 6. Using 200 mA/cm^2 current loads, the output voltage stabilized at 870mV. The good performance may also be attributed to the better interfacial characteristics. For co-fired cells, one of the major concerns is the pyrochlore formation at the cathode-electrolyte interface during sintering. Pyrochlore has much lower conductivity and is considered an insulator, which will significantly increase the cell internal resistance. After co-firing a sample, cathode material was etched away using diluted hydrochloride acid. X-ray diffraction techniques were used to investigate the cathode-electrolyte interface. The analysis indicated no observable pyrochlore at the cathode-electrolyte interface for the second generation co-fired ESC.

SUMMARY

Through optimization of the electrolyte composition and thickness, SOFCo has successfully developed a second generation co-fired electrolyte supported planar solid oxide fuel cell with higher performance and mechanical strength than conventional cells. Co-fired cell power density was increased by 100% and the cell exhibited stable long-term performance. A standard cell size (10cm x 10cm) has sufficient strength and fracture toughness to survive the mechanical and thermal stresses imposed by assembly and operation.

ACKNOWLEDGEMENTS

"This manuscript was prepared with the support of the U.S. Department of Energy, under Award No. DE-FC26-01NT41244. However, any opinions, findings, conclusions, or recommendations expressed herein are those of the author(s) and do not necessarily reflect the views of the DOE." The authors would also like to express their appreciation for the help and support from SOFCo team.

REFERENCES

[1]S. Elangovan, J. Hartvigsen, T.L. Cable, T.A. Morris, and E.A. Barringer, "Status of Planar Solid-Oxide Fuel Cell (pSOFC) Development," *Fuel Cell Seminar Abstracts*, pp.503-506, Portland, OR, Nov. 2000.

[2]E.A. Barringer, T.L. Cable, K.E. Kneidel, T.A. Morris, S. Elangovan, and John Olenick, "AMPS, the SOFCo Planar Solid Oxide Fuel Cell," *Fuel Cell Seminar Abstract*, pp. 299-302, Palm Springs, California, November 18-21, 2002

ELECTROPHORESIS: AN APPROPRIATE MANUFACTURING TECHNIQUE FOR INTERMEDIATE TEMPERATURE SOLID OXIDE FUEL CELLS

Sascha Kuehn and Rolf Clasen
Saarland University, Department of Powder Technology
Geb. 43, 66123 Saarbruecken, Germany

ABSTRACT
Solid electrolytes with good oxygen-ion conductivity are of particular interest for application in high-temperature fuel cells. Gadolinium doped ceria $Ce_{0.9}Gd_{0.1}O_{1.95-\delta}$ (GDC) is candidate material as electrolyte of solid oxide fuel cells (SOFCs) operating in the range of 500 to 700 °C.
A cost-effective manufacturing technique for ceria-based-SOFCs is presented in this paper. The anode-based triple-layer is build up in three steps. A porous nickel anode is electrophoretically impregnated with ceria nano-particles to form an improved anode with a large triple-phase boundary. Subsequently these anodes are coated electrophoretically with a dense GDC layer. The dense layer is sintered afterwards to form a gastight electrolyte. Different sintering atmospheres were applied. The different shrinkage rate between anode and electrolyte had to be compensated and fitted by varying the production parameters. The main parameters are the green density and the sintering-temperature of both layers. The green density of the electrolyte layer is controlled by the suspension and the electrophoresis parameters. The third layer, the cathode, is applied by atmospheric plasma spraying (APS) on the electrolyte layer. The manufacturing process of the SOFC is characterized. Power density measurements of the triple layers are presented.

INTRODUCTION

Gadolinium doped ceria $Ce_{0.9}Gd_{0.1}O_{1.95-\delta}$ (GDC) is the most promising material to be used as electrolyte in solid oxide fuel cells (SOFCs) operating in the range of 500 °C to 700 °C [1]. This is due to its high ionic conductivity [2], its catalytic properties and the good thermal expansion match to steel [3]. The practical application of solid oxide fuel cells is still limited by a number of problems [1]. Among them are the performance of the single components and the high operating temperature most significant.

In this paper a new cost-efficient production method for anode supported triple-layer system, solving those problems, is discussed. Anodes have to fulfill various requirements [4, 5]. First of all, a large triple phase-boundary supplies the anode with sufficient electric and ionic conductivity. Secondly, high gas permeability in the total volume is required [6]. The anode is responsible for the catalysis of hydrocarbons and hydrogen-ionization [7]. Requirements are mechanical stability, thermal expansion fit to the electrolyte and the peripherals as well as chemical resistance and compatibility. All these requirements are demanding a special microstructure of the anode [5]. This design can be achieved by electrophoretic impregnation (EPI) [5, 8].

The electrolyte layer is responsible for the ion diffusion, its high ionic conductivity is inalienable [3, 4, 9, 10]. As a consequence the electronic conductivity ought to be minimal [11]. This layer has to be gastight and mechanically stable [12].

The cathode layer has to be porous again. In our design it will be applied by atmospheric plasma spraying (APS) [13]. APS allows application of the cathode without further sintering steps, that could lead to inter-diffusion between the layers.

EXPERIMENTAL

Nickel (Metatherm, Germany) specimen ($\varnothing$ = 17 mm) were pressed and afterwards impreg nated with gadolinium-doped ceria GDC10 (Rhodia, Germany) according to our former publi cized description [5], to preserve an anode-membrane.

A stabilized suspension of GDC10 is produced by bead milling for five minutes, in order to break up the agglomerated and aggregated particles (Fig.1). The suspensions are prepared with a solids content of 40 wt.-% GDC10.

The aforementioned anode-membrane is saturated in water, clutched with a gasket between the water chamber and the suspension chamber, dividing the electrophoretic unit into two parts according to the membrane method [14]. The distance between the two electrodes averages 6 cm. A direct-current voltage is applied for 30 seconds. Different voltages, varying from 10 to 50 V, are applied to the deposition-cell.

The specimens are dried in air for 24 hours. Then they are sintered in a furnace at 1350 °C as determined in [5]. Two different furnaces were used for this processing step. In the tube furnace the temperature was raised with 5 K/min. The final temperature was kept constant for 30 minutes. In contrast to this, some of the specimens are sintered in an inductive furnace. The specimens are lying between two graphite squares of 20 mm side length and 4 mm thickness. The temperature is raised in 3 minutes to about 1350 °C, whereas the dwell-time is only 5 min utes. SEM and X-ray diffraction were used to characterize the two layer systems afterwards. The furnaces were operated with different gas atmospheres. Inert gas with hydrogen content up to . % was used. The X-ray measurements were performed using a D500 Siemens diffractometer with Cu-$K_{\alpha 1}$ radiation.

On top of the two-layer system, the cathode layer is applied by atmospheric plasma spraying Lanthanum strontium manganate (LSM) powder is calcined [15, 16], ball milled and sieved. The particles used were sized between 45 and 90 µm. Plasma spraying parameters were adjusted to obtain porous layers. The plasma device mainly consists of a power supply with amplifier (METCO, 7MR-50), a control unit (METCO, 7M), a plasma gun (METCO, 9MB) and a powder feed unit (METCO, 4MP).

Power densities were measured clutching the triple layers in a SiO_2-tube. Inert gas (5 % H, 95 % N_2) was used as fuel gas and pure O_2 was used on the cathode side.

RESULTS

The main particle size of the bead milled GDC10 powder was determined to 250 nm by Zetasizer 5000 (Malvern Instruments Ltd.). According to the medium solid content, the suspen sions used for the electrophoretic coating of the anode showed low viscosity. This is advanta geous, as the particle mobility is essential for the later EPD process. The isoelectric point was determined to 9.5 which makes a deposition of GDC10 possible in the whole sour pH range and at pH values above 11 (Fig. 1).

Raising the applied voltage results in a super-proportional increase of the electric current (Fig. 2). This does not correspond to ohm's law because the suspensions resistance does not vary. The strong increase indicates a raise of the number of free charge carriers in the suspen sion. The charge carriers can be disengaged of the charge clouds around the dispersed particles due to the high voltages.

The current decreases afterwards with longer deposition time, indicating a raise of the sus pensions resistance due to the electrophoretic deposition. At 40 V deposition rates of 100 µm per minute were reached.

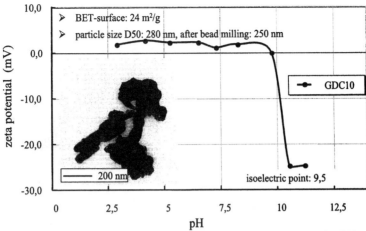

Figure 1: Zeta potential of the GDC10 powder. Bead milling breaks up the GDC10 **agglomer-**
ates being shown in the TEM picture.

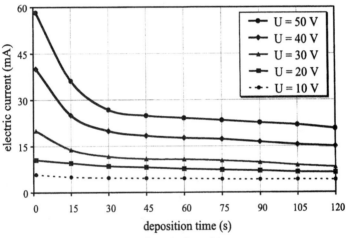

Figure 2: Electric current during electrophoretic deposition.

High green densities were necessary to avoid cracks, because lower shrinkage rates while drying can be achieved. Butt still cracks can appear due to the in-plane shrinkage while drying. By raising the voltage up to 40 V crack free samples were produced. Layers up to 2000 μm thickness were deposited. The voltage, however, cannot be raised infinitely. If the applied voltage is too high, the high current results in warming of the suspension. This leads to defective deposited layers.

The double-layers are sintered afterwards. The different shrinkage rates of the two layers had to be fitted closely to one another. This can mainly be achieved by a high green density of the electrolyte layer. Dilatometer plots of GDC10-greenbodys with different green-densities are presented in Fig. 3. The green-density of the nickel-cermet anodes is lowered to 35 vol.-%. According to this, and because of the good anchoring of the electrolyte layer into the impregnated anode, the two layers were sintered crack-free without changing their shape.

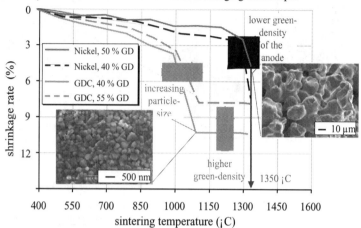

Figure 3: Different shrinkage rates of the two layers.

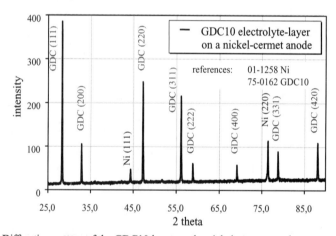

Figure 4: Diffraction pattern of the GDC10 layer on the nickel-cermet anode.

The 30 μm thin GDC10-layers were transparent after sintering, due to their high density. Separate phases were determined by X-ray diffraction. Sintering in reducing atmospheres resulted in a reduction of the CeO_2 to Ce_2O_3. Therefore argon atmosphere was used for the sintering step. In the inductive furnace an reducing atmosphere was less critical due to the short dwelling times at 1350 °C, but argon atmosphere was applied for all experiments.

The cathode is plasma sprayed onto the sintered layers. The plasma-sprayed LSM cathode shows a good adhesion to the electrolyte. The powder-particles sized 45-90 μm are melting in the plasma flame and sputtering to smaller particles that hit the electrolyte layer. The small particles build up a porous membrane. The nickel-membrane in Fig. 5 was only impregnated a mere 5 seconds before deposition of the GDC electrolyte. Adjusting the impregnation process, impregnation depths up to 200 μm are possible.

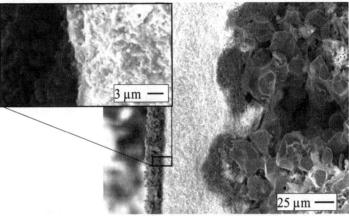

Figure 5: Triple layer: the nickel-cermet anode (on the right) with electrophoretically deposited electrolyte (middle) and plasma sprayed LSM-cathode-layer.

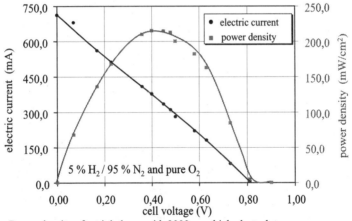

Figure 6: Power density of a triple layer with 2000 μm thick electrolyte.

Afterwards first power density measurements were carried out. The triple layer system with the thickest electrolyte layer was measured first. Power densities of 212 mW/cm^2 were measured at 800 °C with 5 % H_2 on the anode side (Fig. 6).

CONCLUSIONS

Electrophoretic impregnation (EPI) according to the membrane method is an appropriate technique to incorporate nano-sized ceramic particles into a metal membrane and deposit an electrolyte layer on its inner surface. The ceramic second-phase can be applied graded and wi anchor the electrolyte layer for better mechanical stability. EPD is a very fast and cost-effectiv production method for anode-electrolyte bi-layers. In two minutes a metal membrane can be im pregnated with ceramic particles forming a contiguous layer on the inner surface of the mem brane and coated with the electrolyte layer. The cathode can be easily applied after sintering This is a preferred structure for anodes in SOFCs. The produced structure proved high powe densities in preliminary tests. Higher densities are expected for the thinner electrolyte layers. Th use of pure hydrogen is expected to further improve the power density.

ACKNOWLEDGEMENT

The authors gratefully acknowledge the financial support of the German Science Foundatio (Deutsche Forschungsgemeinschaft, DFG) within the scope of the "Graduiertenkolleg: Ad vanced materials for efficient energy conversion".

REFERENCES

[1] B. C. H. Steele, "Materials for IT-SOFC stacks 35 years R&D: the inevitability of gradualness?," *Solid State Ionics*, **134** 3-20 (2000).

[2] V. Butler, C. R. A. Catlow, B. E. F. Fender and J. H. Harding, "Dopant ion radius and ionic conductivity in cerium dioxide," *Solid State Ionics*, **8** 109-113 (1983).

[3] B. C. H. Steele, "Appraisal of $Ce_{1-y}Gd_yO_{2-y/2}$ electrolytes for IT-SOFC operation at 500 °C," *Solid State Ionics*, **129** 95-110 (2000).

[4] T. Ishihara, T. Shibayama, H. Nishiguchi and Y. Takita, "Nickel-Gd-doped CeO_2 cermet anode for intermediate temperature operating solid oxide fuel cells using $LaGaO_3$-based perovskite electrolyte," *Solid State Ionics*, **132** 209-216 (2000).

[5] S. Kühn, J. Tabellion and R. Clasen, "Impregnation of Nickel Foils with Nanocrystalline Ceria as Anodes for Solid Oxide Fuel Cells SOFC," *Ceram. Eng. Sci. Proc.*, **24** [3] 305-310 (2003).

[6] E. Z. Tang, T. H. Etsell and D. G. Ivey, "A New Vapor Deposition Method to Form Composite Anodes for Solid Oxide Fuel Cells," *J. Am. Ceram. Soc.*, **83** [7] 1626-1632 (2000).

[7] P. G. Harrison, I. K. Ball, W. Azelee, W. Daniell and D. Goldfarb, "Nature and Surface Redox Properties of Copper(II)-Promoted Cerium(IV) Oxide CO-Oxidation Catalysts," *Chem. Mater.*, **12** 3715-3725 (2000).

[8] J. Tabellion, C. Oetzel and R. Clasen, "Manufacturing of glass and ceramic matrix composites by electrophoretic impregnation with nanosized powders"; pp. 577-584 in *26th Annual Conference on Composites, Advanced Ceramics, Materials and Structures*, ed. Edited by H.-T. Lin and M. Singh. The American Ceramic Society, Cocoa Beach, Florida, USA, 2002.

[9] D. Y. Wang, D. S. Park, J. Griffith and A. S. Nowick, "Oxygen-ion conductivity and defect interactions in yttria-doped ceria," *Solid State Ionics*, **2** 95-105 (1981).

[10] A. Tsoga, A. Gupta, A. Naoumidis and P. Nikolopoulos, "Gadolinia-doped Ceria and Yttria stabilized Zirconia Interfaces: Regarding their Application for SOFC Technology," *Acta mater.*, **48** 4709-4714 (2000).

[11] C. M. Kleinlogel and L. J. Gauckler, "Mixed Electronic-Ionic Conductivity of Cobalt Doped Cerium Gadolinium Oxide," *J. Electroceram.*, **5** [3] 231-243 (2000).

[12] G. Lindemann, H. Böder and H. Geier, "Preparation and Characterization of ZrO_2 Electrolytes for Planar Solid Oxide Fuel Cells," *Industrial Ceramics*, **18** [2] 99-102 (1998).

[13] G. Fehringer, S. Janes, M. Wildersohn and R. Clasen, "Proton-conducting ceramics as electrode/electrolyte-materials for SOFCs: Preparation, mechanical and thermal-mechanical properties of thermal sprayed coatings, material combination and stacks," *J. Eur. Ceram. Soc.*, **24** [5] 705-715 (2004).

[14] R. Clasen, "Forming of compacts of submicron silica particles by electrophoretic deposition"; pp. 633-640 in *2nd Int. Conf. on Powder Processing Science*, ed. Edited by H. Hausner, G. L. Messing and S. Hirano. Deutsche Keramische Gesellschaft, Köln, Berchtesgaden, 12.-14. 10. 1988, 1988.

[15] S. Kühn, S. Janes and R. Clasen, "Improved Calcination Process of Barium Calcium Niobate BCN18 and its In Situ Application to Thermal Spraying," *Ceram. Eng. Sci. Proc.*, **24** [3] 299-304 (2003).

[16] R. J. Bell, G. J. Millar and J. Drennan, "Influence of synthesis route on the catalytic properties of $La_{1-x}Sr_xMnO_3$," *Solid State Ionics*, **131** 211-220 (2000).

MICROSTRUCTURE-PERFORMANCE RELATIONSHIPS IN LSM-YSZ CATHODES

J.A. Ruud, T. Striker, V. Midha, B.N. Ramamurthi, A.L. Linsebigler, and D.J. Fogelman
GE Global Research
1 Research Circle
Niskayuna, NY 12309

ABSTRACT
Quantitative methods for engineering efficient solid oxide fuel cell (SOFC) composite electrodes are presented. Four strontium-doped lanthanum manganite (LSM)- yttria-stabilized zirconia (YSZ) microstructures were produced by varying batching and sintering conditions. The microstructures were imaged using scanning electron microscopy (SEM) and the phases were identified using scanning Auger analysis. Cathode microstructure was quantified through stereographical analysis of the size distributions and volume fractions of the phases.

Performance of cathodes of various thicknesses was measured using 4-wire symmetric single atmosphere air AC impedance spectroscopy as a function of temperature. All microstructures showed a decrease in total resistance (polarization and ohmic) as the thickness increased up to a critical thickness at which the resistance remained constant. Microstructures with smaller particle sizes had lower resistances, as expected.

The effect of microstructure on performance was modeled analytically based on a particle percolation model. As inputs, the model used temperature-independent microstructural parameters (active area, degree of particle necking), and temperature-dependent parameters (exchange current density, YSZ ionic conductivity, LSM electronic conductivity). Good agreement between the model and the data was observed over a range of microstructures and measurement temperatures. The exchange current density and the active surface area were determined independently from the model fits. The methods can be extended to composite anode microstructures, such as Ni-YSZ or to advanced cathode microstructures.

INTRODUCTION

Optimization of the performance of strontium-doped lanthanum manganite (LSM)-yttria-stabilized zirconia (YSZ) solid oxide fuel cell (SOFC) cathodes requires engineering of a three phase composite microstructure [1]. Oxygen reduction occurs at triple phase boundaries (TPBs) between the gas conducting phase (pore), the electronic conducting phase (LSM) and the ionic conducting phase (YSZ). Active TPBs are those that are connected through continuous pore paths to the air source, connected through continuous LSM paths to the current collector and connected through continuous YSZ paths to the electrolyte. The challenge for cathode microstructure optimization is to engineer the largest volume density of active TPBs, by increasing both the total area and the connectivity.

Prior work in optimizing the LSM-YSZ system concentrated on relating processing conditions to performance [2,3] and on determining the reaction mechanisms [4,5]. The goal of the present work was to relate cathode microstructure to performance quantitatively.

EXPERIMENTAL

LSM-YSZ cathodes were fabricated on 150 μm thick 8 mol% YSZ substrates by screen-printing and sintering. LSM and YSZ powders were mixed with organic binders to form a thick film paste. Cathodes of varying thickness were deposited by printing in multiple layers. Four distinct microstructures were fabricated. Samples of six different thicknesses were produced from a single batch of cathode paste with a 1 hour, 1200 °C sintering treatment (1200-A). Three sets of samples printed at 3 different thicknesses were produced from a second batch of cathode paste sintered at 3 temperatures for 1 hour: 1200 °C (1200-B), 1250 °C (1250-B), and 1300 °C (1300-B). The cathodes were metallized with a Pt organometallic ink, fired and then a Pt wire mesh was attached using a Pt paste with a second firing step.

Samples were measured using AC impedance spectroscopy in a furnace with flowing air. A 10 mV AC amplitude was applied and frequency was scanned between 1 MHz and 0.03 Hz using a Solartron 1260 frequency analyzer and a Solartron 1480 Multistat. Polarization curves were measured at temperatures between 650 °C and 800 °C. The total resistance of the cathode (polarization plus ohmic) was measured by subtracting the ohmic resistance of the YSZ substrate from the low-frequency intercept of the polarization curve with the real axis and dividing by two to account for the double-sided geometry [6].

Images of the LSM-YSZ microstructures were obtained through a Field Emission Scanning Auger Nanoprobe. Phases containing La and Zr were identified on planar cross-sections of the cathode at a magnification of 10kX (14 μm x 11 μm regions imaged at 15 nm/pixel). The pores were determined by difference. Three regions were sampled for each microstructure. The composite microstructure was determined by overlaying the La and Zr maps on a Scanning Electron Micrograph, as shown in Figure 1.

RESULTS

The cathode microstructural parameters, particle size and volume fraction, were determined from the Auger maps and are given in Table I. The particle size was estimated from the average

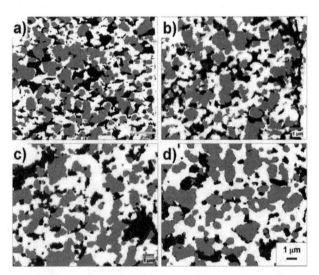

Figure 1. Scanning electron micrographs of LSM-YSZ cathode with overlays determined using Field Emission Scanning Auger Nanoprobe where black is pore, white is YSZ and gray is LSM for a) 1200-A, b) 1200-B, c) 1250-B, and d) 1300-B.

linear intercept. The sample made from the first paste had the smallest particle and pore sizes. For samples made from the second paste, the particle size increased with increasing sintering temperature and the fraction of porosity decreased.

Table I. Cathode microstructural parameters determined from quantitative image analysis.

Sample	Average Linear Intercept (μm)			Volume fraction		
	LSM	YSZ	Pore	LSM	YSZ	Pore
1200-A	0.51± 0.07	0.42± 0.09	0.29± 0.02	0.36± 0.02	0.45± 0.01	0.24± 0.05
1200-B	0.55± 0.05	0.50± 0.02	0.46± 0.02	0.30± 0.06	0.36± 0.03	0.34± 0.03
1250-B	0.58± 0.10	0.63± 0.05	0.44± 0.01	0.30± 0.04	0.45± 0.03	0.25± 0.02
1300-B	0.74± 0.09	0.74± 0.07	0.50± 0.06	0.37± 0.07	0.43± 0.06	0.19± 0.03

The performance of the cathodes is given in Figure 2. All cathode microstructures showed a decrease in cathode resistance with increasing thickness up to a critical thickness at which the resistance leveled off. The 1200-A cathode had the lowest resistance of all the samples, and the cathode resistance increased with increasing sintering temperature. The trends with temperature were consistent for all microstructures.

The cathode performance data can be described by the 1D continuum, microstructural model of Castamagna, Costa and Antonucci [7]. In that model, the cathode is described as a percolated network of monomodal, spherical particles of the electronic conductor (LSM) and the ionic conductor (YSZ). The relative particle diameters and the volume fractions of the two conductors are the inputs. The model is a one-dimensional model based on three differential equations: 1) Ohm's law for the potential distribution through the thickness in the ionic conductor, 2) Ohm's law for the potential distribution through the thickness in the electronic conductor, and 3) the distribution of overpotential based on a linearization of the Butler-Volmer equation. An analytical solution for the total resistance (polarization and ohmic), R, of the cathode was derived [7] as given by Equation 1.

$$R = \frac{a(\rho_{io}^{eff} + \rho_{el}^{eff})\{\cosh(\Gamma) + \Omega[2 + \Gamma \sinh(\Gamma) - 2\cosh(\Gamma)]\}}{\Gamma \sinh(\Gamma)} \tag{1}$$

where a is thickness, ρ_{io}^{eff} and ρ_{el}^{eff} are the effective resistivities of the ionic and electronic conductors, respectively,

$$\Omega = \frac{\rho_{io}^{eff} \rho_{el}^{eff}}{\left(\rho_{io}^{eff} + \rho_{el}^{eff}\right)^2} \tag{2}$$

and

$$\Gamma^2 = \frac{I_0 AF(\rho_{io}^{eff} + \rho_{el}^{eff})a^2}{RT} \tag{3}$$

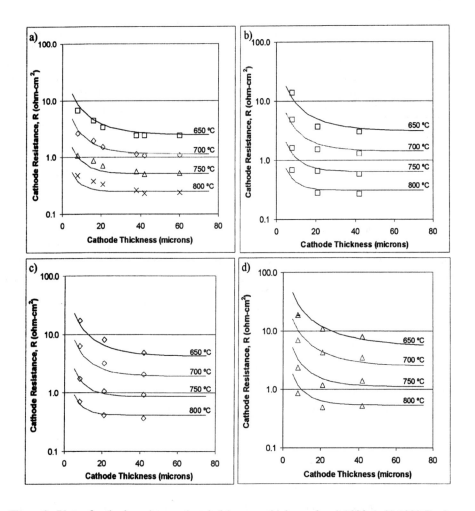

Figure 2. Plots of cathode resistance (symbols) versus thickness for a) 1200-A, b) 1200-B, c) 1250-B, and d) 1300-B as a function of temperature. Solid lines correspond to model fits.

where I_0 is the exchange current density, A is the active TPB area per unit volume, F is Faraday's constant, 96,500 C/mol, R is the gas constant, 8.31 J/mol-K, and T is the temperature.

The effective resistivities of the two phases are given by Equation 4,

$$\frac{1}{\rho^{eff}} = \sigma^{eff} = \frac{\gamma}{(1-n_c)^2}(n-n_c)^2\sigma_0 \qquad (4)$$

where σ_0 is the conductivity of the pure material, n is the number fraction of the ionic or electronically conducting particles, n_c is the critical fraction number set by the percolation threshold, and γ is an adjustable parameter between 0 and 1 that accounts for the degree of necking between particles.

A set of common parameters was used to fit impedance data for all four microstructures at all four temperatures with the only free parameter that was varied among the microstructures being A. The ionic conductivity and the electronic conductivity were temperature-dependent parameters that were determined from the literature [8,9]. The number fraction of the ionic phase was measured to be about 0.55 for all microstructures and the number fraction of the electronic phase was 0.45. The critical fraction number for the ionic phase for each microstructure was determined from the particle sizes in Table I using Equation 5 [7]:

$$n_c^{io} = \frac{Z_c P^2}{Z_0 - Z_c (1 - P^2)} \tag{5}$$

where Z_c is the average coordination number at the percolation threshold for a random assembly of uniform size spheres, taken to be 1.764 after [10], Z_0 is the average coordination number for a binary random packing of spheres, taken to be 6 [11], and P is the ratio of radii of the ionic particles to the electronic particles. A similar equation was used to determine the critical fraction number for the electronic phase [7]. The values are given in Table II.

There were only 3 fitting parameters, γ, I_0, and A. γ was taken to be 0.75 for all microstructures. I_0 was a function of temperature and was fit to the values 100, 300, 1000 and 3000 A/m^2 at the temperatures 650 °C, 700 °C, 750 °C, and 800 °C, respectively. The only fitting parameter that was varied among the microstructures was A, which is given in Table II. The fitted curves are shown in Figure 2. Good agreement with the data was found.

Table II. Microstructure-dependent parameters from the model fits.

Parameters	Microstructure			
	1200-A	1200-B	1250-B	1300-B
n_c^{io}	0.216	0.259	0.330	0.295
n_c^{el}	0.387	0.332	0.260	0.293
A (m^2/m^3)	12x10^4	9x10^4	7x10^4	3.5x10^4

DISCUSSION

All cathodes showed a similar dependence of resistance on thickness. With increasing thickness, the resistance decreased due to a reduction in the polarization component. Increased thickness resulted in an increase in the surface area available for oxygen reduction. Above a critical thickness, the effective surface area for reduction did not increase because of increased ohmic paths to the electrolyte and the current collector.

The microstructure-based model showed good agreement with the data using relatively few fitting parameters. The exchange current density and the active area density were determined by the level of the plateau. The thickness at which the resistance leveled off was determined largely by the effective ionic resistivity of the YSZ, which was the phase with the limiting ohmic resistance. Large ionic resistivity values reduced the critical thickness by reducing the thickness of the active region for oxygen reduction. The effective ionic resistivity of a microstructure was affected by the percolation limit which was determined by the ratio of particle sizes.

CONCLUSIONS

The performance of composite cathode microstructures was described by a model based on the percolation of monomodal spheres. Scanning Auger microscopy was shown to be a useful method for distinguishing between LSM and YSZ phases. The model allowed for the determination of the exchange current density independent of the active area. Extension of the model to other composite electrode structures is possible.

ACKNOWLEDGMENTS

The authors acknowledge the support of this work by the U.S. Department of Energy through cooperative agreement DE-FC26-01-NT41245 under the Solid State Energy Conversion Alliance. Image analysis provided by J. Grande and helpful discussions with M. Thompson, A. Verma, and J. Guan are gratefully acknowledged.

REFERENCES

[1] J. Fleig, "Solid oxide fuel cell cathodes: Polarization mechanisms and modeling of the electrochemical performance," *Annu. Rev. Mater. Res.* **33**, 361-82 (2003).

[2] J.H. Choi, J.H. Jang, and S.M. Oh, "Microstructure and cathodic performance of $La_{0.9}Sr_{0.1}MnO_3$/ytrria-stabilized zirconia composite electrodes," *Electrochimica Acta*, **46**, 867-874 (2001).

[3] M.J. Jorgensen, S. Primdahl, C. Bagger, and M. Mogensen, "Effect of sintering temperature on microstructure and performance of LSM-YSZ composite cathodes," *Solid State Ionics*, **139**, 1-11 (2001).

[4] M.J. Jorgensen and M. Mogensen, "Impedance of solid oxide fuel cell LSM/YSZ composite cathodes," *J. Electrochem. Soc*, **148**, A433-A442 (2001).

[5] E.P. Murray, T. Tsai, and S.A. Barnett, "Oxygen transfer processes in $(La,Sr)MnO_3/Y_2O_3$-stabilized ZrO_2 cathodes: an impedance spectroscopy study," *Solid State Ionics*, **110**, 235-243 (1998).

[6] W.G. Wang, R. Barfod, P. H. Larsen, K. Kammer, J.J. Bentzen, P.V. Hendriksen, and M. Mogensen, "Improvement of LSM cathode for high power density SOFCs"; pp. 400-408 in *Solid Oxide Fuel Cells VIII (SOFC-VIII)*, Edited by S.C. Singhal and M. Dokiya, The Electrochemical Society, Inc., Pennington, New Jersey, 2003.

[7] P. Costamagna, P. Costa, and V. Antonucci, "Micro-modelling of solid oxide fuel cell electrodes," *Electrochimica Acta*, **43** 375-394 (1998).

[8] J.H. Park and R.N. Blumenthal, "Electronic transport in 8 Mole Percent $Y_2O_3 - ZrO_2$" *J. Electrochem. Soc*, **136**, 2867 (1989).

[9] K. Katayama, T. Ishihara, H. Ohta, S. Takeuchi, Y. Esaki, E. Inukai, "Sintering and Electrical Conductivity of $La_{1-x}Sr_xMnO_3$," *Nippon Seramikkusu Kyokai Gakujutsu Ronbunshi*, **97** 1327-33 (1989).

[10] C.H. Kuo and P.K.. Gupta, "Rigidity and conductivity percolation thresholds in particulate composites," *Acta Metall. Mater.* **43**, 397 (1995).

[11] D. Bouvard and F.F. Lange, "Relation between percolation and particle coordination in binary powder mixtures," *Acta Metall. Mater.*, **39**, 3083 (1991).

256

ROLE OF CATHODE IN SINGLE CHAMBER SOFC

Toshio Suzuki, Piotr Jasinski, Fatih Dogan, and Harlan U. Anderson.
Electronic Materials Applied Research Center and
Ceramic Engineering Department
University of Missouri-Rolla
303 MRC Rolla MO 65401 USA
E-mail: toshio@umr.edu

ABSTRACT

The role of the catalytic activity of the cathode materials on the open circuit voltage (OCV) of single chamber solid oxide fuel cell (SC-SOFC) was investigated. Ni-cermet has been proven to be as a very efficient catalyst for fuel and used as anode. On the other hand, typical cathode materials (perovskite oxides) are also known as catalytically active for fuel oxidation and therefore affect the OCV of the cell. Cells with a mixture of doped ceria and (La, Sr)(Co, Fe)O₃ (LSCF) as cathode, and Ag as reference electrode on doped ceria electrolyte (SDC) were prepared and tested in single chamber configuration. The OCV increased as SDC content decreased and reached values as high as 0.1V for pure LSCF. A power density of ~140 mW/cm² was obtained at 600°C using Ni-cermet anode and LSCF-10% SDC cathode.

INTRODUCTION

A high performance fuel cell that is operated in a single chamber (SC-SOFC) has been demonstrated by Hibino et al. [1-3]. In SC-SOFC, both the anode and cathode are exposed to the same mixture of fuel and oxidant gas. Since SC-SOFC is operated in the mixture of fuel and oxidant gas, there are several advantages such as; (i) simplified cell structure since no sealing is needed, (ii) better thermal and mechanical stability as compared to double chamber SOFC, (iii) increasing cell temperature results in reducing cell resistance due to oxidation of fuel at the anode side [3].

Recent studies showed that relatively high power density could be obtained in SC-SOFCs at intermediate operating temperatures using hydrocarbon fuel [1]. Typical SC-SOFC consists of samarium doped ceria (SDC) as electrolyte, Ni-SDC cermet as anode and (Sm, Sr)CoO₃ as cathode [3]. The oxygen activity at the anode and the cathode is not fixed and one of the electrodes has a higher electro-catalytic activity for the anodic oxidation of fuel, whereas the other electrode has a higher electro-catalytic activity for the cathodic reduction of oxygen. As a result, an EMF between two electrodes is generated even in an atmosphere of fuel/air mixture. The following oxidation and/or reforming reactions should occur on the electrodes with a catalyst such as Ni using propane as fuel:

$$C_3H_8 + 3/2O_2 \rightarrow 3CO + 4H_2 \tag{1}$$

$$H_2 + O^{2-} \rightarrow H_2O + 2e^- \tag{2}$$

$$CO + O^{2-} \rightarrow CO_2 + 2e^- \tag{3}$$

$$1/2O_2 + 2e^- \rightarrow O^{2-} \tag{4}$$

The reaction (1) is exothermic evolving heat, which effectively increases the anode temperature. Such reactions can take place not only on the anode but also on the cathode, since typical cathode materials such as perovskites exhibit catalytic activity for hydrocarbon oxidation [4]. The objectives of this study are to determine the catalytic effect of cathode materials on the EMF and performance of SC-SOFC.

EXPERIMENTAL PROCEDURE

Samarium doped ceria (SDC) electrolyte was prepared by pressing of commercially available powders (20 % samarium doped ceria, Daiichiki Genso Co, Japan) and sintering at 1500 °C for 10 hours in air. After screen-printing of 70 wt%Ni-SDC ink as anode material on the electrolyte, the sample was sintered at 1450 °C for 1 hour. Subsequently, mixtures of $La_{0.8}Sr_{0.2}Co_{0.2}Fe_{0.8}O_3$ (LSCF) and SDC powder (0, 10, 30, 50 wt% SDC) were screen printed on the opposite side of the electrolyte and sintered at 1000 °C for 1 hour in air.

To measure the open circuit voltage (OCV) of cathode materials, Ag was used as a reference electrode (Ag paste, Electro-Science Laboratories). Microstructural development of the cell materials was characterized by scanning electron microscopy techniques (Hitachi S4700 field emission gun (FEG)). Typical dimensions of the samples were 0.18 cm^2 electrode area and 0.5 mm electrolyte thickness.

The electrochemical properties of the cells were measured in a tube furnace using a flowing mixed gas with a composition of 10 % propane and 90 % air as shown in Fig. 1. The furnace- and sample-temperatures were monitored simultaneously. Pt- and Au-meshes, in the size of electrode area, were used as current collectors. Furnace temperature was adjusted between 500 and 600 °C. The flow rate of mixed gas was kept constant at 300 cm^3/min using a gas flow controller. The impedance spectroscopy techniques were utilized to investigate the cell performance using Solartron 1470 Battery Tester and 1255B Impedance Gain Phase Analyzer.

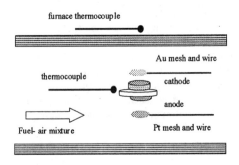

Figure 1: Experimental set up for SC-SOFC measurement

RESULTS AND DISCUSSION

Figure 2 shows SEM images of a cell with Sm-doped CeO_2 (SDC) as electrolyte Ni-SDC as anode and $(La,Sr)(Co,Fe)O_3$ as cathode. The thickness of the electrolyte and electrodes were ~500 μm and ~20 μm, respectively.

Figure 3 shows the open circuit voltage of the cell as a function of SDC content in LSCF vs. Ag electrode. The negative sign of the OCV, measured as high as 0.1 V, is related to the LSCF cathode acting as an anode against the silver reference electrode. This implies that LSCF exhibits a catalytic activity promoting hydrocarbon oxidation.

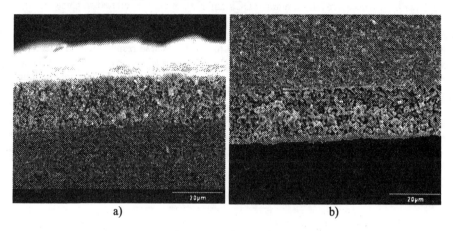

a)   b)

Figure 2: Microstructure of electrolyte supported ceria based fuel cell; a) cathode / electrolyte interface, b) electrolyte / anode interface.

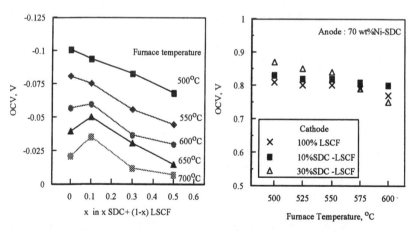

Figure 3: Open circuit voltage as a function of SDC content in LSCF electrode vs. Ag electrode.

Figure 4: Open circuit voltage as a function of SDC content in LSCF electrode vs. Ni-cermet anode.

The OCV decreased as the temperature and the SDC content of the cathode increased. An increase of SDC content decreases the catalytic activity of the electrode leading to a lower reaction rate for Eq. (1), which in turn, results in lower OCV. The decrease of OCV with increasing temperature may be related to the shift of main fuel-oxygen reaction from partial to full oxidation considering Eq. (1). This may lead to a decrease of H_2 concentration resulting in

259

a decrease of the OCV. A similar behaviour of electrode reactions was observed in our previous studies [5-7] and may be explained by determining the catalytic activity of the electrode materials, which are currently under investigation. Note that results shown in Fig.3 can also be affected by the surface area, thickness and size of the cathode material. Nevertheless, the use of LSCF as cathode material results in reduced OCV and lowering of the cell performance.

Figure 4 shows the OCV as a function of furnace temperature for different cathode materials using 70wt.%Ni-SDC anode. It is depicted that the OCV increased with increasing percentage of SDC in LSCF at temperatures up to 550°C. These results are consistent with those shown in Fig. 3. The sample with 30%SDC-LSCF revealed a drop of the OCV at furnace temperatures above 550°C. This may be related to possible damage of the samples during measurements, such as cracking, which usually accompanied with a sudden drop of the OCV.

Figure 5 shows the maximum power density of SC-SOFC with different cathode materials at furnace temperatures from 500 to 600°C. Among different samples, the cell with LSCF-10%SDC cathode reveals the highest power density of 140mWcm^{-2} at 600°C. The role of SDC in LSCF on the cathode overpotential, studied by measurements of the impedance spectra, will be reported elsewhere.

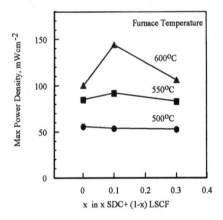

Figure 5: Maximum power density of the electrolyte supported fuel cells as a function of SDC content in LSCF in single chamber SOFC condition. Electrolyte thickness = 0.5 mm.

SUMMARY

Effect of cathode materials on the open circuit voltage (OCV) of single chamber solid oxide fuel cell was investigated. Since, SC- SOCF is operated using mixed gas of air and fuel, the OCV of cell is determined by the difference of catalytic activity between cathode and anode. Cells with a mixture of doped ceria and (La, Sr)(Co, Fe)O$_3$ as cathode material and Ag reference electrode on Sm-doped ceria electrolyte were prepared and tested under single chamber conditions. Using LSCF as anode, the OCV was as high as 0.1 V, which decreased as the SDC content in LSCF increased. It was shown that LSCF exhibits a catalytic activity for oxidation of the fuel and lowers the OCV when it is used in the single chamber SOFC condition. Measurements of SDC-LSCF cathode materials with Ni-cermet anode on SDC electrolyte showed that the OCV increased with increasing SDC content, which is consistent with the results above. Best performance at 600°C among the tested samples was 140 mWcm^{-2} using 10%SDC-LSCF as cathode material on a ~0.5 mm thick SDC electrolyte. This result also indicated a strong influence of SDC content in LSCF on the cathode overpotential.

ACKNOWLEDGMENT

The authors would like to acknowledge the grants from the Research Board of University of Missouri-Rolla and Motorola Advanced Technology Center for the support of this research.

REFERENCES

1. T. Hibino, A. Hashimoto, M. Yano, M. Suzuki, S. Yoshida, M. Sano. *J. Electrochem. Soc.* 149 (2) A133-A136 (2002).
2. T. Hibino, A. Hashimoto, T. Inoue, J. Tokuno, S. Yoshida, M. Sano. *J. Electrochem. Soc.* 148 (6) A544-A549 (2001).
3. T. Hibino, A. Hashimoto, T. Inoue, J. Tokuno, S. Yoshida, M. Sano. *Science.* 288 p. 2031 (2000).
4. R.J.H. Voorhoeve, in *Advanced Materials in Catalysis*, Edited by J.J. Burton and R.L. Garten, Academic Press, p.129 (1977).
5. P. Jasinski, T. Suzuki, Z. Byars, F. Dogan, H.U. Anderson, Electrochemical Society Proc. "Solid Oxide Fuel Cells VIII" PV 2003-07, (2003) 1101.
6. P. Jasinski, T. Suzuki, X.D .Zhou, F. Dogan, H.U. Anderson, Ceramic Engineering and Science Proceedings 24, 3, p.293(2003).
7. P. Jasinski, T. Suzuki, F. Dogan and H.U. Anderson., "Impedance spectroscopy of single chamber SOFC", accepted *Solid State Ionics.*

MORPHOLOGY CONTROL OF SOFC ELECTRODES BY MECHANO-CHEMICAL BONDING TECHNIQUE

Takehisa Fukui and Kenji Murata
Hosokawa Powder Technology Research Institute
1-9 Tajika, Shoudai, Hirakata, Osaka 573-1132,
Japan

Ching Chung Huang
Hosokawa Nano Particle Technology
Center
10 Chatham Road, Summit, NJ 07901

Makio Naito, Hiroya Abe and Kiyoshi Nogi
Joining and Welding Research Institute, Osaka University
11-1 Mihogaoka, Ibaraki, Osaka 567-0047, Japan

ABSTRACT

Nickel Oxide (NiO) - Yttria Stabilized Zirconia (YSZ) composite powders for the fabrication of the cermet anode of Solid Oxide Fuel Cell (SOFC) were processed by a novel processing technique in dry phase. The morphology of the Nickel (Ni)-YSZ cermet anode fabricated from the NiO-YSZ composite powder showed homogeneous porous structure consisted of sub-micron sized Ni and YSZ grains. Moreover, it was found from Auger Electron Spectroscopy observation that several hundred nano-meter sized YSZ grains and fine Ni grains achieved good connections. As a result, the Ni-YSZ cermet anode showed excellent electric performance at lower operating temperatures less than 800°C. The homogeneous morphology consisted of fine Ni and YSZ grains would lead to increasing of the Three Phase Boundary (TPB) length for the electrochemical reaction, which contributed to the performance improvement of the cermet anode.

INTRODUCTION

The SOFC has been widely studied and developed as the most promising fuel cell option for the distributed power market because of its high energy conversion efficiency and the low NO_x and SO_x emissions. Recent trend on SOFC development has focused on the cell operating temperatures less than 800°C to allow for the use of low-cost metallic interconnects and to improve the long-term stabilities of cells and stacks. However, lower operating temperature can cause high polarization losses for both SOFC anodes and cathodes. Therefore, many researches have been conducting with an aim to increase the electrochemical activities for both electrodes at the operating temperatures less than 800°C.

The Ni-YSZ cermet has been used as a conventional anode for SOFC. The electrochemical activity of the cermet anode strongly depends on its morphology, because the electrochemical reaction occurs at the TPB created by Ni grains, YSZ grains, and pores in the anode.[1-3] The electrochemical activity increases with an increase in the TPB length. Therefore, to achieve better anode performance, the cermet anode must have a homogeneous pore structure with highly dispersed fine Ni and YSZ grains in it.

In this paper, a novel processing technique in dry phase, called Mechano-Chemical Bonding (MCB) Technique[4], is applied for the preparation of NiO-YSZ composite powder to control the morphology of the Ni-YSZ cermet anode. This technique is defined as a novel method to create chemical bonds between particles in dry process without any binder. It can be used to control the morphology and composition of the multifunctional composite powders for various applications. During MCB processing, mechanical chemical and thermal interactions are caused to produce strong bonds between same or different kinds of particles. The aim of this study is to make high performance anode, which has homogeneous pore structure with highly dispersed fine Ni and YSZ grains. Relationships between the morphology and performance of the Ni-YSZ cermet anode processed by MCB will be discussed.

EXPERIMENAL PROCEDURE

Nickel oxide (NiO, 99.9%, Yamanaka chemical Co.) and 8-mol% Y_2O_3 stabilized ZrO_2 (YSZ, Tosoh Co., TZ-8Y) were used as starting materials. NiO and YSZ powder were processed by the Mechano-Chemical Bonding (MCB) technique to make multifunctional composite particles. In this experiment, NiO and YSZ weight ratio was selected as 65.6:33.4. NiO-YSZ powder obtained by these operations was observed by a scanning electron microscopy (SEM, Model S-3500N, Hitachi Ltd, Japan) with an energy dispersive analysis of X-ray (EDAX, Model EX-200, HORIBA Ltd., Japan). Particle size distribution was measured by laser diffraction and scattering method (MICROTRAC, Model HRA9320-X100, NIKKISO Co. Ltd., Japan). The powder sample was dispersed in distilled water before particle size measurement.

Then, NiO-YSZ powders were mixed with organic binder. They were printed onto one side of YSZ electrolyte pellet of 25 mm in diameter and 0.2 mm in thickness. The printed body was fired at 1350°C in air to produce the NiO-YSZ anode. After that, (La,Sr)MnO$_3$(LSM)-YSZ powder selected as a cathode material[5] was printed onto the other side of the YSZ electrolyte pellet, and the pellet was fired at 1200°C. Pt wire was used as the reference electrode. The anode polarization between the anode and the reference electrode were measured by the current interruption technique up to one A/cm^2 of current density. The single cell obtained by this experiment was operated at 800 and 700°C in the conditions of $H_2 - 3$ %H_2O for the anode and air for the cathode. A NiO-YSZ anode was reduced at 800°C in $H_2 - 3$ %H_2O to make a Ni-YSZ cermet anode. The morphology of Ni-YSZ cermet anode was observed by SEM and Auger Electron Spectroscopy (680S, ULVAC-PHI Inc.) after the cell test was finished.

RESULTS AND DISCUSSION

Figure 1 shows SEM photographs of NiO starting powder and NiO-YSZ composite powder processed by MCB. Table 1 shows their average particle size and specific surface area. Average particle size of NiO was 0.79 μm, however, the particle size distribution of the NiO powder was broad containing particles lager than a few microns. It is also clear from Fig. 1(a) that NiO powder contains strong agglomerations. On the other hand, average particle size of NiO-YSZ composite powder processed by MCB was 0.42 μm, and its particle size was relatively uniform without coarse

(a) 1µm **(b)** 1µm

Figure 1. SEM photographs of (a) NiO starting powder and (b) NiO-YSZ composite powder

Table 1 Average particle size and specific surface area of NiO, YSZ starting powder and NiO-YSZ composite powder

	NiO powder	YSZ powder	NiO-YSZ composite powder
Average particle size (µm)	0.79	0.08 *	0.42
BET specific surface aera (m^2/g)	5.5	13.0	7.4

*: value calculated from BET surface area

particles. It is also obvious from Fig. 1(b) that the MCB processed powder consists of NiO particles partially covered with fine YSZ particles. It is considered that NiO starting particles are ground down to finer particles, and fine YSZ particles are bonded on the surface of the finer NiO particles during MCB processing.

Figure 2 shows a SEM photograph of Ni-YSZ cermet anode fabricated from NiO-YSZ composite powder processed by MCB. It is evident from this figure that the Ni-YSZ cermet anode fabricated from MCB processed powder shows micron sized homogeneous porous structure, and consists of sub-micron sized Ni and YSZ fine grains. Figure 3 shows the elemental analysis result of AES for the Ni-YSZ cermet anode fabricated from the composite powder processed by MCB. Figure 3(a), (b) and (c) show SEM photograph, Ni elemental mapping and O elemental mapping, respectively. The O elemental

5µm

Figure 2. SEM photograph of Ni-YSZ cermet anode from NiO-YSZ fabricated composite powder processed by MCB.

mapping in Fig. 3(c) means the existence of YSZ. Figure 3(a) and (c) indicate fine YSZ grains less than 500 nm make excellent connections. It is also confirmed from Figure 3 that fine YSZ grains are partially embedded on the surface of Ni grains, and that Ni grains make good connections each other. Therefore, in the Ni-YSZ cermet anode fabricated from

Figure 3. Auger Electron Spectroscopy results of a Ni-YSZ cermet anode fabricated from NiO-YSZ composite powder processed by MCB, (a) SEM image, (b) Ni element mapping and (c) O element mapping.

NiO-YSZ composite powder processed by MCB, it is thought that YSZ grains and Ni grains achieve better connections each other, and that finer YSZ connections are formed on the surface of Ni network.

Figure 4 shows the anode polarization of Ni-YSZ cermet anode at the operating temperature of 700 and 800°C. Anode polarization of Ni-YSZ cermet anode fabricated from the composite powder processed by MCB is excellent in comparison with other published data[6-8] obtained at higher temperature operation (1000°C). Such data shows that the anode electrochemical polarization of 0.1 to 0.14 V could be achieved at the operating temperature of 800°C with current density of 0.5 to 1.0 A/cm^2. Moreover, it is clear form Fig. 4 that the anode electrochemical polarization at the operating temperature of 700°C slightly decreases in comparison with that at 800°C. The anode performance obtained here at 700 and 800°C in Fig. 4 is almost identical to those of other studies[6-8] obtained at 1000°C.

In the Ni-YSZ cermet anode fabricated by MCB technique, homogeneous porous structure brings better gas diffusion, and fine Ni and YSZ grains lead to increasing of the Three Phase Boundary (TPB) length for the electrochemical reaction. Such structure also may bring the better Ni and YSZ connection which form better

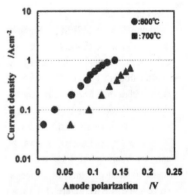

Figure 4. Anode polarization of Ni-YSZ cermet anode at 800°C(•) and 700°C(▲).

electron and ion path. It appears that the Ni-YSZ cermet morphology could result in an excellent anode performance at lower operating temperatures ($\square$800°C).

CONCLUSIONS

NiO-YSZ composite powder consisted of NiO particles partially covered with fine YSZ particles, was successfully processed by a novel processing technique called MCB in dry phase. Average particle size of the NiO-YSZ composite powder was 0.42 μm, and its particle size was relatively uniform without coarse particles. Ni-YSZ cermet anode fabricated by the NiO-YSZ composite powder showed an excellent anode performance at lower operating temperatures ($\square$800°C). The Ni-YSZ cermet anode showed micron sized homogeneous porous structure, and consisted of sub-micron sized Ni and YSZ fine grains. Moreover, in the cermet anode, it was thought that Ni and YSZ grains formed good connections. This advanced anode structure increased the TPB length for anode electrochemical reaction, and led excellent anode performance. The present study suggested that the novel processing method is promising in developing electrodes for lower temperature operation of SOFC.

ACKNOWLAGEMENT

This work was partially supported by Kinki-chiho innovation center.

REFERENCES

[1] T. Kawada, N. Sakai, H. Yokogawa, M. Dokiya, M. Mori and T. Iwata, "Structure and Polarization Characteristics of Solid Oxide Fuel Cell Anodes". *Solid State Ionics*, 1990, **40/41**, 402-408.

[2] N.Q. Minh, "Ceramic Fuel Cells." *J. Am. Cerm. Soc.*, 1993, **76** 563-588.

[3] J. Mizusaki, H. Tagawa, T. Saito, K. Kamitani, T. Hirano, S. Ehara, T. Takagi, T. Hikita, M. Ipponmatsu, S. Nakagawa, and K. Hashimoto, "Preparation of Nickel Pattern Electrodes on YSZ and Their Electrochemical Properties in H_2-H_2O Atmospheres." *J. Electorochem. Soc.*, 1994, **141**, 2129-2135.

[4] T. Fukui, K. Murata, H. Abe, M. Naito and K. Nogi, "Morphology and Performance of Ni-YSZ Cermet Anode Fabricated from NiO-YSZ Composite Powder for SOFCs." *Ceramic Transaction*, in press.

[5] T. Fukui, S. Ohara, M. Naito, and K. Nogi, "Morphology control of the electrode for SOFCs by using Nanoparticles," *J. Nanoparticle Research*, 2001, **3**, 171-174.

[6] H. Itho, T. Yamamoto, M. Mori, T. Watanabe, and T. Abe, "Improved Microstructure of Ni-YSZ Cermet Anode for SOFC with long Term Stability," *DENKI KAGAKU*, 1996, **64**, 549-554.

[7] T. Ioroi, Y. Uchimoto, Z. Ogumi, and Z. Takehara, "Preparation and Characteristics of Ni/YSZ Cermet Anodes by Vapor-phase Deposition," *Denki Kagaku*, 1996, **64** 562-567.

[8] N. Nakagawa, K. Nakajima, M. Sato, and K. Kato, "Contribution of Internal Active Three-Phase Zone of Ni-Zirconia Cermet Anodes on the Electrode Performance of SOFCs," *J. Electrochem. Soc.*, 1999, **146**, 1290-1295.

IMPROVED SOFC CATHODES AND CATHODE CONTACT LAYERS

F. Tietz, H.-P. Buchkremer, V. A. C. Haanappel, A. Mai, N. H. Menzler, J. Mertens,
W. J. Quadakkers, D. Rutenbeck, S. Uhlenbruck, M. Zahid, D. Stöver
Forschungszentrum Jülich GmbH
Institut für Werkstoffe und Verfahren der Energietechnik (IWV)
D-52425 Jülich, Germany

ABSTRACT

SOFC development at Forschungszentrum Jülich is aiming at high power density and high durability to achieve cost reduction in manufacturing and installation. For higher power density the work on materials development has been focused on improving the cathode performance by modification of the component thickness, the microstructure and the choice of materials. All investigations were first carried out with 5 x 5 cm^2 cells and, if appropriate, scaled up to cells having a size of 10 x 10 and 20 x 20 cm^2.

In the case of $(La,Sr)MnO_3$-based cathodes the power output was nearly doubled reproducibly from 0.56 to 1.0 W/cm^2 at 800°C and 0.7 V by varying the $(La,Sr)MnO_3$/YSZ ratio in the composite layer as well as the grain size in and the thickness of both the composite layer and the overlaid pure $(La,Sr)MnO_3$ layer. In addition, SOFCs with cathodes based on $(La,Sr)(Fe,Co)O_3$ have successfully been developed giving a power output of 1.2 W/cm^2 at the same experimental conditions. Similar performance was achieved by depositing the cathode using either wet powder spraying or screen printing. A prerequisite for comparing different cathodes is the reliable and quality-assured fabrication of all other cell components.

For improved durability the development of cathodic contact materials was directed towards a reduced interaction between the ceramic materials and the ferritic steel used as interconnect material. Several materials solutions are available in which the contact resistance remains constant below 70 $m\Omega$ cm^2 within the first 1500 h of testing. Currently, contact materials are under development that do not only show a reliable and low contact resistance but simultaneously act as a diffusion barrier for the chromium that is released from the interconnect.

INTRODUCTION

At Forschungszentrum Jülich the work on materials development in SOFC technology has been focused on improving the cathode performance during the recent two years. Initially, the investigations were related to the optimization of the widely used cathode based on $(La,Sr)MnO_3$ (LSM) by modification of the component thickness and the microstructure. In parallel, also other perovskite materials like $(La,Sr)(Fe,Co)O_3$ (LSFC) were characterized and then used for cell manufacturing. The perovskite materials were synthesized at Forschungszentrum Jülich by spray-pyrolysis of aqueous nitrate solutions [1]. All investigations were first carried out with 5 x 5 cm^2 cells and, if appropriate, scaled up to cells having a size of 10 x 10 and 20 x 20 cm^2.

The attempt to change the processing of the thick layers (anode, electrolyte, cathode) from coating processes like spraying and slip casting to screen printing technology (Table 1) has simultaneously led to a re-investigation of important parameters of the components, because different microstructures can be expected due to a) the change in amount and kind of orga additives in the slurries and pastes and b) the possibility to use much coarser powders with scr printing than with wet powder-spraying (WPS). In the case of cathode processing the composition the composite layer of the cathode was varied as well as the layer thicknesses and the particle size: the two cathode layers composed of LSM and 8 mol% yttria-stabilized zirconia (8YSZ, Tos Japan) and pure LSM, hereafter named cathode functional layer (CFL) and cathode current collect layer (CCCL), respectively. This variation of parameters aimed at reaching or improving performance of SOFCs with wet powder-sprayed (WPS) cathodes [2] by maintaining the sinter conditions. For a comparison with the formerly manufactured components, it is necessary to rely quality-controlled components, which has reached established procedures [3].

Table I. Actual and alternative processing techniques at FZJ [4]

Component	Actual processing technique	Alternative
Cathode	Wet Powder Spraying	Screen Printing
Electrolyte	Vacuum Slip Casting	Screen Printing
Anode	Vacuum Slip Casting	Screen Printing
Anode substrate	Warm Pressing	Tape Casting
Sealing	Automated Paste Dispensing	
Cathodic contact layer	Wet Powder Spraying	
Anodic contact layer	Spot welding	
Interconnect	Machining	Punching & Pressing

CATHODE OPTIMIZATION WITH LSM

The first parameter for optimization of screen-printed cathodes was the LSM/8YSZ ratio in CFL. A set of cells with LSM/8YSZ ratios ranging from 30/70 to 70/30 vol% of the solids w prepared and tested between 900 and 700°C (Figure 1). At 900°C the best performance of 1.8 A/ was found for a ratio close to 50/50. Towards both end members of the series, the current den decreased to about 1.1 A/cm^2 with a remarkable asymmetry along the LSM/8YSZ ratios. At 700 the highest performance of 0.55 A/cm^2 shifted to an LSM/8YSZ ratio of 45/55 but even for inverse ratio the performance remains at 0.46 A/cm^2. The asymmetric shape of the curves in Figu is mainly due to the temperature dependence of the ionic conductivity of 8YSZ and the catal activity of the LSM. At high operation temperatures a good oxygen conversion is achieved v LSM and as soon as the percolation threshold of the ion-conducting phase is reached a sharp incre in performance is observed. At lower temperatures, especially towards 700°C the reaction rat oxygen conversion and the ionic conductivity of the 8YSZ is decreased and its dependence composition is much smoother. Already this parameter variation led to an improvement in performance of more than 30 % compared with formerly manufactured WPS cathodes [2].

In a parallel set of experiments the particle size in the CCCL was varied from a d_{90} value of to about 20 μm. Because the pastes with larger grains contain fragments of hollow spheres due to preparation technique of the perovskite powder, the grain sizes are not satisfactorily described by particle size measurements. However, keeping this uncertainty in mind, the current density at 80 and 0.7 V increased continuously from 1.1 to 1.34 A/cm^2 with increasing d_{90} values from 2.5 to 8 of the LSM powder and an LSM/8YSZ ratio of 60/40. Only for even larger grain sizes with $d_{90} = 16$ μm the current density did not increase further. From the microstructures obtained it can be concluded that the increase of the particle size led to a higher porosity and hence better gas permeability in the cathode. Simultaneously the layers with the larger particles ($d_{90} \geq 8$ μm) showed a decrease in lateral conductivity limiting the current distribution in the CCCL. This is most likely the reason for the reduced performance using large grains. However, by changing the LSM/8YSZ ratio to 50/50 a similarly high performance of 1.46 A/cm² could be achieved using a CCCL paste containing an LSM powder with $d_{90} = 19$ μm.

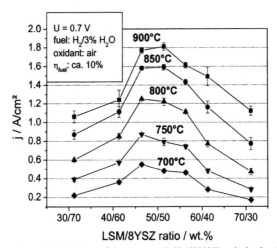

Fig. 1: Dependence of SOFC performance on LSM/8YSZ ratio in the CFL layer.

The layer thicknesses of the CFL and CCCL were varied between 10 and 35 μm and between 20 and 70 μm, respectively, using the LSM/8YSZ paste with LSM/8YSZ 50/50 ratio and the LSM paste with $d_{90} = 19$ μm. For each layer combination only one printing step for the CFL and CCCL was performed. The expected thickness was achieved with screens differing in mesh width and wire thickness. From these experiments it was concluded that
- the CFL thickness should be in the range of 10-25 μm,
- the CCCL thickness should be in the range of 30-70 μm,
- the thickness ratio CCCL/CFL should be in the range of 5:1 to 3:1.

In such cases a power density of about 1.4 A/cm² was obtained. A further increase of performance, however, was not achieved in the course of these experiments.

CATHODE OPTIMIZATION WITH LSFC

Several LSFC compositions were synthesized by spray-pyrolysis and used for cathode preparation. In all cases a $(Ce,Gd)O_2$ interlayer between 8YSZ and LSFC was applied to avoid the formation of $SrZrO_3$ at the interface. For this purpose it is aimed at applying a dense interlayer, but could not be obtained so far due to the constraint sintering conditions. According to the form $La_{1-x-y}Sr_yFe_{1-z}Co_zO_{3-\delta}$ the compounds with x = 0, 0.02, 0.05, y = 0.2, 0.4, and z = 0.2 were test Due to the different compositions the transport properties and the sintering behavior of the powc varied significantly. For the sake of better comparison, the microstructures of the cathodes w therefore adjusted to a similar grain size after sintering [5,6]. Additionally, the sintering of $(Ce,Gd)O_2$ interlayer was kept at 1300°C to avoid unfavorable interaction between 8YSZ $(Ce,Gd)O_2$ [7].

Among the LSFC compositions the best performance was achieved so far v $La_{0.58}Sr_{0.4}Fe_{0.8}Co_{0.2}O_{3-\delta}$ (Figure 2). Compared with the LSM-based cathodes the difference performance at 800°C is much smaller (1.8 vs. 1.46 A/cm^2) than at 700°C or below. The at specific resistance of the cell with this LSFC cathode was 0.18, 0.22, 0.30, 0.50 and 0.94 Ω cm 800, 750,700,650 and 600°C, respectively. These values indicate a remarkable change in apparent activation energy between 600-700 and 700-800°C, which might be related to the ste: increase of oxygen release [8] and oxygen vacancy concentration [9] with increasing temperatu The high performance of $La_{0.58}Sr_{0.4}Fe_{0.8}Co_{0.2}O_{3-\delta}$ and the dependence of electrochemi performance on stoichiometry variations tentatively corresponds with catalytic measurements of uptake and release of oxygen on these perovskites [8,10].

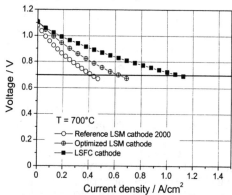

Fig. 2: Current-voltage curves of a formerly used LSM cathode,
an optimized LSM cathode and an LSFC cathode at 700°C.

CATHODE CONTACT MATERIALS

In addition to a cathode with high performance it is necessary to transfer the performance fc cell tests to stack tests and hence to guarantee a good contact between the cathode and interconnect material. Recently the development of cathodic contact materials was directed towarc reduced interaction between the ceramic materials and the ferritic steel used as interconnect mater In a previous work it was shown that even very small compositional changes in the ferritic steel cause significant changes in contact resistance in the long-term [11]. In several previous report was also shown that the use of alkaline earth-containing perovskites led to the formation of alka earth chromates at the ceramic/steel interface, e.g. [12]. Therefore an interlayer of cobalt ox between the metallic interconnect material (either Cr-based alloy or ferritic steel) and the ceramic as already proposed in [12] reduces the interaction, but the contact resistances obtained were rather high (see Figure 3). With the recently developed steel for SOFC application [13], the so-called Crofer22 APU, such interlayers gain new interest, because it is possible to develop cobaltite-based contact layers that do not only show a reliable and low contact resistance within several thousand hours of testing (Figure 3) but simultaneously may act as a diffusion barrier for the chromium that is released from the interconnect (Figure 4).

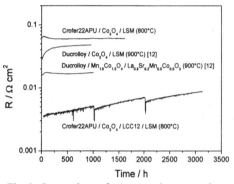

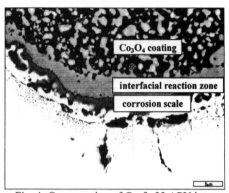

Fig. 3: Comparison of contact resistances using Ducrolloy (taken from [12]) and Crofer22 APU interconnect materials in combination with various spinel interlayers and perovskites.

Fig. 4: Cross-section of Crofer22 APU interconnect material coated with Co_3O_4 and exposed to 800°C for 500 h. Besides a corrosion scale (dark gray) a highly dense interfacial reaction zone (light gray) was formed.

CONCLUSIONS

LSM-based cathodes were improved in performance and reached 1.46 A/cm^2 at 800°C and 0.7 V. LSFC cathodes performed even better and are excellent materials for operation temperatures in the range of 650-750°C. Using the new ferritic steel interconnect material, the application of ceramic interlayers between steel and contact material or cathode resulted in a) low and stable contact resistance and b) a possibility to reduce effectively the release of volatile Cr species.

ACKNOWLEDGEMENTS

The authors gratefully thank all members of Jülich's SOFC development team. Financal support from the German Federal Ministry of Economics and Technology (BMWi) under contract no. O327088C/8-1A and the German Federal Ministry of Education and Science (BMBF) in the frame of the networking project 'Renewable Energies' under contract no. 01SF0039 is gratefully acknowledged.

REFERENCES

[1] P. Kontouros, R. Förthmann, A. Naoumidis, G. Stochniol, E. Syskakis, "Synthesis, Form and Characterization of Ceramic Materials for the Planar Solid Oxide Fuel Cell (SOFC)", *Ionic* 40-50 (1995)

[2] L. G. J. de Haart, I. C. Vinke, A. Janke, H. Ringel and F. Tietz, "New Developments in St Technology for Anode Substrate Based SOFC"; pp. 111-119 in *Proc. 7th Int. Symp. Solid O: Fuel Cells (SOFC-VII)*. Edited by H. Yokokawa and S. C. Singhal. The Electrochemical Soci Pennington, NJ, 2001.

[3] N. H. Menzler, G. Blaß, S. Giesen, H. P. Buchkremer, "Processing and Quality Contro' Planar SOFC Components", *Mat.-wiss. u. Werkstofftech.* 33 367-371 (2002)

[4] F. Tietz, H. P. Buchkremer and D. Stöver, "Components Manufacturing for Solid Oxide Cells," *Solid State Ionics* 152-153 373-381(2002).

[5] A. Ahmad-Khanlou, F. Tietz, I.C. Vinke and D. Stöver, "Electrochemical and Microstruct Study of SOFC Cathodes Based on $La_{0.65}Sr_{0.3}MnO_3$ and $Pr_{0.65}Sr_{0.3}MnO_3$"; pp. 476-484 in *Proc. Int. Symp. Solid Oxide Fuel Cells (SOFC-VII)*. Edited by H. Yokokawa and S. C. Singhal. Electrochemical Society, Pennington, NJ, 2001.

[6] A. Mai, V. A. C. Haanappel, F. Tietz, I. C. Vinke and D. Stöver, "Microstructural Electrochemical Characterisation of LSFC-Based Cathodes for Anode-Supported Solid Oxide Cells"; pp. 525-532 in *Proc. 8th Int. Symp. Solid Oxide Fuel Cells (SOFC-VIII)*. Edited by S Singhal and M. Dokiya. The Electrochemical Society, Pennington, NJ, 2003.

[7] A. Tsoga, A. Naoumidis and D. Stöver, "Total electrical conductivity and defect structur ZrO_2-CeO_2-Y_2O_3-Gd_2O_3 solid solutions", *Solid State Ionics* 135 403-409 (2000)

[8] A. Mai, "Katalytische und elektrochemische Eigenschaften von eisen- und kobalthalti Perowskiten als Kathoden für die oxidkeramische Brennstoffzelle (SOFC)", PhD thesis, R Universität Bochum, 2004, in press

[9] L.-W. Tai, M. M. Nasrallah and H. U. Anderson, "Thermochemical Stability, Electr Conductivity, and Seebeck Coefficient of Sr-Doped $LaCo_{0.2}Fe_{0.8}O_{3-\delta}$", *J. Solid State Chem.* 118 1 124(1995)

[10] A. Mai, F. Tietz and D. Stöver, "Release and uptake of oxygen in mixed-conducting SC cathode materials measured by temperature-programmed methods," *Ionics* 9 189-194 (2003)

[11] R. N. Basu, F. Tietz, O. Teller, E. Wessel, H. P. Buchkremer and D. Stöver, "$LaNi_{0.6}Fe_{0.4}O$ Cathode Contact Material for Solid Oxide Fuel Cells," *J. Solid State Electrochem.* 7 416-420 (200

[12] Y. Larring and T. Norby, "Spinel and Perovskite Functional Layers Between Plansee Met Interconnect (Cr-5 wt% Fe-1 wt% Y_2O_3) and Ceramic ($La_{0.85}Sr_{0.15})_{0.91}MnO_3$ Cathode Materials Solid Oxide Fuel Cells", *J. Electrochem. Soc.* 147 3251-3256 (2000).

[13] J. Pirón Abellán, V. Shemet, F. Tietz, L. Singheiser, W. J. Quadakkers and A. Gil, "Fer Steel Interconnect for Reduced Temperature SOFC"; pp. 811-819 in *Proc. 7th Int. Symp. Solid O: Fuel Cells (SOFC-VII)*. Edited by H. Yokokawa and S. C. Singhal. The Electrochemical Soci Pennington, NJ, 2001.

CHARACTERIZATION OF SOLID OXIDE FUEL CELL LAYERS BY COMPUTED X-RAY MICROTOMOGRAPHY AND SMALL-ANGLE SCATTERING

A.J. Allen and T.A. Dobbins
National Institute of
Standards and Technology
100 Bureau Drive
Gaithersburg, MD 20899

J. Ilavsky
NIST / Department of
ChemicalEngineering
Purdue University
West Lafayette, IN 47907

F. Zhao and A. Virkar
Department of
Civil Engineering
University of Utah
Salt Lake City, UT 84112

J. Almer and F. DeCarlo
X-ray Operations and Research (XOR)
Advanced Photon Source
Argonne National Laboratory
Argonne, IL 60439

ABSTRACT

A combination of advanced x-ray synchrotron-based research tools is presented that shows promise in providing an improved understanding of how solid oxide fuel cell (SOFC) microstructure and chemistry may be controlled through processing, and how their evolution or degradation during service life changes SOFC performance properties. Results are discussed to show how the representative void microstructures may be characterized and quantified through the anode, electrolyte and cathode layers for generic SOFC systems. Structural phase analysis is also demonstrated for sample volumes closely corresponding to those for which the void microstructures are quantified. This paper both summarizes results and elucidates how continued development of such methods could benefit SOFC design.

INTRODUCTION

In solid oxide fuel cell (SOFC) development, control of the microstructure and chemistry in the various component layers, particularly near to and at the interface with the thin electrolyte layer, is of primary importance in determining SOFC performance and cost [1,2]. Since microstructure and chemistry are significantly affected by materials choice and processing, good void and phase microstructure characterization are desirable. The properties of such functional gradient materials depend on a complex superposition of spatial gradients in void morphology and structural phase composition (where the void feature size extends from tens of micrometers down to nanometers). Quantifying these gradients more comprehensively could facilitate an increased exploitation of the underlying phenomena that govern SOFC material properties.

In this connection, a combination of x-ray synchrotron-based research tools is being developed at a 3rd generation hard x-ray synchrotron source, the Advanced Photon Source (APS), Argonne, IL. Ultimately, the aim is to provide a 3D quantitative and *statistically-representative* characterization of the SOFC void and phase microstructures as a function of position throughout the system, and to determine how these change during service life. Specifically, such characterization is achievable using a combination of ultrasmall-angle x-ray scattering (USAXS) [3,4], high-energy small-angle x-ray scattering (HE-SAXS) with associated high-energy wide-angle x-ray scattering (HE-WAXS) [5,6], and computed x-ray microtomography (XMT) [7]. USAXS provides absolute volume fraction void size distributions over much of the scale range of interest, together with surface area information. HE-SAXS offers 5-micrometer spatial resolution for characterization of any fine (nm) features present, as well as surface area determination. HE-WAXS provides complementary phase composition data, also at 5 μm spatial resolution. XMT provides 3D visualization of the coarse features and their interconnectivity; the current feature resolution size is 1.5 μm. In the sections that follow, each technique is briefly described, and results are reported and compared to illustrate how the methods could be used together to obtain a comprehensive picture of the SOFC systems investigated.

EXPERIMENTAL

Materials:

The SOFC samples studied comprised sections cut from complete SOFC's with a porous $La_{0.8}Sr_{0.2}MnO_3$ (LSM) cathode, a denser cathode interlayer, a fully-dense yttria-stabilized zirconia (YSZ) electrolyte layer, and a porous Ni/YSZ cermet anode [2]. A range of electrolyte and cathode interlayer thicknesses, and anode porosities, were investigated. For the USAXS, HE-SAXS and HE-WAXS measurements, 150 μm – 200 μm-thick sections were cut perpendicular to the electrode/electrolyte interfaces. For XMT, coupons of approximate dimension 200 μm x 200 μm x 1 mm were cut with the long dimension parallel to the interfaces.

USAXS Microstructure Characterization:

The small-angle scattering from inhomogeneities in the density of a material (e.g., voids), may be analyzed to give their size, shape and volume fraction [8]. Whereas in conventional SAXS the small-angle scattered intensity is measured on a 2D CCD detector, in USAXS the scattering is determined by measuring the intensity passed through a crystal monolith, placed behind the sample, as it is rotated out of the Bragg condition [3]. Because the angular width of the crystal rocking-curve is much less than that provided by conventional geometric collimation, smaller scattering angles and correspondingly larger scattering features are accessible. In fact, USAXS extends the range of measurable sizes to above one micrometer without compromising the detection of small (nm) scattering features. Here, USAXS data were obtained using the NIST-built USAXS instrument at UNICAT, sector 33-ID of the APS [3.4]. At the x-ray energy used of 16.9 keV, some multiple scattering corrections were required, but the scattering intensities could then be absolutely calibrated with respect to the incident beam intensity without need for a scattering standard. On analyzing the corrected data using the maximum entropy algorithm, MaxEnt [9], absolute volume-fraction size-distributions were obtained over much of the scale range of interest, together with total void volume fractions. From the final slope of the scattering at large scattering angles, Porod surface area values were also obtained [8]. The high brilliance of the 3rd generation synchrotron radiation source permits a continuing improvement in the USAXS instrumental design. In this case, the scattered intensity is sufficient to reduce the vertical slit size so as to allow volume fraction and surface area determination, as a function of position, to be obtained at 10 μm spatial resolution.

Fig. 1a shows USAXS results for the spatial variation perpendicular to the SOFC layers of both the total void surface area and the total void volume fraction for a SOFC with a 20 μm electrolyte layer and a 20 μm thick cathode interlayer. The surface area and volume fraction curves were derived from the absolute-calibrated scattering data within the different SOFC layers using appropriate contrast terms [3,4]. The statistical uncertainties are comparable with the size of the data points. Local microstructure variation may add to these uncertainties. Fig. 1b shows typical MaxEnt-derived void size distributions for the cathode and anode layers [9]. Fig. 1a suggests the measured void volume fraction and surface area would decrease to near zero in the electrolyte with a fine enough beam size and an ideal sample alignment. While increased porosity in the anode away from the interfaces is indicated, the full 48 % anode porosity is not detected because much of this occurs in voids significantly larger than 1 μm in diameter (as indicated below in the XMT results) that are not detectable by SAXS, and hence not included in the size distribution plots of Fig. 1b. However, strong variations in the surface area and measured void volume fraction close to the electrode/electrolyte interfaces are indicated. Such data can be correlated, quantitatively, with variations in the phase composition, and other information, to predict cell properties.

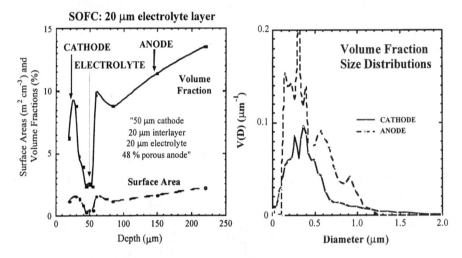

Figure 1. (a) Spatial variation, perpendicular to the SOFC interfaces, of the void volume fraction and total surface area, as measured by USAXS. (b) Typical MaxEnt volume fraction size distributions derived from USAXS data for the cathode and anode layers.

Correlated HE-SAXS Microstructure and HE-WAXS Phase Characterization:

In HE-SAXS, the small-angle scattering intensity is measured using a conventional geometry with a 2D CCD detector and is absolute-calibrated using a scattering standard, which has been defined here through previous USAXS measurement. These experiments were performed at XOR, Sector 1 of the APS, using 80.7 keV x-rays [5]. A sufficient flux of high-energy x-rays is achieved using an undulator and brilliance-preserving monochromator consisting of two bent Si crystals [6]. The high x-ray energy greatly reduces both sample absorption and multiple scattering. Furthermore, the high brilliance enables the beam-defining slits to be reduced to 50 μm wide x 5 μm high. As configured here, HE-SAXS is suited to the characterization of fine features, less than 50 nm in size, and to total surface area measurement. In the present studies, HE-SAXS was used to determine the total surface area of the voids through each layer as a function of position at 5-μm spatial resolution. In Fig. 2a surface areas derived from HE-SAXS data are presented for a SOFC with a 16 μm electrolyte layer and a 20 μm thick cathode interlayer. These data were then compared with the phase composition variation through the layers, as measured by HE-WAXS.

The high x-ray energy of the HE-SAXS instrument means it can be combined with the use of a large MAR345* image-plate detector to obtain diffraction data covering much of the conventional powder diffraction range in a single exposure. Here, characterization of the phase composition by HE-WAXS was obtained, both as a function of position at 5 μm spatial resolution, and at sample positions corresponding to those at which HE-SAXS was measured. Diffraction patterns at a series of such positions through the layers of the SOFC with the 16-mm thick electrolyte layer are presented in Fig. 2b. Some major peaks are identified for the main phases within each SOFC layer. Combined use of HE-SAXS and HE-WAXS allows the spatial variation of the surface area (hence microstructure) and the phase composition to be quantitatively compared through the SOFC layers and across the interfaces between them. Not shown here are variations with position in the uniformity of the Debye-Scherrer rings recorded in the 2D HE-WAXS data, indicating that texture variations may also be significant close to the electrode-electrolyte interfaces. This may be associated with the complex features apparent in Fig. 2b within the electrolyte layer in the vicinity of the YSZ (111) peak. Broader comparisons can also be made between the HE-SAXS / HE-WAXS data and the void size distribution and porosity variations determined by USAXS.

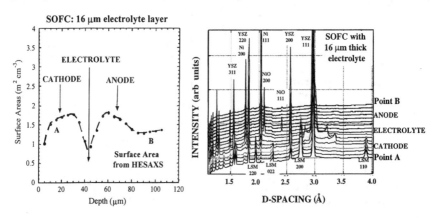

Figure 2. (a) Spatial variation, perpendicular to the SOFC interfaces, of the void total surface area measured by HE-SAXS. (b) HE-WAXS data at successive positions through the SOFC layers with a 20 μm cathode interlayer, a 16 μm electrolyte, and a 48 % porous anode. [Presence of NiO in the anode may be an artefact of sample preparation or cell shutdown.]

Computed x-ray microtomography (XMT):

Coarse features significantly larger than 1 μm are not usually included in SAXS analyses. Furthermore, for purposes of modeling or prediction of the thermo-mechanical or transport properties, the quantitative microstructure and phase data obtained from SAXS and HE-WAXS need to be combined with a 3D density map distribution in order to obtain information on the tortuosity or interconnectivity of the void morphology. To provide this information and characterize the SOFC layer microstructures at length scales significantly above 1 μm, as well as

---

* Information on commercial products is given for completeness and does not necessarily constitute or imply their endorsement by the National Institute of Standards and Technology.

evaluate the integrity of the thin electrolyte layer, XMT studies were also made on the SOFC sections. In XMT measurements, absorption contrast images are obtained for a thin needle-shaped (200 μm x 200 μm) coupon as a function of sample orientation with respect to the incident beam. A visible light camera views and magnifies the x-ray image on a fluorescent screen. Through image reconstruction, a 3D visualization of the sample can then be obtained from the large series of 2D absorption contrast images.

Using the XMT facility at XOR sector 2-BM of the APS [7], the void microstructure through the SOFC layers for voids on the μm-scale and above was visualized. The feature resolution was 1.5 μm for the present measurements, although finer feature resolution will be achievable in future as x-ray optics between the sample and fluorescent screen are progressively introduced to magnify the x-ray image. Figure 3 shows a 3D image reconstruction for a typical section of the SOFC with a 20 μm thick electrolyte and a 20 μm-thick cathode interlayer. The porous anode and cathode regions separated by a dense electrolyte and cathode interlayer are distinguishable. While such images have their own value, increasingly sophisticated image processing algorithms are being developed to allow microstructure parameters (e.g., void surface area, volume fraction and inter-connectivity) to be extracted from the 3D XMT images as a function of position. In principle, image evaluation algorithms such as the "median axis" method [7] can be used to provide these quantitative microstructure data which can then be combined with the SAXS characterization of the finer void microstructures, and with HE-WAXS phase analysis.

CATHODE

20 μm THICK ELECTROLYTE

ANODE

70 μm

Figure 3. 3D XMT image reconstruction of a SOFC section (20 μm electrolyte layer, 20 μm cathode interlayer. Coarse voids are present in the cathode and anode but not in the electrolyte.

DISCUSSION AND CONCLUSIONS

Representative void microstructures have been characterized and quantified through the anode, electrolyte and cathode layers for a selected group of generic SOFC systems using USAXS and HE-SAXS. The corresponding structural phase variations have also been determined by HE-WAXS, with sample volumes and positions closely corresponding to those used for HE-SAXS. These data have also been complemented by XMT imaging of the coarse void morphology. It has been demonstrated that the spatial resolution is sufficient to detect or infer, nondestructively, some of the abrupt microstructure and phase composition changes at the SOFC layer interfaces.

Use of these methods for the quantitative comparison of microstructure and phase in the cathode, as a function of composition, morphology and doping, and their evolution during service life, would be valuable in optimizing cathode design. Similarly, the effects of corrosion in the anode can be explored, together with chemical effects on composition and morphology at the electrode/electrolyte interfaces. It is proposed that the combination of experiments discussed here can provide a sufficiently improved quantitative understanding of the processing – microstructure – property relationships generic to SOFC systems, that it would significantly contribute to improved SOFC designs.

ACKNOWLEDGEMENTS

Use of the Advanced Photon Source (APS) was supported by the U.S. Department of Energy (DOE), Office of Science, Office of Basic Energy Sciences, under Contract No. W-31-109-ENG-38. The UNICAT facility at the APS is supported by the University of Illinois at Urbana-Champaign, Materials Research Laboratory (U.S. DOE, the State of Illinois-IBHE-HECA, and the NSF), the Oak Ridge National Laboratory (U.S. DOE under contract with UT-Battelle LLC), the National Institute of Standards and Technology (U.S. Dept. of Commerce), and UOP LLC.

REFERENCES
[1] N.Q. Minh; "Ceramic Fuel Cells," J. Am. Ceram. Soc.,76, 563-588 (1993).
[2] Y. Jiang, A.V. Virkar; F. Zhao; "The Effect of Testing Geometry on the Measurement of Cell Performance in Anode-Supported Solid Oxide Fuel Cells - The Effect of Cathode Area," J. Electrochem. Soc., 148, A1091-A1099 (2001).
[3] J. Ilavsky, A. J. Allen, G. G. Long and P. R. Jemian; "Effective Pinhole-Collimated Ultrasmall-Angle X-ray Scattering Instrument for Measuring Anisotropic Microstructures," Rev. Sci. Instrum., 73, 1660 - 1162 (2002).
[4] J. Ilavsky, P.R. Jemian, A.J. Allen, G.G. Long; "Versatile USAXS (Bonse-Hart) Facility for Advanced Materials Research," Proc. 8th Int. Symp. Sync. Rad. Instrum., American Institute of Physics, in press (2004).
[5] X-L. Wang, J. Almer, C.T. Liu, Y.D Wang, J.K. Zhao, A.D. Stoica, D.R. Haeffner and W.H. Wang, "In situ synchrotron study of phase transformation behaviors in bulk metallic glass by simultaneous diffraction and small angle scattering," Phys. Rev. Lett., in press (2004).
[6] S.D. Shastri, K. Fezzaa, A. Mashayekhi, W-K. Lee, P.B. Fernandez, P.L. Lee, "Cryogenically Cooled, Bent Double-Laue Monochromator for High-Energy Undulator X-Rays (50-200 keV)," J. Synchrotron Rad. 9, 317-322 (2002).
[7] Y.X. Wang, F. De Carlo, D.C. Mancini, I. McNulty, B. Tieman, J. Bresnahan, I. Foster, J. Insley, P. Lane, G. von Laszewski, C. Kesselman, M.H. Su, M. Thiebaux; "A High-Throughput X-ray Microtomography System at the Advanced Photon Source," Rev. Sci. Instrum., 72, 2062-2068 (2001).
[8] G. Porod; "General Theory," pp. 17-51 in Small-Angle X-ray Scattering, edited by O. Glatter and O. Kratky, Academic Press, London, 1982.
[9] J.A. Potton, G.J. Daniell, and B.D. Rainford; "Particle Size Distributions from SANS Data Using the Maximum Entropy Method," J. Appl. Cryst., 21, 663-668 (1988).

KINETICS OF THE HYDROGEN REDUCTION OF NiO/YSZ AND ASSOCIATED MICROSTRUCTURAL CHANGES

Miladin Radovic, Edgar Lara-Curzio, Beth Armstrong, Larry Walker, Peter Tortorelli and Claudia Walls
Metals and Ceramics, Oak Ridge National Laboratory
Oak Ridge, TN 37831-6069

ABSTRACT

The kinetics of hydrogen (4%H_2-96%Ar mixture) reduction of porous 75mol%NiO/YSZ was investigated by thermogravimetry at 600°C, 700°C, 750°C and 800°C. In addition, the evolution of structural changes during reduction at 800°C were monitored. It was found that reduction of NiO occurs in two stages. In the first stage, linear reduction kinetics with an activation energy of 22 kJ/mol were observed. In the second stage, the extent of reduction with time was found to decrease significantly and the fraction of reduced NiO at which the transition from the first to the second stage occurred was found to increase with increasing temperature. Using optical and scanning electron microscopy and wavelength dispersive X-ray spectroscopy, it was found that the first stage of the reduction process is associated with the displacement of the reduction front across the thickness of the sample, whereas the second stage involves further reduction of NiO grains behind it. The transition between the first and second stages was found to occur when the reaction fronts meet at the middle of the sample. The porosity of the samples before and after reduction was determined by alcohol immersion and it was found that porosity increases linearly with the fraction of reduced NiO.

INTRODUCTION

At present, cermets consisting of interconnecting and interpenetrating Ni-metal and Y_2O_3-stabilized ZrO_2 (YSZ) phases are widely used as anode materials for Solid Oxide Fuel Cells (SOFCs)[1,2]. The porous Ni-YSZ anodes have good electronic conductivity, stability, catalytic properties and compatibility with other constituents of SOFC[1]. Ni metal is vital for the electrochemical reaction on the anode side of the SOFC, while YSZ provides support to the Ni particles and prevents them from sintering during service. Also, by tailoring the phase concentrations the mismatch between the coefficients of thermal expansions of the anode and a YSZ electrolyte can be minimized. Porosity of Ni-based anodes as high as 50 vol% makes it possible to obtain high external current densities and enables hydrogen to infiltrate the cermet and H_2O to escape[3]. In most cases porous Ni-based anodes are fabricated starting with a NiO-YSZ precursor, and then co-sintered with the electrolyte and cathode. When hydrogen fuel gas is supplied to the cell for the first time, NiO-YSZ is converted into a Ni-YSZ cermet. During the reduction process the chemistry, microstructure and properties of the anode change and these changes will induce mechanical stresses in the SOFC. Since stresses introduced during the reduction step would have a significant effect on the reliability and durability of SOFC, it is necessary to understand the kinetics and structural changes associated with this process.

EXPERIMENTAL PROCEDURE

The samples used in this investigation were prepared by mixing 75mol% of NiO (J.T.Baker*, Phillipsburg, NJ) with YSZ (8mol%YSZ, TOSOH Corp.*, Grove City, OH) powder and 30vol% of organic pore former (rice starch, ICN Biomedicals*, Inc Irvine, CA). Green samples were prepared by tape casting layers that were subsequently laminated into ≈1 mm thick specimens. Discs (nominal diameter 25.4 mm) were hot-knifed from the assembled green tapes prior to sintering at 1450 °C in air for 2 hours. The relative porosity of the samples after sintering was found to be 23.5±0.6 vol% as determined by alcohol immersion[4].

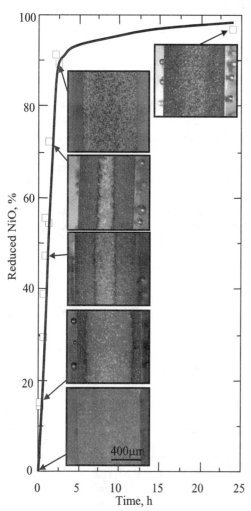

Fig. 1. Fraction of reduced NiO vs. time for reduction at 800 °C. Selected but typical optical micrographs show the microstructure of samples with different fractions of reduced NiO. Dark gray layers on the both sides of the sample correspond to the reduced part of the samples.

The reduction of the samples was carried out in a thermogravimetric unit (Cahn 1000*) with a detection limit of 0.1 mg. Individual sample were placed on a quartz stand that was hung by a Pt suspension wire in the thermobalance. The stand and the sample was introduced into a quartz reaction tube, which in turn was placed in the furnace. The sample was heated under a constant flow of argon (99.999% purity) to the reaction temperature. No mass changes were observed during the heating period. After reaching thermal equilibrium, the gas was changed to a mixture of 4%H_2+96%Ar (min 99.95% pure) which was flowed at a constant rate of 100 cc/min for the duration of the reduction process. The sample mass was monitored continuously using an automatic data acquisition system. The reaction times reported here correspond to the time measured from the beginning of the reaction, i.e. from the moment when 4%H_2+96%Ar gas mixture was introduced into the reaction tube, to the time when the gas flow was changed again to pure argon. Reduction runs were carried out for different times at 800 °C to monitor microstructural changes as a function of reduction time. Also, some samples were fully reduced at 600, 650, 700, 750 and 800°C to investigate the effect of temperature on reduction kinetics. The relative porosity of partially and fully reduced samples was also determined by alcohol immersion[4].

RESULTS AND DISCUSSION

The chemical reaction associated with the reduction of NiO by hydrogen can be represented by:

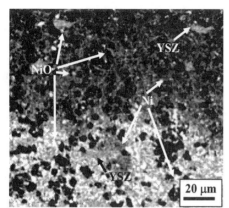

Fig. 2. Elemental surface analysis of a reaction front in partially reduced anode sample obtained using JEOL 8200* Electron Superprobe. Gray is YSZ, white is NiO and dark gray is Ni-metal. Black areas are pores.

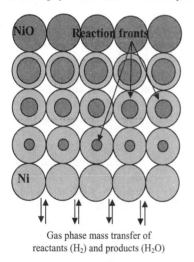

Gas phase mass transfer of reactants (H_2) and products (H_2O)

Fig. 3. Schematic representation of reduction process reaction according to grain model.

$$NiO(s) + H_2(g) \Rightarrow Ni(s) + H_2O(g) \qquad (1)$$

The mass change recorded during the reduction process was converted to the fraction of reduced NiO, r , namely the ratio of the instant mass change ΔW_t, to the theoretical final mass change, ΔW_{th}. The later was calculated for every sample using the following equation, assuming that no other reaction except reaction (1) occurred during the reduction process:

$$\Delta W_{th} = (m_O / m_{NiO}) \cdot W_{NiO} \qquad (2)$$

where W_{NiO} is the initial mass of NiO in the sample. m_O and m_{NiO} are the molecular mass of oxygen (≈ 16 amu) and of NiO (≈ 74.7 amu), respectively.

Fractions of the reduced NiO, r as a function of time from interrupted reduction tests at 800°C are presented as open squares in Fig. 1. Also in Fig.1 results of an uninterrupted test are printed which demonstrate a good reproducibility of the process. It was found that the curves exhibit two stages. In the first stage, the reduction is nearly linear with time until $\approx 85\%$ of NiO has been reduced. Beyond this transition point the amount of mass loss with time decreases gradually. Optical micrographs shown in Fig. 1 illustrate the development of the reduced layer on the surface of the samples during reduction at 800°C. Dark gray layers on both sides of the samples are reduced parts, while light gray areas represent unreduced parts. Those micrographs suggest that the transition occurs when the "reaction fronts" meet at the middle of the sample. However, this does not mean that all NiO is completely reduced within the gray area. The boundaries between reacted and unreacted zones were examined in more detail using scanning electron microscopy and wavelength dispersive spectrometry (WDS). Typical results of surface elemental analysis obtained using WDS are shown in Fig. 2. These analyses suggest that the boundary between reacted and unreacted zone is not sharp, but rather diffusive. NiO grains (white) were found in predominately reduced areas, while Ni-metal (dark gray) was also found in the predominately unreduced zone. These results suggest that reduction kinetics can be explained by grain model of Szekely, Evans, et al.[5,6,7,8] which is schematically presented in Fig. 3, where, for the sake of simplicity, the YSZ grains have been excluded and NiO grains are presented as spherical particles of uniform size. The Ni layer on the surface of the each grain is represented as a

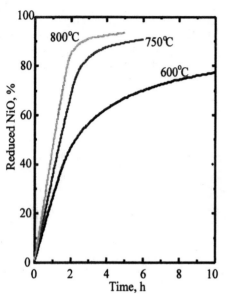

Fig. 4. Fraction of reduced NiO vs. reduction time at different temperatures.

continuous layer although it is likely to be discontinuous due to high volume changes during the reduction of NiO into the Ni.

According to the grain model, the gaseous reactant (i.e. hydrogen) diffuses from the surface of the sample through interstices and reacts with the spherical particles. Then, the reaction between the gas and the solid phase proceeds in accordance with the "shrinking core" model. Gas diffuses through the product layer (Ni) in the each grain and reacts at the spherical reaction interface, Fig. 3. The product gases thus generated diffuse back through the solid product and between the grains into the bulk gas steam. The resultant situation is depicted in Fig.3 showing rows of grains that have undergone a progressively lesser amount of reaction with the depth of penetration of the gas stream.

Results of previous studies[5,6,8,9] indicate that the hydrogen reduction of pure NiO can be successfully explained by the subject grain model assuming that the reaction rate is controlled by the rate of chemical reduction and/or diffusion of the gaseous phase through the pores. It was also shown that the effect of the diffusion of the reactive gas and products of the reaction through the reactive layer on the particle surface can be neglected. The present results suggest that this is the case during the extended first stage, when reaction kinetics are linear, but that diffusion through a tortuous gas path or in the solid state may be controlling thereafter as the amount of further reduction per unit time continually decreases (Fig. 2).

Selected but typical curves showing the fraction of reduced NiO as a function of reduction time at different temperatures are plotted in Fig. 4. The shape of those curves, as well as the fact that the reaction zone is diffusive suggests that the reduction process is controlled probably by both the

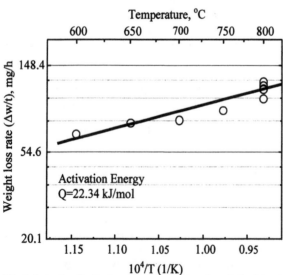

Fig. 5. Arrhenius plot for the reduction of Ni-based anode in 4% and 96% argon gas mixture.

rate of the chemical reaction and the diffusion of the gases through the pores. Additionally, reduction tests at 800°C, which were carried out at a higher gas flow rate of 300 cm³/min resulted in an initial mass loss rate of ≈360 mg/h, which is considerably higher than the weight loss rate of 101-123 mg/h that was obtained from the tests carried out at a gas flow rate of 100 cm³/min. This strongly suggests that the diffusion of the reduction gas through the pores and the gas phase mass transfer must play an important role in the reduction process consistent with the extended time of the first stage[5,9,10].

The initial weight loss rates obtained from the reduction of NiO-YSZ samples in hydrogen are plotted vs. the reciprocial reduction temperatures in Fig. 5. The activation energy of 22.3 kJ/mol calculated from the slope of the straight line on Fig. 5 is close to the activation energies of 21.8 and 18.0 kJ/mol that have been reported for the reduction of pure NiO in the same temperature range[11]. This suggests that the presence of YSZ phase in the examined samples does not have an effect on the mechanism of NiO reduction.

Figure 6 shows a plot of the relative porosity of test specimens, p, as a function of fraction of reduced NiO, r. These results indicate that the relative porosity of the test specimens increases with the fraction of reduced NiO. This is expected since the specific volume of metallic Ni is significantly smaller than that of NiO. Scanning electron micrographs of specimens (not shown here) also confirmed the significant increase in porosity with fraction of reduced NiO. In this case pores that were formed due to shrinkage of NiO particles during their reduction into Ni are much smaller ($\approx$ 1-2 μm) than those initially present in NiO-YSZ samples (10-15 μm).

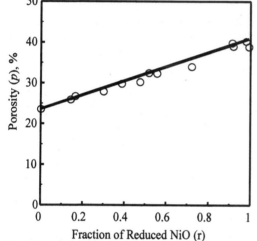

Fig. 6. Porosity of partially reduced samples as a function of fraction of reduced NiO. Solid line corresponds to Eq. (4).

If it is assumed that the decrease in the overall volume of the samples after reduction is negligble then the following expression can be derived for the changes in porosity as a function of fraction of reduced NiO:

$$p = p_o + \rho_o \ \overline{m}_{NiO}^0 (\frac{1}{\rho_{NiO}} - \frac{1}{\rho_{Ni}} + \frac{m_o}{m_{NiO}} \frac{1}{\rho_{Ni}})r \qquad (3)$$

where p_0, ρ_0, ρ_{Ni} and ρ_{NiO} are the average initial porosity, initial density of the samples, density of Ni and density of NiO, respectively, and $\overline{m}_{NiO}^0$ is the initial weight fraction of NiO in the NiO-YSZ. For the examined NiO-YSZ composite the initial weight fraction of NiO was $\overline{m}_{NiO}^0 = 0.587$. For $p_0=0.23$, $\rho_0=4.83$ g/cm³, $\rho_{Ni}=8.88$ g/cm³ and $\rho_{NiO}=6.67$ g/cm³ Eq.(3) yields:

$$p = 0.23 + 0.174 \cdot r \qquad (4)$$

Equation 4, which has also been plotted in Figure 6, confirms the good agreement between experimentally obtained changes of relative porosity after reduction of NiO.

SUMMARY

The reduction of 75mol%NiO/YSZ, with initial porosity of 23vol% in 4%H$_2$+96%Ar gas mixture and 600-800°C temperature range exhibits two stages: the first stage, which takes place through most of the reduction process occurs almost at a constant rate; during the second stage the rate of reduction decrease significantly with time. The transition between the first and second stages was found to occur when the reaction fronts meet at the middle of the sample. The diffusive nature of the reduction front suggests that the grain model can be used to describe the process and that the reduction is probably controlled by the rates of chemical reaction and diffusion of the gaseous phases through pores over most of the time while transport through the reduced phase and or tortuous gas paths plays a role in the latter stages. The value of the activation energy, ≈22 kJ/mol, which was obtained for the first stage of the reduction process is in agreement with published values for the reduction of pure NiO. This finding suggests that the presence of YSZ phase might not have a significant effect on the reduction of NiO. The porosity of the samples was found to increase linearly with the fraction of reduced NiO.

ACKNOWLEDGMENTS

This research work was sponsored by the US Department of Energy, Office of Fossil Energy, SECA Core Technology Program at ORNL under Contract DE-AC05-00OR22725 with UT-Battelle, LLC. The authors are grateful for the support of NETL program managers Wayne Surdoval, Travis Shultz and Donald Collins. The authors are indebted to Shawn Reeves of ORNL for help with thermogravimetric measurements.

REFERENCES

[1]Ming N. Q. and Takahashi T., "Science and Technology of Ceramic Fuel Cells", Elsevier, Amsterdam (1995)

[2]Zhu W.Z. and Deevi S.C., "A review on the Status of Anode Materials for Solid Oxide Fuel Cells", *Mater. Sci. Eng.* A326, 228 (2003)

[3]Morgensen M. and Skaarup S., "Kinetic and Geometry Aspects of Solid Oxide Fuel Cell Electrodes", *Solid State Ionics* 86-88, 151 (1996)

[4] ASTM standard C20-00

[5]Szekely J. and Evans J.W., "Studies in Gas-Solid Reactions: Part I. A Structural Model for the Reaction of Porous Oxides with a Reducing Gas", *Metall. Trans.* 2, 1961 (1971)

[6]Szekely J. and Evans J.W., "Studies in Gas-Solid Reactions: Part II. An Experimental Study of Nickel Oxide Reduction with Hydrogen", *Metall. Trans.* 2, 1969 (1971)

[7]Szekely J. and Evans J.W., "A Structural Model for Gas-solid Reactions with Moving Boundary", *Chem. Eng. Sci.* 25, 1091 (1970)

[8]Evans J.W., Song S. and Loen-Sucre C.E., "The Kinetics of Nickel Oxide Reduction by Hydrogen; Measurements in Fluidized Bed and in Gravimetric Apparatus", *Metall. Trans.* 7B, 55 (1976)

[9]Rashed A.H. and Rao Y.K., "Kinetics of Reduction of Nickel Oxide with Hydrogen Gas in the 230-452°C Range", *Chem. Eng. Sci.* 156, 1-30 (1996)

[10]Szekely J., Lin I. and Sohn H.Y, "A Structural Model for Gas-solid Reactions with Moving Boundary-V: An Experimental Study of the Reduction of Porous Nickel-Oxide Pallets with Hydrogen", *Chem. Eng. Sci.* 28, 1975 (1973)

[11]Sridhar S, Sichen D and Seetharaman S, " Investigation of the Kinetics of Reduction of Nickel Oxide and Nickel Aluminate by Hydrogen", *Z. Metallkd.* 85, 616 (1994)

ELASTIC PROPERTIES, EQUIBIAXIAL STRENGTH AND FRACTURE TOUGHNESS OF 8mol%YSZ ELECTROLYTE MATERIAL FOR SOLID OXIDE FUEL CELLS (SOFCs)

Miladin Radovic, Edgar Lara-Curzio, Rosa Trejo, Beth Armstrong and Claudia Walls
Metals and Ceramics, Oak Ridge National Laboratory
Oak Ridge, TN 37831-6069

ABSTRACT

The mechanical properties of single and multi-layer tape-cast 8mol% yttria-stabilized zirconia (YSZ) were characterized up to 800°C in air. Elastic properties were determined by Impulse Excitation and Resonant Ultrasound Spectroscopy as a function of porosity and temperature. Equibiaxial strength was determined by the Ring-on-Ring test method and the results were analyzed using Weibull statistics. Fracture toughness was determined by double torsion. It was found that the elastic moduli, characteristic strength and fracture toughness of YSZ samples decrease with testing temperature between 20°C and 600°C, but that they increased slightly between 600°C and 800°C. It was also found that the characteristic strength of multi-layer YSZ samples is always lower than that of single-layer samples and that the difference in characteristic strength values between single and multi-layer samples decreased with temperature. Fractographic analyses suggest that a possible explanation for the significantly lower biaxial strength of multi-layer samples is the occurrence of interfacial defects in multi-layer samples.

INTRODUCTION

Solid Oxide Fuel Cells (SOFC) are electrochemical devices that transform chemical energy of the fuel (hydrogen natural gas, etc) directly to electrical energy. As highly efficient and pollution-free energy sources SOFCs have been intensively studied during the past decade. Research activities related to SOFCs have been generally focused on electrochemical, thermal and microstructural properties of the different constituents, as well as on electrochemical performance and processing of different SOFC geometries[1]. Although the reliability of SOFCs is strongly dependent on the ability of SOFC constituents to withstand mechanical stresses that arise during processing and service, limited work has been reported on mechanical properties of SOFC constituents.

Zirconia stabilized with 8mol% of yttria (YSZ) is widely used as to electrolyte of SOFCs. Yttria (Y_2O_3) is added to zirconia (ZrO_2) to: (a) improve ionic conductivity, and (b) stabilize the cubic structure. The objective of this work was to determine the effect of temperature on elastic moduli, biaxial temperature and fracture toughness of the tape cast YSZ. In addition, the effect of porosity on elastic moduli at ambient temperatures, as well as effect of specimen thickness on biaxial strength between 25 and 800 °C were investigated.

EXPERIMENTAL PROCEDURE

The YSZ material examined in this study were prepared using ZrO_2 stabilized with 8 mol% Y_2O_3 (TOSOH Corp., Somerville, NJ) powder. Green samples were prepared by tape casting ≈250 μm thick single layers. Two or four green tapes were additionally laminated to prepare samples of different thickness. Discs for biaxial testing and determination of elastic properties with nominal diameter of 25.4 mm were hot-knifed from the 1-, 2- and 4-layers green tapes and sintered at 1400 °C in air for 2 hours. Rectangular samples for double torsion testing were cut form 4-layers tapes previously sintered also at 1400 °C in air for 2 hours. The relative porosities of all examined samples were determined by alcohol immersion according to ASTM standard C20-00.

Young's, E and shear, G moduli were determined by Impulse Excitation (IE) at ambient temperature and by Resonant Ultrasound Spectroscopy (RUS) in the 25-1000 °C temperature range. IE was performed according to ASTM standard C1259-98. This technique allows for the determination of dynamic elastic moduli from the resonant frequencies of a mechanically excited sample. On the other hand, RUS allows for the determination of dynamic elastic moduli at different temperatures from the resonant spectra of the sample excited by ultrasound waves of constant amplitude and varying frequency[2].

The equibiaxial flexural strength of YSZ specimens was determined by the ring-on-ring test method according to ASTM standard C1499-03. One, two or four layers disk samples with diameter (D) of ≈25 mm were evaluated using a loading ring with a diameter (D_l) of 5.5 mm and a supporting ring with diameter (D_s) of 20 mm. The tests were carried out at a constant crosshead displacement rate of 1 mm/min until fracture.

Fracture toughness, K_{IC}, was determined according to the double torsion test method[3,4]. Initial notches 1-mm wide and 12.5-mm long were cut into one side of the rectangular double torsion samples (20-mm wide, W, 40-mm long, L, and ≈1-mm thick, t) using a circular diamond blade. The notched test specimens were pre-at a rate of 0.02 mm/min. Pre-cracked samples were fast-fractured at a loading rate of 1 mm/min.

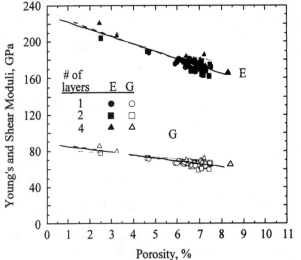

Fig. 1. Young's and shear moduli of YSZ electrolyte at ambient temperature as a function of porosity.

RESULTS

Elastic moduli.
Figure 1 shows the RT elastic moduli E and G as a function of porosity for YSZ. The modulus-porosity relationships determined in this study using linear and exponential relations are shown as separated lines in Fig. 1., while the zero-porosity constants (E_0 and G_0) and porosity dependence constants (b_E and b_G) obtained by least squares regression are given in Table I. These results are in good agreement with previously published data[5].

Young's and shear moduli obtained by RUS for a 8.48 vol% porous, 4-layer YSZ sample are plotted in Fig 2 as a function of temperature. Both, E and G initially decrease with temperature between 25 and 600°C. Above 600°C, E and G increase slightly with temperature up to 1000°C. Similar behavior has been reported [6] for pressureless sintered and hot pressed

Table I. The best fit values for zero-porosity moduli, M_0 and porosity dependant constant, b_M

	E_0, GPa	G_0, GPa	b_E	b_G
Linear	229.85	88.24	3.80	3.69
Expon.	234.64	90.20	4.35	4.51

Linear: $M = M_0(1 - b_M p)$

Expon.: $M = M_0 \exp(-b_M p)$

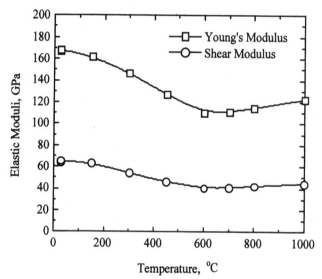

Fig. 2. Elastic moduli of the 8.48 vol% porous YSZ electrolyte as a function of temperature.

6.5 mol% yttria stabilized zirconia. In those cases a sharp decrease in modulus with temperature was also reported in the 25-400°C temperature range, but no changes in modulus were reported between 400 and 1000°C.

Biaxial Strength.

The biaxial strength results for YSZ samples with 1, 2 and 4 layers evaluated at 25°C, 600°C and 800°C were analyzed using Weibull statistics[7]. The results of the Weibull analysis are summarized in Table II, while the characteristic Weibull strength values are plotted in Fig. 3 as a function of temperature. These results show initial decrease in Weibull characteristic biaxial strength with temperature between 25 and 600°C. Above 600°C the characteristic strength

Table II. Results of biaxial tests for YSZ.

	T, °C	#	Characteristic Strength*, MPa	Weibull modulus*	Strength**, MPa
4-layers	25	23	222.1 (200.6,245.1)	3.7 (2.9, 4.6)	201.5±56.5
	600	21	132.8 (123.7,142.1)	5.6 (4.2, 7.3)	123.1±23.5
	800	21	145.1(131.6,159.1)	4.1 (3.0, 5.4)	131.4±37.2
2-layers	25	15	176.2(159.5,193.5)	4.9 (3.4, 6.7)	161.1±39.1
	600	18	109.7(91.0,130.7)	2.9 (1.9, 4.0)	97.4±38.6
	800	12	146.1 (132.4, 160.2)	6.9 (4.1, 10.4)	136.4±23.8
1-layer	25	20	345.3 (314.9, 378.6)	4.2 (3.2, 5.6)	313.7±84.7
	600	15	140.8 (133.5, 148.5)	8.5 (6.1, 11.8)	133.0±18.6
	800	16	177.8 (163.3, 192.8)	5.4 (3.9, 7.3)	164.3±35.2

*Average value (Lower 95% confidence bound, Upper 95% confidence bound)

** Average value ± Standard Deviation

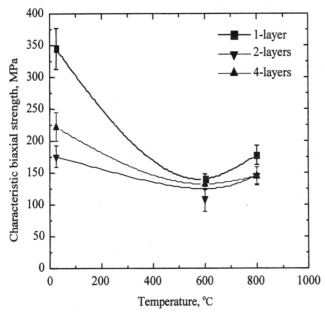

Fig. 3. Biaxial strength of YSZ electrolyte as a function of temperature. Error bars represent 95% confidence bounds.

increases slightly with temperature for all examined thicknesses. Such atypical strength dependence on temperature has also been reported for 6.5mol% YSZ[6] and 8mol% YSZ[8] evaluated in 3-point and 4-point bending respectively.

The results presented in Table II and Fig. 3 also show that the characteristic strength of multi-layered samples at ambient temperature were considerably lower than those of single-layer samples. However, it was found that the differences in characteristic strength between samples of different thicknesses diminished at higher temperatures. Weibull moduli for different thicknesses and temperatures are in ≈3-8.5 range (Table II). Since 95% confidence bounds for Weibull moduli overlap in most cases (Table II) the differences in Weibill moduli can be considered statistically insignificant.

Fracture surfaces after biaxial testing were analyzed using Field Emission Scanning Electron Microscope (FESEM). It was found that the fracture mode was predominately transgranular at ambient temperature and 600°C but mostly intergranular at 800°C. The observed changes in the fracture mode are in agreement with previously published results[6]. Fractographic analyses also indicate that the failure sources in multi-layered specimens are often associated with interfacial defects that are introduced during lamination. Clusters of pores and zirconia agglomerates were other common failure origins in single layered and some multi layered samples.

Fracture Toughness.

The fracture toughness results determined by double torsion for 4-layered samples are listed in Table III and plotted in Fig 4. as a function of temperature. The value of fracture toughness at ambient

Table III. Fracture Toughness of YSZ electrolyte

T, °C	#	Fracture Toughness*, K_{IC}, MPam$^{1/2}$
25	5	1.65±0.03
600	5	1.24±0.27
800	5	1.51±0.13

T-Temperature, # - number of tested samples
*Average value ± Standard Deviation

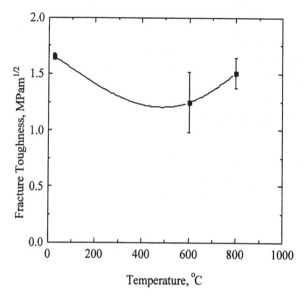

Fig. 4. Fracture toughness, K_{IC} of 4-layers YSZ electrolyte as a function of temperature. Error bars represent 95% confidence bounds.

temperature is comparable to the value of 1.61 MPam$^{1/2}$ that has been reported for tape-cast 8mol% YSZ[9]. As in the case of elastic moduli and biaxial strength, the fracture toughness initially decreases with temperature, and then it increases slightly between 600 and 800°C.

DISCUSSION

The results presented in this study demonstrate atypical temperature dependence of mechanical properties of 8mol% yttria stabilized zirconia. Elastic moduli, biaxial strength and fracture toughness of inherently brittle ceramics usually decrease monotonically with temperature if no phase transformation occurs within the corresponding temperature range. In the case of fully stabilized, cubic, tape cast 8mol% YSZ electrolyte elastic moduli, biaxial strength and fracture toughness decrease with temperature in the 25-600 °C temperature range, and then they increase slightly with temperature up to at least 800 °C.

Mori and co-workers[8] hypothesized that this atypical temperature dependence of the bending strength of 8mol% YSZ could be attributed to the occurrence of small amounts of plastic deformation at elevated temperatures. They suggested that limited grain boundary sliding in fine-grained zirconia may occur even at temperatures as low as 800°C and cause small amounts of plastic deformation. Although the occurrence of plastic deformation can provide a good explanation for the observed increase in fracture toughness, biaxial strength and corresponding changes in fracture mode at temperatures above 600°C, it can't explain the atypical temperature dependence of elastic moduli, which was found in this paper.

Adams[6] suggested that the atypical temperature dependence of the elastic moduli of 6.5 mol% YSZ can be attributed to defect-related phenomena and/or high elastic anisotropy of YSZ ceramics. It is well documented that elastic anisotropy of cubic YSZ increases rapidly with the content of yttria and temperature[10]. Thus, redistribution of stresses at the microstructural scale could provide an explanation for the atypical changes of biaxial strength, fracture toughness and fracture mode with the temperature. However, high-temperature studies using X-ray diffraction and Raman spectroscopy failed to reveal any evidence of changes in the a/c ratio of the lattice or the presence of a phase other than cubic and therefore do not lend support to this argument[11].

It is also worth noting that the lamination process used to synthesize the YSZ samples used in this study is likely responsible for the creation of interlaminar defects and that the decrease in strength with increasing number of layers results from this factor and not from volumetric effects as it would be expected for brittle materials. This argument is supported by the results presented in Fig. 3. If the decrease of the biaxial strength is only due to increase in effective volume of the sample, the biaxial strength of the 4-layer samples should be significantly lower than that of the 2-layers samples, which is obviously not a case here (Fig.3).

SUMMARY

Fully stabilized, cubic, tape cast 8mol% YSZ shows atypical temperature dependence of elastic moduli, biaxial strength and fracture toughness between 25 and 800°C. It was found that these properties initially decrease with temperature between 25 and 600°C, and then they increase slightly with temperature up to at least 800°C. It was also found that synthesis of the test specimens by tape casting can have a deleterious effect on the strength of YSZ.

ACKNOWLEDGMENTS

This research work was sponsored by the US Department of Energy, Office of Fossil Energy, SECA Core Technology Program at ORNL under Contract DE-AC05-00OR22725 with UT-Battelle, LLC. The authors are grateful for the support of NETL program managers Wayne Surdoval, Travis Shultz and Donald Collins. The authors are indebted to Randy Parten from ORNL for help with specimen preparation and characterization and to Tejun Zhen from Drexel University for help with double torsion testing.

REFERENCES
[1]Ming N. Q. and Takahashi T., "Since and Technology of the Ceramic Fuel Cells", Elsevier, Amsterdam (1995)

[2]A. Migliori and J.L. Sarro, "Resonant Ultrasound Spectroscopy: Applications to Physics, Materials Measurements and Nondestructive Evaluation", John Willey & Sons, Inc. New York (1997)

[3]Fuller E. R. Jr, "An Evaluation of Double-Torsion Testing – Analysis" in *Fracture Mechanics Applied to Brittle Materials*, ASTM Special Technical Publication No 678 Edited by Freiman S. W., ASTM, Philadelphia (1997)

[4]Pletka B.J, Fuller E. R. Jr and Koepke B. G. "An Evaluation of Double-Torsion Testing – Analysis", ibid 3

[5]A. Selcuk and A. Atkinson, "Elastic Properties of Ceramic Oxides Used in Solid Oxide Fuel Cells (SOFC)", *J. Euro. Ceram. Soc.* 17, 1523-1532 (1997)

[6]J. W. Adams, " Young's Modulus, Flexural Strength and Fracture Toughness of Yttria-Stabilized Zirconia versus Temperature", *J. Amer. Ceram. Soc.* **80**, 903-908 (1997)

[7]Software package: Weibull++, ReliaSoft, Tucson, AZ

[8]M. Mori, T. Abe, H. Itoh, O. Yamamoto, Y. Takeda and T. Kawahara, " Cubic-stabilized Zirconia and Alumina Composites as Electrolytes in Planar Type Solid Oxide Fuel Cells", Solid State Ionics **74**, 157 (1994)

[9]A. Atkinson and A. Selcuk, "Strength and Toughness of Tape-Cast Yttria-Stabilized Zirkonia", *J. Am. Ceram. Soc.*, **83**, 2029 (2000)

[10]R.W. Rice, "Possible Effects of Elastic Anisotropy on Mechanical Properties of Ceramics", J. Mater. Sci. Lett **13**, 1261 (1994)

[11]Unpublished work, ORNL

SINTERING OF BaCe$_{0.85}$Y$_{0.15}$O$_{3-\delta}$ WITH/WITHOUT SrTiO$_3$ DOPANT

F. Dynys[*], A. Sayir[*1] and P. J. Heimann[*2],
NASA Glenn Research Center[*]/ CWRU[1] /OAI[2]
21000 Brookpark Rd.
Cleveland, OH 44135 USA

ABSTRACT

The perovskite composition, BaCe$_{0.85}$Y$_{0.15}$O$_{3-\delta}$, displays excellent protonic conduction at high temperatures making it a desirable candidate for hydrogen separation membranes. This paper reports on the sintering behavior of BaCe$_{0.85}$Y$_{0.15}$O$_{3-\delta}$ powders doped with SrTiO$_3$. Two methods were used to synthesize BaCe$_{0.85}$Y$_{0.15}$O$_{3-\delta}$ powders: (1) solid state reaction and (2) wet chemical co-precipitation. Co-precipitated powder crystallized into the perovskite phase at 1000 °C for 4 hrs. Complete reaction and crystallization of the perovskite phase by solid state was achieved by calcining at 1200 °C for 24 hrs. Solid state synthesis produce a coarser powder with a average particle size of 1.3 μm and surface area of 0.74 m^2/g. Co-precipitation produced a finer powder with a average particle size of 65 nm and surface area of 14.9 m^2/g. Powders were doped with 1, 2, 5 and 10 mole % SrTiO$_3$. Samples were sintered at 1450 °C, 1550 °C and 1650 °C. SrTiO$_3$ enhances sintering, optimal dopant level is different for powders synthesized by solid state and co-precipitation . Both powders exhibit similar grain growth behavior. Dopant levels of 5 and 10 mole % SrTiO$_3$ significantly enhances the grain size.

INTRODUCTION

The initial pathway for efficient hydrogen production to meet demand from the growing energy sector will come from the existing fossil fuels. To foster the initial growth of a localized distribution, small reformers and electrolyzers will provide on-site hydrogen generation. Reformers based on ceramic membrane technology potentially offer hydrogen production that is comparable to the cost of fossil fuels.

Protonic conducting ceramic with the chemical formula ABO$_3$ offers the promise of highly selective hydrogen separation at intermediate temperature (400-800°C). Among different perovskite-type oxides, barium cerates show promising high protonic conductivity but strong resistance to densification.[1,2] The solid state reaction of barium carbonate with ceria is highly utilized to synthesize BaCeO$_3$ base powders. The reaction is simple and there are no metastable products. Downside to the solid state process is the formation of large particles, high degree of agglomeration, possible inhomogeneities in chemical composition and contamination by the attrition steps. These powder characteristics negatively impact the sintering behavior. The use of excess barium enhances densification but its environmental reaction with CO$_2$ causes mechanical integrity problems.[3] Particle size reduction and cation doping are traditional methods to enhance sintering. Both approaches are used to investigate the sintering behavior of BaCe$_{0.85}$Y$_{0.15}$O$_{3-\delta}$.

Preliminary results on the sintering behavior of SrTiO$_3$ doped BaCe$_{0.85}$Y$_{0.15}$O$_{3-\delta}$ powders is reported. Two different synthesis methods were used to fabricate BaCe$_{0.85}$Y$_{0.15}$O$_{3-\delta}$ powder: (1) solid state reaction and (2) wet chemical co-precipitation utilizing ammonia/ammonium carbonate.[4,5]

EXPERIMENTAL PROCEDURE

Solid state (SS) synthesis was performed by mixing stoichiometric quantities of $BaCO_3$, CeO_2 and Y_2O_3 to yield the composition of $BaCe_{0.85}Y_{0.15}O_{3-\delta}$. The mixture was wet ball milled in alcohol for a period of 48 hrs to enhance homogeneity. The perovskite phase was synthesized by triple calcining: 800 °C for 6 hrs., 1000 °C for 10 hrs and 1200 °C for 24 hrs in Al_2O_3 crucibles. Calcined material was ground in a mortar and pestle to break up agglomerates between calcinations. The crystalline $BaCe_{0.85}Y_{0.15}O_{3-\delta}$ powder was doped with 1, 2, 5 and 10 mole % $SrTiO_3$ powder. The powders were wet ball milled for 24 hrs. to ensure $SrTiO_3$ homogeneity and break down of agglomerates.

Metal salts of $Ba(NO_3)_2$, $Ce(NO_3)_3 \cdot 6H_2O$ and $Y(NO_3)_3 \cdot xH_2O$ were used as precursors. The stoichiometric quantities were dissolved in deionized water at a ratio of 12 ml of water per gram of metallic salt to yield the composition of $BaCe_{0.85}Y_{0.15}O_{3-\delta}$. A precipitant solution was prepared by using a molar ratio of 1:2 of $(NH_4)_2CO_3$ to NH_4OH (28-30.0% aqueous solution). The molar ratio of $(NH_4)_2CO_3$ to $Ba(NO_3)_2$ and H_2O to $(NH_4)_2CO_3$ was held constant at 3 and 119, respectively. Mixed nitrate solution was dripped into a stirred base solution at 2 ml/min using an auto-titrator. Co-precipitation (COP) occurred instantaneously. The precipitate was separated from the solution by filtration and then freeze dried. The perovskite phase was synthesized by calcining at 1000 °C for 4 hrs in Al_2O_3 crucibles. The crystalline $BaCe_{0.85}Y_{0.15}O_{3-\delta}$ powder was doped with 1, 2, 5 and 10 mole % $SrTiO_3$ powder. The powders were wet ball milled for 24 hrs. to ensure $SrTiO_3$ homogeneity and break down of agglomerates.

Green pellets were pressed at 200 MPa using 2 grams of powder in a 15 mm die. Pellets were sintered in air at 1450 °C, 1550 °C and 1650 °C using a heating rate of 5 °C/min. A thin layer of $BaCe_{0.85}Y_{0.15}O_{3-\delta}$ powder was placed between the sample and Al_2O_3 crucible to abate the reaction between the crucible and sample. The densities of sintered pellets were determined by geometrical measurements and mass. Sample cross-sections were polished and etched to reveal the microstructure. Microstructures were examined with an Hitachi S-4700 scanning electron microscope (SEM). The grain size was determined by a linear intercept method.

The surface area of the synthesized powders were measured by nitrogen gas adsorption and analyzed using the 5 point Brunauer-Emmett-Teller method.

The amorphous and crystalline structures were characterized by x-ray diffraction. The x-ray diffractometer was equipped with a Cu K_α source with a wavelength of 0.1540 nm. The operating conditions were 45 KV and 40 mA. Scans were conducted at 3°/min with a sampling interval of 0.02°.

RESULTS AND DISCUSSION

X-ray diffraction (XRD) of the calcined powders is shown in Figure 1. The calcination temperatures used for each synthesis method was sufficient to produce crystalline perovskite powder of $BaCe_{0.85}Y_{0.15}O_{3-\delta}$. The measured surface area of the synthesized powders by SS and COP were 0.74 m^2/g and 14.9 m^2/g, respectively. Figure 2 shows a SEM micrograph of the SS synthesized $BaCe_{0.85}Y_{0.15}O_{3-\delta}$ powder. The particles are irregular shaped and size ranges from 1-5 μm. Average spherical particle diameter calculated from the surface area measurement is 1.3 μm. Grain boundaries are observed on the particle surfaces suggesting that particles are polycrystalline with sub-micron size grains. Figure 3 shows a SEM micrograph of the calcined co-precipitated powder. The powder is highly agglomerated and consists of particles that are approximately 100 nm in size. The calculated average spherical particle size from the measured surface area is 65 nm and correlates well with the SEM observations.

Figure 4 shows the sintered density of $BaCe_{0.85}Y_{0.15}O_{3-\delta}$ as a function of $SrTiO_3$ content for

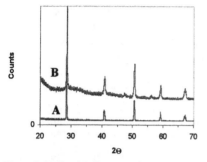

Figure 1. XRD of $BaCe_{0.85}Y_{0.15}O_{3-\delta}$ (A) SS synthesis and (B) COP synthesis.

Figure 2. $BaCe_{0.85}Y_{0.15}O_{3-\delta}$ powder synthesized by SS.

powders synthesized by the SS method. The general trend is that $SrTiO_3$ doping enhanced the sintered density compared to the undoped samples. A 12% improvement in sintered density is observed at 1550 °C and 5.5% at 1650 °C. $SrTiO_3$ appears to be ineffective at 1450 °C. Figure 4 also illustrates that the sintering temperature was more effective in enhancing the sintered density than $SrTiO_3$ doping. Figure 5 compares the sintered microstructures at 1650 °C for

Figure 3. $BaCe_{0.85}Y_{0.15}O_{3-\delta}$ powder synthesized by COP.

Table I

Average Grain Size at 1650 °C

Mole % $SrTiO_3$	Ave. GS (μm) SS	Ave. GS (μm) COP
0	4.0	1.6
1	4.6	1.8
2	4.1	2.5
5	5.5	5.7
10	9.4	12

undoped $BaCe_{0.85}Y_{0.15}O_{3-\delta}$ and 10 mole % $SrTiO_3$ doped $BaCe_{0.85}Y_{0.15}O_{3-\delta}$. The roughness observed within grains in Figure 5 is caused by the preferential etching along crystallographic directions. A larger grain size is observed with the $SrTiO_3$ doping. Table I summarizes the average grain size of SS samples sintered at 1650 °C for $SrTiO_3$ dopant levels of 1 to 10 mole %. The average grain size for the undoped sample is 4 μm, showing very little grain growth when considering the starting particle size. Significant grain size enhancement is observed at 5 and 10 mole % $SrTiO_3$, average grain size of 5.5 μm and 9.4 μm are observed, respectively. In addition, the observed grain size distribution is narrow at 0, 1, 2 and 5 mole % $SrTiO_3$, it significantly broadens at 10 mole %.

Figure 6 shows the sintered density of $BaCe_{0.85}Y_{0.15}O_{3-\delta}$ as a function of $SrTiO_3$ content for powders synthesized by the COP method. The observed sintering behavior is significantly different than the SS samples. The sample sintered at 1650 °C with 10 mole % $SrTiO_3$

showed exceptional bloating, surface bubbles were observed on the surface. The general trend at 1550 °C and 1650 °C is that 1 to 2 mole % SrTiO$_3$ is optimal for sintered density, 5 and 10 mole % caused a reduction. The 1450 °C isotherm behaves similar to the SS material; increasing SrTiO$_3$ content enhances sintered density. The density increased by 19% at 10 mole % at 1450 °C. At 1 mole %, a 11% increased in density is observed at 1550 °C and 1650 °C. The sintered density of the COP samples show a larger dependence upon SrTiO$_3$ doping than upon temperature, opposite to what was observed for the SS samples. Figure 7 compares the sintered microstructures at 1650 °C for undoped BaCe$_{0.85}$Y$_{0.15}$O$_{3-\delta}$ and 10 mole % SrTiO$_3$ doped BaCe$_{0.85}$Y$_{0.15}$O$_{3-\delta}$. Large pores, >10 µm, are observed in the microstructure at 10 mole % SrTiO$_3$. Pore clusters are observed in the undoped sample which can be attributed to agglomeration problem. These large pores and pore clusters are responsible for low sintered densities. Table I shows the average grain for COP material sintered at 1650 °C. All dopant levels of SrTiO$_3$ show enhanced grain size. Significant grain growth occurred at 10 mole % SrTiO$_3$ along with significant broadening of the distribution. The COP grain growth behavior is similar to the SS material.

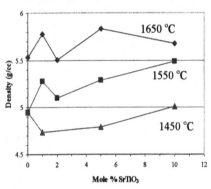

Figure 4. Affect of SrTiO$_3$ on sintered density of SS derived powder.

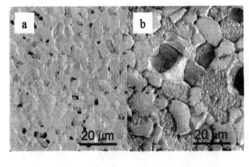

Figure 5. Microstructure of SS derived powder sintered at 1650 °C :(a) undoped and (b) 10 mole % SrTiO$_3$.

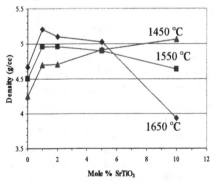

Figure 6. Affect of SrTiO$_3$ on sintered density of COP derived powder.

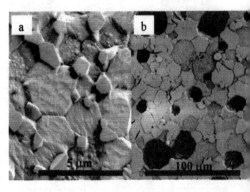

Figure 7. Microstructure of COP derived powder sintered at 1650 °C :(a) undoped and (b) 10 mole % SrTiO$_3$.

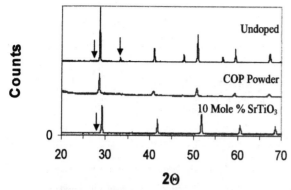

Figure 8. XRD of COP samples sintered at 1650 °C

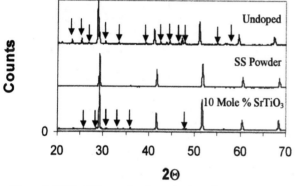

Figure 9. XRD of SS samples sintered at 1650 °C

The surfaces of the sintered samples were analyzed for phases using XRD. Figures 8 and 9 show the XRD patterns for undoped and 10 mole % $SrTiO_3$ doped samples sintered at 1650 °C for COP and SS derived powders, respectively. Both figures contained the XRD of the undoped calcined powders for comparison. The doped and undoped samples do not maintain phase purity after sintering at 1650 °C. The arrows in Figures 8 and 9 indicate peaks due to secondary phase formation. For both powders, XRD indicates that the $SrTiO_3$ goes into solid solution as indicated by the shift in the perovskite peaks. Secondary phase formation is stronger in the SS samples than COP samples. One and two unknown diffraction peaks are observed in the COP samples. A large number of unknown diffraction peaks are observed in the SS samples. The severity of unknown diffraction peaks is more pronounced in the undoped sample than the $SrTiO_3$ doped sample. The unknown diffraction peaks do not correspond to Y_2O_3, CeO_2, $BaCO_3$ and BaO. For each isotherm, the SS and COP samples were sintered together. Contamination from sintering process appears minimal since the COP samples exhibit a few unknown diffraction peaks. Mixed CeO_2 phases were reported in sintered samples by Ma et al.[5] for Ba deficient $BaCe_{0.9}Y_{0.1}O_{3-\delta}$. Ma reported no additional phases in sintered samples with excess Ba. $BaZrO_3$ exhibits similar behavior as reported by Snijkers et al.[6] Both Ma and Snijkers attribute barium loss during sintering for secondary phase formation. Takeuchi et al.[8] have shown that phase transitions occur in Y-doped $BaCeO_3$. Crystal structure is dependent upon Y concentration and annealing atmosphere. Multi-phase materials were observed in atmospheres containing H_2 or H_2O. Calculated XRD from Takeuchi neutron diffraction data does not account for all the extra diffraction peaks. Further characterization is needed to identify the secondary phases.

There is insufficient data to elucidate the difference in sintering behavior between the $SrTiO_3$ doped SS and COP powders. It is suspected that sinter samples that exhibit bloating was caused by gas evolution in combination with a liquid phase. The roundness of the grain edges and boundary curvatures suggest liquid phase but it is undetectable in the SEM work. It is well known that CeO_2 becomes oxygen deficient upon heating ≥ 1200 °C: $CeO_2 \rightarrow CeO_{1.83} \rightarrow CeO_{1.72} \rightarrow Ce_2O_3$. It can be postulated that the reaction of $SrTiO_3$ with $BaCeO_3$ results in the reduction of ceria. More work is needed to understand the material interactions during sintering.

SUMMARY

$BaCe_{0.85}Y_{0.15}O_{3-\delta}$ powders were synthesize by two methods: (1) solid state reaction and (2) wet chemical co-precipitation. Co-precipitation produces a finer particle size and reduces the calcination temperature and time to crystallize the perovskite structure. A 1000 °C calcination for 4 hrs produce an average particle size of 65 nm. Solid state synthesis required multiple calcinations up to 1200 °C for 24 hrs. with an average particle size of 1.3 µm. Powders were doped with 1, 2, 5 and 10 mole % $SrTiO_3$. Samples were sintered at 1450 °C, 1550 °C and 1650 °C. $SrTiO_3$ doping was found to enhance sintering. However, the SS and COP doped powders exhibit different sintering behavior. Both powders exhibit similar grain growth behavior with $SrTiO_3$ doping. Dopant levels of 5 and 10 mole % $SrTiO_3$ significantly enhances the grain size.

ACKNOWLEDGEMENT

This work was supported by NASA Director's Discretionary Fund.

REFERENCES

1. K.D. Kreuer, An. Rev. of Mat. Res., **33**, 333, 2003.
2. A.S. Nowick, Y. Du and K.C. Liang. Solid State Ionics, **125**, 303, 1999
3. D. Shima & S. Haile, Solid State Ionics, **97**, 443, 1997
4. F. Boschini et al., J. Eur. Cer. Soc., **23**, 3035, 2003
5. J. Brzezińska-Miecznik, Mat. Letters, **56**, 273, 2002.
6. G. Ma et al., Solid State Ionics, **110**, 103, 1998.
7. F. Snijkers et al., Scripta Mat., **50**, 655, 2004.
8. K. Takeuchi et al., Solid State Ionics, **138**, 63, 2000.

HIGH TEMPERATURE SEALS FOR SOLID OXIDE FUEL CELLS (SOFC)

Raj N. Singh
Department of Chemical and Materials Engineering
University of Cincinnati, P.O. Box 210012, Cincinnati, OH 45241-0012

ABSTRACT

A variety of seals such as metal-metal, metal-ceramic, and ceramic-ceramic are required for a functioning SOFC. These seals must function at high temperatures between 600-900°C and in oxidizing and reducing environments of the fuels and air. Among the different type of seals probably the metal-metal seals can be readily fabricated using metal joining, soldering, and brazing techniques but the issue of oxidation at these high-temperatures needs to be considered for long-term survivability of all metal seals as well. This cannot be said for the metal-ceramic and ceramic-ceramic seals because the brittle nature of ceramics/glasses can lead to fracture and loss of seal integrity and functionality. Consequently, any seals involving ceramics/glasses require a significant attention and technology development for reliable SOFC operation. This paper is prepared to primarily address the needs and possible approaches for high temperature seals for SOFC.

INTRODUCTION

Solid oxide fuel cell (SOFC) technology is critical to several national initiatives such as Hydrogen Fuel and Clear Skies initiatives to achieve the overall goal of the National Energy Security and Energy Independence, and at the same time to protect our environment [1, 2]. Among the many fuel cell technologies, the SOFC functions at much higher temperatures of 650-900° C and offers unique advantages of utilization of the more abundant fossil-derived fuels (hydrocarbons) as well as hydrogen. Additionally, much higher efficiencies than the low temperature fuel cell are possible for the SOFC when integrated with the combined cycle utilizing the waste heat.

SOFC technology has progressed to a stage where some of the problems associated with the electrochemically-active cell components such as electrolyte and electrode have been adequately addressed. The remaining issues of technology needs such as the cost and long-term performance goals are being addressed through several government and industry supported programs. One area that needs more attention at the present time is related to the seals for SOFC. Reliable seals are essential to the long-term performance and reliability of the SOFC because a poor seal will degrade cell performance and lead to wastage of fuels, and possibly can pose dangers to safety of the fuel cell stacks. Consequently, the seals for SOFC are important and need additional attention. The most recent workshop on SOFC Seals is clearly a timely recognition of the technology needs for this critical area [1, 2].

BACKGROUND ON SEALS FOR ELECTROCHEMICAL SYSTEMS

Seals development for SOFC can take advantage of past experiences with other electrochemical systems such as battery and fuel cells. These may include rigid seals for Na-S battery and wet-seals for Molten Carbonate Fuel Cells (MCFC) [3-9]. A schematic of the Na-S battery cell incorporating some of these seals is shown in Fig. 1 a. A number of seals such as glass-ceramic electrolyte, mildsteel-ceramic, and glass-metal were successfully developed for long-term stability in highly aggressive environments of liquid Na, liquid sulfur, and very high vacuum of a SEM (for in situ studies) [3-5]. The seal between mild steel and alumina was made by thermocompression approach using Al thin foils and the glass-alumina-Na β"-alumina seal utilized a special composition of Boroaluminosilicate glass (see Table 1). These seals are functioning in Na-S battery packs used for load leveling by power generation industry such as American Power [7].

MCFC used technologies for the porous $LiAlO_2$ electrolyte retainer, which formed a wet-seal (Fig. 1 b) between metal bipolar separator and $LiAlO_2$ electrolyte retainer [8,9]. A great deal of ceramic processing optimization was done to develop the porous electrolyte retainer so that it will keep the liquid Li_2CO_3-K_2CO_3 electrolyte at 650°C without flooding the porous electrodes and the seal area. The porosity and pore size distribution were also optimized to provide active liquid electrolyte for making the wet-seals against bipolar metal cell separator [8,9]. A concept akin to MCFC can be advanced and developed to create stress-free seals for SOFC.

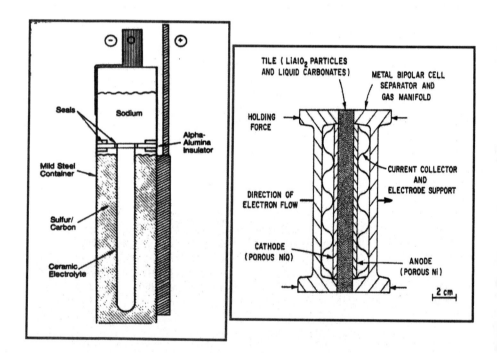

Fig. 1. (a) Schematic of a Na-S cell showing ceramic-ceramic and ceramic-metal seals [3-6] and (b) wet-seals for a Molten Carbonate Fuel Cell (MCFC) [8, 9].

BACKGROUND ON COMPOSITE SCIENCE AND TECHNOLOGY RELEVANT TO
TOUGHENING AND STRENGTHENING OF SEAL MATERIALS

There is a vast knowledgebase on toughening of ceramics and glasses via fiber- or whisker-reinforcement that can be applied to seals for SOFC[10-14]. The ceramic and glass sealant materials can be strengthened and toughened to enhance seal reliability. Towards this end, extensive knowledge of the toughened ceramic matrix composite (CMC) technology (processing and properties) can be employed. An example of toughening a borosilicate glass by fibers is clearly demonstrated in Fig. 2 [10, 11]. In addition, the knowledge base to create conditions in which cracking, cracking pattern, and crack locations can be controlled in CMC (see Fig. 2-b) to take advantage of localizing damage to a predetermined locations to alleviate stress buildup and then providing active sealant at these locations to create active sealing upon cracking of the seals. This is primarily done through control of the fiber-matrix interfacial shear strength and/or via expansion mismatch control of the two materials forming an interface.

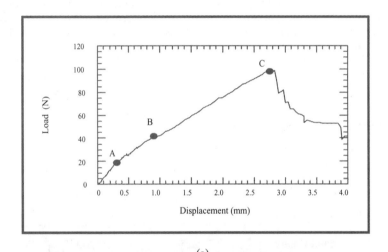

(a)

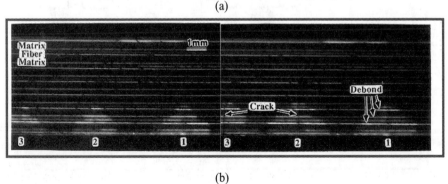

(b)

Fig. 2. (a) Toughening and strengthening of a borosilicate glass by SiC fibers, and periodic cracking and crack arrest in fiber-reinforced glass composite [10, 11].

BACKGROUND ON SEALS FOR SOFC

A variety of seals such as metal-metal, metal-ceramic, and ceramic-ceramic are required for a functioning SOFC. These seals must function for long times from 5000 to 40000 hours, at high temperatures between 600-900°C, and in oxidizing and reducing environments of the fuels and air. Among the different type of seals probably the metal-metal seals can be readily fabricated using metal joining, soldering, and brazing techniques but the issue of oxidation at these high-temperatures needs to be considered for long-term survivability of all metal seals as well. However, this is not possible for the metal-ceramic and ceramic-ceramic seals because the brittle nature of ceramics/glasses can lead to fracture and loss of seal functionality. Consequently, any seals involving ceramics/glasses require a significant attention and technology development for reliable SOFC operation.

Current Status on Selection of Materials for SOFC:

A search for reliable seals for SOFC must of necessity has to start with considerations of selection of materials based on their stability in severe environments of a SOFC. A SOFC operates over a range of temperatures from 600-900°C, in a reducing potential of fuels (anode) and in oxidizing environment of the oxidant (cathode). The fuels for SOFC could be based on diesel, gasoline, hydrogen, propane/natural gas, coal-derived gases, and methanol to name just a few, which provide an important advantage for SOFC over the low-temperature fuel cells because the later cannot tolerate some of these fuels except the purest hydrogen gas. The oxidant

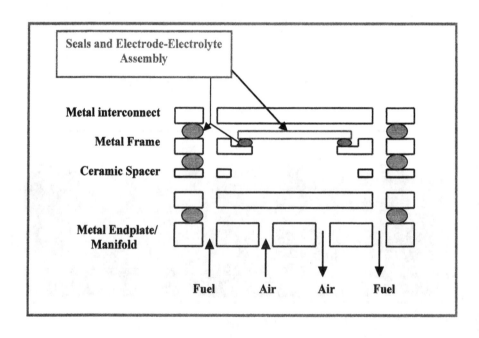

Fig. 3 . Schematic of a SOFC and possible seals and their locations.

can be simply air. Therefore, the sealing materials must be selected based on their survivability in both oxidizing and reducing environments. This may involve selection of metals, ceramics, glasses, and possibly brazing alloys that can survive the severe environment of a SOFC. Figure 3 shows a unit cell of a SOFC and locations of possible seals. The current materials system for the SOFC may consist of 8%-YSZ (yttria stabilized zirconia)-electrolyte, Ni-YSZ anode, doped lanthanide perovskites cathode, doped chromites/alloys as interconnects, insulating seals, and Manifolds made of heat resistant metallic alloys.

Selection of a metallic material for seal's development is very much required so that the seals represent joining of a realistic metallic component to the electrolyte-electrode assembly by ceramic-ceramic-metal seals. To this end, a variety of metallic materials are under consideration for use in SOFC. These may include ferritic alloys such as E-Brite (26 Cr-1 Mo) and 430 stainless steels to name just a few. These alloys form chromia scales and have been studied for their oxidative stability and conductive scale forming ability for the interconnect applications [15]. An external coating of $LaCrO_3$ on T446 stainless steel offered oxidation resistance via reduced chromia scale growth.

Status of Seals for SOFC:
The most delicate component in a SOFC is the electrode-electrolyte assembly because of the brittle electrolyte, which can be as thin as 10-15 μm. Some designs use anode supported electrolyte while other approaches have used a cathode supported electrolyte. The delicate electrolyte-electrode assembly requires a seal to separate the anode from the cathode physically and electrically (insulating) in a functioning SOFC at ~650-800°C (Fig. 2). Consequently, a sealing system must not only transmit the lowest possible stresses to the ceramic components but it must also be stable at high temperatures of 650-850°C over a long period of time.

Weil et. al [16] used alloys like FeCrAlY (Fe 22% Cr, 5% Al, 0.2% Y) as a model metal for sealing by brazing to YSZ. In this study, Ag-CuO braze materials were used. However, additional work is required to determine effect of interdiffusion and durability of the brazed joints at high temperatures in the oxidizing environment of a fuel cell.

A number of more novel sealing concepts are also possible for seals in SOFC. Lewinsohn et. al [17] presented an idea of using polymers of Si-C-N to make seals to SOFC components. Potentially these polymers can be used as a paste to join two surfaces but these materials are inherently unstable against oxidation in air at high temperatures or moist air. They also suggested using ceramic and metallic fillers to control the CTE to match expansion of SOFC components. Loehman [18] has suggested using viscous seals for attaching SOFC components. Chou and Stevenson [19-21] used a novel approach of Mica and Mica-Glass hybrid compressive seals with promising results but these seals will require compression loads. Also, there were issues related to interface reactions and crystallization of some of the glasses. One of the glass composition used was 58% SiO_2, 9% B_2O_3, 11% Na_2O, 6% Al_2O_3, 4% BaO and ZnO, CaO and K_2O. Chou [19] presented research on infiltrated-Mica for seals with promising results after infiltration with Bi-, B-, and glass-forming materials. But, these will still require load and may pose stability issues (evaporation) at high temperatures for use over extended times.

Taniguchi et. al. [22] reported using a combination of glass/YSZ and Fiberfax fiber (essentially alumina fiber) matte to seal the SOFC components. The fiber layer was porous/compliant and provided stress reduction on the electrolyte-electrode assembly via compliance of the fiber layer. It was not apparent from the paper as to how the porous Fiberfax was made impervious or reduced leak rates when in contact with the electrolyte. Donald [23] provided a general review on the glasses and glass-metal seals, which can be used as a reference for guiding materials selection for seals.

A POSSIBLE APPROACH OF SEAL DEVELOPMENT FOR SOFC

These results from open literature clearly indicated that there are promising sealing concepts and materials that have shown some promise for use as seals for SOFC. Table 1 lists coefficient of thermal expansion (CTE) of a variety of realistic metallic, ceramic, and glass materials of interest to seals for SOFC. A seal must somehow accommodate a large mismatch in CTE of metallic and ceramic materials. However, most of the seals work to date have not taken explicitly an integrated approach of selecting materials based on stability considerations, thermophysical properties, modeling of residual stresses upon sealing and during cell operation, and geometrical aspects of seals towards improved reliability. Most of the seals so far have been of rigid nature, which is expected to create detrimental stresses onto delicate electrolyte-electrode assembly leading to fracture of the ceramic components or the seal itself. New concepts, that minimize internal stresses under both steady state and thermal transients, are needed to develop reliable seals for SOFC.

Table 1. Coefficient of Thermal Expansion (CTE) for Some Useful Metallic, Ceramic, and Glass Materials [1, 2, 23, 24].

Materials	Possible Use in SOFC	Temperature Range (°C)	α (ppm/ °C)
Inconel 600	Metallic Hardware	25-1000	16.7
SS 430	Metallic Hardware	25-1000	13
Haynes	Metallic Hardware	25-1000	14-15
Ceramic Perovskite	Interconnect	25-1000	10.6-11.1
Alumina	Insulator	25-1000	8.8
ZrO_2 (Stabilized)	Ionic Conductor	25-1000	10.0
8-YSZ	Electrolyte	25-1000	11
Soda Glass	Sealant	25-1000	9.0
Li_2O-ZnO-Al_2O_3-SiO_2(Glass-Ceramic)	Sealant	25-300	5.5-12 Depending on ZnO content
Boroaluminosilicate Glasses	Sealant for Na-S Battery (Al_2O_3 to Beta"-Alumina)	25-300	5.7-6.4

An integrated approach/concept for seal development is shown in Fig. 4. It consists of materials selection, analytical modeling, processing, property measurements, and seal testing. Since the seals for SOFC must function in oxidizing environment, it will be most desirable to select sealing materials, which are thermochemically stable in air between 600-1000°C. Most suitable sealing materials must of necessity be either crystalline oxides, oxide glasses, or materials that upon active oxidation form stable crystalline oxides or oxide glasses. The second most important requirement on materials is that they should be thermochemically compatible with the YSZ-electrolyte and metal/cermet/ceramic-electrodes/current collectors. In addition, the selected sealing materials must have reasonably close match of the coefficient of thermal expansion (thermomechanical compatibility) with the components being sealed to avoid significant residual stresses that can lead to premature seal failures.

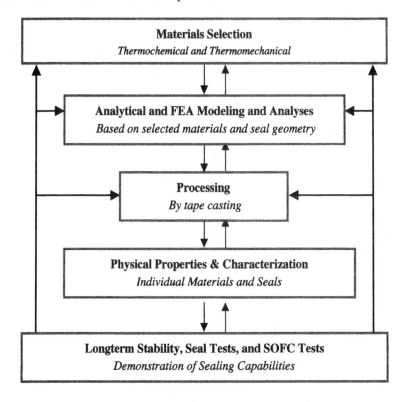

Fig.4. A flow chart of the integrated approach of seals for SOFC.

The selected materials must be analyzed for residual stresses using sealing materials and components being sealed such as electrolyte-electrode assembly with the metallic components. The analytical and FEA models can be readily used for this purpose so that the selected materials must form a seal with the least amount of the stresses. This analysis must be done using materials properties, realistic processing temperatures, and thermal transients encountered upon cell operation. This step will lead to selection of the best possible materials for seal's development, which will be used to develop processing methods for making seals between selected components. A variety of sealing concepts can be used at this point to develop appropriate processing methods for making seals such as rigid, viscous, wet, and/or flexible seals. The sealing materials thus processed must be characterized for physical, mechanical, and thermophysical properties over a range of temperatures. These properties can then be used as inputs into the analytical models for assessing residual stresses in the seals and adjoining components. These inputs then can be used as a feedback mechanism to select better materials for creating reliable seals for SOFC. Long-term stability of the sealing materials and seals themselves needs to be evaluated at elevated temperatures in order to assess stability of the materials systems and seals. Subsequent to these steps the seals will be evaluated into a working SOFC to determine the functionality and long-term survivability of the seals. It is hoped that this integrated approach for seal's development for SOFC will lead to successful seals in a functioning SOFC.

REFERENCES

1. Wayne Surdoval, " Fossil Energy Fuel Cell Program," SECA Workshop/Meeting, July 8-9, Sandia Labs. (2003).
2. Prabhakar Singh, " Sold Oxide Fuel Cell Power Generation: Technology Status," SECA Workshop/Meeting, July 8-9, Sandia Labs. (2003).
3. R.N. Singh, " Asymmetric Polarization Behavior of Sodium Beta"-Alumina Electrolyte," J. Am. Ceram. Soc.,70[4], 221 (1987).
4. R.N. Singh and N. Lewis, " Role of Electrolyte-Sodium Interface Behavior to the Degradation of a Sodium-Sulfur Cell," Solid State Ionics, 9/10, 159 (1983).
5. R.N. Singh, R.H. Ettinger and N. Lewis, Rev. Sci. Inst., " Hot-Stage for In Situ Operation of a Battery in a Scanning Electron Microscope," 55[5], 773 (1984).
6. R.N. Singh, " Etched Beta"-Alumina Ceramic Electrolyte," US Patent # 4381968 (1983).
7. American Power, Private communication (2002).
8. R.N. Singh, J.T. Dusek, and J.W. Sim, "Fabrication and Properties of Porous Lithium Aluminate Electrolyte Retainer for Molten Carbonate Fuel Cells," J. Am. Ceram. Soc., 60[6], 629 (1981).
9. R.N. Singh and J.T. Dusek, " Porous Electrolyte Retainer for Molten Carbonate Fuel Cells," US Patent # 4389467 (1983).
10. Y. Sun and R.N. Singh, " Generation of Multiple Matrix Cracking and Fiber-Matrix Interfacial Debonding in a Glass Composite," Acta Mater., 46[5], 1657 (1998).
11. Y. Sun and R. N. Singh, " Determination of Fiber bridging Stress Profile By Debond Length Measurement," Acta Mater., 48, 3607 (2000).
12. S. Kumar and R.N. Singh, "Effects of Fiber Coating properties on the Crack Deflection and Penetration in Fiber-Reinforced Ceramic Composites," Acta Materialia, 45[11], 4721 (1997).
13. Y.L. Wang, U. Anandkumar, and R.N. Singh, "Effect of Fiber Bridging Stress on the Fracture Resistance of Silicon-Carbide-Fiber/Zircon Composites," J. Am Ceramic Soc., 83[5], 1207 (2000).
14. J.E. Web and R.N. Singh, " Thermal Shock behavior of Two-Dimensional Woven Fiber-Reinforced Ceramic Composites," J. Am. Ceram. Soc., 79[11], 2857 (1996).
15. G. Meier, " Fundamental Study of the Durability of Materials for Interconnct in Solid Oxide Fuel Cells," SECA Core Technology Program Review, October 1 (2003), Albany, NY.

16. K. S. Weil, J.S. Hardy, and J.Y. Kim, " Use of a Novel Ceramic-Metal Braze for Joining in High Temperature Electrochemical Devices," The Joining of Advanced and Specialty Materials, V, ASM (2003).

17. C. Lewinsohn, S. Quist, and S. Elangovan, " Novel Materials for Obtaining Compliant, High Temperature Seals for Solid Oxide Fuel Cells," SECA Core Technology Program Review, October 1 (2003), Albany, NY.

18. R. Loehman, " Development of High Performance Seals for Solid Oxide Fuel Cells," SECA Core Technology Program Review, October 1 (2003), Albany, NY.

19. Y.S. Chou, " Compressive Seals Development," SECA Core Technology Program Review, October 1 (2003), Albany, NY.

20. Y.S. Chou, J.W. Stevenson, and L.A. Chick, " Ultra Low Leak Rate of Hybrid Compressive Mica Seals for SOFC," J. Power Source, 112, 130 (2002).

21. Y.S. Chou and J.W. Stevenson, " Mid-Term Stability of Novel Mica-Based Compressive Seals for SOFC," J. Power Source, 115, 274 (2003).

22. S. Tanaguchi, M. Kadowaki, T. Yasuo, Y. Akiyamu, Y. Miyaki and K. Nishio, " Improvement of Thermal Cycle Characteristics of a Planar-Type SOFC by Using Ceramic Fiber as a Sealing Material," J. Power Sources, 90, 163 (2000).

23. I.W. Donald, " Review: Preparation, Properties, and Chemistry of Glass and Glass-Ceramic-to Metal Seals and Coatings," J. Mater. Sci., 28, 2841 (1993).

24. W.D. Kingery, H.K. Bowen, and D.R. Uhlman, "Introduction to Ceramics," Wiley, 2nd Edition (1976), p. 590-598.

EVALUATION OF SODIUM ALUMINOSILICATE GLASS COMPOSITE SEAL WITH MAGNESIA FILLER

K.A. Nielsen, M. Solvang, F.W. Poulsen, P.H. Larsen
Materials Research Department, Risø National Laboratory,
Frederiksborgvej 399, DK-4000 Roskilde, Denmark

ABSTRACT

Glassy seals are used for short term testing, less than 3-4000 hours, of SOFC materials and components in the laboratory. A composite glass seal where Magnesia particles have been suspended in a Sodium Aluminosilicate glass matrix was investigated in order to evaluate the sealing performance towards metallic interconnects of Fe22Cr-steel. Seal compressibility limited the amount of particles, which could be suspended, thus limiting the range of thermal expansion (<11.7 ppm K^{-1}) for the composite material to values at the lower end of the range of technological interest (12-13 ppm K^{-1}). Excellent wetting and strong bond developed between seal and interconnect materials at temperatures above 850°C.

INTRODUCTION

Glasses are obvious candidates for SOFC sealing because it is possible to tailor the physical and chemical properties within a wide range. Different glass and glass-ceramic compositions have been tried within the groups of Alkali Silicate, Alkali Aluminosilicate, Alkaline Earth Silicate, Alkaline Earth Aluminoborosilicate, Phosphate and Borosilicate glasses (1-6). However, reports on composite seals with crystalline filler materials dispersed into base glass, e.g. Alkali Borosilicate glass (7) or in Sodium Aluminosilicate glass (NAS) (2) show promising results in terms of exact matching the coefficients of thermal expansion (CTE) between the sealant and the seal surfaces, and at the same time attain suitable values of Tg and sufficient wetting of the surfaces. High chemical stability and slow crystallisation behaviour has been reported for the NAS glass (8). In this study, Magnesia filler particles were mixed with a NAS glass, (Tg: 515°C) to reach CTE in the range 12-13 ppm K^{-1} to match the CTE for the two interconnect materials, A: 12.9 ppm K^{-1} (Sandvik OYC44) and B: 12.2 ppm K^{-1} (Crofer 22APU), and for the fuel cell itself, 12.8 ppm K^{-1}. Development of the sealing composite started by determination of the compressibility and thermal expansion as a function of the filler concentration followed by experiments on selected compositions to monitor the influence of temperature and time on the sealing properties.

The set up for laboratory testing of a single stack element (short stack) at temperatures between 700-850°C has been shown schematically in Figure 1. The relatively robust design with large spacing between cells and interconnects provides easy assembly routines with large freedom for building in additional probes and features. Prior to the performance testing, fuel cells and stacks are subjected to an initial sequence at higher temperatures, the sealing temperature, in order to consolidate the sealing, attain electrical contact between the different stack components and initiate the anode material. The range of sealing temperatures for this study is limited to maximum values at 950°C in order to protect other stacking components from excessive degradation. Thus, the sealing has to be sufficiently soft at the sealing temperature, wet the sealing surfaces and then reach the desired properties of thermal expansion.

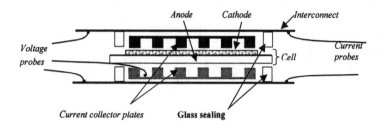

Figure 1: Stacking components schematically.

EXPERIMENTAL

Sample preparation

The NAS glass was melted from analytical grade chemicals SiO_2, Al_2O_3 and Na_2CO_3 in an Alumina crucible at 1650°C for 2 hours reaching a final composition: Na_2O: 17.8mole%, Al_2O_3: 9.4mole%, SiO_2: 72.8mole%, crushed and re-melted in order to ensure homogeneity. Finally, the glass was milled into fine powder, $d_{50}=24\mu m$. For the composite sealing material, the Magnesia powder was screened to grain sizes from 90-200µm prior to mixing in a ball mill with the glass powder and Stearic acid added as binder. Test bars for CTE-measurements were prepared by hydraulic pressing of small samples, which were then sintered at 800°C for 2 hours in air. The seals were prepared similarly, except that the powder mixtures were filled into graphite moulds and sintered at $4*10^{-4}$Pa. The sintered seals were finally machined to obtain parallel, smooth surfaces. Steel samples were machined to dimensions, etched by an aqueous HNO_3/HF-solution in an ultrasonic bath, rinsed in Ethanol and dried.

Analytical techniques

Particle sizes were measured on a Beckman coulter LS particle size analyser. Coefficients of thermal expansion were measured in a Netzsch DIL 402C ramping at 2 °C/minute until a maximum temperature, usually equal to the dilatometric softening temperature (Ts) where the relative shrinkage exceeds a preset limit at a pressure of 1.9 kPa. XRD analysis was done in a STOE STADIP diffractometer (Cu-α) from 10°2Θ to 80°2Θ. Microscopy and element analysis were done from polished samples in a JEOL JSM 5310 LV microscope with EDS detector.

Heat treatment

Symmetrical samples of steel plates and sealing materials were assembled as shown on figure 1 by exchanging the central fuel cell with the one type of steel, type B, and with the outer interconnect plates of the second type of steel, type A. The width of the current collector plates was reduced on both anode- and cathode sides in order to accommodate the samples of sealing material inside the outer glass seals, which were similarly prepared, except for the filler material of Zirconia (YSZ) (2). A linear transducer monitored seal compression during heating sequence. In order to investigate the thermal stability of the sealing, samples were heated at 180°C per hour and exposed for 120-140 hrs to both cathode (air) and anode environments (9 vol% H_2, 3 vol% H_2O, in Ar) at three different temperatures 750°C, 850°C, 950°C and to cathode environments for 720 hrs at 750°C. The perpendicular sealing pressure was 410 kPa and 16 kPa, respectively.

310

RESULTS AND DISCUSSION

The coefficient of thermal expansion for the NAS-MgO composite increased monotonously with the concentration of Magnesia filler, as seen from figure 2, and covers the interval of interest at concentrations from 40vol% to 75vol% Magnesia. The dilatometric softening temperature, Ts, at which the seals started to yield at low sealing pressure (1.9 kPa) increased similarly, and reached the maximum value of interest for the stack sealing temperature (~950°C) at a filler concentration of approximately 32vol% Magnesia. Not too surprisingly, compressibility of the composites disappeared above approximately 60vol% Magnesia filler. Obviously, this result leaves further study with a choice between desired CTE and mechanical performance, and the subsequent selection of composites with 30 vol% Magnesia filler content for evaluation of sealing performance weighted the mechanical performance over an exact matching of CTE.

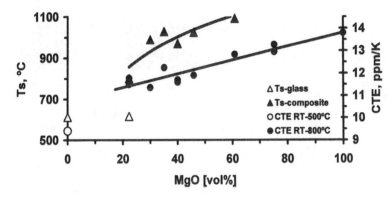

Figure 2: Coefficient of thermal expansion (CTE) and dilatometric softening temperature, Ts, of NAS glass composites with MgO filler. MgO data (9)

When heated at 180°C per hour under a high sealing pressure (410 kPa) seal composites with 30 vol% Magnesia filler started to yield already at 585-600°C, approximately 80°C above Tg, and were compressed 20% linearly at 670°C and 30% at 820°C, which meets the technological specifications for sealing performance. The sealing was tested gas tight at temperatures above 850°C.

Indications from the corrosion pattern as seen from Figure 3 are that the seal microstructure remains open when heated at low sealing pressure until about 750-800°C in contrast to heating at high sealing pressure. The lighter grey areas in the glassy part on the right half of the Figure 3 are dominated by Chromia, which most likely evaporated from the interconnect into the open pores in the glass composite until eventually, the seal wetted the surface and prevented further evaporation.

As evaluated from cross sections of the heat-treated samples and summarised in table 1, the glass adhered far better to the interconnect, type B, than to the type A, which partly may be due to the higher surface roughness on the former and partly to differences in the chemistry and microstructure of the corrosion inhibiting scale. However, further analysis is needed to clarify the latter point. In cases where glass sealing and interconnect materials had separated after heat treatment,

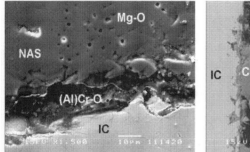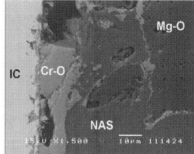

Figure 3: Interface between the glass and the interconnect, type B, after exposure to 750°C in air at different sealing pressure, 410 kPa (left) and 16 kPa (right)

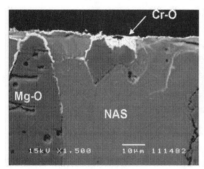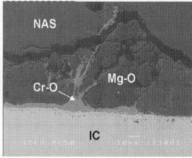

Figure 4: Interface between the glass sealing and interconnects, type A (left) and type B (right), after heat treatment at 850°C for 120 hrs in air. The interconnect A is fully separated from the glass sealing and not visible in the photo.

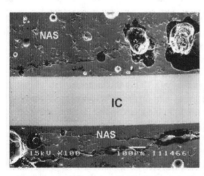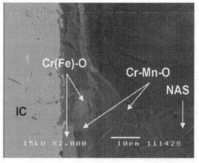

Figure 5: Interface between the glass and the interconnect type B after exposure to 950°C to both anode (left, lower) and cathode environments (left, upper). Left figure shows a detailed view of the interface on the cathode side.

the crack went along the glass-interconnect interface for the type A but followed a transversal route in the glass composite for the interfaces to interconnect, type B. Also, the degree of separation of glass composite and interconnect materials showed a correlation to the CTE-mismatch as interconnect, type A, separated more readily from the glass sealing than did interconnect, type B.

The Magnesia particles in the composite seem to stay inert at low temperature but develop more rounded contours as the temperature increases, although no crystalline reaction products were identified. Cracks between the glass matrix and larger Magnesia particles were visible, and although they seem to get less pronounced at higher temperatures, a more mechanically stable microstructure could be expected on reducing the fraction of the largest particles.

From low temperature samples, no pronounced differences between anode and cathode side of the glass-interconnect interfaces were observed but from samples exposed to 950°C, the Chromia content in the amorphous interface layer was significantly higher (3-7at%) on the cathode side than on the anode side (<1at%), where only isolated Chromia particles were observed. This difference may be due to the generally higher vapour pressure of Cr^{VI}-species on the cathode side during the initial part of the temperature profile, until the glass adheres to the interconnect surfaces and prevents further oxidation from gaseous Oxygen.

The glassy part of the composite seal appeared to be stable over time and after 720 hrs at 750°C the only clear intensity from primary crystallisation in the glass composite, as monitored by XRD, was from the Na-Al-Si-oxides, Albite and Nepheline. Same type of primary crystallisation (Albite and Nepheline) has been identified from a similar type of glass, composed like stoichiometric Albite, after nearly 4000 hours of service in SOFC-testing at 850°C. However, additional crystallisation from Cr-Mn-oxides was observed in the glass at the interconnect-glass interfaces, cf. figure 5, but these phases have not yet been identified by XRD.

At 750°C and at 850°C Chromium oxide seems to diffuse from the interconnect surface into flaws in the glass composite with very little dissolution into the glass matrix, cf. figure 3 and figure 4. The protective layer of Chromia, which usually develops on the interconnect itself after oxidation at high temperatures, seems to remain intact at 850°C but almost disappears after exposure to 950°C, where the interface layer develops into a glassy Cr-Al-Si-Oxide and forms a very strong bond between the glass matrix and the interconnect, cf. figure 5. Interconnect surfaces develop a normal corrosion resistive microstructure and the glass corrosion does not seem to change the metal durability.

Table 1: Summary of observations on seal adhesion and reactivity.

Properties of NAS-MgO glass composite after120 hrs at temperature.	750°C	850°C	950°C
Wetting	Poor	Excellent	Excellent
Adhesion to Interconnect A Adhesion to Interconnect B	Poor Some	Fair Good	Some Excellent
Corrosion between glass and interconnect	CrO_x diffuses into glass pores	CrO_x diffuses into glass pores	Cr-Al-Si-O interface in air
Glass composite microstructure	Glass wets the MgO	Glass wets MgO	Rounded MgO grains
Primary crystallisation after 720 hrs at temperature	Albite, Nepheline	n.a.	n.a.

CONCLUSION

Additions of filler material improved the mechanical stability and increased the thermal expansion of the glass composite until <11.7 ppm K^{-1}, close to the desired range, 12-13 ppm K^{-1}. The mechanical performance of the glass composite set an upper limit to the filler concentration at approximately 32 vol% Magnesia, which may be increased until 40-50 vol% and thereby also increasing the thermal expansion of the composite if sealing pressures higher than 410 kPa could be applied. The base glass, having relatively high Sodium content, shows an excellent wetting behaviour and the Tg at 515°C ensured a sufficiently low viscosity at the working temperatures. The glass composite developed gas tight sealing performance above 850°C and adhered better to the type of interconnect steels, B, on which a layer of Chromia-Manganese-spinel usually forms in air, than to the second, A, on which a dense Chromia layer usually limits corrosion in air. The glass matrix appeared stable over time and showed only small amounts of primary crystallisation even after long time exposure.

ACKNOWLEDGEMENTS

The Danish Power Utility Company, ELKRAFT, and the Ministry for Energy are acknowledged for funding to the DK-SOFC project under contract no: 103594 (FU3403). The authors also gratefully acknowledge the inspiring collaboration with both Haldor Topsøe A/S and the staff at Materials Research Department, Risø National Laboratory.

LITERATURE

(1) J.G. Larsen, P.H. Larsen and C. Bagger, "High Temperature Sealing Materials". US Patent Ser. no. 60/112039 pending.

(2) P.H. Larsen, "Sealing Materials for Solid Oxide Fuel Cells", Ph.D. thesis, Materials Research Department, Risø National Laboratory, Roskilde, Denmark. (1999).

(3) S-B. Sohn, S.-Y. Choi, G.-H. Kim, H.-S. Song and G.-D. Kim, "Stable sealing glass for planar solid oxide fuel cell", J. of Non-Crystalline Solids, 297 (2002), p 103.

(4) R.E. Loeman, H. Hofer and H. Dumm, "Evaluation of Sealing Glasses for Solid Oxide Fuel Cells", 26th Annual International Conference on Advanced Ceramics & Composites, Jan. 13-18, 2002, Cocoa Beach, FL. USA.

(5) S.B. Adler, B. T. Henderson, M. A. Wilson, D. M. Taylor and R. E. Richards, "Reference Electrode Placement and Seals in Electro-chemical Oxygen Generators", Solid State Ionics, 134 (2000), p. 35.

(6) Yang,Z., Stevenson, J.W., and Meinhardt, K.D., "Chemical Interactions of Barium-Calcium-Aluminosilicate-based Sealing Glasses with Oxidation Resistant Alloys". Solid State Ionics, 160 (2003), p213.

(7) X. Qi, F.T. Atkin, Y.S. Lin, "Ceramic-Glass High Temperature Seals for Dense Ionic-Conducting Ceramic Membranes", J. of Membrane Sci. 193 (2001), p. 185.

(8) W. Höland and G. Beall, "Glass-Ceramic Technology", The American Ceramic Society, Westerville, OH, (2002).

(9) Shaffer, P.T.B. "Handbook of High Temperature Materials" No. 1, Materials index. Plenum Press, New York, (1964).

DURABLE SEAL MATERIALS FOR PLANAR SOLID OXIDE FUEL CELLS

C. A. Lewinsohn*, S. Elangovan, and S.M. Quist
Ceramatec Inc.
2425 South 900 West
Salt Lake City
Utah, 84119
USA

ABSTRACT

Planar solid oxide fuel cells require a seal to separate air and fuel and to bond components together. The development of seals with acceptable lifetime and performance has proven to be technically challenging. In this paper, a novel method to form durable seals that meet the requirements for commercial application will be described. The seal material is a composite with a preceramic polymer-derived matrix and particulate fillers. Partial pyrolysis is used to control shrinkage during pyrolysis of the seal material. Fillers are selected to adjust the seal material's physical properties. Seal compliance, and its effect on stack stresses, will also be discussed.

INTRODUCTION

Solid Oxide Fuel Cells (SOFC) convert chemical energy to electrical energy directly from a variety of fuels, and thus offer the potential for high-efficiency stationary and mobile power generation with lower emissions than current, commercial power systems. Planar, SOFC designs offer high power density per unit volume and lower manufacturing costs than other designs. In planar SOFC designs a seal is required to prohibit fuel and air from mixing and decreasing the oxygen gradient required for operation. These seals must be thermomechanically stable at high temperatures (700-850°C), be highly impermeable (in order to prevent mixing of the reducing and oxidizing atmospheres), be chemically compatible with the other SOFC materials, have a similar coefficient of thermal expansion (CTE) to the materials that they seal, and be electrically insulating. Glass-ceramic, glass, high-temperature cement, and mechanical (compressive) seals have been reported (1-7). Current seals do not meet the performance criteria for commercially viable SOFC systems. In particular, seal materials and designs that are capable of allowing cells and stacks to survive planned and unplanned thermal cycles, are compatible with SOFC component materials and environments, are mechanically and chemically stable for the projected lifetime of a commercial SOFC (40,000 h for stationary systems, or at least 5,000 h and 3,000 thermal cycles for transportation systems), and can be fabricated cost-effectively must be developed in order for systems utilizing SOFCs for power generation to be viable.

Non-oxide materials offer the potential for chemically stable and mechanically durable seals, however elevated temperatures are usually required to densify these materials. Furthermore, non-oxide materials typically have significantly lower values of thermal expansion than SOFC materials. Certain silicocarbon polymers, however, can be pyrolysed at relatively low temperatures to form extremely stable, covalently bonded, amorphous non-oxide materials. These materials have compositions in the field of Si-O-B-C-N. It has been shown that by incorporation of certain fillers into these polymers that the thermal expansion of the resulting composite material can be made to match that of SOFC materials (8). Preliminary results indicated that composite seal materials, derived from polymer precursors, could be used to make seals with little change in leak rate after a number of thermal cycles (8). In this work, experiments aimed at investigating the effect of various fillers and process parameters on the mechanical properties of the seal materials were performed.

In addition to minimizing the mismatch in thermal expansion among SOFC components, stresses in critical cell components (i.e. electrolyte, anode, and cathode) can be reduced by imparting mechanical compliance to the cells or stack. The seals between interconnects or interconnects and cells offer the potential for imparting compliance via adjusting the seal material properties or cell and stack design. In this study, finite element analysis was used to investigate the effect of seal parameters on stresses in cells.

EXPERIMENTAL

Two commercially available precursor polymers were used in this study: an allyl-hydridopolycarbosilane (aHPCS, Starfire Systems, Inc.), and a methyl siloxane (Oxycarbide-A, Starfire Systems, Inc.). Powders were prepared by partially pyrolysing the polymers at temperatures below those that lead to full conversion from the silicocarbon polymer to a covalently bonded ceramic. The resulting solid material was crushed into powder and milled. The milled powder was blended with fresh polymer and filler powders and formed into seals or pressed into bars and pellets for additional characterization. In order to obtain compositions with thermal expansion properties similar to those of other fuel cell components, filler powders were selected based upon criteria of a melting temperature above 1000°C and a coefficient of thermal expansion (CTE) above 10×10^{-6} °C^{-1}. The details of the processing routes are proprietary.

Experiments were performed to determine whether the materials developed would affect SOFC performance and the capability of the seal materials to withstand thermal cycling. Strontium and magnesium doped, lanthanum gallate cells were sealed to zirconia tubes (Fig. 1) and tested in button cell test apparatus (Fig. 2). The cells were cycled three times to investigate the effect of thermal cycling. Each cycle consisted of heating the cells to 800°C in 8 h and cooling them to room temperature in 8 h. The open circuit voltage was measured as a function of temperature to indicate the seal performance.

To evaluate the effect of the filler material on the mechanical properties of candidate seal materials, bar shaped specimens were made by pressing mixtures of polymer precursor, powder from partially pyrolysed precursor, and filler materials. The bars were then completely pyrolysed in an inert atmosphere. The strength of the resulting bars was

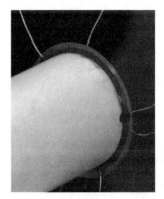

Figure 1   Photographs of doped, lanthanum gallate fuel cells sealed to zirconia tubes.

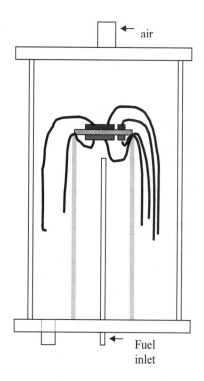

Figure 2   Schematic illustration of a button cell test apparatus.

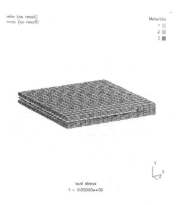

Figure 3.   Finite element model of a planar solid oxide fuel cell (light) and interconnects (dark).

tested in four-point bending according to ASTM C1161.

Finite element methods were used to evaluate the effect of seal properties on stresses in the electrolyte and seal material. A simple model of a 10 cm^w x 10 cm^l stack was used (Fig. 3). A single cell of the stack was analysed and, due to symmetry, one quadrant of a repeating unit was meshed. Boundary conditions were placed on the model to simulate the effects of the rest of the stack: the top surface of the stack was unconstrained, however the displacements of the bottom surface were fixed in the vertical direction. The stresses caused by cooling the stack from the sealing temperature, 1000°C, were determined.

RESULTS & DISCUSSION

As shown in Table I, preliminary results indicated that composite seals, consisting of polymer-derived material and high thermal expansion filler materials, had thermal expansion behavior within the range of other SOFC materials. Table II shows that seals made from the composition exhibiting acceptable thermal expansion behavior allowed fuel cell operation and showed very little degradation after three thermal cycles. These results, as well as those reported previously, demonstrate the feasibility of using materials derived from silicocarbon polymer precursors to form durable seals for planar SOFCs.

Results of FEA modeling are shown in Table III. The elastic modulus of the seal material was varied between 20 and 200 GPa and the resultant stresses in the electrolyte and seal were calculated.   The elastic modulus of materials derived from non-oxide, preceramic polymer precursors is expected to be in the range of 100-150 GPa. The results indicate that the stresses in the electrolyte are not strongly influenced by the compliance of the seal material. Likewise, the maximum principal (tensile) stress in the seal is strongly influenced by the seal compliance. On

Table I
CTE values

Composition	Temperature Range (°C)	CTE (ppm/ °C)
8 mol% yttria-doped zirconia	25-1000	10.6-11.1
Seal 1 (aHPCS + filler)	200-700	10.0

Table II
OCV values

Temperature (°C)	Seal 1
600	1.096 V
900	1.008 V
cooled to 50°C	
600	1.085
900	0.949
cooled to 50°C	
600	1.05
900	0.948
cooled to 50°C	
600	1.042
900	0.961

Table III
FEA Results

Seal Modulus (GPa)	Electrolyte		Seal	
	Maximum Principal Stress (MPa)	Maximum Shear Stress (MPa)	Maximum Principal Stress (MPa)	Maximum Shear Stress (MPa)
20	24.4	147	48.4	36.4
100	26.6	145	45.8	75.1
200	27.1	143	45.8	138

the other hand, the shear stress in the seal is a strong function of the seal compliance. Although the magnitude of the stresses predicted by the finite element model should not be considered accurate, the results indicate that a seal with a low elastic modulus will prevent high shear stresses from occurring within the seal. In addition, the shear strength of most silicate glasses, a popular seal material, is typically around 30 MPa (9); pyrolysed preceramic polymer materials, 60 MPa.

Bend bars made from polymer blended with partially pyrolysed polymer powder were extremely weak: the four-point, flexural strength of this material was 8.4 MPa with a 95% confidence level of 1.8 MPa. Bars made with the filler powders used to obtain acceptable thermal expansion and thermal cycling behavior, however, exhibited a flexural strength of 157 MPa, with a 95% confidence level of 11.5 MPa. This is in contrast to the tensile strength of silicate glasses, which ranges from 60-100 MPa (9). These results, and those of the finite element modeling, indicate that a non-oxide material, derived from preceramic polymer precursors, with a low elastic modulus offers the potential for an extremely reliable seal for planar solid oxide fuel cells.

SUMMARY

It has been demonstrated that by the addition of appropriate fillers, non-oxide, preceramic polymer precursors can be pyrolysed to form materials with similar thermal expansion to SOFC materials. Furthermore, the materials have been shown to provide sealing capability to button cell tests and can survive thermal cycling. Furthermore, the seal materials exhibit promising flexural strength values. The results reported in this paper indicate that a non-oxide material, derived from preceramic polymer precursors, with a low elastic modulus offers the potential for an extremely reliable seal for planar solid oxide fuel cells.

ACKNOWLEDGEMENTS

The authors would like to thank Ms. K. Cameron, Mr. D. Larsen, and Mr. J. Pearce for their assistance preparing and characterizing specimens, and Mr. M. Timper for performing fuel cell measurements. This work was funded by the U.S. Department of Energy (DOE) under Phase I Small Business Innovations Research Projects granted to Ceramatec, Inc..

REFERENCES
1. N.Q. Minh, "Ceramic Fuel Cells," J. Am. Ceram. Soc., 76 [3], pp. 563-588 (1993).
2. F. Tietz, Ionics, 5, p. 129 (1999).
3. N. Lahl, L. Singheiser, K. Hilpert, K. Singh, D. Bahadur, Solid Oxide Fuel Cells-VI, S. Singhal, M. Dokiya (Eds.), Electrochemical Society, Pennington, NJ, PV 99-19, p. 1057, 1999.

4. Y. Sakai, M. Hattori, Y. Esaki, S. Ohara, T. Fukui, K. Kodera, Y. Kubo, Solid Oxide Fuel Cells-V, U. Stimming, S. Singhal, H. Tagawa, W. Lehnert (Eds.), Electrochemical Society, Pennington, NJ, PV 97-40, p. 652, 1997.
5. C. Gunther, G. Hofer, W. Kleinlein, , Solid Oxide Fuel Cells-V, U. Stimming, S. Singhal, H. Tagawa, W. Lehnert (Eds.), Electrochemical Society, Pennington, NJ, PV 97-40, p. 746, 1997.
6. P. Larsen, C. Bagger, M. Mogensen, J. Larsen, , Solid Oxide Fuel Cells-IV, vol. 69, M. Dokiya, O. Yamamoto, H. Tagawa, S. Singhal (Eds.), Electrochemical Society, Pennington, NJ, PV 95-1, p. 652, 1995.
7. S.P. Simner, J.W. Stevenson, "Compressive Mica Seals for SOFC Applications," J. Pow. Sources, 102, 310-316 (2001).
8. C.A. Lewinsohn, P. Colombo, I. Reimanis, and O. Unal, "Stresses During Joining Ceramics Using Preceramic Polymers," J. Am. Ceram. Soc., 84 [10], 2240-2244 (2001).
9. D.C. Boyd and D.A. Thompson, "Glass," pp. 807-880 in Kirk-Othmer Encyclopedia of Chemical Technology, Volume 11, 3rd Edition, J. Wiley & Sons, 1980.

DEVELOPMENT OF A COMPLIANT SEAL FOR USE IN PLANAR SOLID OXIDE FUEL CELLS

K. Scott Weil and John S. Hardy
Pacific Northwest National Laboratory
P.O. Box 999
Richland, WA 99352

ABSTRACT

We have developed a deformable seal for planar solid oxide fuel cells (pSOFCs) that can accommodate significant thermal mismatch between the adjoining components and still remain hermetic. Essentially composed of a thin stamped metal foil bonded to both sealing surfaces, the seal offers a quasi-dynamic mechanical response to thermally generated stresses. It remains well adhered to both faying surfaces, i.e. non-sliding, but readily yields or deforms under modest thermo-mechanical loading, thereby mitigating the transfer of these stresses to the adjacent ceramic and metal components. Initial thermal testing demonstrates that the seal retains its original hermeticity and strength after a number of rapid cycles between ~75°C and 750°C.

INTRODUCTION

One of the critical issues in designing and fabricating a high performance planar solid oxide fuel cell (pSOFC) system is developing an appropriate means of hermetically sealing the metal and ceramic components in the stack. At present, there are essentially two standard methods of sealing: (1) by forming a rigid joint or (2) by constructing a compressive "sliding" seal. Each type of seal has its own set of advantages and design constraints. For example, rigid glass joining is a cost-effective and relatively simple method of bonding ceramic to metal. However, the softening point of the glass limits the maximum operating temperature to which the glass joint may be exposed. In addition, because the resulting glass-ceramic is a brittle material and forms a non-dynamic, low-yielding seal, it is imperative that the temperature dependent coefficient of thermal expansion (CTE) of each joining component, i.e. the ceramic cell, the seal, and the metal separator, be approximately equal. If not, high thermal stresses can develop within the components during stack heat-up and/or cool-down, causing fracture of the cell or seal. Currently only a narrow range of high temperature glass compositions within the borate- or phosphate-doped aluminosilicate families display coefficients of thermal expansion that closely match those of the ferritic stainless steels commonly employed in device separators and housings. Unfortunately, these glasses typically display signs of devitrification within the first few hours of exposure at operating temperature.[1] As the glass begins to crystallize, its carefully engineered thermal expansion properties change significantly, ultimately limiting the number of thermal cycles and the rate of cycling at which the resulting joint is capable of surviving.

Alternatively, in compressive sealing a compliant high-temperature material is captured between the two sealing surfaces and compressed, using a load frame external to the stack, to deliver hermetic sealing in the same way rubber gaskets are used in everyday appliances.[2] Because the seal conforms to both sealing surfaces and is under constant compression during use, it forms a dynamic seal. That is, the sealing surfaces can slide past one another without disrupting the hermeticity of the seal and CTE matching is not required between the ceramic cell and the metal separator. At present however, this technology remains incomplete due to the lack of a reliable high-temperature sealing material that would form the basis of the compliant seal. A number of materials have been considered, including mica, nickel, and copper, but each has been found deficient for any number of reasons, ranging from oxidation resistance in the case of the metals to poor hermeticity and through-seal leakage with respect to the mica.[3]

We are currently developing a third sealing alternative for pSOFCs, the bonded compliant seal, which in concept incorporates the advantages of both rigid and compressive sealing. That is, the seal should provide excellent hermeticity (comparable to that observed with glass-ceramic seals) and high mechanical integrity under thermal cycling and mechanical vibration (i.e. the seal is non-brittle and quasi-dynamic), as well as mitigate any mismatch stresses that arise between adjacent components by "trapping" much of the stress as elastic or plastic strain within a thin, deformable sealing membrane. One of the primary advantages that this type of seal offers is that a much wider range of alloy compositions can be considered for use in the pSOFC interconnect. At present, the candidate list is severely limited to those that offer good CTE matching with the ceramic cell, namely the chromia scale-forming ferritic stainless steels, which in general display significantly poorer mechanical, oxidation, and through-scale electrical properties than their nickel-based counterparts.[4]

EXPERIMENTAL

Materials

The sealing technology that we have developed relies on an elasto-plastically deformable metal foil membrane that is bonded to the adjacent ceramic and metal component surfaces. A room temperature analog of this concept is an elastomeric gasket coated on either side with a sticky adhesive such that the gasket adheres to both sealing surfaces. Because the gasket is made from a low stiffness material, it readily deforms in response to stresses generated at the interfaces with the neighboring components, minimizing the transfer of stresses from the sealing gasket to the adjacent materials. In the compliant seal design envisioned here, the stiffness of the seal is minimized not only by proper selection of the foil membrane material, but also by appropriate geometric design of the membrane.

There are potentially a number of high temperature alloys that can be considered in this application, including: nickel- and cobalt-based superalloys and other aluminum- and/or chromium-containing non-ferrous alloys; stainless steel and iron-based superalloys; and noble metals such as silver, gold, palladium, and platinum. For the initial proof-of-concept development effort, our materials selection process focused on four key properties: high oxidation resistance, low stiffness, high ductility, and acceptable cost. After an initial materials screening analysis based on these factors, we chose to use a commercially available alumina-forming ferritic steel as the foil membrane: DuraFoil (22% Cr, 7% Al, 0.1%La+Ce, bal. Fe, manufactured by Engineered Materials Solutions, Inc.). Supplied as 50μm thick sheet, the DuraFoil was sheared into 3 cm x 3 cm samples, annealed in vacuum at 900°C for 2hrs, and stamped into cap-shaped washers using a die designed specifically for this purpose, as shown in Figures 1(a) and (b). The stamped foils were ultrasonically cleaned in soap and water, then flushed with acetone to remove the lubricant from the stamping operation.

Each foil washer was bonded to a NiO-5YSZ bilayer disc that was prototypical in composition, layer thickness, and material density to the anode-supported membranes commonly used in mid-temperature pSOFC stacks. Formed by tape casting and lamination, the NiO and 5YSZ layers were approximately 600μm and 7μm in thickness, respectively, while the overall sintered discs measured ~25mm in diameter. The other substrate to which the foil seal was bonded was a 6.2mm thick Haynes 214 washer with an outer diameter of 44mm and an inner diameter of 15mm. As an alumina-scale forming nickel-based superalloy, Haynes 214 displays excellent oxidation resistance at temperatures in excess of 1000°C, but also exhibits an average CTE of 15.7 μm/m·K, which is nearly 50% higher than that of the anode-supported bilayer (CTE = 10.6 μm/m ·K). The stamped DuraFoil component was joined to the Haynes 214 washer using BNi-2 braze tape (Wall Colmonoy, Inc.). A second brazing operation using an in-house prepared braze, 4mol% CuO in silver,[5] was conducted at 1000°C in air for 15min to join the top side of the foil to the YSZ side of the ceramic bilayer disc, thereby forming the test specimen.

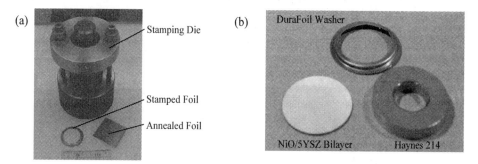

FIGURE 1   (a) The stamping die and (b) sealing specimen components used in fabricating the foil seals.

Testing and Characterization

Rupture strength testing was employed to facilitate quantitative comparison of the RAB seal joint strengths.[6] Figure 2 shows a schematic of the rupture test equipment. The test sample is placed within a fixture that consists of a bottom and top flange, a coupling that secures and centers the two flanges, and an o-ring that is squeezed against the bottom surface of the Haynes 214 washer. Compressed air was used to pressurize the back-side of the washer specimen up to a maximum rated pressure of 150psi. A digital regulator allows the pressure behind the joined bilayer disk to be slowly increased to a given set point. This volume of compressed gas can be isolated between the specimen and a valve, making it possible to identify a leak in the seal by a decay in pressure. In this way, the device can be used to measure the hermeticity of a given seal configuration without causing destructive failure of the seal. Alternatively, by increasing the pressure to the point of specimen rupture, we can measure the maximum pressure that the specimen can withstand. Thermal cycle testing was performed by heating the specimens in air to 750°C in ten minutes using an infrared furnace, holding at temperature for ten minutes, and cooling to ≤70°C in forty minutes before re-heating under the same conditions for a given number of cycles. A minimum of six specimens was tested for each test condition. Microstructural analysis of the joints was conducted on polished cross-sectioned samples using a JEOL JSM-5900LV scanning electron microscope (SEM) equipped with an Oxford energy dispersive X-ray analysis (EDX) system that employs a windowless detector.

RESULTS AND DISCUSSION

Shown in Figure 3(a) is a composite cross-sectional micrograph of an as-joined rupture specimen. The sample was well sealed, as determined by hermeticity testing conducted prior to metallography. The entire seal is approximately 1.1mm thick, although it is expected that this can be readily reduced simply by altering the geometry of the DuraFoil stamping. As seen in the sequence of higher magnification micrographs in Figures 4(a) – (c), the two brazes each account for 50 – 100μm of the overall seal thickness. Also apparent from Figure 3(a) are changes to the geometry of the foil washer that occurred during joining. For example, the BNi-2 braze causes the outer periphery of the foil to curl due to a mismatch in CTE between the two materials. While this does not degrade the perfomance of the rupture test specimen, it may affect the performance of the seal in the stack. A conceptual drawing of the seal in an actual application is shown in figure 3(b). Sealing occurs around the gas manifold holes, which are incorporated into both the ceramic ell and the separator plate.

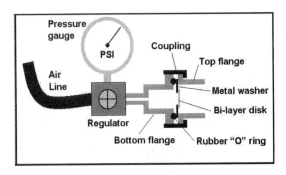

FIGURE 2   Rupture strength test schematic.

On the ceramic side of the seal, the silver braze region is thicker in-board of the specimen than out. As seen in Figures 4(b) and (c), this braze appears to form a robust joint between the 5YSZ and the alumina scale of the DuraFoil. While EDX analysis shows no indication of a reaction zone at the 5YSZ/braze interface, a 10 – 15μm thick zone appears to have formed on the DuraFoil due to reaction between the Al_2O_3 scale and the CuO in the braze. The dominant product in this interfacial region is a mixed oxide phase, $2CuO \cdot Al_2O_3$.

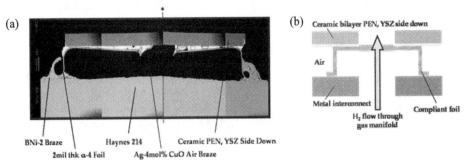

FIGURE 3   (a) A composite cross-sectional micrograph of an as-joined rupture test specimen. (b) A schematic of the bonded compliant sealing concept in use in pSOFC gas manifold sealing.

Results from rupture strength testing of the sealed specimens are shown in Figure 5. All of the tested specimens were found to be hermetic up to the maximum pressure (60 psi) tested

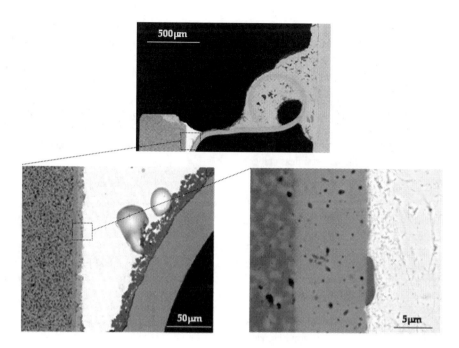

FIGURE 4   Cross-sectional SEM micrographs of the YSZ-silver braze region of an as-sealed rupture specimen at the following magnifications: (a) 52X, (b) 450X, and (c), 4000X.

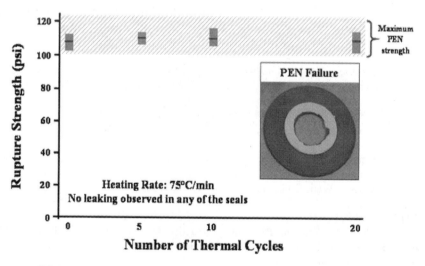

FIGURE 5   Rupture strength of the bonded compliant joint specimens in the as-joined and as-cycled conditions. Inset: typical mode of failure observed in the rupture specimens, within the center of the ceramic disc.

during initial leak testing. More extensive pressure testing up to the point of rupture indicated no failure in any of the seals, even in the specimens that underwent as many as twenty rapid thermal cycles. Instead in each case, failure occurred within the center of the ceramic disc (of the type shown in the inset in Figure 5), revealing this as the weakest component in the current specimen configuration. By way of comparison, similar rupture specimens constructed of the same NiO/5YSZ bilayer discs joined directly to 3mm thick 430 stainless steel washers using a typical barium calcium aluminosilicate pSOFC sealing glass displayed an average strength of only 10.5psi in the as-joined condition, which degraded to nearly zero upon rapid thermal cycling out to 10 cycles.[6] Failure in these glass sealed specimens occurs along the interface between the sealing glass and the 430 stainless steel.

CONCLUSIONS

One of the major challenges remaining before planar SOFCs can be commercialized is developing a robust cell-to-separator seal that maintains its hemeticity under varied operating conditions over the lifetime of the stack. Of particular interest is designing a seal that allows us to use better performing, but higher CTE, nickel-based superalloys in the separator plate. In this regard, our work on a new pSOFC sealing concept, the bonded compliant seal, shows some promise. In the present embodiment, a thin, oxidation-resisitant metal foil is employed as an elasto-plastically deformable membrane that when bonded to the ceramic cell and metal separator forms a hermetic seal. Aggressive thermal cycle testing indicates that this seal retains its original hermeticity and bond strength even when it is used to join materials with a significant difference in thermal expansion behavior.

ACKNOWLEDGMENTS

The authors would like to thank Nat Saenz, Shelly Carlson, and Jim Coleman for their assistance. This work was supported by the U. S. Department of Energy as part of the SECA Program. The Pacific Northwest National Laboratory is operated by Battelle Memorial Institute for the United States Department of Energy (U.S. DOE) under Contract DE-AC06-76RLO 1830.

REFERENCES

[1] K. Eichler, G. Solow, P. Otschik, and W. Schaffrath, "BAS (BaO·Al$_2$O$_3$·SiO$_2$) Glasses for High Temperature Applications," *Journal of the European Ceramic Society*, **19** [6-7] 1101-4 (1999).

[2] J. Kim and A. Virkar, "The Effect of Anode Thickness on the Performance of Anode-Supported Solid Oxide Fuel Cells," *Proceedings of the Electrochemical Society*, **99-19**, 830-6 (1999).

[3] S.P. Simner and J.W. Stevenson, "Compressive Mica Seals for SOFC Applications," *Journal of Power Sources*, **102** [1-2] 310-6 (2001).

[4] Z. Yang, K.S. Weil, D.M. Paxton, and J.W. Stevenson, "Selection and Evaluation of Heat-Resistant Alloys for SOFC Interconnect Applications," *Journal of the Electrochemical Society*, **150** [9] A1188-201 (2003).

[5] K.S. Weil, J.Y. Kim, and J.S. Hardy, "Reactive Air Brazing: A New Method of Sealing Solid Oxide Fuel Cells and Other High Temperature Electrochemical Devices," *Journal of Materials Research*, in review.

[6] K.S. Weil, J.E. Deibler, J.S. Hardy, D.S. Kim, G.G. Xia, L. A. Chick, and C.A. Coyle, "Rupture Testing as a Tool for Developing Planar Solid Oxide Fuel Cell Seals, *Journal of Materials Engineering and Performance*, in review.

A COMPARISON OF THE ELECTRICAL PROPERTIES OF YSZ PROCESSED USING TRADITIONAL, FAST-FIRE, AND MICROWAVE SINTERING TECHNIQUES.

Michael Ugorek and Doreen Edwards
School of Engineering
New York State College of Ceramics
Alfred University
Alfred, NY 14802

Dr. Holly Shulman
Ceralink Inc.
Alfred, NY 14802

ABSTRACT

Yttria-stabilized zirconia (YSZ) is an oxygen-conducting electrolyte used in solid oxide fuel cells (SOFCs). In this work, three different sintering techniques – traditional (low heating rate), fast-fire, and microwave – were investigated as a means of sintering 8-mol% YSZ. YSZ powders were isostatically pressed into pellets and sintered at $1350^{\circ}C$ – $1500^{\circ}C$ with rates ranging from $3^{\circ}C/min$ (traditional) to 60 - $100^{\circ}C/min$ (microwave and fast-fire). Sintered pellets were characterized using scanning electron microscopy (SEM) to determine microstructure and impedance spectroscopy to determine high-temperature electrical properties. The electrical conductivity of the sintered samples ranged from 0.08 to 0.20 S/cm at $1000^{\circ}C$. While the dependency of density and grain-size on sintering temperature was different for the three techniques, there were no significant differences in the relationships of density and electrical conductivity to grain size.

INTRODUCTION

Yttria-stabilized zirconia (YSZ) is widely used as an electrolyte in solid oxide fuel cells (SOFC). Microwave sintering has been considered as an alternative to conventional sintering of YSZ because of benefits related to lower sintering temperature, lower energy requirements, and reduced thermal-mismatch stresses.[1-3]

In this work, three different sintering techniques – traditional (low heating rate), fast-fire, and microwave – were used to sinter ZrO_2 doped with 8 mol% Y_2O_3. The sintered samples were characterized to determine the relationships between density, grain-size, and electrical conductivity for samples processed using the three different methods.

EXPERIMENTAL PROCEDURE

One-gram, ½-inch diameter pellets were prepared from Tosoh TZ-8YS powder by unaxial pressing at 80 MPa followed by cold isostatic pressing at 240 MPa. The pellets were sintered to 1350 - $1500^{\circ}C$ via traditional ($3^{\circ}C/min$), microwave (60-$100^{\circ}C/min$), and fast-fire (60-$100^{\circ}C/min$) techniques with no dwell time at peak temperature. Microwave sintering was conducted using a 1.3 kW, 2.45 GHz variable-power system (ThermWAVE, Research Microwave Systems, USA). Samples were placed between two SiC susceptors inside a refractory-board container and heated using 50% power output until $500^{\circ}C$, followed by 100% power output to the desired sintering temperature. Fast-fire sintering was conducted in a vertical tube furnace with computer-controlled sample feed, which allowed the heating profiles to be programmed to mimic the heating profile measured during the microwave heating runs. For both techniques, temperature was measured using a thermocouple placed near, but not touching, the sample. The thermocouple used during microwave heating was shielded.

Pellet density was measured using Archemedes' method. SEM was conducted on the pellet surfaces after polishing to 1 μm. Grain size was determined using a circular-intercept method.[4] Impedance spectroscopy was conducted on samples electroded with platinum paste. Measurements were conducted from 100°C to 1000°C in air, using a Solatron 1260 equipped with custom control software. The samples were heated at 10°C/min, with a dwell time of 10 min at each measurement temperature. An excitation potential of 1 V was used, and frequency was scanned from 10MHz to 5Hz. The resulting spectra were analyzed using the Zview software package (Scribner and Associates, USA).

RESULTS AND DISCUSSION

Figure 1 shows the density and average grain size of the YSZ samples as a function of sintering temperature. The densities of traditionally processed samples were higher than the corresponding densities of the either the microwave-processed or fast-fired samples over the entire temperature range investigated. This difference is attributed primarily to the longer sintering times associated with the slower heating rates.

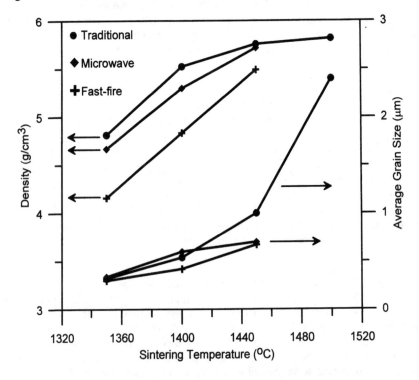

Figure 1. Density and average grain size of YSZ as a function of sintering temperature.

Below 1450°C, the density achieved at a given sintering temperature depends strongly on the method used. At 1450°C, the densities of the samples sintered using traditional (low-rate) and microwave heating are similar to each other, but greater than that of the fast-fired sample. This demonstrates that microwave sintering can achieve equivalent-density YSZ in a shorter time than can be achieved by either traditional (low-rate) sintering or fast-fire sintering. While the higher density of the microwave-sintered sample compared to that of the fast-fired sample processed with the same heating profile may suggest the existence of a so-called "microwave effect", it may also reflect differences in the average temperature of the two samples. Although the thermocouples were placed within a few millimeters of the sample surface for both techniques, the internal temperature of the fast-fired samples is expected to lag behind the surface temperature, whereas the internal temperature during microwave sintering may be higher than that at the surface due to volumetric heating effects.

At 1350°C, there is no notable difference in the grain size of the samples. With increasing sintering temperature, the traditionally fired samples show a larger increase in grain size than do the samples processed with the other two methods. As with the differences in density, this increase in grain size can be attributed in part to the longer sintering time associated with the lower heating rate.

Figure 2 shows representative impedance data collected at three different temperatures. The

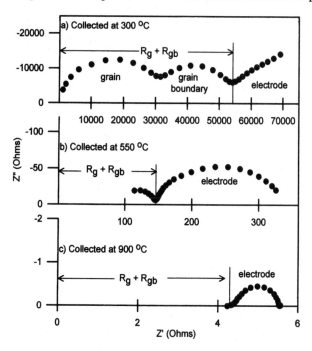

Figure 2. Representative impedance spectra collected at a) 300°C,
b) 550°C, and c) 900°C for YSZ .

spectrum collected at 300°C (Figure 2a) shows three arcs, which correspond to grain, grain-boundary, and electrode contributions. For the spectrum collected at 550°C (Figure 2b), only two arcs are present, corresponding to grain-boundary and electrode contributions. Because the frequency response of the grain interior lies outside the measurement range, the grain contribution was modeled as an offset resistance. For the spectrum collected at 900°C (Figure 2c), only one arc, corresponding to the electrode contribution, is observed. In this case, the sample resistance (grain plus grain-boundary) is taken as the offset-resistance. The sample resistance (determined by combining the contributions of the grain and grain boundary resistances) were extracted from the spectra and used to calculate the sample conductivity.

Figure 3 is a representative plot of the conductivity data for a set of YSZ samples sintered at a given temperature using the three different methods. For ionic conductors, the conductivity is expected to depend on temperature as

$$\sigma = \frac{\sigma_o}{T} e^{-Ea/kT} \tag{1}$$

where σ is conductivity, σ_o is a constant, T is measurement temperature in Kelvin, E_a is activation energy, and k is Boltmann's constant. Values for the activation energies and σ_o were extracted from the plots of ln (σT) vs. 1000/T and are summarized in Table I, along with the statistics of the linear least-squares fit to the data. The calculated activation energies range from 0.87 to 0.95 eV. For a given sintering method, the activation energy increases with decreasing sintering temperature, which is attributed to the larger number of grain boundaries in the samples processed at lower sintering temperatures. The electrical conductivity of the samples at

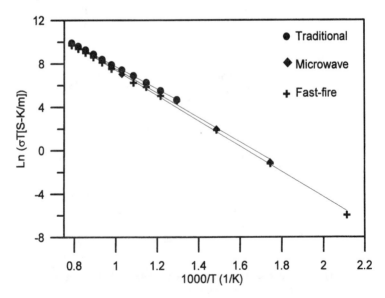

Figure 3. Electrical conductivity of YSZ as a function of measurement temperature.

1000°C ranged from 0.08 to 0.20 S/cm, which is comparable to conductivity values reported previously for microwave sintered YSZ.[3]

Table I. Electrical Conductivity Parameters of YSZ*

Sintering Method	Sintering Temp. (°C)	σ_0 (S/cm)	E_A (eV)	R^2
Traditional	1350	7.5×10^5	0.93	0.999
	1400	6.9×10^5	0.89	0.999
	1450	6.4×10^5	0.88	0.995
	1500	6.7×10^5	0.86	0.999
Microwave	1350	6.5×10^5	0.92	0.995
	1400	8.4×10^5	0.92	0.997
	1450	6.0×10^5	0.87	0.999
Fast-fire	1350	6.7×10^5	0.95	0.999
	1400	8.5×10^5	0.95	0.993
	1450	8.4×10^5	0.92	0.991

* Conductivity at any temperature can be calculated as $\sigma = (\sigma_0/T) \exp(-E_A/kT)$

Figure 4 shows the dependence of density and conductivity at 1000°C on the grain size for all three sintering techniques. All three techniques exhibit a similar density-*vs.*-grain-size

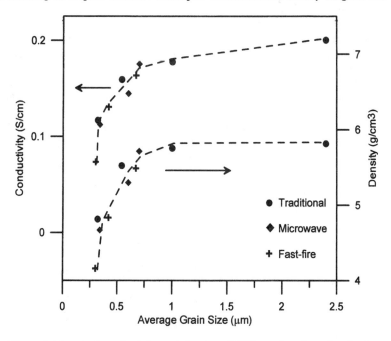

Figure 4. Electrical conductivity and density of YSZ as a function of grain size.

relationships, which suggests that the relative contribution of lattice, grain-boundary, and surface diffusion in 8 mol% YSZ does not change significantly with heating rate or with the sintering method used.

The conductivity of 8 mol% YSZ increases with increasing grain size, reflecting the significant contribution of grain boundaries to sample resistance. The conductivity-*vs.*-grain-size relationship is similar for all three methods, which suggests that the local electrical properties of the grain and grain-boundary regions do not depend on the sintering method used.

CONCLUSIONS

The microstructure and electrical properties of 8 mol% YSZ sintered using three different techniques were compared. Although the dependency of density and grain size on sintering temperature is different for the three techniques, the relationship of density and electrical conductivity to grain size is similar. This implies that any of the three techniques can be used to achieve 8 mol% YSZ with desired characteristics, but that the time required to achieve the desired microstructure will be different.

ACKNOWLEDGEMENTS

This work was supported, in part, by the Center for Advanced Ceramic Technology (CACT) at the New York State College of Ceramics at Alfred University.

REFERENCES

[1]S.A. Nightingale, H.K. Worner, and D.P. Dunne, "Microstructural Development during the Microwave Sintering of Yttria-Zirconia Ceramics," *Journal of the American Ceramic Society,* **80** [2] 394-400 (1997).

[2]M.A. Janney, C.L. Calhoun, and H.D. Kimrey, "Microwave Sintering of Solid Oxide Fuel Cell Materials: I, Zirconia-8 mol% Yttria," *Journal of the American Ceramic Society,* **75** [2] 341-46 (1992).

[3]F.T. Ciacchi, S.A. Nightingale, and S.P.S. Badwal, "Microwave Sintering of Zirconia-Yttria Electrolytes and Measurement of their Ionic Conductivity," *Solid State Ionics,* **86-88** 1167-72 (1996).

[4]Z. Jeffries, E. Met, A.H. Kline, and E.B. Zimmer, "The Determination of Grain Size in Metals," *Transactions,* American Institute of Mining and Metallurgical Engineers, **54** 594-607 (1917).

ENHANCEMENT OF YSZ ELECTROLYTE THIN FILM GROWTH RATE FOR FUEL CELL APPLICATIONS

Zhigang Xu and Jag Sankar
Center for Advanced Materials and Smart Structures
North Carolina A&T State University
Greensboro, NC 27411

ABSTRACT

Some of the factors that control the film growth rate in combustion chemical vapor deposition (CCVD) are the concentration of species available near the surface of the substrate to be coated, film growth mode, and the fluid mechanics of the diffusion boundary layer next to the substrate surface. We have proved that by increasing the precursor concentration in an appropriate range, the film growth rate can be increased linearly. The film growth mode can be controlled by substrate temperature. At a lower temperature regime, film growth was in a surface reaction controlled mode; while in the higher temperature regime, film grew in a gas phase diffusion controlled mode. Moreover, the effect of thermophoresis on the film growth was also analyzed. An increase in the temperature gradient near substrate surface enhances the momentum of species providing a faster film growth. The films were studied by scanning electron microscopy.

INTRODUCTION

Solid oxide fuel cells have many advantages over other kinds of fuel cells. However, the high cost of the SOFCs obstructs their commercialization. Many scientist and researchers are making endeavors to reduce the cost. Application of thin film electrolyte is a major measure besides the invention of new cell materials[1]. SOFCs with thin film electrolytes are expected to work at higher efficiencies and can be operated at lower temperatures. By lowering of the operation temperature, cheap materials can be used to construct the fuel cells. Therefore, both the manufacturing and running cost of the SOFCs can be reduced.

Combustion chemical vapor deposition (CCVD) has been shown to be a versatile technique to process metal oxide coatings for various applications[2]. Recently, it has also been used to process complex oxide films like yttria-stabilized zirconia (YSZ) electrolyte thin films for applications in solid oxide fuel cells[3]. Low-cost processing is of vital importance to the development fuel cells. The advantage of low capital investment of the CCVD system and its open-air operating feature provides a potential of low cost thin film processing. High film growth rates are also highly desired for cost purposes. In this paper, experimental studies and results towards the enhancement of YSZ thin films will be presented.

EXPERIMENTALS

The liquid fuel CCVD system and experimental procedures have been introduced elsewhere[4]. Zirconium 2-ethylhexanoate (Zr-2EH) and Yttrium 2-ethylhexanoate (Y-2EH) were chosen as reagents, which were dissolved in toluene separately. The solutions were delivered by a HPLC pump to the nebulizer with a precise flow control. At the exit of the nebulizer, aerosol of the mixture of the solutions and high-speed oxidant flow was produced, and was ignited by the pilot flame. Films were deposited on the substrates that were placed downstream of the aerosol flame.

Mirror-polished Si(100) single-crystal wafer was cut into a size of 10×10 mm as substrates. Substrate temperature was varied between 900°C and 1300°C while keeping the total metal concentration at 1.25×10^{-3} M to study its effect on the film growth rate. Investigation on the film growth rate and film morphology was conducted at the total metal concentration ranging from 0.5×10^{-3} to 5×10^{-3} M while the substrate temperature was kept at about 1200°C. Deposition experiments were also performed at different substrate to nozzle distances to study the effect of the thermophoresis on the film growth rates. The substrate to nozzle distance was varied in a range of 51 to 83 mm. The morphology and thickness of the films were characterized with scanning electron microscopy (SEM).

RESULTS AND DISCUSSION
    The substrate temperature is a significant parameter that controls the film growth modes and quality of the deposited films in chemical vapor deposition. The temperature affects the absorption/desorption, the reactivity and diffusivity of adatoms or clusters on the substrate surface, and hence affects the film growth rate and its morphology. Within the range of temperature employed in the experiments, two different film growth regimes were determined (as shown in Figure 1). In both of the regimes, film growth rate was increased as the substrate temperature was raised but with different increasing rate. Moreover, at the higher substrate temperatures, X-ray characteristic peaks were stronger and the particles were more faceted at one predominant orientation[3].

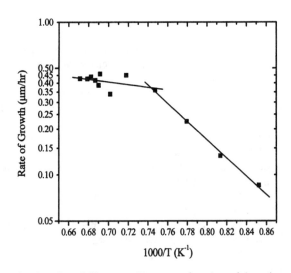

Figure 1. Arrhenius plot of film growth rate as a function of the substrate temperature.

    From this result, it can be inferred that in order to obtain higher film growth rate and well-crystallized particles in the film, higher substrate temperature is desired, e.g. higher that 1100 °C.
    Figure 2 shows morphology of the YSZ films deposited at different total metal concentrations. By observing the morphologies of the samples, the films had well crystallized and faceted particles as long as the total-metal-concentration was not more than 4.25×10^{-3} M.

Beyond this limitation, the film was in a cauliflower-like structure. At the lowest concentration, the film was the smoothest and with smallest particle size among all the films. With the increase in concentration, the particle size was significantly increased. By observing the cross-sections of the coated samples, the thicknesses of the films were obtained and then converted into film growth rates. Within the range of the concentration employed in our experiments, a linear relationship was obtained between the film growth rate and the concentration (as shown in Figure 3.). In order to maintain a reasonably high growth rate of the YSZ film, higher total metal concentration is necessary and also is practical with the technique of CCVD.

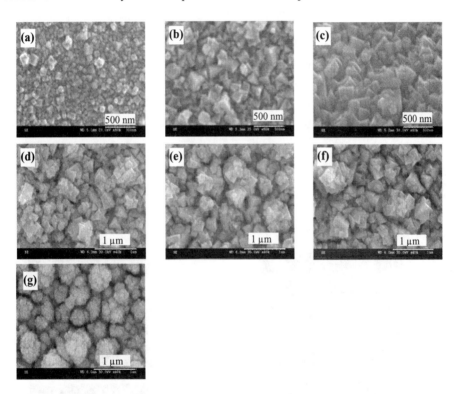

Figure 2. Morphological SEM micrographs of the YSZ films deposited at substrate temperature of 1200°C, O_2 to solution ratio of 1600:2.0 cm^3/min, S-N distance of 51 mm, ratio of Y/Zr of 16 %, and at different total metal concentrations (a) 5×10^{-4}M, (b) 1.25×10^{-3}M, (c) 2×10^{-3}M, (d) 2.75×10^{-3}M, (e) 3.5×10^{-3}M, (f) 4.25×10^{-3}M and (g) 5×10^{-3} M.

By comparing the cross-sections of the films shown in Figure 4, the effect of the total metal concentration on the microstructures of can be further confirmed. At low concentrations, the films are relatively thin, but dense. When the concentration was increased, i.e. 3.5×10^{-3}M, columnar growth feature is very obvious. Some voids or gaps among the columns can be seen, which are believed to be the disadvantage of the high concentration. The situation was worse when the concentration was increased to 5×10^{-3} M. The film becomes non-uniform in thickness, and the columns are composed of very small crystallites (as small as 100 nm). This phenomenon can be possibly interpreted as a result of relatively short time for the adsorbed species to migrate to a proper lattice position comparing to their adsorption rate. However, a reasonably high concentration can still be used for obtaining a high film growth rate; because it is has been proved in our research that the voids or gaps can be removed by appropriate thermal annealing procedures.

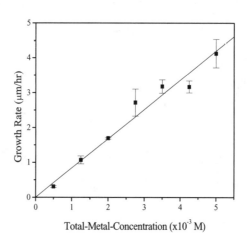

Figure 3. Film growth rate versus total-metal-concentration in the liquid solution.

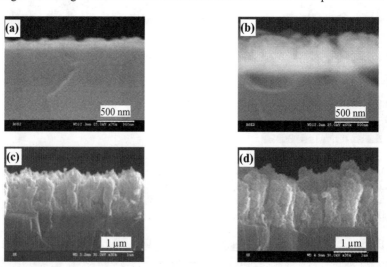

Figure 4. Cross-sectional SEM micrographs of the selected samples from Figure 2. Graph a) corresponds the total metal concentrations $5x10^{-4}$M, b) $2x10^{-3}$M, c) $3.5x10^{-3}$M, and d) $5x10^{-3}$ M, respectively.

Thermophoresis occurs when a temperature gradient exists in the diffusion boundary layer of a flow. This gradient affects the kinetic energy of gas molecules. There is a net force to transfer gas molecules or clusters in the diffusion boundary layer to move from the hotter region to a cooler region. Thermophoresis can be used to enhance the growth rate of thin films[5].

Since at the positions closer to the nozzle of the nebulizer, the flame temperature is higher, if the substrate temperature is kept constant, the temperature gradient across the diffusion boundary layer next to the substrate surface will be larger when the substrate is moved closer to the nozzle. Therefore, the change of the substrate to nozzle distance is, in fact, equivalent to changing of the temperature gradient across the diffusion boundary layer next to the substrate surface, and hence the effect of the thermophoresis. The film thickness measured on SEM images is plotted versus the S-N distance in Figure 5. The data can be fitted by an exponential equation.

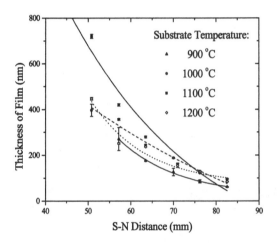

Figure 5. YSZ film thickness measured from the cross-sections as a function of S-N distance at different substrate temperatures. The films were deposited for 20 min.

From the curves in Figure 5, a very obvious trend of increasing in the film growth rate is noticed when the substrate to nozzle distance is decreased. Therefore, it is advisable to maintain a small S-N distance in order to acquire a high YSZ film growth rate. However, Figure 5 shows that the lower substrate temperatures do not give the greater effect of thermophoresis on the film growth rate, even though the temperature gradients were higher. For example, substrate temperature of 900°C results in the lowest growth rate; substrate temperature of 1000°C gives rise to higher growth rate. It can be understood that the thermophoresis is not the only factor to affect the film growth rate. When the substrate temperatures are low, e.g. 900°C and 1000°C, the film growths are in the surface reaction controlled regime, which results in low film growth rates (see Figure 1), although the effect of thermophoresis is strong in these circumstances. Higher film growth rate at the substrate temperature of 1000°C is also consistent with the results as shown in Figure 1. The substrate temperature of 1100°C combines the advantages of the thermophoresis effect and high film growth rate in the gas phase diffusion controlled regime, which results in the highest growth rate in our study.

CONCLUSIONS

The deposition experiments on polished substrates demonstrated that the film growth rate could be enhanced by choosing relatively higher substrate temperature and higher total metal concentration of the reagent solution. The effect of thermophoresis was observed in the experiments of film depositions at various substrate-to-nozzle distances. An increasing of the temperature gradient in the flame next to substrate can generally increase the film growth rate. It can be further stated that the incorporation of the effect of thermophoresis and high film growth rate in the gas-phase controlled reaction regime will be more effective to enhance to film growth rate. All of the effects can be utilized to enhance the film growth rate and reduce the manufacturing cost of SOFC.

ACKNOWLEDGEMENT

This research was sponsored by NSF through the NSF Center for Advanced Materials And Smart Structures (NSF-CAMSS).

REFERENCES

[1] B. Steele, "Ceramic ion conducting membranes and their technological applications", *Comptes Rendus de l'Académie des Sciences - Series IIC - Chemistry,* **1**(9) 533-43, 1998.

[2] J.M., Hampikian and W.B. Carter, "Combustion chemical vapor deposition of high temperature materials", *Materials Science and Engineering A,* 267(1), 7-18,1999.

[3] Z. Xu, J. Sankar, and Q. Wei, "Preparation and properties of YSZ electrolyte thin films via liquid fuel combustion chemical vapor deposition", *Ceramic Engineering and Science Proceedings,* **23**(3) 711-18, 2002.

[4] Z. Xu, J. Sankar, and Q. Wei, "Combustion Chemical Vapor Deposition of YSZ Thin Films for Fuel Cell Applications", Paper IMECE2001/MD-24800, pp1-8 in *Effects of Processing on Properties of Advanced Ceramics,* published in proceedings of ASME Congress, New York, 2001.

[5] C. He, and G. Ahmadi, "Particle deposition with thermophoresis in laminar and turbulent duct flows", *Aerosol Science and Technology,* **29**(6) 998, 525-46 1998.

SYNTHESIS OF YTTRIA STABILIZED ZIRCONIA THIN FILMS BY ELECTROLYTIC DEPOSITION

Zhigang Xu, Samuel Tameru, Jag Sankar
NSF Center for Advanced Materials and Smart Structures
North Carolina A&T State University, Greensboro, NC 27411, USA

ABSTRACT

Yttria stabilized zirconia (YSZ) thin film coatings have been deposited on different substrates, such as nickel and porous strontium doped lanthanum manganite (LSM) that is a cathode material for solid oxide fuel cells (SOFCs), by electrolytic deposition. The films were processed by cathodic deposition in the solution of the inorganic salts of yttrium and zirconium. A mixture of methanol and water was used as the solvent. The continuous coatings were developed by multiple iterations of ultra thin film depositions, room temperature drying, and intermediate temperature thermal treatment. The cracks in the films after drying were alleviated by reducing the thickness of individual layers and utilizing two different additives, such as hydrogen peroxide (H_2O_2) and cationic polyelectrolyte poly(dimethyldiallylammonium chloride) (PDDA). The desired thickness and the continuity of the coating were obtained by controlling the metal ion concentration in the solution, deposition current density, deposition time and thermal treatment temperature. The structures of YSZ coatings were characterized using scanning electron microscopy.

INTRODUCTION

YSZ is a fast ion-conductive material that is widely used as an electrolyte for SOFCs and other electrochemical devices[1]. To prepare thin film electrolyte for solid oxide fuel cells, low-cost and efficient processing is desired. In the past years, chemical vapor deposition, electrochemical vapor deposition, sol-gel, and state-of-the-art tape casting or screen-printing techniques have been studied for processing thin electrolyte layers for fuel cell application. However, it is difficult for them to achieve cost and quality requirements.

Electrolytic deposition has been developed as a new technology in ceramic processing for processing of nanostructured thin films and powders[2]. Cathodic electrolytic deposition is achieved via hydrolysis of metal ions or complexes by an electro-generated base to form oxide, hydroxide or peroxide deposits on cathodic substrates[3]. Hydroxide and peroxide deposits can be converted to corresponding oxides by thermal treatment.

The interest in this technique stems from a variety of advantages, such as low cost of equipment, rigid control of processing thickness and uniformity as well as the possibility of forming coatings on substrates of complex shape. The uniformity of ceramic coatings results from the insulating proprieties of deposits and the electric field dependence of the deposition rate.

In this paper, the experimental studies of deposition YSZ complex ceramic coating on Ni substrates will be presented. The technique and process of crack filling in the deposited film was the major effort in the experiments. The application of the deposition technique on porous LSM substrates that are the typical cathode material for SOFCs was also studied experimentally.

EXPERIMETNAL

Commercial zirconium dichloride oxide hydrate, $ZrOCl_2 \cdot 8H_2O$, (99.9%-Zr, Strem Chemicals, Inc.) and yttrium nitrate hexahydrate, $Y(NO_3)_3 \cdot 6HO_2O$ (99.9%-Y, Strem Chemicals, Inc.) were used as the source materials to form metal oxides and hydroxides. Hydrogen peroxide, H_2O_2, (Fisher Scientific) and Poly(diallyldimethylammonium chloride) (low molecular weight, 20 wt. % in water, Sigma-Alrich) were used as additives. The mixture of methanol and deionized water with a volume ratio of 3:1 was used as solvent. The total metal ion concentration of Zr^+ and Y^+ in the solution was 1×10^{-3} M. The atomic ratio of Y^+ to Zr^+ was 16:100 with which to prepare electrolyte materials close to the composition of $(Y_2O_3)_{0.8}(ZrO_2)_{0.98}$ for SOFCs. Concentration of 5×10^{-3} M for H_2O_2 and 0.5 g/l of PDDA were used. Cathodic deposits were obtained by the galvanostatic method on Ni and LSM substrates at current densities ranging from 1-10 mA/cm^2. The electrochemical cell for deposition included a cathodic substrate centered between two parallel platinum counterelectrodes. After drying at room temperature, the electrolytic deposit was thermally annealed followed by characterization and subsequent deposition, drying and thermal treatment cycles in order to achieve a crack-free coating. Powder scraped from some room temperature dried samples were used for thermogravimetric analysis (TGA). The thermoanalyzer was operated in air between room temperature and 850°C at a heating rate of 5°C/min. The morphologies of the coating were characterized by scanning electron microscopy.

RESULTS AND DISCUSSION

Thermogravimetric analysis of the room temperature dried deposits from solutions with different additives is shown in Fig.1. In the case of deposition with H_2O_2 as additive, the weight loss of the deposits increases monotonically during heating and most of the weight loss (~38%) occurs below 400°C, after which there is no significant weight change. This weight loss is associated with the dehydration of the hydroxide. According to X-ray diffraction of the thermally treated deposit of 0.005 M $ZrOCl_2 \cdot 8H_2O$ with H_2O_2 as additive, small peaks attributed to tetragonal phase of zirconia appeared at 400°C[4]. The TGA data of the deposits with 0.5g/l PDDA as additive appear in a very different way. Three marked reductions of the sample weight can be

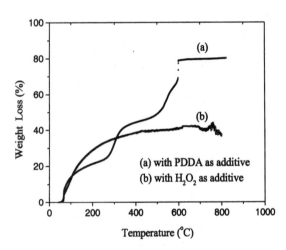

Figure 1. Thermogravimetric analysis data for deposits obtained from 0.001 M ($ZrOCl_2$ + $Y(NO_3)_3$) and additives a) PDDA , b) H_2O_2, respectively.

340

seen in Fig.1, the first one at ~100°C, the second at ~ 300°C and the third at ~600°C. There was no significant weight loss above 600°C. The total weight loss was approximately 79%. The weight loss is believed to be associated with dehydration of the hydroxide and the polymer combustion.

Fig. 2 shows the deposition weight gain versus deposition time for the prepared solutions. When the current density was maintained at 3 mA/cm^2, as shown by the curves Fig.2(c) and (d), the deposition weight gain in the PDDA added solution is lower than that with the H$_2$O$_2$ added solution within 8 min. The deposition weight gain in the solution with PDDA increases monotonically for the duration of the experiment, while in the solution with H$_2$O$_2$, the deposition weight gain increases up to 4 min, then start decreasing gradually thereafter. The decrease of the deposition weight gain in this case may be interpreted by the result of spallation. There is no spallation for the deposits from the solution with PDDA addition, because of the binding effect of the polymeric electrolyte. When the current was held at 8 mA/cm^2, the deposition rates in both solutions were higher than at the lower current. Unfortunately, it was not possible to deposit for longer than 6 min, due to limited maximum power supply voltage (60 V).

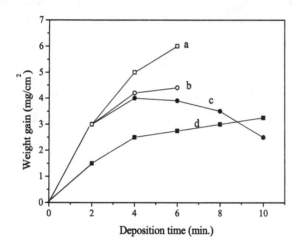

Figure 2. Deposition weight gain in the solution of 0.001M of ZrOCl$_2$ and Y(NO$_3$)$_3$ versus deposition time at 8 mA/cm^2 with a) 0.5 g/l PDDA, b) 0.005 M H$_2$O$_2$ as additives, and at 3 mA/cm^2 with c) 0.005 M H$_2$O$_2$ and d) 0.5 g/l PDDA as additives.

Fig. 3 shows morphological evolution of the YSZ films coated on Ni substrates at a current density of 10mA/cm^2. For a deposition duration time of 0.5 min, the film was very thin (see Fig.3(a). The solvent in the film can easily reach the surface during drying at room temperature. The drying rate of the film through the thickness is roughly uniform, therefore, the film is smooth and there are only small cracks. When the deposition time was increased, the film became thicker, and the drying rate of the surface was higher than within the film, thus, potentially creating residual stresses for crack formation. In order to reduce the cracking, the deposition time was limited to 1 min in the later experiments.

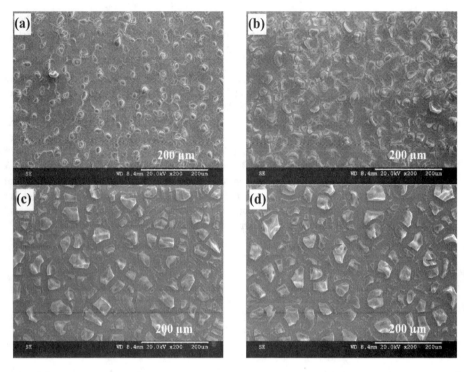

Figure 3. SEM images of the YSZ films coated on Ni substrates at current destiny of 10 mA/cm^2, total metal ion concentration:1×10^{-3} M, with 5×10^{-3} M of H_2O_2 as additive. The deposition temperature was 1.5 °C and the deposition times were a) 0.5 min, b) 1.0 min, c) 1.5 min, and 2.0 min.

Crack formation is an intrinsic problem in liquid phase deposition, especially when water is used as all or part of the solvent. According to the principle of the cathodic deposition, the cathode as a substrate must be conductive. The substrates used in this experiment were all conductive metals or oxides. The deposited zirconium and yttrium hydroxides are relatively less conductive, and their oxides are non-conductive. Therefore, as the deposit builds up, the resistance of the film increases and therefore, the deposition rate decreases. This mechanism can be used to fill the cracks. The operation steps include the repetition of deposition, thermal treatment. The previous deposit needs to be annealed to become nonconductive; therefore, the subsequent deposition will only take place in the conductive areas, i.e. the cracks. Fig. 4 shows an example of crack filling accomplished by this method. In Fig.4(a), the film is fragmented into isolated flakes on the substrate after thermal treatment. The subsequent 3 min deposition fills into the cracks as shown in Fig. 4(b). After a total deposition time of 15 min, the cracks were filled and the width of the small crack is lass than 0.5 μm on the surface as shown in Fig.4(e). The coating in the solution with PDDA as binding phase had a very different appearance from those depositions in the solution without binder. On the surface of coating with PDDA binder, there are significantly less and smaller cracks. In the majority of the coated area, cracks were hardly found even after thermal treatment at 650°C for 30 min. Fig.4(f) was taken from a cracked area to demonstrate the crack filling for this instance.

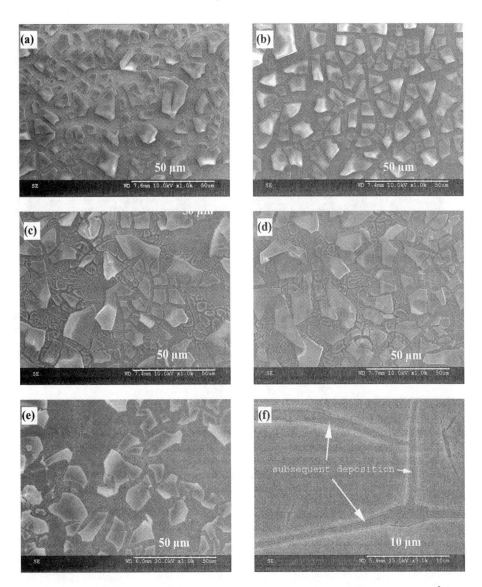

Figure 4. SEM images of YSZ coatings on a Ni substrate at a current density of 5 mA/cm² in the solution of 1×10^{-3} M $ZrOCl_2$ and $Y(NO_3)_3$. For a) to e), H_2O_2 at 5×10^{-3} M was used as an additive and a different number of deposition cycles was applied: a) 1 cycle, b) 2 cycles, c) 3 cycles, d) 4 cycles, and e) 5 cycles. A cycle corresponds to 1 min of deposition + 30 min air drying + 1 min of deposition + 30 min air drying + 1 min of deposition + 48 hr air drying + 30 min of thermal treatment at 650°C. For f), PDDA at 0.5 g/l was used as additive with deposition of 4 cycles.

Fig. 5 shows the coatings on porous LSM substrates in the solutions either with H_2O_2 or PDDA as additives. As demonstrated by the images, this deposition technique has the ability to form continous films on porous substrates, which is of significant interest to the applications in fuel cells and electrochemical sensors.

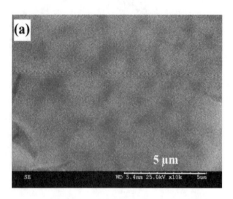 

Figure 5. SEM images of the 650°C annealed coatings on porous LSM substrates for 6 deposition cycles in the solution with a) H_2O_2 and b) PDDA as additives. Other processing conditions refer to those for the coatings shown in Fig. 4.

CONCLUSIONS

Through the experimental studies, the electrolytic deposition technique demonstrated its potential to prepare oxide coatings at very low cost and ease. Continuous films can be obtained by optimization of the time in each deposition combining with thermal treatment and proper number of processing repetitions. The use of PDDA as an additive appears to be more promising than the use of H_2O_2.

REFERENCES

[1] L.J.M.J. Blomen, and M.N. Mugerwa, pp.473. in *Fuel Cell Systems*, Plemun Press, New York, 1993.

[2] I. Zhitomirsky, "Cathodic Electrosynthesis of Titania Films and Powder", *Nanostructured Materials*, **8** 521-28 (1997).

[3] S-K. Yen, "Mechanisms of Electrolytic ZrO2 coating on Commercial Pure Titanium", *Mater. Chem. Phys.*, **63** 256-62 (2000)

[4] I. Zhitomirsky and L. Gal-Or, "Cathodic Electrosynthesis of Ceramic Deposits", *J. Eur. Ceram. Soc.*, **16** 819-824 (1996)

SINTERING AND STABILITY OF THE $BaCe_{0.9-x}Zr_xY_{0.1}O_{3-\delta}$ SYSTEM

Z. Zhong[*+], A. Sayir[*+] and F. Dynys[*]
NASA GRC[*]/CWRU[+]
21000 Brookpark Road
Cleveland, OH 44135 USA

ABSTRACT

Protonic separation membranes for the hydrogen industry require thermo-chemical stability and high conductance. The perovskite $BaCe_{0.9}Y_{0.1}O_{2.95}$ exhibits good proton conduction at high temperatures, but shows poor thermo-chemical stability. Substituting Zr for Ce in $BaCe_{0.9}Y_{0.1}O_{2.95}$ improves the thermo-chemical stability but reduces proton conduction. The objective of this work was to study the optimization of protonic conductance and thermo-chemical stability by changing the ratio of Ce to Zr in $BaCe_{0.9-x}Zr_xY_{0.1}O_{2.95}$. To elucidate the dopant effect, a coprecipitation method has been developed to produce single phase perovskites of $BaCe_{0.9-x}Zr_xY_{0.1}O_{2.95}$ ($0 \leq x \leq 0.9$). The coprecipitation method has been optimized to yield high purity and homogeneous powders with a particle size of 50-100 nm in diameter. The sintering characteristics were studied in the temperature range of 1400-1650°C. $BaCe_{0.9}Y_{0.1}O_{2.95}$, $BaCe_{0.7}Zr_{0.2}Y_{0.1}O_{2.95}$ and $BaCe_{0.5}Zr_{0.4}Y_{0.1}O_{2.95}$ can be sintered to high density at 1650°C. Sintered $BaCe_{0.5}Zr_{0.4}Y_{0.1}O_{2.95}$ and $BaCe_{0.3}Zr_{0.6}Y_{0.1}O_{2.95}$ show good chemical stability against water and CO_2. Electric conductivity decreases with Zr content.

INTRODUCTION

Since the discovery of proton conduction in sintered oxides by Iwahara et al. in the early 1980s,[1, 2] there is a tremendous interest in research and the application of this type of material. These sintered metal oxides, which exhibit good proton conduction between 300 °C and 1000 °C, are generally known as high temperature protonic conductors (HTPC)[3-7]. Applications of HTPC membranes include hydrogen and steam sensors, electrolytes for fuel cell, hydrogen purification, hydrogen pump and steam electrolyzer.[5, 8-11] One important aspect of HTPC is the chemical stability to the environment. $BaCeO_3$-based ceramics show the highest proton conductance, but it is thermodynamically the least stable among all HTPC materials.[7, 12-17] It is known that sintered pellets disintegrate and decompose upon contacting boiling water.[12, 16] Additionally, the material shows reactivity with carbon dioxide. On the other hand, $BaZrO_3$-based ceramics show high stability towards water and carbon dioxide[17, 18] and exhibits poor protonic conductivity. Solid solutions of $BaCeO_3$-$BaZrO_3$[19-22] have been investigated to improve environmental durability and maintain high protonic conductivity. Zirconium substitution enhances environmental durability but deteriorates protonic conductivity. The composition has to be compromised between environmental durability and protonic conduction.

$BaCeO_3$ and $BaZrO_3$ are notoriously difficult materials to densify. Sintering temperatures >1600°C are needed to achieve densities ≥90% theoretical. Traditionally, solid state methods are used to prepare coarse (1-5 μm) HTPC powders. It is advantageous to process at lower temperatures. Reduction in particle size can reduce sintering temperatures. Co-precipitation of oxalate precursors has been reported to prepare fine powders of Nd-doped $BaCeO_3$.[23, 24] Sintered densities ≥90% were achieved at 1300°C. This work reports on coprecipitation of $BaCe_{0.9-x}Zr_xY_{0.1}O_{2.95}$ ($0.1 \leq x \leq 0.9$) solid solutions.

Crystalline fine powders generated from the coprecipitation method were sintered at various temperatures to evaluate the densification behavior. Powders and sintered samples were tested for chemical stability in boiling water and CO_2 at 900°C. Electrical conductivity of sintered samples was measured by impedance spectroscopy.

EXPERIMENTAL

Metallic salts of $Ba(NO_3)_2$, $Ce(NO_3)_3 \cdot 6H_2O$, $Y(NO_3)_3 \cdot xH_2O$, $ZrO(NO_3)_2 \cdot xH_2O$ (99.9+%, Alfa Aesar) were used as precursors. The precursors were weighed to yield $BaCe_{0.9-x}Zr_xY_{0.1}O_{2.95}$ compositions (x = 0, 0.2, 0.4, 0.6, 0.9). The salts were dissolved in deionized water. In a separate beaker, $(NH_4)_2CO_3$ (99.999%, Alfa Aesar) and NH_4OH (28-30.0% aqueous solution, Alfa Aesar) were mixed in 1:2 molar ratio in the amount that is sufficient to precipitate the metals in the solution. The precipitant solution was placed in an ultrasonic bath and the mixed nitrate solution was added using a plastic dropper. The ultrasonic bath provided a turbulent mixing environment that eliminated concentration and pH gradients rapidly. All the metals in the salt solution precipitated instantaneously forming a homogenous mixture. The precipitate was separated from the solution by filtration and freeze-dried.

Thermogravimetric analysis (TGA) of the dried precipitates was performed in an open platinum pan by heating at 5°C/min from room temperature to 1200 °C. Based on the TGA results, the freeze dried powders were calcined in static air for 4 hours at 1000 °C to yield the perovskite phase.

Powder x-ray diffraction data were obtained using a Phillips APD3600 automatic powder diffractometer with Cu K_α radiation at 3° per minute for continuous scan from 20 to 70° (2θ). Chemical analysis using ICP was carried out to determine the compositional purity of the products. Powder size measurements were determined by BET surface area, laser scattering particle size analyzer, and scanning electron microscopy (SEM).

Green pellets were pressed with 1.2 gram of powders using a 15 mm diameter die using 200MPa. Green pellets were sintered in an atmosphere of air or oxygen at temperatures ranging from 1200 °C to 1650 °C. The densities of sintered pellets were determined by geometrical measurements and the water immersion (Archimedes) method. Microstructures of the polished cross-section were acid etched and then examined by a Hitachi S-4700 SEM.

For electrochemical measurement, both sides of the pellets were polished with SiC polishing paper. Platinum (Heraeus CL11-5349) paste was painted on the polished surfaces and fired at 1100°C for 60 minutes. A ProboStat™ measurement cell[25] was used as sample holder. The measurement system included a Solartron 1260 frequency response analyzer and a Solartron 1287 electrochemical interface connected to a personal computer. ZPlot/ZView electrochemical impedance software and was used for data acquisitions and analysis. The measurements were carried out in the frequency range of 0.1 Hz to 60 kHz at 500, 700 and 900°C.

RESULTS AND DISCUSSION
Powder Properties and Sintering

The TGA results indicate the weight loss (i.e. the conversion of barium carbonate in the precursors) is completed by 1000°C for all compositions studied. Figure 1 shows the XRD patterns for different calcination temperatures for the composition of $BaCe_{0.9}Y_{0.1}O_{2.95}$ calcined for four hours in air. The as-precipitated powder is amorphous and shows very weak crystallinity when calcined at 350 °C. By 700 °C, crystalline phases appear. Formation of the perovskite phase, $BaCe_{0.9}Y_{0.1}O_{2.95}$, is completed at temperatures ≥ 900°C as observed by XRD. The perovskite phase $BaCe_{1-x}Zr_xO_3$ ($0.1 \leq x \leq 0.9$) was attained at a calcination temperature of 1000 °C as shown in Figure 2.

$BaCe_{0.9-x}Zr_xY_{0.1}O_{2.95}$ powders were calcined in ZrO_2 crucibles. Chemical analysis by ICP was performed on the composition of $BaCe_{0.9}Y_{0.1}O_{2.95}$. Exceptional compositional purity was achieved. The zirconium level was within 50ppm of the nominal composition, similar to the powder calcined in a platinum crucible. Calcining the powder in zirconia crucibles does not compromise the purity of the product. The phosphorus and calcium levels were 200 ppm and 170 ppm respectively; the source of the contaminations was not identified. All the other impurities were <100 ppm. The ICP results gave the composition of the product as $Ba_{0.995\pm0.008}Ce_{0.900\pm0.006}Y_{0.101\pm0.002}O_{3-\delta}$. Similar purity levels were observed for the other compositions in the $BaCe_{0.9-x}Zr_xY_{0.1}O_{2.95}$ (x=0.2, 0.4, 0.6, 0.9) system. The hafnium impurity level was found to be dependant on zirconium concentration. The hafnium level in $BaZr_{0.9}Y_{0.1}O_{2.95}$ was 200 ppm, which was presumably originally from $ZrO(NO_3)_2 \cdot xH_2O$ per Alfa Aesar data sheet.

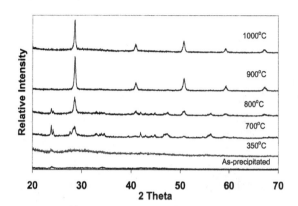

Fig. 1 Phase formation for $BaCe_{0.9}Y_{0.1}O_{2.95}$.

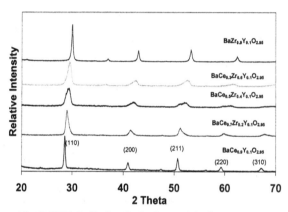

Fig. 2 XRD for $BaCe_{1-x}Zr_xO_3$ ($0.1 \leq x \leq 0.9$ after calcining at 1000 °C for 4 hours.

The particle size of the freeze dried powders ranged from 50 to 100 nm per SEM analysis. Figure 3 shows an SEM image of $BaCe_{0.9}Y_{0.1}O_{2.95}$ powder after calcination at 1000°C for 4 hours. Calcination at 1000°C does not cause particle sintering or coarsening. BET surface area for calcined $BaCe_{0.9-x}Zr_xY_{0.1}O_{2.95}$ (x = 0, 0.2, 0.4, 0.6, 0.9) ranged from 11.67~17.94 m^2/g. This corresponds to equivalent particle sizes of 54~83 nm, assuming a theoretical density of 6.2 g/cc.

A thin layer of $BaCe_{0.9-x}Zr_xY_{0.1}O_{2.95}$ powder was used between the sample and crucible to abate sample reaction with the crucible. Samples were sintered in air between 1250°C to 1650°C for 4 hours as shown in Fig. 4. No significant sintering occurs at 1250°C and 1350°C. Sintered densities of 80 to 90% were achieved in the temperature range of 1400 to 1650°C. Examination of the microstructures showed pore clusters were inhibiting densification. Pore clusters were attributed to particle agglomeration. The powder was de-agglomerated by wet ball milling in isopropanol for 4 hrs. De-agglomeration treatment improved the sintering behavior; sintered density increased to 92.5% at 1650°C. Isostatic pressing did not significantly improve the sintered density. The source of agglomerates occurs either at precipitation or at calcination or both. It was found that changing calcination atmosphere influenced agglomeration formation. Calcination of the $BaCe_{0.9}Y_{0.1}O_{2.95}$ precursor in flowing argon at 1000°C was sufficient to change the densification behavior. The reduction in oxygen in the atmosphere can inhibit the formation of hard agglomeration.[26] Further work is needed to elucidate the mechanism. Sintered density of 98% was achieved at 1650°C using the powder calcined in argon.

Substitution of Zr for Ce has a strong influence upon the sintering behavior. Using the calcinations in argon, $BaCe_{0.7}Zr_{0.2}Y_{0.1}O_{2.95}$ and $BaCe_{0.5}Zr_{0.4}Y_{0.1}O_{2.95}$ were sintered above 95% density at 1650°C. The material became difficult to sinter above 80% density when the Zr molar fraction increased to 0.6. The density for $BaCe_{0.3}Zr_{0.6}Y_{0.1}O_{2.95}$ sintered at 1650 °C was 80%. $BaZr_{0.9}Y_{0.1}O_{2.95}$ sintered to 65% density at 1650 °C regardless of calcination atmosphere of air, argon, or hydrogen. Ball milling the powder did not improve the sintering behavior of the powder. The use of hydrogen or oxygen environment at 1550°C did not improve the densification characteristic of $BaZr_{0.9}Y_{0.1}O_{2.95}$. No effort was made to sinter it at a higher temperature than 1650°C.

Fig. 3 SEM of $BaCe_{0.9}Y_{0.1}O_{2.95}$ powder, calcined for 4 hours at 1000 °C.

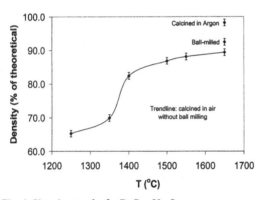

Fig. 4 Sintering results for $BaCe_{0.9}Y_{0.1}O_{2.95}$.

Fig. 5 SEM image of sintered $BaCe_{0.7}Zr_{0.2}Y_{0.1}O_{2.95}$.

Figure 5 shows an SEM image of the sintered $BaCe_{0.7}Zr_{0.2}Y_{0.1}O_{2.95}$ pellet. The polished cross-section was acid etched to reveal the grain structure. The roughness observed within the interior of the grains in Figure 5 is due to preferential etching along crystallographic orientation.

The Chemical Stability and Electric Conductivity.

The chemical stability of $BaCe_{0.9-x}Zr_xY_{0.1}O_{2.95}$ (x = 0, 0.2, 0.4, 0.6, 0.9) powders to water was tested in boiling water. The powders tested were calcined at 1000°C for 4 hours in air. After being boiled in water for 3 hours, the powders were filtered and air-dried overnight. XRD results show that all the powders reacted with boiling water in 3 hours. $BaCe_{0.9}Y_{0.1}O_{2.95}$ and $BaCe_{0.7}Zr_{0.2}Y_{0.1}O_{2.95}$ decomposed significantly. $BaZr_{0.9}Y_{0.1}O_{2.95}$ demonstrated the highest stability among all, but was also attacked by boiling water.

Sintered pellets of $BaCe_{0.9}Y_{0.1}O_{2.95}$, $BaCe_{0.7}Zr_{0.2}Y_{0.1}O_{2.95}$ and $BaCe_{0.5}Zr_{0.4}Y_{0.1}O_{2.95}$ with densities >95% were tested in boiling water for six hours. The low density sample of $BaCe_{0.3}Zr_{0.6}Y_{0.1}O_{2.95}$, 80%, was also tested in boiling water. None of them crumbled as reported in the literature for $BaCeO_3$-based HTPC.[16] Stability of the sintered pellets to exposure to CO_2 atmosphere was also tested at 900 °C for 2 hrs. Visible inspection of the sample and XRD were used to evaluate the environmental stability. $BaCe_{0.9}Y_{0.1}O_{2.95}$ and $BaCe_{0.7}Zr_{0.2}Y_{0.1}O_{2.95}$ exhibit poor stability in the water and CO_2 tests. Environmental testing of $BaCe_{0.5}Zr_{0.4}Y_{0.1}O_{2.95}$ and $BaCe_{0.3}Zr_{0.6}Y_{0.1}O_{2.95}$ shows good stability to exposure to water and CO_2. This is consistent with the reported results.[21] The polished surface of the pellet of $BaCe_{0.9}Y_{0.1}O_{2.95}$ became matte after being boiled in water, however, the polished surface of the pellet of $BaCe_{0.5}Zr_{0.4}Y_{0.1}O_{2.95}$ remained glossy

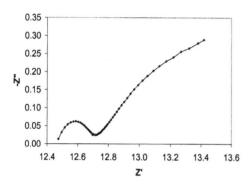

Fig. 6 Impedance data for $BaCe_{0.7}Zr_{0.2}Y_{0.1}O_{2.95}$ at 700°C.

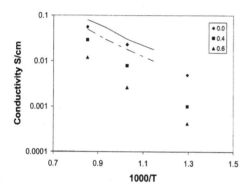

Fig. 7 Conductivities of $BaCe_{0.9-x}Zr_xY_{0.1}O_{2.95}$ measured in wet air (1.22×10^3 Pa of water) in comparison with the results of Katahira et al.[21], where ♦ for x=0.0, ■ for x=0.4, ▲ for x=0.6, solid line for x=0.0 and dashed line for x=0.4 from the results of Katahira et al.

after being boiled in water. Sintered material shows significantly higher stability than its powder form is not unique in the $BaCe_{0.9-x}Zr_xY_{0.1}O_{2.95}$ (x = 0, 0.2, 0.4, 0.6, 0.9) system.

The overall electrical impedance for $BaCe_{0.9}Y_{0.1}O_{2.95}$ at 900°C is very low (~10 Ω·cm), and the bulk and the grain boundary resistance cannot be resolved in the impedance spectrum. This characteristic in the impedance spectra was observed for all compositions studied at 900 °C. However, at T≤700°C, the impedances from different components were separated. Figure 6 is a representative impedance spectrum for $BaCe_{0.7}Zr_{0.2}Y_{0.1}O_{2.95}$ at 700°C in a complex plane format (Nyquist diagram). The impedances of the bulk, the grain boundary and the interfacial (charge transfer) contribution were separated. This is in agreement with previous results.[23] Figure 7 is the plot of the bulk conductivities for $BaCe_{0.9-x}Zr_xY_{0.1}O_{2.95}$ (x = 0, 0.2, 0.4, 0.6) measured in wet air (1.22×10^3 Pa of water), which are slightly lower than those with the same compositions made by solid-state method.[21] The samples made by the solid-state method were also sintered to >95% at a higher temperature and the reported data were compatible with theoretical estimation.[3, 18] However, the grain boundary resistances are low. (see Fig. 6) The improved grain boundary conductivity measured in this work are possible due to better stoichiometric control and/or improved homogeneity of microstructure using coprecipitation method, which reduces formation of secondary phase around grain boundaries. It can also be caused by the differences in measurement method, as the inconsistent results in the electric conductivity of HTPC were frequently reported.

CONCLUSION

High purity, homogeneous, and fine powders of $BaCe_{0.9-x}Zr_xY_{0.1}O_{2.95}$ (x = 0, 0.2, 0.4, 0.6, 0.9) have been prepared by a co-precipitation method. The sintering characteristics of $BaCe_{0.9-x}Zr_xY_{0.1}O_{2.95}$ were strongly dependent on the Zr content. The density of $BaCe_{0.9-x}Zr_xY_{0.1}O_{2.95}$ ceramics processed at 1650 °C exceeded 95% for $0 \leq x \leq 0.4$. For $0 \leq x \leq 0.2$, the chemical stability was poor against boiling water and carbon dioxide. $BaCe_{0.5}Zr_{0.4}Y_{0.1}O_{2.95}$ can be sintered to good density and also shows good chemical stability. The electric conductivity decreases with the increase of zirconium content as $BaZrO_3$-based materials has lower electric conductivity than $BaCeO_3$-based materials. $BaCe_{0.9-x}Zr_xY_{0.1}O_{2.95}$ ceramics made using the coprecipitation method had higher conductivity than conventional solid-state sintering.

ACKNOWLEDGEMENTS
This research was partly funded by NASA strategic research fund through NASA cooperative agreement NCC3-850 and by AFOSR under Grant No. FA9620-01-1-0500. We thank Professor Eric D. Wachsman for helpful discussion during the meeting.

REFERENCES
[1] H. Iwahara, T. Esaka, H. Uchida, and N. Maeda, Solid State Ionics, 3-4, 359 (1981)
[2] H. Uchida, N. Maeda, and H. Iwahara, Solid State Ionics, 11, 117 (1983)
[3] T. Norby, Solid State Ionics, 125, 1 (1999)
[4] K.D. Kreuer, Chem. Mater., 8, 610 (1996)
[5] H. Iwahara, Solid State Ionics, 77, 289 (1995)
[6] H. Iwahara, Solid State Ionics, 86-88, 9 (1996)
[7] N. Bonanos, Solid State Ionics, 145, 265 (2001)
[8] W.G. Coors, J. Power Sources, 118, 150 (2003)
[9] G. Zhang, S.E. Dorris, U. Balachandran, M. Liu, Solid State Ionics, 159, 121 (2003)
[10] T. Yajima, K. Koide, N. Fukastu, T. Ohashi, H. Iwahara, Sensors and Actuators B, 13-14, 697 (1993)
[11] T. Yajima, K. Koide, H. Takai, N. Fukastu, H. Iwahara, Solid State Ionics, 79, 333 (1995)
[12] C.W. Tanner, A.V. Virkar, J. Electrochem. Soc., 143, 1386 (1996)
[13] F. Chen, O. Toft-Sørensen, G. Meng, D. Peng, J. Mater. Chem., 7, 481 (1997)
[14] Z. Wu, M. Liu, J. Electrochem. Soc., 144, 2170 (1997)
[15] J. Guan, S.E. Dorris, U. Balachandran, M. Liu, J. Electrochem. Soc., 145, 1780 (1998)
[16] S.V. Bhide, A.V. Virkar, J. Electrochem. Soc., 146, 2038 (1999)
[17] S.V. Bhide, A.V. Virkar, J. Electrochem. Soc., 146, 4386 (1999)
[18] K.D. Kreuer, Solid State Ionics, 125, 285 (1999)
[19] S. Wienströer, H. Wiemhöfer, Solid State Ionics, 101-103, 1113 (1997)
[20] K.H. Ryu, S.M. Hail, Solid State Ionics, 125, 355 (1999)
[21] K. Katahira, Y. Kohchi, T. Shimura, H. Iwahara, Solid State Ionics, 138, 91 (2000)
[22] K.D. Kreuer, S. Adams, W. Münch, A. Fuchs, U. Klock, J. Maier, Solid State Ionics, 145, 295 (2001)
[23] S.D. Flint, R.C.T. Slade, Solid State Ionics, 77, 215 (1995)
[24] F. Chen, P. Wang, O. Toft-Sørensen, G. Meng, D. Peng, J. Mater. Chem., 7, 1533 (1997)
[25] S. Gallini, M. Hänsel, T. Norby, M.T. Colomer, J.R. Jurado, Solid State Ionics, 162-163, 167 (2003)
[26] C. Liu, B. Zou, A.J. Rondinone, Z.J. Zhang, J. Am. Chem. Soc., 123, 4344 (2001)

MICROSTRUCTURE AND ORDERING MODE OF A PROTONIC CONDUCTING COMPLEX $Sr_3(Ca_{1+x}Nb_{2-x})O_{9-\delta}$ PEROVSKITE

Marie-Hélène Berger
Ecole des Mines de Paris
BP 87
91003 Evry Cedex – FRANCE

Ali Sayir
NASA GRC / CWRU
21000 Brookpark Road
Cleveland, OH 44135 - USA

ABSTRACT

Complex perovskites are developed for applications which require high temperature protonic conducting (HTPC) ceramics. The solid-state sintering of the HTPC perovskites leads to limited protonic conductivities possibly due to the grain-boundary blocking effect. A $Sr_3(Ca_{1+x}Nb_{2-x})O_{9-\delta}$ HTPC perovskite has been produced using the melt growth process with the aim of reducing the grain-boundary blocking effect. Complex perovskite compositions of $Sr_3Ca_{1+x}Nb_{2-x}O_{9-\delta}$ have microstructures characteristic of cellular growth. Each cell has distinct core and shell regions. A composition gradient in bivalent to pentavalent cation ratio is observed from the core to the shell regions. Nano-scaled domains have been revealed inside the cells using high-resolution transmission electron microscope. The domains of **1:1** order of the cations on the **B** sites and domains of oxygen cage tilting octahedral are explained. The implications of structural and chemical analysis to the protonic conduction characteristic are discussed.

INTRODUCTION

The complex perovskite systems are envisioned as high temperature protonic conductors (HTPC) for hydrogen separation or solid oxide fuel cells (SOFC). Complex perovskites provide a wide range of possibilities for tailoring electronic to protonic conductivities and for maintaining a high degree of chemical stability, compared to simple perovskites.

Protonic conducting membranes or electrolytes usually consist of layered structure ceramics produced by solid-state sintering. The protonic conductivities and chemical stabilities obtained are often lower than those expected mainly due to the grain-boundary blocking effect.[1] Solid-state sintering produces high-energy grain boundaries that promote segregation and the formation of second phases. These grain boundary phases are detrimental for chemical stability at high H_2O and CO_2 partial pressures and may act as electrical barriers. The present work focuses on the melt processing of $Sr_3(Ca_{1+x}Nb_{2-x})O_{9-\delta}$ composition with the objective of reducing grain-boundary blocking effect. Melt processing of ceramics yields elongated microstructures with coherent and clean interfaces. Melt processed materials are expected to attain increased protonic conductivity. We report the structural and chemical analysis and its relation to processing conditions.

MELT PROCESSING

High purity polycrystalline $SrCO_3$ (99.999%), CaO (99.99%) and Nb_2O_5 (99.99%) powders were obtained from Alpha Aesar. The powders were blended in the ratio giving $Sr_3Ca_{1+x}Nb_{2-x}O_{9-\delta}$ x=0.18, and milled without milling agents for 40 h. The powder mixture was then calcined at 800 °C and 1000 °C in air for 6 and 10 hours respectively. The calcined powder was cold isostatically pressed into a cylindrical rod and sintered in air at 1500 °C. Laser heated melt processing technique[2] was used to directionally solidify the $Sr_3Ca_{1+x}Nb_{2-x}O_{9-\delta}$ rod in air at a rate of 15 mm/min, Fig.1.

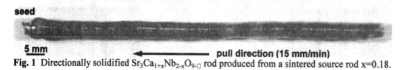

seed

5 mm ⟵——————— **pull direction (15 mm/min)**

Fig. 1 Directionally solidified $Sr_3Ca_{1+x}Nb_{2-x}O_{9-\square}$ rod produced from a sintered source rod x=0.18.

RESULTS AND DISCUSSION

Cellular Morphology – Core shell structure

Cross sections of the directionally solidified rod were mechanically polished with an alcohol-based lubricant. The observation in back scattering electron-scanning electron microscope (BSE-SEM), Fig 2, revealed a dense, cellular microstructure with cell widths of 5 to 10 μm. Each of the cells exhibited a core-shell structure. The cores of the cells were richer in calcium and strontium whereas the shells were richer in oxygen and niobium as revealed by wavelength dispersive x-ray (WDX) chemical mappings, Fig 3, and line profiles, Fig 4. The calcium content was lower than expected, Table 1. The Sr/(Nb+Ca) ratio was higher than 1, (1.06 in the cell cores), which could indicate some substitution of Sr on the B sites. Intergranular strontium rich second phases were observed at triple junction lines and to a minor extent between cells.

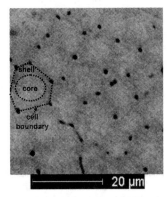

Fig 2. Cross-section of $Sr_3Ca_{1+x}Nb_{2-x}O_{9-\delta}$ in BSE-SEM. Schematic of drawing showing core, shell and cellular boundary are superimposed on the actual BSE-SEM image

Intra - Cell Domains

Figure 5 presents a bright field transmission electron microscope (TEM) image[*] of a cell exhibiting irregular domains with wavy contours. Figure 6 shows selected area diffraction (SAD) patterns of the cell in <100> and <110> zone axes. Compared to patterns of an ideal perovskite structure, they show extra reflections at $\pm h/2\ k\ l$, $\pm h/2\ \pm k/2\ l$ and $\pm h/2\ \pm k/2\ \pm l/2$. The extra reflections at 1/2 1/2 1/2 are the strongest and originate from a 1:1 ordered distribution of the Ca and Nb cations on the B site. Figure 7 presents a corresponding dark field image with curved anti phase boundaries separating 1:1 ordered domains. The fainter extra reflections are typical of perovskites with tolerance factor, $t=(R_A+R_O)/\sqrt{2}$ (R_B+R_O), lower than 0.985.[3] For such perovskites, the ionic radii of the A site species are too small to occupy the available volume fully. Oxygen octahedra rotate in order to reduce the size of the cuboctahedral interstices of the oxygen sublattice.[4] The reflections at $\pm h/2\ \pm k/2\ l$ are due to in phase tilting of the oxygen octahedra. Reflections at $\pm h/2\ k\ l$ are associated anti-parallel shift of Sr cations.

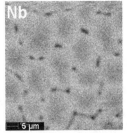 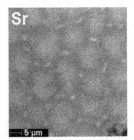 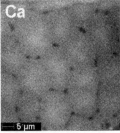

Fig 3 WDX chemical mapping: cores are Sr- and Ca- rich, shells are Nb-rich.

$Sr_3Ca_{1+x}Nb_{2-x}O_{9-\delta}$ x=0.18, δ=3x/2	O	Sr	Ca	Nb
At%	59.27	20.37	8.01	12.36

Table 1: Nominal starting composition with the assumption that $A=Sr^{2+}$, $B=(Ca^{2+},Nb^{5+})$.

---

[*] FEI Tecnai F20 ST

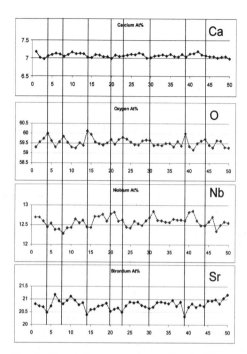

Fig. 4. Chemical concentration profiles (at %) across several cells (vertical lines indicate cell boundaries). Oxygen-rich boundaries are revealed.

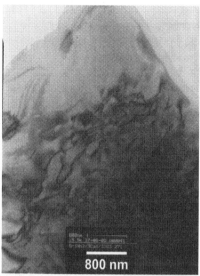

Fig. 5. Bright field TEM of a cell: intra-cell domains.

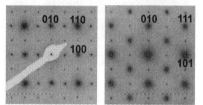

Fig. 6. SAD patterns in <100> and <110>zone axes.

Figure 8 is a dark field image on the corresponding reflection showing elongated domains with irregular contours. Figure 9 presents a high-resolution image recorded across an interface between diffracting and non diffracting domains and exhibiting a modulation in the <100> direction with a doubling of the parameter which is absent in the bottom. A contribution of antiphase tilt to the $\pm h/2 \pm k/2 \pm l/2$ reflection in addition to the 1:1 order contribution is not excluded.

Chemical Analysis of the Intra-Cell Domains

Figure 10 presents a STEM dark field image of a shell region. Domains with a maximum width of 150 nm are revealed. A chemical profile of 1.5 μm in length (evidenced by the contamination line) has been carried out in STEM-EDX across the domains and a secondary phase (a), with 60 analyses separated by 25 nm and a probe size of 3 nm. The decrease of the count numbers for Sr, Nb and Ca from the left to the right of the curves of Fig 11 is due to a decrease in thickness (and a preferential thinning of the second phase). Therefore, mainly qualitative information on the variation of the cation count ratios is attainable. The plot of the Nb/Ca count ratio shows some variation from one domain to another. Analyses in larger domains (b, c, d) exhibit smoother variations. The Sr/(Nb+Ca) ratio is more constant except in the transition between domains (at the boundary). Additional analyses need to be performed to confirm this result.

DISCUSSION

The incorporation and diffusion of protons in the complex perovskite is related to the concentration and distribution mode of the oxygen vacancies in the ceramic. The oxygen non-stoichiometry is fixed by the ratio of the bivalent to pentavalent cations (B'^{2+}/B''^{5+}) on the B-site of the perovskite. Local variations in B'^{2+}/B''^{5+} with respect to the nominal composition induce local variations in the conduction behaviour.[5] This study has shown that a melt grown $Sr_3Ca_{1+x}Nb_{2-x}O_{9-\delta}$ presents variation of B'^{2+}/B''^{5+} at two scales, micro- and nano-scale.

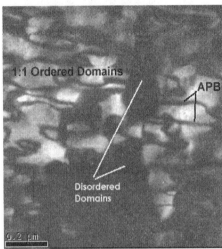

Fig. 7. Dark field from a 1/2, 1/2, 1/2 reflection. 1:1 ordered domains separated by APBs.

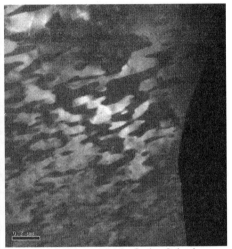

Fig. 8. Dark field from a 1/2 1 0 reflection. Ordered domains with anti-parallel shift of Sr cation.

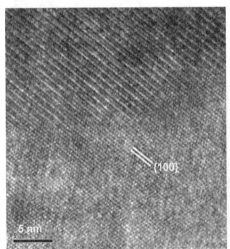

Fig. 9. High resolution image across a diffracting (top) / non diffracting (bottom) domain of Fig 8. Modulation in <100> in the diffracting part.

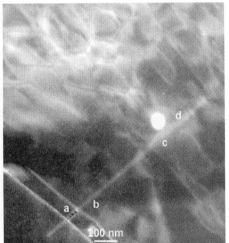

Fig. 10. STEM dark field images of a shell region. Domains and Sr rich second phase (a). Contamination line of 1.5 μm in length shows the location of the profile of Fig 11.

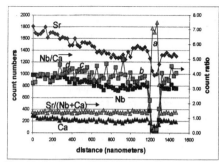

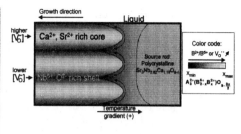

Fig. 11. Qualitative STEM-EDX profile (letters from Fig 10).

Fig. 12. Schematic diagram of solidification front.

The micro-scale variations are attributed to the cellular growth characteristics of microstructure. The first solid formed, which corresponds to the cell cores, is richer in bivalent cations than the liquid composition, as shown in the schematic diagram of Fig 12. Under steady-state growth conditions, the rate of rejection of Nb^{5+} at the center of cell-front to the cell boundary is probably equal to the rate of rejection of Ca^{2+} and Sr^{2+} cations at cell boundary to the cell core. For the particular growth rate of 15 mm/min studied here, lateral concentration gradients set-up ahead of the liquid-solid interface (Fig.12) produce a steady state cell-spacing of around 10 μm (Fig.2).

The slow diffusion rate in the solid does not permit a homogenization of the composition. We do not know the concentrations across the interface in the liquid ahead of the cell-front (at the tip) and in front of cell-boundary. The liquid at the center of the cell-front is richer in Nb^{5+} and has an equilibrium freezing temperature lower than the freezing temperature of single phase Sr_3 (Ca Nb_2) O_9. This causes a distinct spatial distribution of B'^{2+}/B''^{5+} ratio within the cellular structure. The B'^{2+}/B''^{5+} ratio is highest at the center of the cellular structure (core) and decreases monotonically at the shell region (Fig. 4). The spatial deviation of B'^{2+}/B''^{5+} ratio from its nominal value is balanced through a decrease in the concentration of the oxygen vacancy from the core to the boundary of the cells. As a consequence, the shell can act as resistive barriers for protonic conduction. The description is corroborated by the results shown in Figs. 3 and 4 which show deviation from oxygen stoichiometry. These substantial deviations from the oxygen stoichiometry should be accompanied by segregation of point defects in the anion sublattice which can contribute to fast proton transport in the core of the cell. Such a configuration could increas the protonic conductivity parallel to the rod axis and this is a subject matter of current research efforts.

Fig. 13 Effect of oxygen cage tilting on OH⁻ rotation.

The nano-scale variations could reflect a miscibility gap with the formation of nano-domains of distinct B'/B''. The distribution of the B' and B'' cations in a $A_3(B'_{1+x}B''_{2-x})O_{9-\delta}$ perovskite can be random or ordered and differ as a function of the nature of the cations, amount of $B'' \rightarrow B'$ substitution (x) and synthesis method. Du and Nowick[5] have shown that ordered and disordered domains can coexist in HTPC complex perovskites. In the melt grown perovskite, a 1:1 ordered distribution of the B cations on the $\{111\}_P$ planes of the perovskite unit cell ($_P$) has been revealed by the strong extra reflection at 1/2 1/2 1/2. The doubling of the parameter in <111>$_P$ can be explained by two distribution modes; a ratio of $B''/B' = 2$ and a $B'/B'' = 1$.

The first distribution mode maintains a ratio of $B''/B'=2$. The distribution B' and B'' cations on the $\{111\}_P$ is then described as follows: a plane with a random mixture of B' and B'' with $B'/B''=2$ (B', B', B'') alternates with a plane entirely filled by B" ions (B'', B'', B''). This structure has been described for Sr_3 (Ca Nb$_2$) O$_9$ by Hervieu and Raveau[6] The second distribution mode induces a B'/B'' ratio of 1. In this case $\{111\}_P$ planes of B' alternate with $\{111\}_P$ planes of B''. Du and Nowick assume that this model describes the $Sr_3Ca_{1+x}Nb_{2-x}O_{9-\delta}$ HTPC perovskites.[5] In both cases ($B'/B''=2$ or $B'/B''=1$) the ordered domains do not have the mean composition of the rod, so that they have to coexist with disordered nanoscale domains to balance the global stoichiometry and charge neutrality. The composition gap is however lower in the first case and might better fit the Nb/Ca variations observed using STEM-EDX shown in Fig.11. Furthermore, oxygen cage tilting is recognized for the first time for the $Sr_xCa'_{1-x}Nb'_{2-x}O_{9-x}$. Oxygen cage tilting is reported for $Sr_3(ZnNb_2)O_9$ similar perovskites.[4]). This tilting reduces the rotation angle of the O-H group permitting the protonic diffusion from one octahedron to the other octahedron[7] (Fig 13). The oxygen cage tilting is also expected to lower the activation energy needed for the rotation.

The anti-phase boundaries are the channels with low activation energy for protonic diffusion. Diffusion coefficients along these defects can be 6 – 8 orders of magnitude higher than in the bulk material.[8] The engineering of these structures is our main objective and our current efforts are concentrating on the production of HTPC ceramics that have sufficient concentration of anti-phase domain structures to form a continuous network. The formation of microdomains of different composition and structure (Figs. 7 through 10) is possible to produce if there is a miscibility gap leading to the observed microheterogeneties. The effectiveness of melt processing in mitigating the issues of protonic transport rate and environmental durability will be addressed in a subsequent publication. The similarity of perovskite related domains structures produces coherent interphase-interfaces, lowering the elastic part of free energy of the system. The reaction of such structures to water and their high temperature mechanical performance has been addressed in the accompanying papers.[9]

CONCLUSION

Melt processing can produce microstructure architectures that cannot be achieved by solid-state sintering. Melt processing solves the membrane fabrication problems: (1) it is capable of producing fully dense materials, (2) homogenization eliminates secondary phases. Thus, melt processing provides an effective means to mitigate the issues of density, protonic transport rate and environmental durability of HTPC ceramics.

REFERENCES
[1] A.S. Nowick, Y. Du and K.C. Liang. Solid State Ionics 125 303-311 (1999).
[2] A. Sayir and S. C. Farmer, Mat. Res. Proc., 365 11-21 (1995).
[3] I. M. Reaney E.L. Colla and N. Setter Jpn. J. Apll. Phys., 33 3984-3990 (1994).
[4] E. L. Colla I.M. Reaney and N. Setter J. Appl. Phys., 74 (5) 3414 (1993).
[5] Y. Du and A. Nowick , J Am Cer Soc., 78 [11] 3033-3039 (1995).
[6] M. Hervieu and B. Raveau, J Solid State Chem., 28, 209-222, (1979).
[7] E. Matsushita, T. Sasaki Solid State Ionics, 125 31 (1999).
[8] E. Goldberg, A. Nemudry, V. Boldyrev and R. Schollhorn., 110 223 (1998).
[9] M.J. L. Robledo, J. M. Fernández, A.R. A. López, A. Sayir, in this proceeding.

NUCLEAR MICROPROBE USING ELASTIC RECOIL DETECTION (ERD) FOR HYDROGEN PROFILING IN HIGH TEMPERATURE PROTONIC CONDUCTORS

Pascal Berger
Laboratoire Pierre Süe, CEA-
CNRS, CEA Saclay, 91191
Gif sur Yvette - FRANCE

Ali Sayir
NASA GRC / C. W. R. U.
21000 Brookpark Road
Cleveland, 44135 OH - USA

Marie-Hélène Berger
Ecole des Mines des Paris,
BP 87
91003 Evry Cedex – FRANCE

ABSTRACT

The interaction between hydrogen and various high temperature protonic conductors (HTPC) has not been clearly understood due to poor densification and unreacted secondary phases. The melt-processing technique is used in producing fully dense simple $SrCe_{0.9}Y_{0.1}O_{3-\delta}$ and complex $Sr_3Ca_{1+x}Nb_{2-x}O_{9-\delta}$ perovskites that can not be achieved by solid-state sintering. The possibilities of ion beam analysis have been investigated to quantify hydrogen distribution in HTPC perovskites subjected to water heat treatment. Nuclear microprobe technique is based on the interactions of a focused ion beam of MeV light ions (^1H, ^2H, ^3He, ^4He, ..) with the sample to be analyzed to determine local elemental concentrations at the μm^3 scale. The elastic recoil detection analysis technique (ERDA) has been carried out using $^4He^+$ microbeams and detecting the resulting recoil protons. Mappings of longitudinal sections of water treated $SrCeO_3$ and $Sr(Ca_{1/3}Nb_{2/3})O_3$ perovskites have been achieved. The water treatment strongly alters the surface of simple $SrCe_{0.9}Y_{0.1}O_{3-\delta}$ perovskite. From Rutherford Back Scattering measurements (RBS), both Ce depletion and surface re-deposition is evidenced. The ERDA investigations on water treated $Sr_3Ca_{1+x}Nb_{2-x}O_{9-\delta}$ perovskite did not exhibit any spatial difference for the hydrogen incorporation from the surface to the centre. The amount of hydrogen incorporation for $Sr_3Ca_{1+x}Nb_{2-x}O_{9-\delta}$ was low and required further development of two less conventional techniques, ERDA in forward geometry and forward elastic diffusion $^1H(p,p)^1H$ with coincidence detection.

INTRODUCTION

High temperature protonic conducting (HTPC) ceramics offer the promise of highly selective hydrogen separation possibly at intermediate temperature range 400-600°C. Both simple and complex perovskites can be protonic and electronic conductors with the general formula $A_3B'_{1+x}B''_{2-x}O_{9-\delta}$ (A=Sr^{2+}, Ba^{2+}; B'=Ca^{2+}, Sr^{2+}; B''=Nb^{5+}, Ta^{5+}; x= 0 – 0.2). The B-site occupancies of B' and B'' can be ordered or disordered depending on the ionic radius of A ion and the ionic radius difference of the B' and B'' ions. Several obstacles remain to realizing the full potential of these ceramic membranes: (1) difficulty of attaining high density even at processing temperatures exceeding 1600°C, (2) thermodynamic instability in H_2O or CO_2 containing environment and (3) the lack of measurement techniques and unambiguous model for conductivity.

The solid-state sintering approach has not resolved the challenges associated with processing to produce full density materials. The segregation at grain boundaries and inadequate characterization of oxygen vacancy concentration made transport measurements questionable. The interaction between hydrogen and various high temperature protonic conductors (HTPC) has not been clearly understood and the role of hydrogen has been described in the literature as *challenging to understand*.[1] The common approach for the study of the incorporation of hydrogen into the HTPC structure is through thermogravimetric methods. The thermogravimetric techniques provide information about the response of the whole sample but a precise understanding of the transport mechanisms requires local hydrogen profiling of fully dense ceramic HTPC ceramic.

The approach for this work employed a melt-processing developed at NASA GRC to produce HTPC ceramics. The advantages of melt processing over solid-state sintering are the fabrication of full density materials and preferred crystal growth to form textured microstructures. The nuclear microprobe of the Laboratoire Pierre SÜE / France has been used to investigate quantitatively the spatial distribution of hydrogen after water heat treatment. The approach exploits the Elastic Recoil Detection Analysis (ERDA) combined with Rutherford back scattering (RBS) measurements.[2] The melt-processing technique and nuclear microprobe analysis are both employed for the first the time for the HTPC's. From the broad compositional range of perovskite structures, we limited our efforts on $SrCe_{0.9}Y_{0.1}O_{3-\delta}$ and $Sr_3Ca_{1+x}Nb_{2-x}O_{9-\delta}$, x=0.18. $SrCe_{0.9}Y_{0.1}O_{3-\delta}$ and $Sr_3Ca_{1+x}Nb_{2-x}O_{9-\delta}$, x=0.18 compositions. $SrCe_{0.9}Y_{0.1}O_{3-\delta}$ and $Sr_3Ca_{1+x}Nb_{2-x}O_{9-\delta}$ have not been studied previously and represents simple- and complex perovskite structures respectively.

EXPERIMENTAL

Melt processing

For preparation of the source rod[4-5], high purity (99.999% pure) polycrystalline $SrCO_3$ powder, high purity (99.999% pure) polycrystalline CeO_2, $CaCO_3$, Nb_2O_5 and Y_2O_3 (99.99% pure) were obtained from Alfa Aesar. These powders were blended and milled without milling agents for 40 hours. The powder mixture is then calcined at 800 °C in air for 6 hours. The calcined samples are crushed and sieved and the calcining step is repeated. The powders are cold isostatically pressed into the cylindrical rods and sintered in air at 1500 °C.

Fig.1. $SrCe_{0.9}Y_{0.1}O_{3-\delta}$ rod produced by melt processing. Ø = 8

For the heating source, a coherent CO_2-laser beam (PRC 2200; Landing, NJ 0785) was used to melt process. The laser was split into two beams, 180° apart from each other. Each beam was then focused with zinc-selenide lenses onto the molten zone at the top of a polycrystalline source rod in the center of the processing chamber. The maximum available laser power was 2200 watts, but only a small fraction of this power (800 to 1100 watts) was needed because most of the radiation from the CO_2-laser (wavelength=10.6 μm) is absorbed in the oxides. The molten zone temperature was measured in the infrared radiation region with a custom made infrared thermal monitor. Absolute temperature was not measured because the emissivity of the molten Sr-Ce-Y-O and Sr-Ca-Nb-O systems were not known. All experiments were conducted in air. Neither the crystal nor the source rod was rotated and the external appearance is shown in Fig. 1.

Microstructure and sample preparation

SEM and WDX have been performed on longitudinal sections of $SrCe_{0.9}Y_{0.1}O_{3-\delta}$ and $Sr_xCa'_{1-x}Nb'_{2-x}O_{9-x}$ rods, mechanically polished with an alcohol-based lubricant. Proton incorporation experiments were performed by exposing the samples to wet air at 500 °C for 10h followed by a slow cooling (100°C%h). Wet air was obtained by bubbling N_2+21%O_2 gas through a flask with pure water. The nuclear microanalysis has been performed on water treated samples at high temperature, either on as-treated surfaces, cross-sections or thin foils (25-30 μm).

Nuclear microprobe

Nuclear microprobe techniques were used to analyze the surface composition, to evidence, for instance, possible hydrolysis or hydrogen incorporation into the lattice structure. The measurements employed the elastic recoil detection analysis (ERDA) technique combined with Rutherford Back Scattering (RBS) technique. The nuclear microprobe uses the interactions of a focused ion beam of MeV light ions with the sample to determine local elemental concentrations at the μm³ scale. We utilized ^1H and ^4He for the characterization of HTPC ceramics. Several analytical techniques are performed, based on the spectrometry of the induced X-and γ-rays, and of the particles scattered or produced by nuclear reactions. The particle induced X-ray emission (PIXE) technique is currently most accepted technique and is the complementary to the electron induced X-ray emission technique, for elements with Z ≥ 11. The PIXE technique has very low background due to the reduced bremsstrahlung and permits trace element determination. It has been used here only for control since trace determination was not needed.

The RBS technique is pertinent for medium and high Z elements determination and has depth resolution capabilities ranging from a few nanometers to a few micrometers, whereas the ERDA technique is specifically used for very light elements (^1H and ^2H mostly). In order to improve the sensitivity of the detection of hydrogen, less conventional techniques have been undertaken; ERDA in transmission mode and elastic diffusion with coincidence detection which will be discussed further in relation to results with HTPC ceramics. The nuclear reaction analysis (NRA) technique is suitable for light elements, especially from Li to F with isotopic selectivity and depth profiling capability. A good accuracy for the determined compositions may be then easily obtained without reference samples whose composition is close to that of the sample to be analyzed. Both for RBS, ERDA and for NRA, the compositions are derived from the raw spectra with the use of pertinent simulation programs. To determine the analyzed volumes and derive depth concentration profiles, the rate of energy loss in the sample of the particles, dE/dx, can be calculated from the weighted mean of the dE/dx values for the constituent elements with only very slight dependence on the chemistry (Bragg rule).[5] As these stopping powers are known with a good accuracy of approximately 3% for energy level between 1 to 4 MeV,[6] the nuclear microprobe gives reliable depth information.

RESULTS

Simple perovskite $SrCe_{0.9}Y_{0.1}O_{3-\delta}$

Scanning electron micrograph (SEM) images of $SrCe_{0.9}Y_{0.1}O_{3-\delta}$ show a cellular topology. A representative SEM image of the microstructure is shown in Fig. 2 with three different regions marked as outer-, intermediate-, and centre region. The outer region is around 500 μm from the surface and it's microstructure is composed of large cells with only a small amount of second phase inclusions alternating periodically with regions containing elongated thin second phases. Each cell has a width of around 100 μm. The thickness of the intermediate region is around 400 μm. The intermediate region is composed of elongated grains (length ≈200 μm width ≈25 μm), separated by a second phase. The centre region of the rod is approximately 700 μm and composed of grains with lower aspect ratios and wavy boundaries. The intergranular second phases are thicker. A small amount of misalignment of the cells with respect to the rod axis is noticeable. The radial growth markings reveal solid-liquid interface shape and a cellular topology.

	Sr	Ce	O	Y	Al	F	Total
10 analyses on 50*50 μm² (outer region)							
Mean, w %	29.70	46.14	16.65	3.45	0.07	0.41	96.42
Std. Dev., ±	0.34	0.67	0.18	0.13	0.04	0.10	0.55
Mean, At. %	19.13	18.59	58.73	2.19	0.14	1.21	
Std. Dev., ±	0.13	0.38	0.29	0.08	0.09	0.30	
10 analyses on 50*50 μm² (intermediate region)							
Mean, w %	29.58	44.95	17.36	3.55	0.25	0.41	96.09
Std. Dev., ±	0.17	0.87	0.28	0.13	0.10	0.09	0.61
Mean, At. %	18.61	17.69	59.82	2.20	0.51	1.17	
Std. Dev., ±	0.17	0.48	0.40	0.06	0.21	0.27	
10 analyses on 50*50 μm² (centre region)							
Mean, w %	29.52	44.22	17.62	3.74	0.43	0.41	95.94
Std. Dev., ±	0.38	0.83	0.40	0.18	0.17	0.09	0.50
Mean, At. %	18.38	17.22	60.06	2.29	0.86	1.18	
Std. Dev., ±	0.37	0.54	0.67	0.09	0.33	0.26	

Table 1   Overall electron microprobe analysis using WDX mode.

	Sr	Ce	O	Y	Al	F	Total
Second phase							
Weight %	32.00	34.11	23.66	4.85	3.97	0.20	98.81
Atomic %	15.88	10.58	64.30	2.37	6.39	0.46	
Grain in outer region							
Weight %	30.46	46.59	16.31	3.65	0.09	0.50	97.59
Atomic %±	19.64	18.78	57.58	2.32	0.19	1.48	
Grain in centre region							
Weight %	30.60	47.71	16.26	3.49	0.002	0.40	98.48
Atomic %	19.76	19.27	57.51	2.22	0.04	1.19	

Table 2   Local electron microprobe analysis using WDX mode.

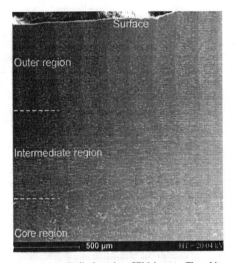

Fig. 2. Longitudinal section. SEM image. The white contrast of the boundaries is due to a preferential etching of the boundaries during polishing.

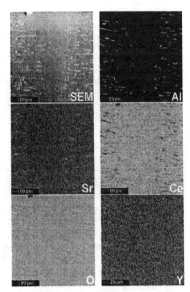

Fig. 3. Outer region (WDX)

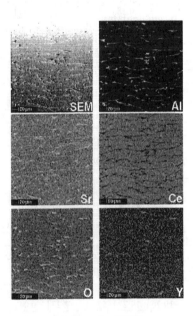

Fig. 4. Intermediate region (WDX)

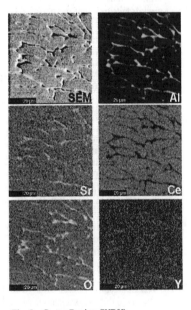

Fig. 5. Centre Region (WDX)

360

Secondary electron images showing the spatial distributions of Sr-, Ce-, Y-, Al- and O cations are depicted in Figs. 3, 4 and 5. Quantitative chemical analyses were completed using electron microprobe (wavelength dispersive x-ray analysis; WDX). Table 1 summarizes the average value of 10 measurements in WDX mode in an area of 50x50 μm^2 positioned at the outer-, intermediate and centre regions (cf. Figs.3 to 5). The total weight percentage reaches ~96%. The weight to atomic conversion is calculated on the basis of a total atomic % of 100. It is suspected that topographic irregularities on the measurement windows are responsible for the 4% deficiency. The spot analyses results are summarized in Table 2 for the center of the grain and second phase. These analyses have been completed on a large grain and intergranular phase located in centre region and on a large crystalline phase in outer region. The transmission electron microscope (TEM) analysis show that intergranular second phase is amorphous.

Nuclear microprobe was first applied to investigate simple perovskite $SrCe_{0.9}Y_{0.1}O_{3-\delta}$. The experimental conditions were chosen as follow: $^4He^+$, energy: 3.06 MeV, beam size 2 x 3 μm^2, surface barrier particle detectors: standard at 30° for ERDA (required) and annular at 170° for RBS. Investigations were

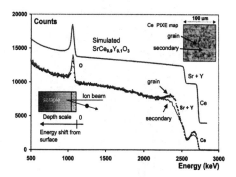

Fig. 6. RBS spectra of $SrCe_{0.9}Y_{0.1}O_{3-\delta}$ following water treatment (included PIXE map of Ce).

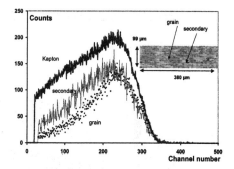

Fig. 7. ERDA spectra of a kapton® foil, grains and intragranular phase in $SrCe_{0.9}Y_{0.1}O_{3-\delta}$.

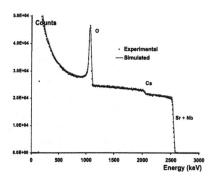

Fig. 8. Comparison of RBS spectra for $Sr_3Ca_{1+x}Nb_{2-x}O_{9-\delta}$ and simulated (solid line).

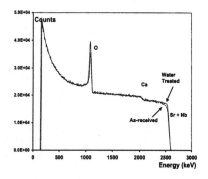

Fig. 9. Experimental RBS spectra of as-received and water treated $Sr_3Ca_{1+x}Nb_{2-x}O_{9-\delta}$.

at the center region of the rod (same region as shown on Fig.5) following the water treatment. Modifications of the surface are clearly evidenced (Figure 6). Indeed, in a standard RBS spectrum, the height of the steps and their intensity variations with the energy provide information on the surface concentration and in-depth profile, respectively. As the energy shift from the right side of the step corresponds to a depth scale from the surface, in-depth modifications of the concentrations are unambiguously revealed, Fig. 6. A homogeneous $SrCe_{0.9}Y_{0.1}O_{3-\delta}$ sample would give rise to the simulated spectrum shown as a solid line in Fig. 6. The anticipated relative heights of the Ce and Sr+Y steps (Sr and Y are not mass separated) are also included.

Complex Perovskite $Sr_3Ca_{1+x}Nb_{2-x}O_{9-\delta}$

The complex perovskite structure is able to accommodate a broad spectrum of chemical substitution making an ideal platform for probing correlations between chemistry, order and protonic conductivity. Due to their large valence difference, the distribution of the B'(2+) and B"(5+) cations on the B site is often ordered. Two types of order schemes are described, giving rise to the so-called 2:1 and 1:1 structures. The TEM investigation of $Sr_3Ca_{1+x}Nb_{2-x}O_{9-\delta}$ has revealed a more complex structure than that described by Du and Nowick[7] and described extensively in the accompanying paper.[8]

The reaction of complex perovskite $Sr_3Ca_{1+x}Nb_{2-x}O_{9-\delta}$ with water was studied by RBS and ERDA techniques. $Sr_3Ca_{1+x}Nb_{2-x}O_{9-\delta}$ had only slight surface alteration subsequent to water treatment as evidenced from RBS spectra shown in Fig.8. The simulation based on $Sr_3Ca_{1+x}Nb_{2-x}O_{9-\delta}$ composition of x=0.18 also depicted in Fig. 8 as solid line. The simulation is in good agreement with the observed data. The comparison of the RBS spectra of as-polished (3 weeks before measurement) and water treated samples shows a more important surface alteration for storage at room temperature than for a short water heat treatment, Fig. 9. The ERDA spectrum also demonstrate more surface hydrolysis for the as-stored sample than mechanically polished specimen suggesting strong time dependence of $Sr_3Ca_{1+x}Nb_{2-x}O_{9-\delta}$ surface composition with water.

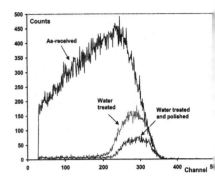

Fig. 10. ERDA spectra of as-received, water treated, polished $Sr_3Ca_{1+x}Nb_{2-x}O_{9-\delta}$.

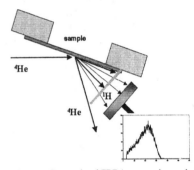

Fig. 11. Conventional ERDA at a grazing angle.

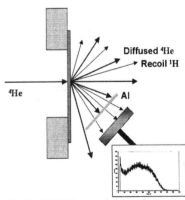

Fig. 12. ERDA in transmission mode.

DISCUSSION

Simple Perovskite $SrCe_{0.9}Y_{0.1}O_{3-\delta}$

The comparison of Table 1 and 2 shows that the grains have a Sr/Ce ratio of 1.05-1.02. The average Sr/Ce ratio for outer-, intermediate and centre regions were 1.03, 1.05 and 1.07 respectively. This ratio is lower than the nominal composition of simple perovskite (1.1). The Sr/Ce ratio in second phase is 1.5. The partitioning of Sr between the grains and second phase is expected to account for the difference between the nominal composition and measured values. Preferential evaporation of SrO during the growth may have occurred but is expected to be small since the Sr/Ce ratio for the interior and outer regions are close to the average value. The Al-contamination is expected to change the Sr partitioning and aid the glass formation at the intergranular phase. The origin of the Al contamination is the use of alumina crucibles during calcining and sintering.

The TEM diffraction analysis indicates that the grains are well developed and their diffraction pattern is consistent with the perovskite structure with a reticular parameter of 8.53 ± 0.04 Å. This parameter is within the range of simple perovskites. The chemical compositions of the grains in the core and in the outer region are comparable as shown through WDX analysis, Table 1 and 2. Apart from Al contamination, Figs. 2 through 5 and Table 1 demonstrate a homogeneous distribution of cations throughout the structure within the resolution WDX and X-ray maps. The microstructure changes modestly from the surface to the interior up to thicknesses 1500 µm. The full density and superior

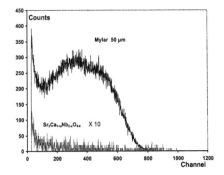

Fig. 13. ERDA in transmission mode data on a 50 µm mylar and 30 µm $SrCe_{0.9}Y_{0.1}O_{3-\delta}$.

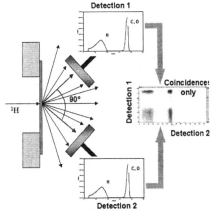

Fig. 14. Forward elastic scattering $^1H(p,p)^1H$ with coincidence detection.

homogeneity and of melt-processing structures are exemplified in an enhanced thermodynamic stability. Simple perovskite $SrCe_{0.9}Y_{0.1}O_{3-\delta}$ sintered at 1500°C disintegrated in laboratory environment (25 °C) within a day. The melt processed $SrCe_{0.9}Y_{0.1}O_{3-\delta}$ was physically stable for a prolonged period of time (months) and time dependence is currently under investigation. The nuclear microprobe analysis using RBS spectra of $SrCe_{0.9}Y_{0.1}O_{3-\delta}$ confirm a strong alteration of the surface, Fig. 6. At the surface the Ce/(Sr+Y) ratio is close to 0.2 and the Ce concentration decreases from the surface towards the interior of the sample. This is interpreted as a selective leaching of Ce accompanied by a re-precipitation, since both a Ce depletion from the bulk composition and a surface enrichment are observed. The whole thickness of the reaction layer may be estimated from the spectrum and is approximately 300 to 500 nm. At higher depths, the composition remains unchanged as demonstrated by electron microprobe measurements. The PIXE maps confirm Ce depleted composition of the intergranular secondary phase and the corresponding RBS spectra show a differentiated alteration of the grains and of the boundaries (Fig. 6).

363

The ERDA spectra, as shown in Fig. 7, have to be interpreted similarly to the RBS spectra. The slope is analogous to the known standard kapton® polymer foil. Thus, the $SrCe_{0.9}Y_{0.1}O_{3-\delta}$ reveal homogeneous hydrogen distribution within both the grains and secondary phase given that both have monotonous decrease of counts with respect to channel number. The comparison of ERDA spectra acquired from

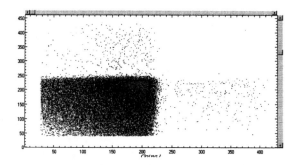

Fig. 15 Coincidence map obtained on 25 µm $Sr_3Ca_{1+x}Nb_{2-x}O_{9-\delta}$.

grains and secondary intergranular phase confer high hydrogen content in the altered layer, especially for the secondary phase. The probe depth is limited to the first few 100 nm and measurements represent the response of the hydrogenation of the reaction layer. A transverse cut after water heat treatment would be necessary to probe bulk hydrogen transport properties.

Complex Perovskite $Sr_3Ca_{1+x}Nb_{2-x}O_{9-\delta}$

The ERDA mapping of the water treated $Sr_3Ca_{1+x}Nb_{2-x}O_{9-\delta}$ along the transverse cross-section does not confirm any difference of chemical reactivity from the exposed surface to the center region. The RBS measurements show a good chemical stability with water treatment, but due to strong time dependence (Fig. 10) the long-term stability requires further study. The measured amount of hydrogen using ERDA at grazing incidence was low and measurement sensitivity reduced the depth resolution. To minimize these difficulties, two alternate techniques were developed to investigate thin foils (25-30 µm) of $Sr_3Ca_{1+x}Nb_{2-x}O_{9-\delta}$ in transmission mode; ERDA in forward geometry and forward elastic diffusion $^1H(p,p)^1H$ with coincidence detection. The principles of hydrogen detection in transmission mode are known, but few teams within the ion beam analysis community practice these kinds of techniques. Both of these new methods have been exploited to detect hydrogen in $Sr_3Ca_{1+x}Nb_{2-x}O_{9-\delta}$.

ERDA in forward geometry

ERDA method is based on the detection of the recoil protons, the sample hit by the ^4He ions of the beam. Both backscattered ^4He ions and recoil protons are collected using a particle detector. An aluminum filter is placed before the detector to stop ^4He so that only the protons are detected. In conventional ERDA, the incident ^4He beam is oriented at a grazing angle from the surface of the sample, Fig. 11. Thin samples enable forward geometry. The particles are collected behind the sample as shown schematically in Fig. 12. The foundation for using the forward geometry method is the expected benefit from higher cross-section (sensitivity) and an increased depth analysis. The mass of the incident ^4He of the beam is relatively high and their range in the target is expected to be low. In view of this fact, very thin samples of $Sr_3Ca_{1+x}Nb_{2-x}O_{9-\delta}$ were required. In order to optimize sample thickness, the approach was to use an $^4He^{++}$ beam instead of $^4He^+$. With the same voltage of the ion accelerator, the velocity of the incident ion could be doubled and their range in the sample increased. The challenge was to extract enough $^4He^{++}$ ions from the ion plasma source of the accelerator which was not designed originally for this purpose. The feasibility has been demonstrated using 5 MeV $^4He^{++}$ but the reduced intensity of the beam limits the size of the spot to 50 x 100 µm². Although the sensitivity is evidenced by the measurements on mylar foils, the hydrogen signal is very weak on a 30 µm thick $Sr_3Ca_{1+x}Nb_{2-x}O_{9-\delta}$; Fig. 13. The origin of this unexpected low hydrogen content for $Sr_3Ca_{1+x}Nb_{2-x}O_{9-\delta}$ is not known. The altered surface of the treated sample might behave as a barrier for hydrogen incorporation. Massive hydrogen losses during the preparation of the thin sample seems unlikely.

Forward elastic diffusion ^{1}H(p,p)^{1}H with coincidence detection

Although backscattering offers the best combination of mass and depth resolution for analytical purposes, forward scattering remains more probable and the related analytical techniques are expected to be more sensitive. Like for transmission ERDA, Fig 12, forward scattering is set on a thin sample with the detector behind it. In the case of a proton beam, the collected particles include both scattered incident and recoil protons from the hydrogen atoms hosting within the lattice. However, the interference with the high magnitude scattering from the other elements of the target was difficult and has necessitated further engineering.

To achieve a selective detection, a modification was structured that consisted in placing not only one but two detectors, both 45° from beam axis, Fig 14. The kinematics of the impact of protons on protons implies a 90° angle between the scattered and the recoil. Both of them may be then detected in coincidence between the two detectors. A simple scattering event is rejected because only one detector records an event. The other recoil nuclei with higher mass are not detected, either because they do not escape from the sample (recoil energy too low) or because they are not scattered in the same direction. Only scattering on hydrogen atoms enables the coincidence detection on both detectors.

These tests with 3.65 MeV protons have shown that forward elastic diffusion ^{1}H(p,p)^{1}H with coincidence detection technique is the most promising in terms of sensitivity. With standard collimated detectors, the counting rates were high on mylar foils and huge with the $SrCe_{0.9}Y_{0.1}O_{3-\delta}$ perovskite. The intensity of the beam had to be drastically reduced, lower than the usual range for an analytical microbeam. This is promising to map the hydrogen content with a good lateral resolution (reduced beam size within a few μm^2). For this preliminary experiment, the minimum coincidence time (maximum delay between two successive events above which they are considered as independent) tested was still high (600 psec.). Because of the fluctuations of the intensity of the beam, this value does not prevent from the observation of "false" coincidences (i.e., two independent events on both detectors, occurring in less than 600 psec.). Figure 15 shows the coincidence map obtained on a 25 μm $Sr_3Nb_2CaO_9$ sample. Due to the geometry and the high coincidence time, most of the recorded events corresponded to false coincidences and represent $\approx$ 1 % of the whole detected particles. Hence, our current efforts are focusing on the further development of the detection device. An order of magnitude improvement of the coincidence time could be gained by using a rejection device that can be inserted before the acquisition system.

CONCLUSIONS

Simple $SrCe_{0.9}Y_{0.1}O_{3-\delta}$ and complex $Sr_3Ca_{1+x}Nb_{2-x}O_{9-\delta}$, perovskite solid structures up to 8 mm in thickness has been produced using melt processing. The melt-processing facilitate in producing fully-dense materials that can not be achieved by solid-state sintering. The microstructural homogeneity of both simple $SrCe_{0.9}Y_{0.1}O_{3-\delta}$ and complex $Sr_3Ca_{1+x}Nb_{2-x}O_{9-\delta}$ perovskite structures has been demonstrated. The full-density and superior homogeneity produced an enhanced thermodynamic stability. The chemical homogeneity could be further improved by controlling the partitioning of Sr in $SrCe_{0.9}Y_{0.1}O_{3-\delta}$ between the grains through minimizing the contamination.

The possibilities of ion beam analysis have been investigated to quantify hydrogen distribution in HTPC perovskites subjected to water treatment. The elastic recoil detection (ERDA) has been carried out using ^{4}He$^+$ microbeams and detecting the resulting recoil protons. Mappings of longitudinal sections of water treated $SrCe_{0.9}Y_{0.1}O_{3-\delta}$ and $Sr_3Ca_{1+x}Nb_{2-x}O_{9-\delta}$ perovskites have been achieved. The water treatment strongly alters the surface of simple $SrCe_{0.9}Y_{0.1}O_{3-\delta}$ perovskite. From RBS measurements, both Ce depletion and surface re-deposition is evidenced. In spite of depth profiling capabilities, ERDA measurement directly from the water treated surface may only probe the altered layer because of the probed depth limited to the few first 100 nm. To investigate bulk properties of hydrogen transport, further measurements on cross-sections, following controlled water treatment will be essential. The ERDA investigations on water treated $Sr_3Ca_{1+x}Nb_{2-x}O_{9-\delta}$ perovskite did not exhibit any spatial difference for the hydrogen incorporation from the surface to the center. The amount of hydrogen incorporation was for $Sr_3Ca_{1+x}Nb_{2-x}O_{9-\delta}$ low and required further development of less conventional techniques. Two new methods, ERDA in forward geometry and forward elastic diffusion $^1H(p,p)^1H$ with coincidence detection, have been developed for hydrogen detection in complex perovskite systems. The tests of ERDA in forward geometry using 5 MeV $^4He^{++}$ and forward elastic diffusion $^1H(p,p)^1H$ with coincidence detection using 3.65 MeV protons have shown promising sensitivity. Further work is necessary to optimize detector devices to elucidate the chemistry of structure at the elemental level and correlate with the efforts on conductance to achieve a comprehensive understanding.

ACKNOWLEDGEMENTS

This research was funded by NASA cooperative agreement NCC3-850 and through European Office of Aerospace Research and Development by AFOSR under Grant No. FA8655-03-1-3040.

REFERENCES

[1] K.-D. Kreuer, Chem. Mater., **8** 610 (1996).

[2] J.R. Tesmer and M. Nastasi Eds. *Handbook of Modern Ion Beam Materials Analysis* Materials Research Society, Pittsburg, Pennsylvania (1995).

[3] A. Sayir and S. C. Farmer, "The Effect of the Microstructure on Mechanical Properties of Directionally Solidified Al2O3/ZrO2(Y2O3) Eutectic," Acta Mat., **48** 4691-4697 (2000).

[4] A. Sayir, "Directional Solidification of Eutectic Ceramics," pp.197-211in Computer-Aided Design of High-Temperature Materials. Eds. A. Pechenik, R.K. Kalia, P. Vashista, Oxford University Press (1999)

[5] W.H. Bragg and R. Kleeman, Phil. Magazine **10** 318 (1905).

[6] J.F. Ziegler, J.P. Biersack and U. Littmark *The Stopping and Range of Ions in Matter.* Vol. 1 Pergamon Press, New York (1985).

[7] Y. Du and A. Nowick , "*Journal of the American Ceramic Society*" **78** [11] 3033-3039 (1995)

[8] M.-H. Berger and A. Sayir, in this proceeding.

IONIC CONDUCTIVITY IN THE Bi_2O_3-Al_2O_3-M_xO_y (M=Ca, Y) SYSTEM

Yi-Tzuen Liu, and Tzer-Shin Sheu
Department of Materials Science and Engineering
I-Shou University, Kaohsiung, Taiwan

ABSTART

A two-phase composite of $BiAlO_3$ plus Bi_2O_3 was sintered at 700-850°C from the mixed oxide powders and the co-precipitated powders. With the mixed oxide powders as the starting materials, the sintered a nd heat-treated B i_2O_3-containing samples contained $Bi_2Al_4O_9$, γ-Bi_2O_3, r-Bi_2O_3 and δ-Bi_2O_3, but not $BiAlO_3$. However, the sintered specimens obtained from the co-precipitated powders o nly had a very small amount of $BiAlO_3$. It indicated that B $i_2Al_4O_9$ was much easier to form than $BiAlO_3$. From the electrical conductivity measurement, these solid electrolytes with δ-Bi_2O_3, $BiAlO_3$, or $Bi_2Al_4O_9$ all had an excellent ionic conductivity at high temperatures, approximately $4*10^{-2}$ S/cm at 700°C. Except for a major content of Bi_2O_3 and Al_2O_3, these solid electrolytes also had a minor Y_2O_3 or CaO content. The correlations among ionic conductivity, phase existence, and the minor dopant content were also discussed.

INTRODUCTION

Bismuth-oxide-containing ceramics have been extensively studied for their excellent oxygen ionic conductivity.[1-11] Pure Bi_2O_3 contains two well-known equilibrium solid phases, δ-Bi_2O_3 at high temperatures and α-Bi_2O_3 at low temperatures.[1-3,10] It have been reported that δ-Bi_2O_3, the face-centered cubic phase, has a best ionic conductivity at high temperatures among most oxygen ionic conductors.[5-6,10] For stabilizing δ-Bi_2O_3 at lower temperatures, some researchers have found that the stabilized phases in the Bi_2O_3-rich phase region, including the stabilized δ-Bi_2O_3 phase, closely correlate with the type and the ionic size of cation dopants.[4-6,11]

In this study, a two-phase composite δ-Bi_2O_3 +$Bi_2Al_4O_9$ or a single phase $BiAlO_3$ is alloy-designed to improve physical and chemical properties of Bi_2O_3-containing solid electrolytes. From t he a spect o f t he coefficient o f t hermal e xpansion, t hese e lectrolytes a re expected t o b e thermally compatible with some commercial electrodes for solid oxide fuel cells. Currently, the formation of $BiAlO_3$, phase existence, and ionic conduction of the $BiAlO_3$ plus Bi_2O_3 solid electrolytes are studied. The relationships among phase existence, ionic conductivity, alloying, fabrication processes, and heat treatments are discussed.

EXPERIMENTAL PROCEDURES

The starting powders were Bi_2O_3 (99.5% pure, Aldrich), Y_2O_3 (99.99% pure, Aldrich), Al_2O_3 (99.9% pure, Aldrich), and CaO (99% pure, Aldrich). These starting powders were well mixed together in ethanol, according to the compositions as listed in Table I. Subsequently, the mixed powders were dried to remove ethanol. A batch of the dried powders was cold-pressed uniaxially to form a cylindrical green powder compact. The green powder compacts were then sintered at 750°C or 850°C for 30min. The sintered specimens at 850°C

Table I. Phase existence of the sintered and heat-treated specimens.

Sample index	Starting Composition	Initial* Powders	Sintering temperature(°C)	Heat treatments	Phases
M1a	$BiAlO_3$	M	850	No	$\gamma\text{-}Bi_2O_3 + Bi_2Al_4O_9$
M1b	$BiAlO_3$	M	850	750°C,10h	$\gamma\text{-}Bi_2O_3 + \delta\text{-}Bi_2O_3 + Bi_2Al_4O_9$
M1c	$BiAlO_3$	M	850	600°C,60h	$r\text{-}Bi_2O_3 + \delta\text{-}Bi_2O_3 + Bi_2Al_4O_9$
M1d	$BiAlO_3$	M	850	500°C,60h	$\gamma\text{-}Bi_2O_3 + Bi_2Al_4O_9$
M2a	$Bi_{0.9}Y_{0.1}AlO_3$	M	850	No	$\delta\text{-}Bi_2O_3 + Bi_2Al_4O_9$
M2b	$Bi_{0.9}Y_{0.1}AlO_3$	M	850	750°C,10h	$\gamma\text{-}Bi_2O_3 + Bi_2Al_4O_9$
M2c	$Bi_{0.9}Y_{0.1}AlO_3$	M	850	600°C,60h	$\gamma\text{-}Bi_2O_3 + Bi_2Al_4O_9$
M2d	$Bi_{0.9}Y_{0.1}AlO_3$	M	850	500°C,60h	$\gamma\text{-}Bi_2O_3 + \delta\text{-}Bi_2O_3 + Bi_2Al_4O_9$
M3a	$Bi_{0.9}Ca_{0.1}AlO_{2.95}$	M	850	No	$\delta\text{-}Bi_2O_3 + Bi_2Al_4O_9$
M3b	$Bi_{0.9}Ca_{0.1}AlO_{2.95}$	M	850	750°C,10h	$\delta\text{-}Bi_2O_3 + Bi_2Al_4O_9$
M3c	$Bi_{0.9}Ca_{0.1}AlO_{2.95}$	M	850	600°C,60h	$\delta\text{-}Bi_2O_3 + Bi_2Al_4O_9$
M3d	$Bi_{0.9}Ca_{0.1}AlO_{2.95}$	M	850	500°C,60h	$r\text{-}Bi_2O_3 + \delta\text{-}Bi_2O_3 + Bi_2Al_4O_9$
M4a	$Bi_{0.9}Y_{0.05}Ca_{0.05}AlO_{2.975}$	M	850	No	$\delta\text{-}Bi_2O_3 + Bi_2Al_4O_9$
M4b	$Bi_{0.9}Y_{0.05}Ca_{0.05}AlO_{2.975}$	M	850	750°C,10h	$r\text{-}Bi_2O_3 + \delta\text{-}Bi_2O_3 + Bi_2Al_4O_9$
M4c	$Bi_{0.9}Y_{0.05}Ca_{0.05}AlO_{2.975}$	M	850	600°C,60h	$\delta\text{-}Bi_2O_3 + Bi_2Al_4O_9$
M4d	$Bi_{0.9}Y_{0.05}Ca_{0.05}AlO_{2.975}$	M	850	500°C,60h	$Bi_2Al_4O_9$
M5a	$Bi_{0.8}Y_{0.1}Ca_{0.1}AlO_{2.95}$	M	850	No	$\delta\text{-}Bi_2O_3 + (Bi_2Al_4O_9)$
M5b	$Bi_{0.8}Y_{0.1}Ca_{0.1}AlO_{2.95}$	M	850	750°C,10h	$\delta\text{-}Bi_2O_3 + Bi_2Al_4O_9$
M5c	$Bi_{0.8}Y_{0.1}Ca_{0.1}AlO_{2.95}$	M	850	600°C,60h	$\gamma\text{-}Bi_2O_3 + Bi_2Al_4O_9$
M5d	$Bi_{0.8}Y_{0.1}Ca_{0.1}AlO_{2.95}$	M	850	500°C,60h	$\delta\text{-}Bi_2O_3 + Bi_2Al_4O_9$
C1	$BiAlO_3$	C	750	No	$(BiAlO_3) + \gamma\text{-}Bi_2O_3 + Bi_2Al_4O_9$
C2	$Bi_{0.9}Y_{0.1}AlO_3$	C	750	No	$(BiAlO_3) + \delta\text{-}Bi_2O_3 + Bi_2Al_4O_9$
C6	$Bi_{0.95}Y_{0.05}AlO_3$	C	750	No	$(BiAlO_3) + Bi_2Al_4O_9$
C7	$Bi_{0.85}Y_{0.15}AlO_3$	C	750	No	$(BiAlO_3) + Bi_2Al_4O_9$

*M, the mixed oxide powders; C, the coprecipitated powders.

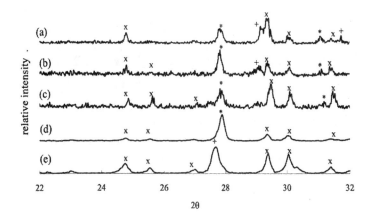

Figure 1. X-ray diffraction patterns for samples (a) M2d, (b) M2a, (c) M1a, (d) C1, and (e) C2. Symbol "x" is for $Bi_2Al_4O_9$; Symbol "*" is for γ-Bi_2O_3, and symbol "+" for δ-Bi_2O_3.

were further heat-treated at 500°C, 600°C, and 750°C for 10-60h to observe any phase transformation.

For obtaining the co-precipitated powders, each starting material was separately dissolved in HNO_3 to obtain an aqueous solution. These aqueous nitrate solutions were then mixed together, according to the each specific chemical composition as listed in Table I. Subsequently, the mixed nitrate solutions and ammonium hydroxide were droplet-like added into a beaker together to obtain hydroxide precipitates, under the continuously stirring conditions and pH=10. After co-precipitation, the precipitate/solutions were either dried (for the Ca-containing systems) or filtered (for the non-Ca-containing systems) to collect the hydroxide precipitates. Dried hydroxide powders were then calcined at 550°C for 30 min to obtain oxides. Calcined oxide powders were sometimes ground, and then the −325-mesh of powders w ere collected for cold pressing. The cold-pressed green powder compacts were sintered at 700 -750°C for 30min.

Phase existence of the sintered and heat-treated samples was determined by the X-ray diffraction method at room temperature. Sintering curve of the green powder compact was sometimes measured by a dilatometric unit at 25-800°C, with a heating rate of 10°C/min. Differential thermal analysis (DTA, Perkin Elmer DTA7) was used to observe any phase transformation or chemical reaction at high temperatures. Electrical conductivity was measured by using an LCR meter (HP model 4284B) at 25°-700°C in air.

(a)   (b)

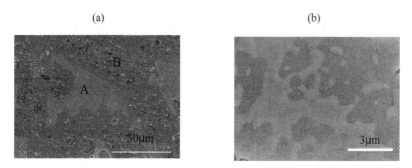

Figure 2. SEM micrographs of thermal etched specimens for samples (a) M2d, and (b) C6.

RESULTS AND DISCUSSION

Phase Existence

The phase formation studied using X-ray analysis is summarized in Table I. For the mixed oxide powders, its sintered and heat-treated samples contained δ-Bi_2O_3, γ-Bi_2O_3, r-Bi_2O_3 and $Bi_2Al_4O_9$. However, the $BiAlO_3$ phase was not being observed, even though some samples had an exact composition of $BiAlO_3$. With a long heat treatment at 500°C or 600°C for 60h, the $BiAlO_3$ phase was not being found in the heat-treated samples. The corresponding SEM micrograph of one sintered and heat-treated specimen is shown in Fig. 2(a). Through the composition analysis from electron dispersive X-ray spectrum (EDXS), the gray area marked by the letter "A" was identified to be the δ-Bi_2O_3 phase; the dark area marked by the letter "B" was the $Bi_2Al_4O_9$ phase. However, the fine microstructures inside the $Bi_2Al_4O_9$ phase were not being carefully examined yet; therefore this sample probably had another phase in addition for δ-Bi_2O_3 and $Bi_2Al_4O_9$.

For the co-precipitated powders, its sintered specimens also contained γ-Bi_2O_3 and $Bi_2Al_4O_9$ phase, as listed in Table I. A small amount of $BiAlO_3$ was also occasionally observed in these sintered samples from the coprecipitated powders. From SEM micrographs of the thermal etched specimens, as shown in Fig. 2, sample C6 had much smaller grains than sample M2d. It indicated that the sintered samples from the coprecipitated powders had sub-mircon grains. This phenomenon could be further confirmed from the width of diffracted peaks shown in Fig. 1, in which the sintered samples from the coprecipitated powders had broadening diffraction peaks.

Electrical Conductivity

The relationships between electrical conductivity and temperature for the sintered and heat-treated specimens are shown in Fig. 3 (a). The electrical conductivity of these samples had two temperature-dependent regimes, 25-200°C and 300-700°C. At high temperatures, at

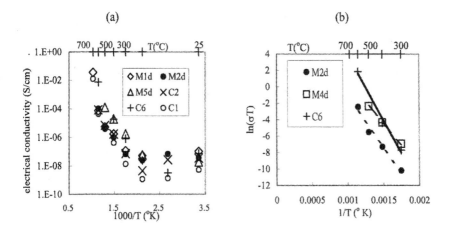

Figure 3. The relationships between electrical conductivity (σ) and temperature (T) for (a) σ vs. 1/T, and (b) ln(σT) vs. 1/T.

T=300-700°C, samples C6 and M5d had a higher electrical conductivity. Except for the major constituents of Al_2O_3 and Bi_2O_3, these two samples contained the oxide dopants Y_2O_3 and CaO. Whether the oxide dopant played an important role in the $BiAlO_3$ plus Bi_2O_3 composite was needed to further study. Among these sintered and heat-treated specimens, electrical conductivity at 700°C was $4*10^{-2}$ S/cm. However, electrical conductivity at high temperatures did not show any significant differences for the sintered samples from the mixed oxide powders and from the coprecipitated powders.

From the relationships between ln(σT) and 1/T shown in Fig. 3(b), activation energy of a charge carrier in these solid electrolytes was 0.6-1.5eV. Therefore, the charge carrier was believed to be oxygen vacancy.

CONCLUSIONS

A two-phase composite of $BiAlO_3$ plus Bi_2O_3 was sintered at 700-850°C from the mixed oxide p owders a nd t he c o-precipitated p owders. With t he m ixed o xide p owders, t he s intered and heat-treated specimens contained γ-Bi_2O_3, r-Bi_2O_3, δ-Bi_2O_3 and $Bi_2Al_4O_9$, but not $BiAlO_3$. From SEM micrograph observations and the composition analysis, these sintered and heat-treated samples had two different phases, δ-Bi_2O_3 and $Bi_2Al_4O_9$. However, the sintered specimens from the mixed oxide powders had much larger grains that those from the coprecipitated powders, >30μm in the former one but ~0.3μm in the latter case. The sintered specimens from the coprecipitated powders also contain $Bi_2Al_4O_9$, δ-Bi_2O_3 and γ-Bi_2O_3. A small amount of $BiAlO_3$ was occasionally observed in these sintered samples.

371

Electrical conductivity of the $BiAlO_3$ plus Bi_2O_3 solid electrolytes was divided into two temperature-dependent regimes, 25-200 °C and 300-700 °C. At high temperatures, solid electrolytes with the oxide dopants Y_2O_3 and CaO had a higher ionic conductivity, $4*10^{-2}$ S/cm at 700°C. From the measurement of activation energy, the charge carrier at high temperatures was believed to be oxygen vacancy for these Bi_2O_3- and Al_2O_3-containing solid electrolytes.

ACKNOWLEDGEMENT

A special thank to Taiwan National Science Counsel for providing a financial support of this research.

REFERENCES

[1] C. N. R. Rao, G . V. Subba Rao, and S. R amdas, "Phase Transformation and Electrical Properties of Bismuth Sesquioxide," *Journal of Physics and Chemistry of The Earth*, **73**, 672-5(1969)

[2] V.Joshi, S. Kulkarni, J. Nachlas, J. Diamond, N. Weber, "Phase Stability and Oxygen Transport Characteristics of Yttria- and Niobia- Stabilized Bismuth Oxide, " *Journal of Materials Science*, **25** 1237-45 (1990)

[3] K. Z. Fung and A. V. Virkar, "Phase Stability, Phase Transformation Kinetics, and Conductivity of Y_2O_3-Bi_2O_3 Solid Electrolytes Containing Aliovalent Dopants, " *Journal of American Ceramic Society*, **74** [8] 1970-80 (1991)

[4] R. K. Datta and J. P. Meehan, "The System Bi_2O_3-R_2O_3 (R=Y,Gd), " *Z. Anorg Allg. Chem.*, **383**, 328-37 (1971)

[5] T. Takahashi, T. Esaka, and H. Iwahara, "High Oxide Ion Conduction in the Sintered Oxides of the System Bi_2O_3-Gd_2O_3," *J. App. Electrochem.*, **5**, 197-202 (1975)

[6] T. Takahashi, H. Iwahara, and T. Arao, "High Oxide Ion Conduction in the Sintered Oxides of the System Bi_2O_3-Y_2O_3, " *J. App. Electrochem.*, **5**, 187-95(1975)

[7] D. Liu, Y. Liu, S.Q. Huang, and X. Yao, "Phase Structure and Dielectric Properties of Bi_2O_3-ZnO- Nb_2O_5-Based Dielectric Ceramics, " *Journal of American Ceramic Society*, **76** [8] 2129-32 (1995)

[8] K. Z. Fung, J. Chen, and A. V. Virkar, "Effect of Aliovalent Dopants on the Kinetics of Phase Transformation and Ordering in RE_2-Bi_2O_3 (RE=Yb,Er,Y, or Dy) Solid Solutions, " *Journal of American Ceramic Society*, **76** [10] 2403-18 (1993)

[9] P. Su and A. V. Virkar, "Ionic Conductivity and Phase Transformation in Gd_2O_3-Stabilized Bi_2O_3, " *J. Electrochem. Soc.*, **139**, [6] 1671-77 (1992)

[10] P. Su and A. V. Virkar, "Cubic-to-Tetragonal Displacive Transformation in Gd_2O_3- Bi_2O_3 Ceramics, " *Journal of American Ceramic Society*, **76** [10] 2513-20 (1993)

[11] T. Ishihara, H. Matsuda, and Y. Takita, "Doped LaGaO$_3$ Perovskite Type Oxide as a New Oxide Ionic Conductor," *Journal of American Ceramic Society,*116 [10] 3801-3 (1994)

A PERFORMANCE BASED MULTI-PROCESS COST MODEL FOR SOFCS

Mark Koslowske, Heather Benson, Isa Bar-On
Worcester Polytechnic Institute
100 Institute Road
Worcester, MA 02459

R. Kirchain
Massachusetts Institute of Technology
77 Massachusetts Avenue
Cambridge, MA 02139

ABSTRACT

Cost effective manufacturing is a major concern for the development and commercialization of solid oxide fuel cells (SOFC). Costs are frequently compared for layer structures that differ in materials selection, manufacturing processes, and cell design. A meaningful cost model needs to consider all these parameters when deriving manufacturing cost estimates. Modeling tools are needed to aid in the selection of the appropriate process combination prior to making expensive investment decisions.

This paper describes the development of a performance based multi-process cost model. This model permits the comparison of manufacturing cost for different processing combinations and various materials while at the same time considering the effect of power density and operating temperature. The model consists of three parts: a performance model, a process yield model, and a process based cost model.

The results are summarized in cost performance maps. These maps indicate cost regimes for combinations of operating temperature and power density. The results show that SECA cost targets can be met for a judicious selection of performance and process parameters.

INTRODUCTION

Process based cost models (PBCM)[1] are a valuable tool for estimation of manufacturing costs. More importantly, they aid in making strategic decisions concerning choice of processing route, investment in equipment, and in process and equipment development. This tool is especially useful in the absence of high volume manufacturing experience when costs cannot be based on prior manufacturing experience as in the case of solid oxide fuel cells (SOFC).

Traditional cost models are based on detailed information of the materials and manufacturing processes. In addition, they use geometric design data, such as size, thickness and shape complexity. For fuel cells the performance parameters of power density and operating temperature have a direct effect on the layer thickness of the parts (and vice versa) and on the number of cells needed per kilo Watt. If this relationship is ignored it becomes difficult to compare costs for varying geometries and design assumptions [2]. The availability of cost models in the open literature is limited[3] and the relationship between geometry, electro-chemical performance and cost is not detailed enough.

This paper describes a process based cost model for planar SOFCs. The model correlates power density and operating temperature levels and tolerances to manufacturing process cost considering explicitly the capability of each process as expressed in process yield values.

MODEL DESCRIPTION

The model consists of three modules as depicted in Figure 1: the cost model, the process yield module, and the cell performance module. The cost model is developed based on the technical and economic data of the processes under consideration: tape casting, screen printing and sputtering. The capabilities of the process and the layer thickness tolerance determine the yield characteristics for different layer thicknesses. The layer thickness can either be inserted as an input, or it can be calculated from a performance model that calculates the required layer thickness and layer thickness tolerance based on the desired power density and temperature range and temperature range variation.[456]

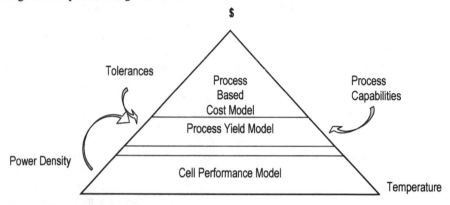

Figure 1: Cost Model Components

Two processing routes are treated in this model, a co-sintering and a layer by layer route. These routes are described in Figure 2 for the manufacture of planar SOFC's. Both routes consider materials preparation and sintering. For the co-sintered route the layers are tape cast individually and then laminated and blanked. In the layer by layer route the layers are cast (screen printed) and sintered successively.

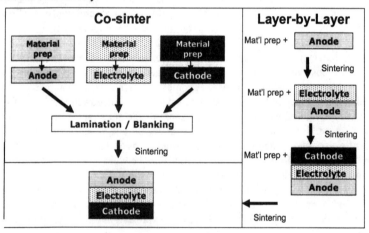

Figure 2: Alternative processing routes for a unit cell.

The baseline cell for this study consists of a 700 μm nickel cermet anode supporting a 20 μm yttria stabilized zirconia (YSZ) electrode and a 50 μm Lanthanum Strontium Manganese (LSM) cathode. The cell operates at nominally 735°C - 750°C and at 1W/cm², unless noted otherwise.

Interconnects are not included. The data for the cost model are based on interviews with equipment manufacturers and on in house processing expertise.

RESULTS
Production Volume and cost elements

Figure 3a shows unit cell cost as a function of production volume for two processing routes: three layers tape cast and co-sintered (TC-CS), and anode tape cast, electrolyte screen printed and sintered, then cathode screen printed and sintered (TC-SP-SP). Equipment size is adapted to production volume within the range of currently available equipment. The 'bumps' in the graphs correspond to the addition of equipment. It can be seen that for production volumes below 65,000 units/year the TC-SP-SP process combination is less costly than TC-CS. For production volumes above 100,000 units/year TC-CS is less costly. The reason for this crossover can be seen in Fig 3a which shows a comparison of the cost elements at two production volumes for each of the processing routes.

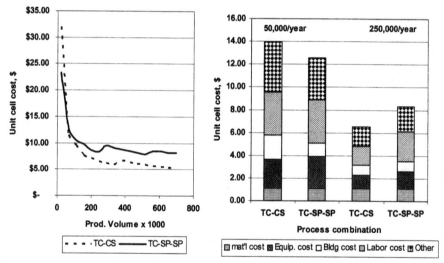

Figure 3a: Unit cell cost as a function of yearly production volume

Figure 3b: Cost elements for two processing routes at two production volumes

For the higher production volume the contribution of the cost elements to unit cell costs decreases in general. Only material cost is assumed constant due to insufficient information on high volume powder pricing. At 250,000 units/year the contribution of labor, building and equipment cost has changed in such a way that TC-CS is more cost effective. For TC-SP-SP each added layer is sintered separately requiring more sintering capacity and leading to a process yield of approximately 60% compared to an 84% yield for the co-sintered process. Also, currently available equipment is not optimized for a given production volume of fuel cells.

Effect of process yield

Process yield in the micro-electronic industry is usually treated on a per part basis. For the wet processes considered here slurry preparation is included in the considerations and yield is calculated based on material usage. This explains also the low yield numbers since scrap is not reused.

Figure 4 shows a comparison of process yield as a function of layer thickness for tape casting (TC), screen printing (SP), and sputtering (SPT). The yield for TC and SP drops off significantly for layer thicknesses below 20 μm. At greater thicknesses the yield for SP is better than that for TC, while this is inverted at lower thickness. The yield for sputtering is constant at above 90% over the range of thickness values considered.

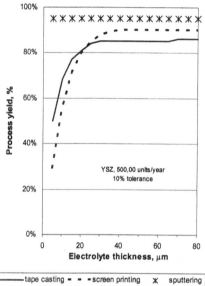

Figure 4: Process yield as a function of electrolyte thickness for TC, SP, SPT.

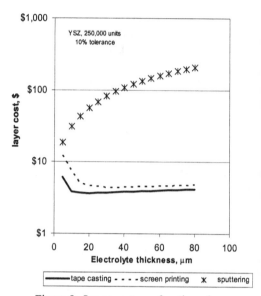

Figure 5: Layer cost as a function of electrolyte thickness for TC, SP, SPT.

Figure 5 shows the corresponding layer cost. SP is slightly more costly than TC for the range considered despite the better yield numbers for SP. In comparison, sputtering is more than 20 times as costly as TC but becomes interesting at very low thickness. If film quality is important, sputtering might be attractive for very thin layers. SP is expensive as it requires more labor than TC, while the equipment for TC is costly. SP and SPT have advantages over TC in terms of layer quality and the ability to incorporate features. Investment in automation and SOFC specific equipment might make these processes more competitive.

Layer tolerance

Layer thickness tolerance and its effect on yield is considered in addition to layer thickness and the results are shown in Figure 6 comparing TC, SP and SPT for two levels of thickness tolerance: 5% and 10%. Process yield is higher for lower thickness tolerance values (empty shapes) for TC and SP. Process yield is not affected for sputtering in the thickness and tolerance range investigated.

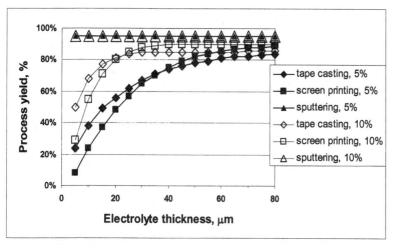

Figure 6: Effect of layer thickness tolerance on process yield for different processes

Unit cell cost

Cost/kW changes either with operating temperature or with a change in power density. As the power density changes a different number of cell units is needed. A change in operating temperature will correspond to a different electrolyte thickness. Cost/kW is shown for two process combinations assuming a 50% efficiency loss in the stack.

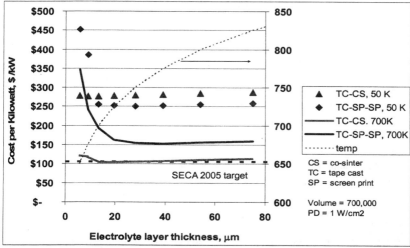

Figure 7: Cost per kW for cells operating at 1 W/cm^2 as a function of electrolyte layer thickness for different process combinations.

It can be seen that the 2005 SECA cost value can be met the TC-CS process route. This should not differ much for any production volume above 300,000 units/year.

Cost Maps

The dependence of cost on operating temperature and power density is shown in Fig. 8 for TC-CS and for TC-TC-SP. TC-TC-SP is selected here as it is likely more cost competitive than TC-SP-SP. TC-CS is in general less costly, but this ignores the quality of the layer and its ability to withstand power densities over an extended period of time. It shows that SP might become cost competitive with further investment in automation and process development.

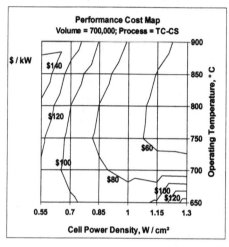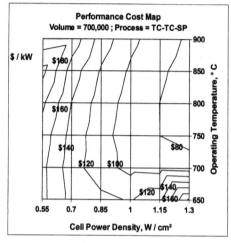

Figure 8: Cost – performance maps for two process combinations.

CONCLUSIONS

A performance based multi-process cost model has been used to compare manufacturing costs of tape cast units with tape cast and screen printed cells. The effect of yield has been investigated for different layer thicknesses. The model shows that tape cast units are currently the most cost effective for large production volumes. Screen printing could become competitive if the process can be developed to change its characteristics somewhat.

REFERENCES

[1] Randolph Kirchain and Frank R. Field III, "Process-Based Cost Modeling: Understanding the Economics of Technical Decisions", in Encyclopedia of Materials Science and Engineering, Elsevier, 2001.
[2] Bar-On, I., Kirchain R., Roth, R.,"Technical Cost Analysis for PEM Fuel Cells", J. of Power Sources, 109 (2002) 71-75.
[3] Arthur D. Little, "Assessment of Planar Solid Oxide Fuel Cell Technology, Report to DOE FETC, Ref, 39463, 1999, also www.netl.doe.gov/publications/proceedings/01/seca/adlstack.pdf
[4] W.Kim, A.V.Virkar, SOFC VI, 830-839 (1999).
[5] W.Kim, A.V.Virkar, K.Z.Fung, K.Mehta, S.C.Singhal, Journal of the Electrochemical Society 146 (1), 69-78 (1999).
[6] H.Chan, K.A.Khor, Z.T.Xia, Journal of Power Sources 93, 130-140 (2001).

DEVELOPMENT OF A TRI-LAYER ELECTROCHEMICAL MODEL FOR A SOLID OXIDE FUEL CELL

Badri Ramamurthi
GE Global Research Center
K1-4C30 1 Research Circle
Niskayuna NY 12309

Vikas Midha
GE Global Research Center
K1-4C17 1 Research Circle
Niskayuna NY 12309

James Ruud
GE Global Research Center
MB165 1 Research Circle
Niskayuna NY 12309

Mark Thompson
GE Global Research Center
K1-4B4 1 Research Circle
Niskayuna NY 12309

ABSTRACT

We report the development of a 1D continuum model to describe the performance of a Solid Oxide Fuel Cell (SOFC). The model consists of two coupled components: 1) Charge transport, and 2) Species transport. The effective transport properties that feed into these components were computed from percolation theory.

Ionic and electronic potential distributions in the cell were obtained by solving for transport of charged species (oxygen ions and electrons), the production/loss of charged species being modeled by Butler-Volmer Kinetics. A multi-component diffusion model was employed to account for the diffusion of multiple neutral species through the porous electrode and solve for species concentration within the cell. Good agreement was observed between the model and experimental I-V characteristics of SOFCs under a series of operating conditions.

INTRODUCTION

Over the last few years Solid Oxide Fuel Cells (SOFCs) have received considerable attention as a potentially viable source for stationary power generation. In order to achieve the goals that could make an SOFC commercially viable, a fundamental understanding of the physical processes in a fuel cell and their effect on cell performance is critical.

The basic structure of models that link cell properties to cell performance already exists in the literature [1,2]. Kim et al. [1] provide a 0D model which can be used to generate performance curves for a cell by computing polarization resistances due to aggregate ohmic and charge transfer losses. Some of the deficiencies of a 0D model lie in the fact that this model does not account for multicomponent diffusion and reaction within the cell, which occurs with internal reforming. Furthermore, the total overvoltage in a 0D model is treated as a sum of individual overvoltages due to activation, ohmic and mass transport processes. Since the charge transfer process is non-linear with respect to the overvoltage, a 1D model which calculates the voltage as a function of the distance through the cell is required to accurately compute the total overvoltage. Chan and Xia [2] have reported a 1D model accounting for ionic/electronic and species transport through the electrodes. However, they assume equimolar counter-current diffusion to compute individual species fluxes. This assumption is valid only in the case of gas-phase binary diffusion being dominant over Knudsen diffusion, which is typically not the case for state-of-the-art SOFCs with high surface area, small pore size electrodes.

The model presented in this article addresses the deficiencies of earlier models and is applicable to a cell consisting of multiple layers with different microstructures. In order to solve for ionic and electronic transport through the different layers of the cell, it is convenient to explicitly solve for ionic and electronic potentials in the cell as opposed to solving for the overvoltage. This approach lends itself to being applicable to arbitrary cell geometries with multiple layers, as long as the appropriate boundary conditions are employed.

MODEL DESCRIPTION

The physical processes occurring in an operating fuel cell can be summarized as follows:

1) Ionic and electronic transport through the corresponding phases of the composite electrode, which are driven by their corresponding potentials.
2) Charge transfer reactions occurring along the triple phase boundaries which lead to charge exchange between ions and electrons.
3) Transport of neutral species such as H_2, H_2O, N_2 and O_2 through the porous electrode to reaction sites.

The model presented in the following section addresses the processes shown above with the following assumptions:

1) The fuel cell operates under isothermal conditions.
2) All effective properties such as triple phase boundary area and electronic/ionic conductivity are uniform within a layer of the cell.
3) The electrode microstructure is assumed to consist of uni-modal spherical particles and a uni-modal pore size distribution.

Table I. Model equations describing charged and neutral species transport through the porous electrodes.

Description	Equation	
Electron Transport:	$\dfrac{d}{dx}\left(\sigma_{el}^{\,eff}\dfrac{dV_{el}}{dx}\right)=\pm Ai_n$	(1)
Ion Transport:	$\dfrac{d}{dx}\left(\sigma_{io}^{\,eff}\dfrac{dV_{io}}{dx}\right)=\mp Ai_n$	(2)
Charge Transfer Current:	$i_n = i_0\left\{\dfrac{C_r}{C_r^{ref}}\exp\left(\dfrac{\beta n_e F\eta}{RT}\right)-\dfrac{C_p}{C_p^{ref}}\exp\left(-\dfrac{(1-\beta)n_e F\eta}{RT}\right)\right\}$	(3)
Species Transport:	$\varepsilon\dfrac{\partial c_i}{\partial t}+\nabla\cdot\mathbf{N}_i=\pm\dfrac{Ai_n}{nF}$	(4)
Species Flux:	$\mathbf{N}_i=\mathbf{J}_i+x_i\mathbf{N}_T$	(5)
Multicomponent Diffusive Flux:	$\displaystyle\sum_{j\neq i}\dfrac{x_j\mathbf{J}_i-x_i\mathbf{J}_j}{\Delta_{ij}}=\dfrac{-p}{RT}\nabla x_i-\dfrac{x_i}{RT}\left(1-\dfrac{1/D_i^{eff}}{\sum_s x_s/D_s^{eff}}\right)\nabla p$	(6)
Total Flux:	$\displaystyle\mathbf{N}_T=\sum_i\mathbf{N}_i=\dfrac{\sum_{s=1}^{N_c}\dfrac{\mathbf{J}_s}{D_s^{eff}}}{\sum_{s=1}^{N_c}\dfrac{x_s}{D_s^{eff}}}-\dfrac{1}{RT}\left(\dfrac{Bp}{\mu}+\dfrac{1}{\sum_{s=1}^{N_c}\dfrac{x_s}{D_s^{eff}}}\right)\nabla p$	(7)

The model solves for transport of charged and neutral species through the thickness of an anode supported SOFC in which there are four layers: 1) a support anode; 2) an electrochemical or active anode; 3) an electrolyte and 4) a cathode. The model equations are shown Table I. Equations (1) and (2) describe the transport of electrons and ions with their corresponding

382

potentials. Equation (3) describes the kinetics of charge transfer given by the Butler-Volmer equation [3]. C_r^{ref} and C_p^{ref} correspond to the reference state concentrations of reactants and products respectively. The reference state was assumed to be the state corresponding to OCV condition when no current is drawn from the cell. Equations (4-7) describe the transport and reaction of neutral species through the porous electrode. The multicomponent fluxes are given by the dusty gas model [4] accounting for species diffusion through a porous medium, in which the kinetic theory of gases is applied to a mixture of the gas and the medium itself which is treated as large, solid gas molecules with fixed positions. The effective properties of the porous medium such as the ionic/electronic conductivities σ_{io}^{eff}, σ_{el}^{eff} and the triple phase boundary area, A, were computed from a percolation model [5].

The boundary condition for species transport was incorporated by specifying a fixed value of the surface concentration of H_2, H_2O and O_2. For the electrons, an electronic potential equal to the cell operating voltage was specified on the anode boundary and a reference potential of zero was assumed at the cathode. For the ions, zero current density (or zero ion flux) was specified at both the boundaries.

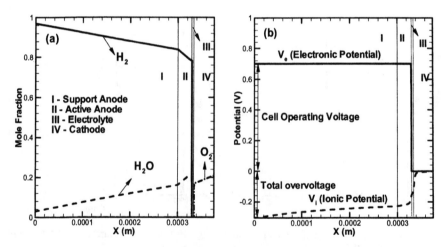

Fig 1. a) Species and b) Potential profiles for baseline conditions shown in Table II. The cell operating voltage was fixed at 0.7 V. The discontinuity in electronic potential is a result of the assumption that the electrolyte (Region III) is purely ion conducting.

RESULTS AND DISCUSSION

The baseline condition for the results in Figs. 1 and 2 is shown in Table II. Fig. 1a shows the species profile for the baseline condition. Since the net electrochemical reaction involves the conversion of H_2 and O_2 into H_2O, a gradient in H_2 and O_2 is set up. For the fuel cell operating at steady state, the net flux of H_2 and O_2 is equal to J/2F and J/4F respectively, where J is the total current density and F is Faraday's constant. The gradient of H_2 is seen to be discontinuous at the interface between the support and active anode due to different microstructures in the two layers. Fig. 1b shows the distribution of ionic and electronic potentials across the electrode. The electronic potential is almost uniform due to high electronic conductivity. Furthermore, it is discontinuous across the electrolyte due to the fact that the electrolyte is purely ion conducting. The ionic potential on the other hand assumes a continuous profile across the electrode. Near the cathode/electrolyte interface, the ionic potential is much lower than the electronic potential, leading to significant charge transfer reaction and the production of oxygen ions. The ionic potential is linear in the electrolyte due to the assumption of pure ionic conductivity. At the anode/electrolyte interface, ions exchange charge with electrons and a further drop in ionic potential is obtained. The total cell overvoltage is given by the value of the ionic potential on the left boundary of the anode. This overvoltage signifies the total polarization loss (activation, ohmic and mass transport) incurred by the cell to generate current. In contrast to a 0D model, a cell overvoltage distribution can be obtained through the thickness of the cell..

Figs. 2a and 2b show the extent of charge transfer reaction and the current density distributions respectively. It can be seen from Fig. 2a that the reaction depth is typically ~10 μm, implying that the conversion of ionic to electronic current and vice-versa occurs very close to the electrolyte/electrode interface as seen in Fig. 2b.

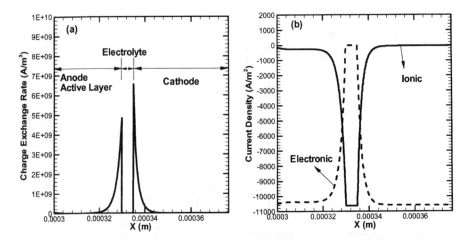

Fig 2. a) Charge transfer reaction rate and b) electronic/ionic current densities along the thickness of the cell for baseline conditions shown in Table II.

In order to compare performance curves generated by the model with experiment, tests were performed on "button cells" under low utilization conditions. Anode-supported cells were fabricated by tape casting and sintering three layers: 1) a highly porous, low surface area NiO/YSZ support anode, 2) a less porous, higher surface area NiO/YSZ active anode and 3) a YSZ electrolyte. A strontium-doped lanthanum manganate (LSM) cathode was deposited on the cell by screen-printing and sintering. Pt wire mesh was attached to the anode and cathode surfaces using a Pt paste. The anode area was 2.85 cm^2 and the cathode area was 1.36 cm^2. The anode-side of the cell was bonded to an alumina tube using a high-temperature ceramic cement, and the anode was reduced at temperature in situ. For testing, the cell temperature was held constant at 750 °C. Since the H_2 composition was varied, N_2 was used as a diluent with a total fuel flow rate of 200 sccm. Air flow rate was maintained at 500 sccm. I-V curves were generated using a potentiostat with a current scan rate of 10 mA/s.

The model was compared with two sets of experimental curves as shown in Fig. 3. The conditions in Fig. 3a correspond to an electrochemical anode thickness of 30 μm. The adjusted parameter values in the model are shown in Table III. It is important to note that a single set of parameter values was used to compare the model prediction with all of the experimental data. Note that the exchange current density, I_0, was varied independently of A, the exchange current area, where A was calculated from the percolation model; however, the product could also be used as a single adjustable parameter.

Table II. Baseline condition used in the model

Parameter	Value
Support thickness	300 μm
Active layer thickness	30 μm
Electrolyte thickness	5 μm
Cathode thickness	40 μm
Cell voltage	0.7 V
H2 composition	97%
H2O composition	3%
O2 composition	21%

Table III. Adjustable parameters used in the model to compare with experiments shown in Fig. 3.

Adjustable Parameters	Value
I_0A (Active Anode)	1230 A/cm^3
I_0A (Cathode)	3269 A/cm^3
Pore to particle size ratio (Anode)	0.35
Pore to particle size ratio (Cathode)	0.8
Tortuosity	6

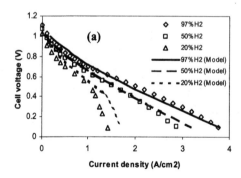 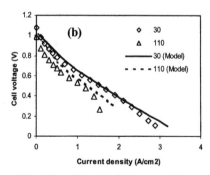

Fig. 3. Comparison between model and experiment for a) Baseline electrode thicknesses shown in Table II and b) Active anode thicknesses of 30 and 110 μm. Fuel and air flow rates were adjusted to ensure low utilization conditions.

At moderate fuel utilizations (<50%), good agreement is observed between the model and experimental data (for reasonable values of pore to particle size ratios and tortuosity) even in the region of potential mass transport limitation (high current densities). The fit at high current densities contrasts with the assertion that surface diffusion mechanisms must be incorporated to match the tail of a performance curve [6]. At high utilizations (high current densities), there is a deviation between the model and experiment. This deviation is believed to be a consequence of the change in average surface concentrations at high utilizations.

Fig. 3b shows the comparison between model and experiment for active anode thicknesses of 30 and 110 µm. At low current densities (<0.5 A/cm^2), the predicted and experimental current densities are in good agreement. At higher current densities, the experimental and predicted curves for the thick active anode are consistent with an increasing effect of mass transport limitation. Note that there is a difference in the magnitude of the predicted and experimentally observed voltages, which may be a result of the initial OCV of the experiment versus that assumed in the model.

CONCLUSIONS

A continuum 1D model was developed to describe the transport of charged and neutral species through the porous electrodes of a Solid Oxide Fuel Cell. The model was validated with experiments performed on "button cells" under relatively low utilization conditions. With reasonable adjustable parameters in the model, good agreement was observed between model and experiment. The model is a useful design tool in investigating microstructures and geometries that result in better cell performance.

ACKNOWLEDGMENTS

The experimental work in this article was supported by the U.S. Department of Energy through cooperative agreement DE-FC26-01-NT41245 under the Solid State Energy Conversion Alliance (SECA).

REFERENCES

[1] J.W. Kim, A.V. Virkar, K.Z. Fung, K. Mehta and S.C. Singhal, "Polarization Effects in Intermediate Temperature, Anode-Supported Solid Oxide Fuel Cells," *Journal of the Electrochemical Society*, **146** [1] 69-78 (1999).

[2] S.H. Chan and Z.T. Xia, "Anode Micro Model of Solid Oxide Fuel Cell," *Journal of the Electrochemical Society*, **148** [4] A388-A394 (2001).

[3] J.S. Newman, "Electrochemical Systems," Prentice Hall, Englewood Cliffs NJ, (1991).

[4] R. Jackson, "Transport through Porous Catalysts," Elsevier Scientific, New York (1977).

[5] P. Costamagna, P. Costa and V. Antonucci, "Micro-modelling of Solid Oxide Fuel Cell Electrodes," *Electrochimica Acta*, **43** [3] 375-394 (1998).

[6] R.E. Williford, L.A. Chick, G.D. Maupin and S.P. Simner, "Diffusion Limitations in the Porous Anodes of SOFCs," *Journal of the Electrochemical Society*, **150** [8] A1067-A1072 (2003).

REDUCTION AND RE-OXIDATION OF ANODES FOR SOLID OXIDE FUEL CELLS (SOFC)

Jürgen Malzbender, Egbert Wessel and Rolf W. Steinbrech, Lorenz Singheiser
Forschungszentrum Jülich GmbH
Institute for Materials and Processes in Energy Systems
52425 Jülich, Germany

ABSTRACT

The influence of reduction and re-oxidation on microstructure and residual stress has been studied for anode supported planar SOFCs. The transition in oxidation state shows a non-reversible behaviour. Shrinkage caused by the reduction of NiO to Ni is over-compensated in the re-oxidation step. Microstructural SEM observations reveal that after re-oxidation the NiO has a higher porosity and requires a larger volume than in the initial, oxidised state. The residual stress development has been analysed via monitoring of the curvature changes of unconstrained half-cells. Re-oxidation is a dynamic process that starts at the free surface of the anode and proceeds towards the interface with the electrolyte. Accordingly the residual stresses in the anode and electrolyte change. Ultimately, the tensile residual stress in the electrolyte exceeds tensile strength, resulting in cracks which lead to deterioration of the entire cell function.

INTRODUCTION

Robust mechanical performance is an important prerequisite for reliable operation of thermal cycling resistant planar SOFC components. However, since in an SOFC brittle materials are rigidly bonded as a multilayer composite, there are problems regarding mechanical integrity during thermal cycling. Thermal mismatch, layer geometry and elastic moduli of the materials affect the development of residual stresses[1]. In addition stresses arise from the constrained fixation of the cells in a SOFC stack[1]. Various mechanical properties of SOFC materials, cells and interfaces within cells and stacks have been characterised[1,2]. The investigations predominantly considered the standard conditions of an initially oxidised and then reduced anode before operation. Although the re-oxidation of cells can be detrimental, it has only received limited attention[3]. Re-oxidation might occur at high temperatures due to a lack of fuel gas, e.g. in the case of sealant damage. Cell fracture is assumed to follow the rapid oxidation and volume expansion of the Ni particles in the YSZ matrix[3]. Macroscopically, the volume increase during re-oxidation causes an expansion of the anode as compared to the initial oxidised situation[5]. The anode expansion leads to electrolyte cracking[4]. However it has also been reported that cyclic reduction and oxidation resulted in an increase of the polarisation resistance which was attributed to the formation of cracks within the anode[5].

The present work is aimed to gain further insight into the re-oxidation of planar SOFCs. Global curvature studies of unconstrained half cells provide macroscopic information on the re-oxidation development and allow an analytical treatment of the residual strain/stress behaviour. In an additional local approach based on SEM observation, the microstructural changes are resolved with respect to strain and used to explain re-oxidation related electrolyte cracking.

EXPERIMENTAL

The re-oxidation tests were carried out with SOFC half-cells supplied by Research Centre Juelich (FZJ), Netherlands Energy Research Foundation (ECN) and Risoe National Laboratories

(RNL) in the EU project "Component Reliability in Solid Oxide Fuel Cell Systems fo
Commercial Operation" (CORE-SOFC). All half-cell variants (total thickness between ~ 0.2
and ~ 1.5 mm) had a porous Ni-YSZ as anode current collector (ACC) which supported an anod
functional layer (AFL) and an YSZ electrolyte film (thickness ~ 10 μm). The microstructur
studies were restricted to FZJ cells with a ~ 5 μm thick AFL of Ni-YSZ. Starting from as
received larger cells, specimen strips of ~ 35 mm × 10 mm × cell thickness were cut using
diamond saw.

The thermoelastic behaviour was tested between RT and 800°C in a quartz chamber und
either oxidising (air) or reducing atmosphere (Ar / 4% H_2). The radius of curvature of th
specimen was measured in-situ using a far field microscope system (Questar M100). Th
microstructure was analysed before and after reduction/re-oxidation using a Gemini scannin
electron microscope (SEM). The elastic moduli (Table I) were determined by depth sensitiv
indentation (Fischerscope H100C). The thermal expansion coefficients were taken fro
literature[2].

Table I. Thermal expansion coefficients and elastic moduli of SOFC materials.

	oxidized anode	reduced anode	Electrolyte
$\alpha / 10^{-6} \, K^{-1}$	12.6	13.2	10.9
E / GPa	98	39	205

THEORY

The impact of thermal expansion mismatch on curvature of unconstrained multilayer
composites was used to determine the residual stresses in the half cells. The stresses arising fro
the difference in strain at any position y within a layer i can be expressed by[6]:

$$\sigma_{res,i,y} = \left(\frac{1}{r} - \frac{1}{r_0}\right)\frac{E_i}{(1-v_i)}(y - y_{na}) + \left(\frac{E_i}{1-v_i}\sum_{j=1}^{n}\frac{E_j}{1-v_j}t_j(\varepsilon_j - \varepsilon_i)\right)\left(\sum_{j=1}^{n}\frac{E_j}{1-v_j}t_j\right)^{-1} \tag{1}$$

where r is the radius of curvature, E the elastic modulus, v the Poisons ratio, t the thickness
the strain and r_0 is the radius at ε_0. With the neutral axis y_{na}[6]:

$$y_{na} = \sum_{i=1}^{n} E_i w_i t_i \left(2\sum_{j=1}^{i-1}t_j + t_i\right) \Big/ \left(2\sum_{i=1}^{n}E_i w_i t_i\right) \tag{2}$$

the change in curvature of a multilayer specimen becomes[6]:

$$\frac{1}{r} - \frac{1}{r_0} = \frac{\sum_{i=1}^{n}\left(E_i t_i\left(2\sum_{j=1}^{i-1}t_j + t_i\right)\sum_{j=1}^{n}\frac{E_j}{1-v_j}t_j(\varepsilon_j - \varepsilon_i)\Big/ 2\sum_{i=1}^{n}E_i w_i t_i\right)}{\frac{1}{3}\sum_{i=1}^{n}E_i\left(\left(\sum_{j=1}^{i}t_j - y_{na}\right)^3 + \left(y_{na} - \sum_{j=1}^{i-1}t_j\right)^3\right)} \tag{3}$$

388

Equations (1) to (3) permit a calculation of the stress at any position y within a layer i due to a change in strain in one or more layers. To determine the strain during re-oxidation these equations have to be solved.

For simplification first a three layer system will be assumed. Layer one is the re-oxidised material near the free surface of the ACC, layer two is the still reduced layer between re-oxidised ACC and electrolyte and layer three refers to the electrolyte. Fitting the theoretical curvature plots to the experimental data provides a value for the unknown strain. However, as will be demonstrated in the results section, the influence of the fourth layer (AFL) on curvature cannot be neglected. There are also some boundary conditions that have to be considered. The strain only reflects the expansion of the re-oxidised material and the thickness of the re-oxidised material increases from zero to the total thickness (ACC plus AFL). The latter seems reasonable if the width of the specimen is larger than the thickness.

RESULTS AND DISCUSSION

Planar SOFCs with YSZ electrolyte are typically co-fired at temperatures above 1200°C and operated at around 800°C. Thus, due to the difference in thermal expansion coefficients, the electrolyte is under compression and the anode experiences on average tension at operation temperature. This is associated with an initial curvature of an unconstrained half cell towards the anode. Subsequent reduction and re-oxidation leads to changes in curvature as a result of the shrinkage and expansion, associated with the transition from NiO to Ni and vice versa (Fig. 1).

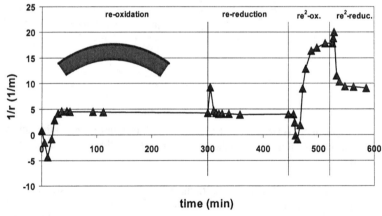

Fig. 1: Curvature changes during re-oxidation, re-reduction cycles (0.27 mm half cell, 800°C).

During re-oxidation the half cell first bends towards the electrolyte and then again back to the anode side. However, the final curvature is larger than the initial one. Further re-reduction / re-oxidation cycles increase the maximum curvature (Fig. 1).

The initial bending towards the electrolyte can be explained assuming that re-oxidation starts at the free surface of the ACC. The kinetics result in two effects. Due to the increase in stiffness of the re-oxidised material as compared to the still reduced material the neutral axis moves initially towards the free anode surface. Also the expansion (Ni to NiO) of the re-oxidised as compared to the reduced material supports curvature towards the electrolyte. Once the re-oxidised layer has progressed above the neutral axis the expansion causes a moment into the opposite direction, changing the curvature again as well as moving the neutral axis further away from the free anode surface. The process, however, does not explain why the final curvature is larger in the re-oxidised than in the initial oxidised state.

Based on equations (1) to (3) the radius of curvature as a function of the thickness of the re-oxidised material can be calculated. The theoretical curve in Fig. 2 is fitted to the experimental results obtained for a 0.27 mm thick half-cell. Note that a linear proportionality between re-oxidation time and thickness was assumed.

It can be seen that the experimental data are initially well described but a large deviation exists when the re-oxidation front approaches the electrolyte. However, agreement between theory and experiment becomes better, if the AFL (Fig. 3) is also taken into consideration. Although identical in composition to the ACC, the AFL has a significantly lower porosity. The AFL will experience higher strain when the Ni expands during re-oxidation. Since this occurs at the end of the re-oxidation process it causes a larger final curvature (Fig. 2). Qualitatively similar curvature results are also obtained for cell variants with larger total thickness.

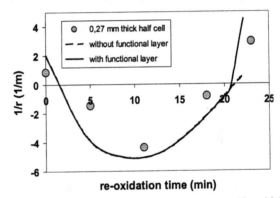

Fig. 2: Curvature of half cell (0.27 mm) as a function of progressing thickness of re-oxidise material (800°C). Experimental data are compared with theoretical curves for three and four laye model.

From the analytical modelling with three layers (re-oxidised, reduced anode and electrolyte) can be suggested that re-oxidation leads to a strain of 0.15 ± 0.01 %, which is equivalent to maximum stress of ~ 250 MPa in the electrolyte. This value is close to the modulus of rupture of an ACC (416 MPa at RT, 265 MPa at 900°C[1]). However, since in addition a residual compressiv stress exists in the electrolyte due to thermal mismatch (~ 550 MPa at room temperature and 21 MPa at 800°C) a tensile stress of 250 MPa appears not to be critical. If the anode functional laye is also taken into account (strain 1.5 %) the re-oxidation tensile stress in the electrolyte of ~ 50 MPa exceeds at 800°C the sum of compressive stress and fracture strength (modulus of rupture Thus fracture phenomena in the electrolyte become likely and are in fact observed (Fig. 4).

In a local approach complementary to the global curvature measurements, micrographs were taken from the very same location of a cell, comparing the initial (oxidised), the reduced and the re-oxidised state (Fig. 3.a to c).

The reduction results in material shrinkage when the NiO particles are converted to Ni (Fig. 3a and 3b). After re-oxidation the morphology of the NiO has changed compared to the initial oxidised microstructure (Fig. 3a and 3c). The re-oxidised particles exhibit a higher porosity. The volume increase leads to additional compressive stresses in ACC and AFL and, more important, to higher tensile strain and stress in the electrolyte.

The micrographs in Fig. 3 allow a direct determination of the strain. Measuring the distance between identical features of the electrolyte, e.g. pores, a change in strain of about 0.5 % from oxidised to re-oxidised state can be estimated. This corresponds to a tensile re-oxidation stress of ~ 1000 MPa, sufficient to fracture the electrolyte despite the presence of the compressive residual stress. This result agrees with the prediction made on the basis of the curvature changes.

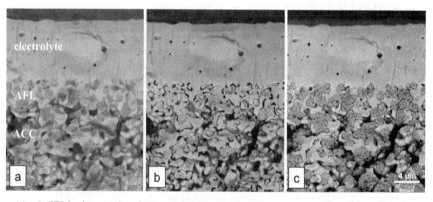

Fig. 3: SEM micrographs of 1.5 mm thick half cell. Cross section displays from top to bottom electrolyte, anode functional layer (AFL) and anode current collector (ACC). The same location is shown in oxidised (a), reduced (b) and re-oxidised state (c).

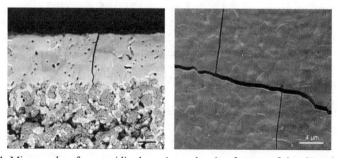

Fig. 4: Micrographs of a re-oxidised specimen showing fracture of the electrolyte, a) cross-section, b) electrolyte surface.

CONCLUSIONS

The re-oxidation of planar SOFC half cells has been analysed using in-situ measured radius curvature behaviour and micrographs of the cells in different oxidation states. The re-oxidation results in an expansion of the NiO as compared to the initial NiO in as fabricated anodes. The expansion causes a large tensile strain and respective stress in the electrolyte that leads fracture. The present analysis highlights the importance of a higher density anode functional layer which has a significant impact on the stress in the electrolyte.

In principle the stress in the electrolyte can be reduced if the thickness of the anode is reduced. However, an increase in the porosity of the anode functional layer will have a stronger effect since the strain during the expansion of the Ni is better accommodated. A reduction of t sintering temperature of the anode, as suggested in literature[5], will only be beneficial if der electrolytes can still be obtained. Investigations are ongoing to study the influence of the Ni particle size on the re-oxidation phenomena.

ACKNOWLEDGEMENTS

The authors would like to thank F. Tietz, Research Centre Juelich, M. Mogensen, Ris National Laboratories, N. Christiansen, Haldor Topsoe, and G. Rietveld, Netherlands Ener Research Foundation, for their generous support of the re-oxidation studies. The work w funded by the European Union as part of the project "Component Reliability in Solid Oxide F Cell Systems for Commercial Operation (CORE - SOFC)".

REFERENCES

[1]A. Selcuk and A. Atkinson, "Strength and toughness of tape-cast yttria-stabilized zircon* *Journal of the American Ceramics Society*, **83** [8] 2029-2035 (2000).

[2]J. Malzbender, R.W. Steinbrech and L. Singheiser, "Determination of the interfacial fract energies of cathodes and glass ceramic sealants in a planar solid-oxide fuel cell design," *Jour. of Materials Research*, **18** [4] 929-934 (2003).

[3]G. Stathis, D. Simwonis, F. Tietz, A. Moropoulou and A. Naoumides, "Oxidation a resulting mechanical properties of Ni/8Y(2)O(3)-stabilized zirconia anode substrate for so oxide fuel cells," *Journal of Materials Research*, **17** [5] 951-958 (2002).

[4]G. Robert, A. Kaiser, K. Honegger and E. Batawi, "Anode Supported Solid Oxide Fuel C with a Thick Anode Substrate," pp. 116-122 in *Proceedings of European Solid Oxide Fuel (Forum V*, Edited by J. Huijsmans, Lucerne, 2002.

[5]D. Fouquet, A.C. Müller, A. Weber and E. Ivers-Tiffee, "Kinetics of Oxidation a Reduction of Ni/YSZ Cermet," pp. 467-474 in *Proceedings of European Solid Oxide Fuel (Forum V*, Edited by J. Huijsmans, Lucerne, 2002.

[6]J. Malzbender and R.W. Steinbrech, "Mechanical methods to determine layer complian within multilayered composites," *Journal of Materials Research*, **18** [6]: 1374-1382 (2003).

NUMERICAL CHARACTERIZATION OF THE FRACTURE BEHAVIOR OF SOLID OXIDE FUEL CELL MATERIALS BY MEANS OF MODIFIED BOUNDARY LAYER MODELING*

Ba Nghiep Nguyen, Brian J. Koeppel, Prabhakar Singh and Mohammad A. Khaleel
Pacific Northwest National Laboratory
P.O. Box 999, Richland, WA 99352, USA

Said Ahzi
University Louis Pasteur, IMFS
2 Rue Boussingault
67000 Strasbourg, France

ABSTRACT

A modified boundary layer modeling approach is used to predict the fracture toughness and crack resistance behaviors of solid oxide fuel cell (SOFC) materials. In this approach, a pre-existing sharp crack inside a layer or at an interface between two different layers is assumed under plane strain conditions. Fracture is allowed to occur in a small process window situated at the crack tip. The process window is contained in a circular region, which can involve one or two different materials and their interfaces. Elastic asymptotic crack-tip fields are prescribed as remote boundary conditions. Special attention is focused on the cracking of the interface between the glass seal and the electrolyte material.

INTRODUCTION

Two fracture mechanisms for the SOFC stacks have been identified: *interlaminar cracking* (delamination) along the interface and *transverse cracking* across the layers [1]. It is essential to develop suitable materials and to design stacks so that SOFCs are able to sustain thermal cycling without cracking. The fracture issue can be addressed through experimental and numerical characterizations of the fracture toughness and crack resistance behaviors of the constituent layers and their interfaces. This paper proposes a *numerical characterization* of the fracture behavior by means of a *modified boundary layer (MBL) modeling* approach.

In the MBL model, a pre-existing crack inside a layer or at an interface between two different layers under plane strain conditions is assumed. The crack propagation is allowed to occur in a small process window situated at the crack tip. In this paper, the crack is assumed to be sharp, and elastic brittle materials are considered. Thus, the debonding between the elements containing the initial crack plane can be described by a simple traction-separation law obeying a critical stress criterion. The debond option and the critical stress criterion in the finite element code ABAQUS were used for simulating crack propagation. The process window is contained in a circular region whose outer radius must be much greater than the size of the process window so that fracture o ccurring i nside i t d oes n ot a ffect t he r emote f ields. T he a dvantage o f t he M BL

* This manuscript has been authored by Battelle Memorial Institute, Pacific Northwest Division, under Contract No. DE-AC06-76RL0 1830 with the U.S. Department of Energy. The United States Government retains and the publisher, by accepting the article for publication, acknowledges that the United States Government retains a non-exclusive, paid-up, irrevocable, world-wide license to publish or reproduce the published form of this manuscript, or allow others to do so, for United States Government purposes.

analysis is that it can be used efficiently to characterize the material interface toughness and crack resistance behavior without knowing the exact geometry of the structure and the associated boundary conditions. These, however, are represented here through the asymptotic crack tip fields [2]. In this study, the crack-tip fields in an elastic solid and for an interface crack between dissimilar elastic materials are prescribed as remote boundary conditions.

Finally, the effect of surrounding material properties on the interface toughness is studied. In this paper, the properties of the electrolyte material as a function of its porosity are computed using the Eshelby-Mori-Tanaka approach [3], [4] in which the pores are assumed to have spherical shapes and behave as inclusions with zero stiffness.

MODIFIED BOUNDARY LAYER MODELING APPROACH

Figure 1 presents a schematic description of the modified boundary modeling approach. Figures 1a and 1b show an internal crack inside a material and an interface crack, respectively. The crack is generally subject to complex loading conditions represented by the stress field σ_{ij}^{∞} applied to the structure. The problem consists of determining the stress intensity factor that causes an initial crack to propagate (*fracture toughness*). This problem can be solved in terms of a MBL analysis in which a small circular region containing the initial crack-tip is analyzed. This kind of analysis is defined as *small-scale fracture* analysis. It relies on the assumption that fracture is allowed to develop in a process zone, which is much smaller than all relevant specimen dimensions so that a boundary layer problem can be formulated with the crack-tip fields in an undamaged solid as the remote boundary conditions applied on the outer contour of the circular region containing the process zone. The crack-tip displacement field in an elastic solid, e.g. for Mode I loading is

$$u_1 = \frac{K_1}{2\mu}\sqrt{\frac{r}{2\pi}}\cos(\frac{\theta}{2})[\kappa - 1 + 2\sin^2(\frac{\theta}{2})], \quad u_2 = \frac{K_1}{2\mu}\sqrt{\frac{r}{2\pi}}\sin(\frac{\theta}{2})[\kappa + 1 - 2\cos^2(\frac{\theta}{2})], \quad (1)$$

where r and θ are polar coordinates centered at the crack tip, K_1 denotes the Mode I loading stress intensity factor, μ is the shear modulus, and κ is given by: $\kappa = 3 - 4\nu$ (with ν, the Poisson ratio) for the plane strain conditions assumed in the analysis. The crack-tip displacement field for an interface crack between two dissimilar elastic materials is given by [5], [6]:

$$\mu^{(j)}(u_1 + iu_2) = \frac{|K|r^{1/2}}{2\sqrt{2\pi}\cosh(\pi\varepsilon)}\left\{ \frac{(3 - 4\nu^{(j)})e^{i\theta/2+\varepsilon(\theta-\pi)-i\psi_0}}{1 - i2\varepsilon} \right.$$
$$\left. - \frac{e^{-i\theta/2-\varepsilon(\theta-\pi)-i\psi_0}}{1 - i2\varepsilon} - i\sin\theta\, e^{i\theta/2+\varepsilon(\theta-\pi)+i\psi_0} \right\}, \quad (2)$$

where the superscript j (j=1, 2) denotes the associated material, $i = \sqrt{-1}$, ψ_0 is a measure of the mode mixity, and ε is the oscillation index given by:

$$\varepsilon = \frac{1}{2\pi}\ln\left(\frac{\mu^{(1)} + \mu^{(2)}(3 - 4\nu^{(1)})}{\mu^{(2)} + \mu^{(1)}(3 - 4\nu^{(2)})}\right) \quad (3)$$

394

The remote loading for an interface crack problem is thus specified by $|K|$ and ψ_0. In this paper, the stress intensity factor is prescribed as the value of loading that is increased during the incremental process. Hence, the critical $|K_c|$ at the onset of crack propagation can be determined.

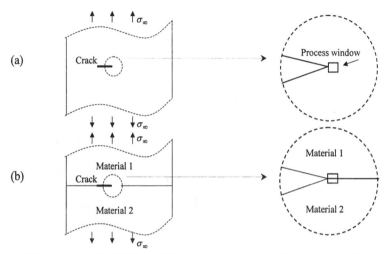

Figure 1. Schematic of the MBL modeling approach: (a) intralaminar and (b) interface crack problems.

EFFECT OF THE POROSITY ON THE ELASTIC PROPERTIES

The Mori-Tanaka approach [3] based on the Eshelby equivalent inclusion method [4] can be used to determine the homogenized (*effective*) properties of a composite material from those of its constituents. To do so, the inhomogeneous material is replaced by an equivalent material for which the strain field inside the inclusion is adjusted by the so-called *eigenstrains* to ensure the same response to an applied strain. The eigenstrains are then expressed in terms of the perturbed strains (due to the presence of the inclusion) via the Eshelby tensor, which accounts for the inclusion shape and the elastic properties of the matrix material. The stiffness tensor of a porous medium can be approximated by the Mori-Tanaka method by assuming that the pores have spherical shapes and behave as the inclusions with zero stiffness. Accordingly, it reads

$$\overline{C} = \lim_{C^{\text{inclusion}} \to 0} C_m [I + pB(I+pE)^{-1}]^{-1}$$

(4)

where C_m is the stiffness tensor of the bulk material (matrix material), and B and E are the tensors dependent on the Eshelby tensor and the stiffness of the inclusion. In this paper, Equation (4) is used to compute the stiffness tensor of the YSZ material in terms of its porosity p (pore volume fraction), and its elastic properties are then obtained accordingly.

NUMERICAL APPLICATIONS

This section will first focus on the validation tests for the MBL modeling approach to predict the fracture toughness values of fuel cell materials. As an example, the YSZ material studied in [7] is used here for the numerical tests. Since fracture is allowed to occur in a small process

395

window, the mesh size in this region strongly determines the accuracy of the prediction. A mesh sensitivity study was carried out to determine a mesh refinement criterion. This investigation indicated that the element mesh size d_e should be comprised between the *grain size* d_g and r_c, *the distance ahead the crack tip at which fracture occurs* ($r_c < d_e < d_g$). Figures 2a and 2b illustrate the prediction of the onset of cracking for the YSZ material used in [7]. Table I presents the predicted fracture toughness values for this material at 20°C and 900°C, which are in good agreement with the experimental results by Selçut & Atkinson [7] also shown in Table I.

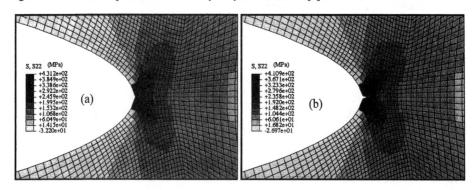

Figure 2. Distributions of the crack opening stress prior to (a) and at the onset (b) of crack propagation. The element size in the process widow is of 0.005 mm (while the average material grain size d_g and the distance r_c are of ~0.008 and 0.0023 mm, respectively).

Table I. Experimental and predicted values of K_I^c for the YSZ material used in [7]

Fracture Toughness (MPa.m$^{1/2}$)	Selçut & Atkinson [7]	MBL Modeling Prediction
K_I^c (20°C)	1.61 ± 0.12	1.72
K_I^c (900°C)	1.02 ± 0.05	1.1

Next, the crack resistance behavior of the YSZ material used by Kumar and Sørensen [1] was predicted. Figures 3a shows a propagation stage for a crack advance about 0.046 mm while Figure 3b presents the predicted crack resistance curves expressed in terms of the energy release rate as a function of the crack advance for this material at 20°C. Two different meshes were used for the analyses. The experimental results show no crack resistance ability for this material. Such a behavior was modeled by imposing the stresses to be relaxed immediately after the failure criterion had been satisfied. The average experimental curve from [1] is also presented in Figure 3b that shows a good agreement of results both in tendency and numerical values.

The mode mixity ψ_0 has a strong effect on the fracture toughness of the interface between dissimilar materials as pointed out in e.g. [8]. The MBL modeling is used here to numerically establish the fracture toughness/mode mixity relationship that is highly important for design of fuel cell stacks, in particular the design of glass seal/electrolyte interfaces. It is noted that ψ_0 is linked to a reference length whose choice can be based on a material length scale such as the size of the fracture process zone, the zone of dominance of the K-field, etc.

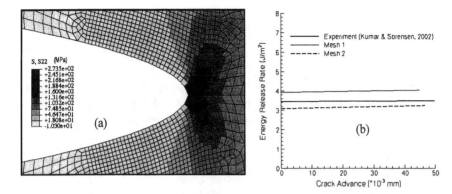

Figure 3. (a) Contour of the crack opening stress for a crack advance of 0.046 mm. (b) Predicted and experimental energy release rates versus crack advance for the YSZ material used in [1].

The Young's modulus and Poisson's ratio of glass used in this application were 76 GPa and 0.25, respectively while the corresponding elastic properties for the YSZ material were: $E=200$ GPa and $\nu = 0.315$. Figure 4a illustrates the variation of the normalized fracture toughness with ψ_0 for an interface crack between the glass and YSZ material. The minimum value of $|K_c|$ occurring at $\psi_0 = 23°$ for which the contribution of Mode II is minimum was used as the reference value for the normalization. For $-35° < \psi_0 < 23°$ and $23° < \psi_0 < 90°$, there is strong increase in toughness with increasing contribution of Mode II to the loading. Figure 4b shows the opening stress profile around the crack tip for the case $\psi_0 = 69°$ where the stress field is strongly governed by Mode II.

The elastic moduli of an YSZ material versus its porosity, which were obtained from Equation (4), are illustrated in Figure 5a. The computed results were then used in the MBL modeling to determine the effect of the YSZ porosity on the fracture toughness of the YSZ/glass interface (Figure 5b). In these simulations, the elastic properties of glass are identical to those used in the previous application. The mode mixity was kept at ~23° while the elastic properties of YSZ were varied as a function of its porosity. Figure 5b shows the interface toughness values normalized by the reference value at zero porosity. Overall, the effect of the YSZ porosity is negligible. However, it is noted that this approach does not account for local defects or voids situated in the immediate vicinity of the crack tip, which can decrease the fracture toughness.

CONCLUSION

The MBL modeling approach used in this paper is efficient and practical since it allows the fracture toughness and crack resistance behavior of SOFC materials to be directly estimated. In particular, the numerically established interface toughness/mode mixity relationship is very useful for the design of sealing materials in SOFC stacks. The approach only requires a few material data inputs such as the elastic properties and failure stresses, although it does require knowledge of the grain size for the mesh size criterion for the process window.

REFERENCES

[1]A.N. Kumar and B.F. Sørensen, "Fracture Energy and Crack Growth in Surface Treated Yttria Stabilized Zirconia for SOFC Applications," *Materials Science & Engineering A*, **A333** 380-9 (2002).

[2]B.N. Nguyen, P.R. Onck and E. van der Giessen, "Crack-Tip Constraint Effect on Creep Fracture," *Engineering Fracture Mechanics*, **65** 467-90 (2000).

[3]J.D. Eshelby, "The Determination of the Elastic Field of an Ellipsoidal Inclusion and Related Problems," *Proceedings of the Royal Society London*, **A 241** 376-96 (1957).

[4]T. Mori and K. Tanaka, "Average Stress in Matrix and Average Elastic Energy of Materials with Misfitting Inclusions," *Acta Metallurgica*, **21** 571-74 (1973).

[5]J.R. Rice, "Elastic Fracture Mechanics Concepts for Interfacial Cracks," *Journal of Applied Mechanics*, **55** 98-103 (1988).

[6]V. Tvergaard and J.W. Hutchinson, "Effect of Strain Dependent Cohesive Zone Model on Predictions of Interface Crack Growth," *Journal de Physique IV*, **6** 165-71 (1996).

[7]A. Selçut and A.Atkinson, "Strength and Toughness of Tape-Cast Yttria-Stabilized Zirconia," *Journal of the American Ceramic Society*, **83** [8] 2029-35 (2000).

[8]J.W. Hutchinson and Z. Suo, "Mixed Mode Cracking in Layered Materials," pp. 65-191 in *Advances in Applied Mechanics*, Volume 29 Edited by J.W. Hutchinson and T.Y. Wu, Academic Press, New York, 1991.

ACKNOWLEDGEMENT

This work was funded by the U.S. Department of Energy's Office of Energy Efficiency and Renewable Energy through the Hydrogen, Fuel Cells & Infrastructure Technologies Program. The support by John Garbak, Technology Development Manager is gratefully acknowledged.

(a) (b)

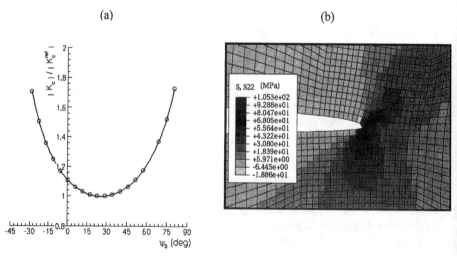

Figure 4. (a) Glass/YSZ interface toughness as a function of the mode mixity ψ_0. (b) Opening stress profile at the onset of crack propagation for the case $\psi_0 = 69°$.

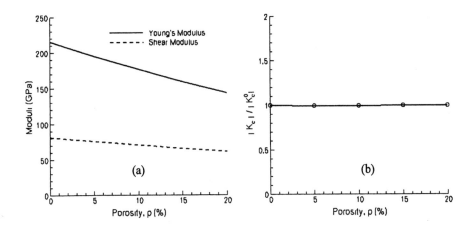

Figure 5. (a) Moduli of YSZ versus its porosity. (b) Normalized glass/YSZ interface toughness versus the YSZ porosity.

CHROMIUM POISONING OF CATHODES BY FERRITIC STAINLESS STEEL

T. D. Kaun, T. A. Cruse*, and M. Krumpelt
Argonne National Laboratory
Chemical Engineering Div.
9700 S. Cass Avenue,
Argonne, IL 60439

ABSTRACT

Chromium poisoning of cathodes has been reported by several groups developing solid oxide fuel cells (SOFC) with metallic bipolar plates. Typically, chromium is presumed to migrate from the bipolar plates into the cathode via a volatile oxyhydroxide species. In this paper we report results of experiments where cells measuring 2.5 cm by 2.5 cm were operated with 430 SS and E-Brite bipolar plates and with two different cathode materials. At 700°C, the SOFC degradation rate (percent loss of peak current generating capability) was about 3 times faster than at 800°C. In addition, samples of cathode material coated onto the stainless steel were equilibrated at temperatures of 700°C and 800°C in 25% humid air and analyzed for chromium content. It appears that chromium oxide covering the metal can react chemically with the cathode materials.

INTRODUCTION

Chromium contamination of SOFC cathodes has been observed by several groups of researchers developing cells with metallic bipolar plates. These cells exhibited significant performance declines leading to speculation that chromium contamination may "poison" the cathode performance. Hilpert et al. have attributed the chromium transport to the formation of volatile oxyhydroxide species that forms when chromium containing steels are exposed to oxygen and water at elevated temperature (1, 2). It has been noted that formation of volatile chromium species (oxyhydroxide) is more prevalent at 700°C than at 800°C (3) and that the cathode acts as a nucleation site for gaseous Cr deposition (4). In addition, one may ask whether the two most common cathode materials, lanthanum manganite (LSM) and lanthanum ferrite (LSF) can react chemically with the chromia scale that protects the steels. We have examined the phase stabilities of these systems using the HSC thermodynamic data base. Details of these calculations will be presented at a later date, but it appears that both cathode materials can partially react with chromia when in direct contact. With LSM cathodes, the HSC code predicts that 50% of the lanthanum in LSM forms lanthanum chromite when in contact with chromia, and for LSF, 25% is present in that form.

In this paper we will present results of work addressing the degradation of the cell performance in the presence of different steels and of the chemical reactivity of steels with cathode materials

EXPERIMENTAL APPROACH

The primary approach to investigating the chromium poisoning of the SOFC cathode is to operate SOFC cells at a constant potential of 0.7 volts and observe performance degradation. SOFCs of 2.5 cm X 2.5 cm using 430 SS, E-Brite, and Crofer 22 APU interconnects were tested.

To enhance the chromium poisoning effect, particles of the metal interconnect were placed on top of the cathode. Next, a Pt current collector was placed on top of the cathode and particles. A plate of interconnect, with slits cut in it, was then placed on top of the Pt current collector. Separate cells with either LSF or LSM were operated at 700°C and 800°C. The air contained 2% humidity. The cells were operated until only 50% of the initial current was supported at 0.7 volts. Thick cathode samples (10 microns) were examined by scanning electron microscopy (SEM) to identify Cr distribution and/or Cr gradient using energy dispersive x-ray spectroscopy (EDS).

A second approach used coupon samples that were prepared with 10-20 micron thick cathode material coated onto prospective interconnect materials in a controlled-atmosphere furnace test (air at 250 ml/min). The coupons were stacked with alumina felt to give access to the flowing humid air within the furnace tube, and then equilibrated for 400 hours at either 700°C or 800°C in air containing 25% humid air. The furnace tests were a means of accelerating reactivity due to the elevate level of humidification (25%) that was used. These test samples examined two types of Cr alloy, 430 SS with 17% Cr, and E-Brite with 26% Cr. Either LSF (LaSrFeO$_3$) or LSM (LaSrMnO$_3$) were applied. To examine the reactivity between chromia and the cathode materials (LSM and LSF), a layered powder pellet of chromia and LSF or LSM was pressed and exposed to ambient air at 800°C for 400 h. After equilibration, the chromium content was analyzed at selected locations by SEM and EDS.

RESULTS OF SOFC TESTS

Results of the cell tests are shown in figures 1a and 1b. Fig 1a shows polarization curves of a cell with an LSM cathode and a 430 SS bipolar plate at 700°C after 2, 48, 144, 168, and 216 hours of operation. Clearly, the slope of the polarization curve declines quite rapidly. In Fig. 1b, the cell current at 0.7 volts is plotted versus operating time at 800°C and at 700°C. The cell current at 700°C is significantly smaller than at 800°C as it should be, considering the higher resistance of the electrolyte and cathode interface at the lower temperature. With time, the current through both cells decreases significantly. Although there is considerable scatter, the decline at 700°C appears more rapid than at 800°C, implying a more severe poisoning reaction at the lower operating temperature.

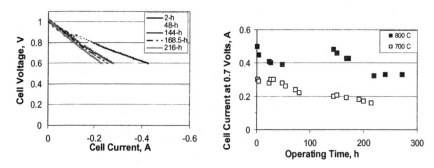

Fig.1. a) polarization curves of InDec cell #17 with LSM and 430 SS at 700°C and 2% humid air showing ongoing performance decline, b) Cell current at 0.7 V for two SOFCs with LSM cathode, 430 SS and humid 2% air; one at 700°C and the other at 800°C

SEM was used to examine physical condition of the degraded cells. EDS was used to determine Cr content through the cathode thickness, e.g., concentrated at the cathode/electrolyte interface or evenly dispersed. Fig.2a shows the chromium content of the cathode and bipolar plate for the cell operated at 700°C at 5 locations. In the cathode, chromium levels of 0, 1.5, and 0.15 % were found and in the bipolar plate, the chromium content at location 1 and 2 was depleted compared to the nominal concentration of 17 %. Fig. 2b shows the equivalent data for the cell operated at 800°C. The amount of Cr detected in the cathode by EDS is significantly less. Only 0.4% Cr is found at the outer-most part of the cathode at 800°C. Although not shown, the 430 SS formed a chromia scale when cell tests were carried out at 800°C. Tests of SOFCs with LSF have generally yielded higher levels of Cr in the cathode along with greater rates of performance decline than with LSM cathode. Post-test analysis conducted by SEM/EDS indicated that the level of Cr was 3.5wt% for the LSF cathode.

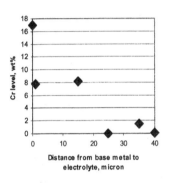 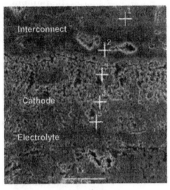

Fig. 2a. SEM of LSM Cathode and 430 SS Interconnect from SOFC (#17) at 700°C, Cr Content at Sites by EDS, Sites 1-2: Cr depletion of 430 SS: 7.8 and 8.2 wt% Cr from 17 wt% Cr, Sites 3-5 Cr in cathode: 0, 1.5, 0.15 wt% Cr at electrolyte interface

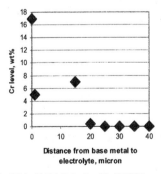 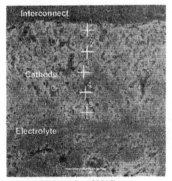

Fig. 2b. SEM of LSM Cathode with 430 SS Interconnect from SOFC (#16) at 800°C, Cr Content at Sites by EDS, Sites in Cathode 1-5: 0.4, 0, 0, 0 wt% Cr at electrolyte interface.

RESULTS OF SAMPLES TO EXAMINE CATHODE REACTIVITY

The interactions of cathode and interconnect metals were examined with cathode-coated coupon samples. These samples were subjected to a harsher environment than SOFC cell tests by increasing the water content of the air to 25%. The Cr distribution results of SEM/EDS examination are shown in Fig. 3 for the 430 SS coupon samples. The amount of Cr found in both the LSM and LSF cathode layers (10-20 micron thick), was distinctly lower at 700°C than at 800°C. The chromium content decreases away from the metal interface.

A different behavior is observed with E-Brite samples. As shown in Fig. 4, the chromium content is higher in absolute terms and higher at 700°C than at 800°C. An indication for the different behavior of the two types of steels may be seen in Fig. 5, showing the thickness of the chromia layers on the steels. The bare 430SS surface had a chromia layer of 10 micron thickness after equilibration at 800°C, and only 1 micron at 700°C. 430SS in contact with LSF formed a chromia coating at 800°C but not at 700°C. In fact, the chromium content of the steel became completely depleted of chromium in the surface down to 10 micron thickness. E-Brite, which has significantly higher chromium content than 430 SS also formed chromia coatings on the bare metal, and in contrast to 430SS, retained a thin layer of chromia when in direct contact with LSF or LSM.

When a layer of chromia powder was equilibrated with a layer of various cathode materials at 800°C, only a trace of Cr was observed at the outer most edge of the LSM layer. However, the stoichiometric LSF layer showed 2.3 at % Cr near the chromia interface and abruptly trailed off. While a La-deficient LSF layer showed 6.5 at % Cr near the chromia interface and remained at 0.6 at % for several hundred micros away from the interface.

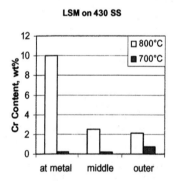
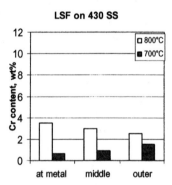

Fig.3. EDS results of Cr found in cathode layer of 430 SS coupon samples at 800°C and 700°C. SEM shows a 1 micron chromia scale at 800°C and Cr depletion at 700°C.

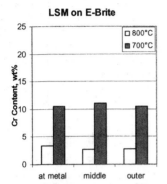

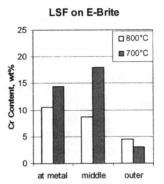

Fig.4. EDS results of Cr found in cathode layer of E-Brite coupon samples at 800°C and 700°C. SEM shows a 1-10 micron chromia scale at 800°C and 1 micron at 700°C

DISCUSSION

The results suggest that there is not a single mechanism responsible for the interaction between SOFC cathodes and steels. As this work continues to reproduce the preliminary results presented here, we may revise some observations. With this caveat, the results are interpreted as follows:

1) Formation of the chromia scale on either of the two steels is much slower at 700°C than at 800°C, but the degradation of the cell performance appears to be faster at 700°C, suggesting that the volatile oxyhydroxide species may form on virtually bare metal. This is consistent with the results in Fig. 5a showing the depletion of chromium from the surface of 430 SS.

2) Chromia does react chemically with cathode materials as suggested by thermodynamics and confirmed in the equilibration of chromia powder with cathode materials and by the Fig. 4a and b, showing high chromium content in cathodes contacting E-Brite. The latter forms chromia more readily than 430SS due to the higher chromium content, as also shown in fig. 5b.

3) The solid state reaction between chromia and cathode materials is influenced by kinetics. The fact that sub-stoichiometric LSF contains more chromium than stoichiometric LSF, and LSM contains even less when equilibrated with chromia suggests that oxide ion conductivity of the cathode material may be influencing the rate of reaction.

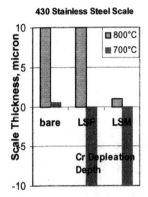
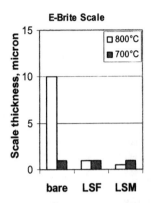

Fig.5. SEM results of Cr oxide scale found on 430 SS or E-Brite after 400h at either 800°C or 700°C. In a number of cases, a cathode layer significantly alters the scale formation, e.g. bare 430 SS has minor scale, but addition of cathode results in no scale and 10 micron of total Cr depletion.

CONCULSIONS

While ferritic stainless steel based bipolar plates are a potentially attractive component for planar solid oxide fuel cells, the protective chromia scale that can form interacts with the cathode material and must be stabilized. Approaches to stabilize metal interconnect may be formation of spinels or perovskites of chromium, or finding another electrically conductive coating.

ACKNOWLEDGEMENTS

We would like to thank the U.S. Department of Energy, Office of Fossil Energy, Solid State Energy Conversion Alliance (SECA) Program, for funding this research under Contract W-31-109-ENG-38 and our program manager Lane Wilson.

REFERENCES

(1) Hilpert, K., D. Das, M. Miller, D.H. Peck and R. Wei, *J. Electrochem. Soc.*, vol. 143, no. 11, pp. 3642-3647 (1996).
(2) C. Gindorf, L. Singheiser, K. Hilpert, Steel Research , 72, pp.528-533, (2001)
(3) G. Fryburg, F. Kohl,and C. Stearns, J. of Electrochemical Soc.121, pp.952-958, (1974)
(4) S. P. Jiang, et al., J .of European Ceramic Society, 22, pp. 361-373, (2002)

406

EFFECT OF IMPURITIES ON ANODE PERFORMANCE

C. A.-H. Chung, K. V. Hansen and M. Mogensen
Materials Research Department
Risø National Laboratory
4000 Roskilde
Denmark

ABSTRACT

Large differences in anode performance were observed between several batches of Ni-YSZ cermet anodes made from nominally identical batches of anode slurries and on electrolyte tapes made from different, but nominally identical, batches of YSZ powder. Impedance spectroscopy measurements were performed at 1000°C and 850°C in a H_2/H_2O atmosphere. In spite of everything being nominally equal, a large difference in the polarization resistances was observed between the batches, with cells from one batch exhibiting polarization resistances of 300-525 mΩ cm^2 and cells from another batch, made from the very same anode slurry, exhibiting polarization resistances of only 42-63 mΩ cm^2, at 1000°C.

Investigation into the processing of the anode cermet did not provide an explanation for the differences, but X-ray Photoelectron Spectroscopy (XPS) analysis showed that there were higher levels of Si and Y on the surface of the YSZ tape from which the cells with higher polarization resistances were made. It was found that co-sintering of YSZ tape and Ni-YSZ cermets alleviated the problem with silica-impurities substantially.

INTRODUCTION

It is generally recognized that it is difficult to fabricate SOFC cells with good reproducibility over time, even though everything is kept as constant as possible. Here some extreme cases of a lack of reproducibility in the fabrication of symmetric cells, with YSZ electrolyte and Ni-YSZ cermet anodes, are reported. This problem was investigated by analyzing the cell components. Furthermore, the effect of cell fabrication procedure (separate sintering versus co-sintering) was examined in order to determine if this was a factor that could reduce the deleterious effect of silica impurities.

EXPERIMENTAL

Sample preparation

Several batches of two-electrode symmetrical cells, with equal Ni-YSZ electrodes on each side of a ~170 μm thick YSZ tape, were prepared. The YSZ tapes were prepared by tape casting TZ8Y (Tosoh Corporation, ZrO_2 stabilized with 8 m/o Y_2O_3) and the Ni-YSZ anode cermets were prepared by spray painting.[1] Both the NiO-YSZ slurries and the YSZ electrolyte tapes used were nominally equal for all of the batches.

For all batches, except S4270, the anode slurry was hand-sprayed onto both sides of sintered YSZ tapes in layers of 10 - 15 μm thickness, sintering at 1300°C for 2 hours in-between each layer. For the S4270 batch the anode was robot-sprayed, using the spraying robot in the pre-pilot plant at Risø, onto both sides of a green YSZ tape. The anode and electrolyte were subsequently co-sintered at 1310°C. The batch details are given in Table 1. YSZ tape batch 2 was produced out of house, and batches 1 and 3 were produced at Risø. Samples S2901, S2902 and S2910 were made from the same anode slurry, and E0017 and S4270 were made from different, but nominally equal slurries.

Table I. Sample batch details. S2901, S2902 and S2910 were made from the same anode slurry, E0017 and S4270 were made from different anode slurries

Sample ID	YSZ tape batch	Method of Spraying	Number of anode layers	Sintering
E0017	1	Hand	3 (10, 15, 15 μm)	Separate for YSZ and anode
S2901	2	Hand	3 (10, 15, 15 μm)	Separate for YSZ and anode
S2902	2	Hand	3 (10, 15, 15 μm)	Separate for YSZ and anode
S2910	3	Hand	1 (10 μm)	Separate for YSZ and anode
S4270	3	Robot	1 (10 μm)	Co-sintered

Electrochemical testing

Electrochemical testing was carried out on samples of approximately 4 mm by 4 mm. To ensure satisfactory current collection, an additional 20-40 μm current collect layer in the form of a Ni/YSZ (90/10 v/o) slurry, was applied to the anodes. A 3-cell set-up was used, in which the cell being tested was sandwiched between 2 auxiliary cells, with the electrodes of the test cell and the auxiliary cells contacting opposite sides of Pt meshes. Current was passed between the outer electrodes of the auxiliary cells and the potential was measured between the 2 electrodes of the test cell. In this way gas was supplied and removed close to the working electrodes, thus minimizing stagnant layer gas diffusion.[2]

Impedance spectroscopy was carried out around OCV in a one-atmosphere set-up. The set-up used could house 4 samples for testing at a time. Samples were heated to 1000°C at 4 °C min^{-1} in air and then reduced in wet H_2 (H_2 + 3% H_2O) for 3 hours prior to making measurements. A total gas flow rate of 100 ml min^{-1} (at 25°C) of dry gas was used. 4-lead, 2-electrode impedance was measured in the frequency range 2 MHz to 0.2 Hz, with an applied voltage of 28 mV, using a Solartron 1260 FRA.

Data handling

The measured impedance spectra contained two or three arcs. The data were fitted with the EQUIVCRT software by B. A. Boukamp[3] using the equivalent circuit $LR_s(RQ)_{1\alpha}(RQ)_{1\beta}$ or $LR_s(RQ)_{1\alpha}(RQ)_{1\beta}(RQ)_2$, where L is an inductance, R a resistance and Q a constant phase element with an admittance $Y^* = Y_0(j\omega)^n$. Y_0 is an admittance factor, j is the imaginary unit, ω is the angular frequency and n is the frequency exponent. The index numbering of the arcs has been chosen to be consistent with previous work.[4,5]

Different sets of R and Y_0 values may only be compared directly if the frequency exponents (n-values) are constant. The n-values for Ni/YSZ cermet anodes have previously been derived by fitting spectra from a large number of anodes under various conditions.[6] In previous work two arcs were generally observed for symmetrical anode cells, with n-values of 0.8 and 0.75 for the high frequency arc (f_{summit} 5-10 kHz) and the low frequency arc (f_{summit} approx. 100 Hz), respectively.[7] It was found that the spectra in this study could be satisfactorily described (within 2% relative error) using n-values ($n_{1\alpha}$, $n_{1\beta}$, n_2) = (0.8, 0.8, 0.75), with summit frequencies in the range of 40-200 kHz, 1-5 kHz and 50-150 Hz, respectively.

XPS measurements

The surface compositions of YSZ tapes from batches 2 and 3 were examined by XPS. The measurements were performed on a Sage 100 from Specs. A Mg Kα X-ray source was used and an electron flood gun was used to minimize charging of the sample. The analyses were performed at a 90° angle to the surface. For the quantification, 20 narrow scans were performed with 0.2 eV steps and a pass energy of 23 eV for each element. The SDP software from XPS International was used to analyze the spectra.

RESULTS

Typical impedance spectra for cells from batches E0017, S2902 and S2910 at 1000°C are shown in Figure 1. A summary of the polarization resistances at 1000°C and 850°C for all of the sample batches and for batch S3 (from previous work[4]) is given in Table II. In some of the spectra for the S2901 samples a negative resistance and negative constant phase element were observed at low frequency (a low frequency inductive arc), indicated with an asterisk in Table II. Typical sets of data obtained by fitting with the EQUIVCRT software, along with the summit frequencies, are given in Table III.

In some cases (see spectra for S2902 and S2910 in Figure 1) the low frequency part of the spectra could not be satisfactorily fitted to a (RQ) component. Therefore, the very low frequencies of these spectra were not fitted. The discrepancy between the R_p values obtained by

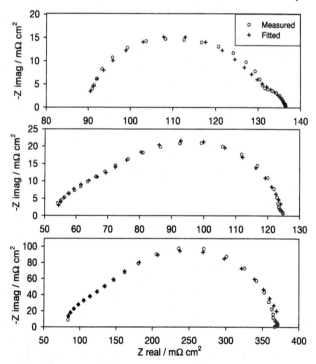

Figure 1   Typical impedance spectra at 1000°C for (from top to bottom) E0017, S2902 and 2910

Table II. Summary of polarization resistances (R_p) at 1000°C and 850°C

Sample	R_p (1000°C) / mΩ cm²	R_p (850°C) / mΩ cm²	YSZ tape batch Reference 4	Anode slurry batch Reference 4
S3	124	227		
E0017	48-50	-	1	A
S2901	42*	72-77*/173-182	2	B
S2902	70-73	227-237	2	B
S2910	300-524	-	3	B
S4270	64-90	-	3	C

*A negative resistance and constant phase element were observed at low frequencies

Table III. Typical data sets, at 1000°C, obtained through fitting with the EQUIVCRT program

Sample	$R_{1\alpha}$ / mΩ cm²	$Y_{0,1\alpha}$ / mSs$^{0.8}$ cm^{-2}	$n_{1\alpha}$	$R_{1\beta}$ / mΩ cm²	$Y_{0,1\beta}$ / mSs$^{0.8}$ cm^{-2}	$n_{1\beta}$	R_2 / mΩ cm²	$Y_{0,2}$ / mSs$^{0.75}$ cm^{-2}	n_2	$f_{sum,1\alpha}$ / kHz	$f_{sum,1\beta}$ / kHz	$f_{sum,2}$ / kHz
S3	-	-	-	70	0.32	0.8	54	0.17	0.75	-	6	0.09
E0017	-	-	-	41	1.81	0.8	7	314	0.75	-	4.1	0.10
S2901	15	0.73	0.8	35	3.06	0.8	-8*	-212*	0.75	46	2.6	0.13
S2902	16	0.50	0.8	57	1.92	0.8	-	-	-	68	2.5	-
S2910	63	0.06	0.8	271	0.29	0.8	-	-	-	182	3.8	-
S4270	7	0.86	0.8	58	2.8	0.8	9	446	0.75	97	1.5	0.05

*In this case R_2 and Q_2 ($Y_{0,2}$) are not ascribed to gas diffusion but to an activation phenomenon

Table IV. XPS data for YSZ tape batches 2 and 3, corresponding to samples S2901/2 and S2910 respectively

Elemental ratio	Batch 2 (S2901/2)	Batch 3 (S2910)	Theoretical for 8YSZ
Y/Zr	0.26	0.30	0.17
Si/Zr	0.1 - 0.13	0.24 - 0.26	0

fitting with or without the low frequency part of the spectra was 1-5 mΩ cm^2.

The XPS measurements of the Y/Zr and Si/Zr ratios on the surfaces of YSZ tapes from batches 2 and 3 (which were used to make samples S2901/2 and S2910 respectively) are given in Table IV. The theoretical ratios for 8YSZ are also shown for comparison.

DISCUSSION

From the data in Tables II and III, and from Figure 1, it is clear that there are large differences in behavior between different batches of samples. Firstly, the S2901, S2902, S2910 and S4270 samples exhibit a high frequency arc (1_α), whereas the S3 and E0017 samples do not. It has been indicated that the high frequency process is sensitive to anode structure, and is related to the resistance to charge transfer from Ni to YSZ, and of oxygen ions in the anode cermet.[4] Thus, the extra 1_α arc indicates that an extra process related to the cermet structure is occurring. The $R_{1\beta}$ at 1000°C is much greater for the S2910 samples than for the other batches.

The low frequency arc, R_2 is not observed for all of the samples. It is known that R_2 is caused by diffusion impedance relating to the gas directly above the anode structure.[2] Where R_2 is present its value is small (<10mΩ cm^2), except in the case of the S3 samples. The S3 samples exhibit larger R_2 values because the 3-cell set-up, designed to minimize the diffusion impedance, was not used for testing these samples.

In some cases the S2901 and S2902 samples exhibit a low frequency arc with a negative resistance and a negative constant phase element. This behavior is probably due to some sort of activation process.[8]

From examination of the anode microstructures and electrolyte-anode interfaces by SEM, no significant difference can be seen to account for these results. However, XPS analysis of the YSZ surfaces, Table IV, reveals that there is a significant difference in the Si/Zr ratio between YSZ tape batches 2 and 3. The Y/Zr ratio on the surface of the YSZ tapes is considerably greater than the theoretical ratio. It is well known that Y and impurities, such as Si, in YSZ migrate to the external surfaces during sintering.[9] Several studies on the effect of impurities on electrochemical performance have been made, and it has been demonstrated that the level of impurities, in particular silica, at Ni-YSZ interfaces can significantly affect the specific resistance.[10,11] The S2910 samples exhibiting high R_p values were made from the YSZ batch with higher levels of impurities. This indicates that the impurities at the YSZ-anode interface have a blocking effect and lead to high polarization resistances. By this reasoning the S2470 samples, which were made from the same YSZ batch as S2910, would also be expected to exhibit high polarization resistances. However, for the S2470 samples the anode and YSZ were co-sintered, so the impurities in the YSZ tape could not migrate to form a blocking layer on the YSZ surface before the anode cermet was applied.

In all cases raw materials with similar purity specifications were used for fabrication of electrolytes and electrodes, but evidently there must have been important differences in the impurity content for some of the electrolyte slurry components. Much more work is required in this area in order to establish knowledge about the types and levels of impurities that are acceptable in the ingredients of a YSZ slurry. Unfortunately, this cannot be determined from the present work.

CONCLUSIONS

Differences in the amount of silica that migrated to the surface of YSZ tapes resulted in large differences in Ni-YSZ anode performance in the case of separate sintering of the electrolyte and anode. The measured difference in the surface concentration of silica must originate from differences in raw materials, even though these were nominally equal. Thus, much more work is required in order to establish knowledge of the levels of impurities that are allowable for different types of SOFC cell fabrication.

In the case of co-sintering of the electrolyte together with the anodes, using the very same "impure" material batches that resulted in poor anode performance for samples in which the electrolyte and anodes were sintered separately, much lower degradation in anode performance was observed.

ACKNOWLEDGEMENTS

This work was carried out within the Danish solid oxide fuel cell project, DK-SOFC b Long Term R&D, j. nr. 1443/02-0001. The authors wish to thank the Danish Energy Agency for financial support, and colleagues in the fuel cell group for their help.

REFERENCES

[1] C. Bagger, "Improved production methods for YSZ electrolyte and Ni-YSZ anode for SOFC"; pp. 241-244 in *1992 Fuel Cell Seminar*, Courtesy Associates, Inc., Washington, D.C., 1992

[2] S. Primdahl and M. Mogensen, "Gas Diffusion Impedance in Characterization of Solid Oxide Fuel Cell Anodes," *Journal of the Electrochemical Society*, 146 [8] 2827-2833 (1999)

[3] B. A. Boukamp, "A non-linear Least Squares Fit procedure for analysis of immittance data of electrochemical systems," *Solid State Ionics*, 20 [1] 31-44 (1986)

[4] S. Primdahl and M. Mogensen, "Oxidation of Hydrogen on Ni/YSZ-Cermet Anodes," *Journal of the Electrochemical Society*, 144 3409-3419 (1997)

[5] M. Brown, S. Primdahl and M. Mogensen, "Structure/performance Relations for Ni/YSZ Anodes for SOFC," *Journal of the Electrochemical Society*, 147 [2] 475-485 (2000)

[6] M. Mogensen, S. Primdahl, J. T. Rheinländer, S. Gormsen, S. Linderoth and M. Brown, "Relations between Performance and Structure of Ni-YSZ-cermet SOFC Anodes"; pp. 657-666 in *Solid Oxide Fuel Cells IV*, edited by M. Dokiya, O. Yamamota, H. Tagawa and S. C. Singhal. **PV 95-1**, The Electrochemical Society, Inc., Pennington, NJ, 1995

[7] S. Primdahl and M. Mogensen, "Gas Conversion Impedance, a Test Geometry Effect in Characterization of Solid Oxide Fuel Cell Anodes," *Journal of the Electrochemical Society*, 145 2431-2438 (1998)

[8] T. Jacobsen, B. Zachau-Christiansen, L. Bay, M. Juhl Jørgensen, "Hysteresis in the solid oxide fuel cell cathode reaction", *Electrochimica Acta*, 46 1019–1024 (2001)

[9] S. P. S. Badwal and A. E. Hughes, "The Effects of Sintering Atmosphere on Impurity Phase Formation and Grain Boundary Resistivity in Y_2O_3-Fully Stabilized ZrO_2," *Journal of the European Ceramic Society*, 10 115-122 (1992)

[10] K. V. Jensen, R. Wallenberg, I. Chorkendorff and M. Mogensen, "Effect of impurities on structural and electrochemical properties of the Ni-YSZ interface," *Solid State Ionics*, 160 27-37 (2003)

[11] M. Mogensen, K. V. Jensen, M. J. Jørgensen and S. Primdahl, "Progress in Understanding SOFC Electrodes," *Solid State Ionics*, 150 123-129 (2002)

Ceramics in Environment Applications

COMPARISON OF CORROSION RESISTANCE OF CORDIERITE AND SILICON CARBIDE DIESEL PARTICULATE FILTERS TO COMBUSTION PRODUCTS OF DIESEL FUEL CONTAINING FE AND CE ADDITIVES

D. O'Sullivan, S. Hampshire
Materials Ireland Research Centre,
University of Limerick,
Limerick, Ireland

M.J. Pomeroy
Materials and Surface Science Institute,
University of Limerick,
Limerick, Ireland

M.J. Murtagh
Corning Incorporated,
Corning,
New York 14831 , USA

ABSTRACT

Cordierite porous honeycomb structures have been used successfully as particulate filters in heavy duty diesel engine applications for nearly two decades. However, some cracking and melting of cordierite based products have been reported under some uncontrolled regeneration conditions in passenger car exhaust systems. As a result silicon carbide based diesel particulate filters have emerged in recent years for application in passenger cars, particularly in Europe.

The present paper describes a series of experiments to investigate chemical interactions between cordierite or silicon carbide filter substrate materials and synthetic ash compositions expected to be deposited on the surfaces of the ceramic filter and within its pore structure as a result of the combustion of diesel fuel containing catalytic additives. Analysis of the experimental data allows prediction of the behaviour of filters under regeneration conditions and sets upper temperature limits to be defined for effective use of the two filter materials.

INTRODUCTION

Tightening of environmental legislation has led to significant effort to reduce motor vehicle emissions. Attention has focussed on solid particulate matter (PM) emissions, identified as carcinogens, from diesel vehicles. PM consists mainly of carbon soot from incomplete fuel combustion as well as inorganic contaminants such as calcium, zinc and phosphorous from the engine lubricating oil and iron from engine abrasion. Ceramic diesel particulate filters (DPFs) are structures that trap PM from the diesel exhaust gases while allowing "clean" gases to exit.

DPFs become loaded with PM and, to avoid unacceptable counter pressure in the engine, regeneration through incineration of the carbon soot is necessary. Controlled regeneration temperatures are typically of the order of 600°C but the use of catalytic fuel additives such as ferrocene ($(C_5H_5)_2Fe$) or cerium carboxylate ($Ce(COOH)_4$) can reduce this by up to 250°C[1]. These additives introduce solid iron or cerium species in the exhaust stream, which along with the inorganic components of PM, are deposited on and in the walls of the filter. During regeneration the inorganic contaminants are oxidised and the oxides remain on the filter as an ash.

Carbon burn-off is a highly exothermic reaction and if uncontrolled, as may occur during prolonged idling of the engine, can lead to temperature peaks over 1000°C within the filter and strong thermal gradients across the honeycomb. At such elevated temperatures, reactions between components of the ash and the ceramic filter structure may occur leading to local melting or cracking of the filter. Degradation may occur by the formation of low eutectic surface liquids and corrosion induced crystalline and vitreous phases resulting in potentially incompatible thermal expansion coefficients. In addition, formation of cracks in the DPF could occur from thermal stresses generated as a result of the local exothermic nature of the oxidative reactions.

Cordierite ($Mg_2Al_4Si_5O_{18}$) is particularly suitable for use in honeycomb structures because of its low coefficient of thermal expansion, high thermal shock resistance and thermal stability, coupled with low cost and ease of component manufacture. As a result it has been routinely used as a DPF material in heavy duty applications for 20 years[2]. However, cracking and melting of current designs of cordierite based DPF products have been reported under some uncontrolled regeneration conditions in passenger car exhausts[3]. In recent years, silicon carbide (SiC) based DPFs have been used in passenger cars. Properties that make SiC suitable for use include high strength, good thermal conductivity, refractoriness, and chemical durability[4]. A common perception is that SiC based filters survive in uncontrolled regeneration because of their ability to withstand high temperatures. However, it is more likely that their high volumetric heat capacity allows them to act as large heat sinks absorbing heat created by the exothermic regeneration reaction and thus minimising interior filter temperatures[5].

This study reports an investigation carried out to characterise reactions occurring (a) between individual ash components when iron and cerium based fuel additives were used and (b) between ash components and the two studied ceramic filter substrate materials.

EXPERIMENTAL

Analysis[6] of ash recovered from a diesel engine filter running with a ferrocene additive indicated that the major species present and their relative proportions by weight were:

$$Fe_2O_3 : CaO : P_2O_5 : SO_3 : ZnO = 1 : 0.35 : 0.4 : 0.45 : 0.20.$$

Using this data and assuming that all ferrocene was oxidised to Fe_2O_3 and all cerium carboxylate to CeO_2, the ratios of substrate material to the various ash components (CaO, P_2O_5, SO_3 and ZnO) and oxides resulting from fuel additives (Fe_2O_3 or CeO_2) were calculated for an in-service lifetime of 160,000 km.

In order to monitor reactions between individual ash components in the presence of carbon and between the ashes and the substrate materials, three ash compositions were prepared. These compositions corresponded to ashes predicted to form on a DPF as a result of the combustion of diesel fuel containing no fuel additive (A), diesel fuel containing the iron based fuel additive (FeA) and diesel fuel containing the cerium based fuel additive (CeA).

Powder compacts of the ashes and the substrate-ash systems were prepared by pressing milled mixtures of crushed DPF, carbon, oxides of iron, cerium, calcium, phosphorous and zinc and sulphates of iron and zinc. The carbon content of the ashes corresponded to the predicted soot present on a DPF if one regeneration cycle was missed.

Simultaneous differential thermal analysis/thermogravimetric analysis (DTA/TGA) was carried out on the pressed pellets to identify reactions occurring in the ashes and the substrate-ash compositions using a Stanton Redcroft STA 1640 with a heating rate of 10°C/min to 1400°C and an air + 200 ppm SO_2 gas atmosphere. The reactions identified by thermal analysis were further investigated by 24 hour heat treatments of (a) ash compacts for 24 hours at 900°C, 970°C, 1100°C and 1250°C, (b) SiC-ash mixed pellets at 900°C, 1100°C and 1250°C and (c) cordierite-ash mixed pellets at 900°C, 1000°C, 1100°C and 1250°C.

To simulate the effect of an ash deposit on the filter surface, SiC and cordierite filter samples were sprayed with salt solutions to obtain surface coatings with the predicted ash compositions (A, FeA and CeA). These samples were heat treated at the same temperatures used for the equivalent powder compact samples. Only the 900°C heat treatment was carried out for 24 hours with the other heat treatments carried out for 30 minutes. These times were chosen to reflect the total estimated time that a diesel particulate filter would experience these temperatures as a result of repeated regenerations during its in-service lifetime.

All samples were heat treated in the air + 200 ppm SO_2 gas atmosphere using heating rates of 50°C/min with furnace cooling. After heat treatment, the pellets were visually inspected and shrinkage on heat treatment was measured. X-ray diffraction (XRD) was carried out on powders of the heat treated samples using a Philips X'Pert system employing Cu K_α radiation.

RESULTS AND DISCUSSION
Ash Behaviour

Thermal analyses of the three ashes showed exotherms at approximately 800°C associated with weight losses corresponding to carbon burnout. Additional weight losses and endotherms at about 950 and 1000°C were due to sulphate decomposition, although the significance of the endotherms for the A and FeA ashes may also indicate partial liquefaction of the samples. Sintering of baseline ash (A) pellets in the sulphurous gas mixture indicated that significant radial shrinkage (16%) occurred at 970°C and that almost complete liquefaction occurred at 1100°C. Such observations would be consistent with the $Ca_3(PO_4)_2$ - $Zn_3(PO_4)_2$ phase diagram[7], where the lowest liquidus is 1048°C for Zn rich compositions, and previous analyses which showed that phosphorus is responsible for inducing low melting liquid formation[8]. The significant sintering of ash A at 970°C is thought to arise from suppression of the liquidus by sulphur trioxide. Subsequent XRD analyses clearly indicated that calcium phosphate and glass were the only two phases present at temperatures above 950°C and thus the glass comprised calcia, zinc oxide and phosphorus and also probably sulphur trioxide. The addition of iron shifted the temperature required for significant sintering to higher values. Thus, only 9% radial shrinkage of compacts occurred at 1100°C, indicating that less liquid was formed in the ash compacts when iron was present. Crystalline phases present after heat treatment at 1100 and 1250°C were zinc ferrate, calcium phosphate and Fe_2O_3 indicating that Zn was removed by Fe additions, thus probably suppressing liquid formation due to $Ca_3(PO_4)_2$ - $Zn_3(PO_4)_2$ interactions. Accordingly less liquid was formed due to (a) Zn uptake in zinc ferrate formation and (b) the dilution effect of the iron oxide addition to the ash. That said, at 1250°C significant liquid formation appeared to occur as evidenced by pellet slumping. Ceria appeared to decrease liquid contents more appreciably such that, at 1100 and 1250°C, 4% and 13% radial shrinkages were observed, respectively.

The data obtained for the ash clearly indicates that the baseline ash undergoes appreciable sintering at 970°C and that such a process could cause DPF pore blockage even if the DPF material itself remains unattacked by the ash deposit. Ashes containing Fe_2O_3 or CeO_2 would be expected to be less likely to block filters unless temperatures of 1100°C were exceeded. Ceria based ashes had better resistance to sintering due presumably to Ce removing phosphorus from the system as $CePO_4$ resulting in less liquid formation.

Silicon Carbide – Ash Interactions

Thermal analysis showed that SiC undergoes exponential weight increases once temperatures of 800°C are exceeded and this is consistent with oxidation of SiC. For each of the SiC - ash systems (SA, SFeA and SCeA) the weight changes, after carbon burnout, were in excess of those observed for SiC alone, indicating an accelerated oxidation mechanism induced by contact with the Ca-Zn-P-S-O ash, similar to observations of Ogbuji and Opila[9] in relation to SiC oxidation in contact with alkali-metals. The contamination of the silica scale formed on SiC allows greater oxygen transport rates giving rise to more rapid oxidation. Exothermic events in addition to carbon burnout were observed in all three systems. In addition the carbon burnout exotherm was shifted to a lower temperature for the SCeA system indicating improved catalytic activity.

Radial shrinkage analyses of SiC - ash pellets heat treated at 900, 1100 and 1250°C indicated that negligible sintering had taken place. XRD analyses of the pellets following heat treatment

(Table I) are compared with those for heat treated sprayed filters. SiC pellets oxidised to give a phase assemblage of α-SiC and cristobalite (and quartz). The sprayed filters also oxidised. For the baseline ash system (SA), calcium sulphate was formed in pellets at 900°C while at higher temperatures, glass phase was present. For the filters, calcium sulphate and tricalcium phosphate accompanied α-SiC and cristobalite in the phase assemblage at 900 and 1100°C whilst for a heat treatment temperature of 1250°C, calcium sulphate was no longer present and evidence for quartz formation and glass was observed. When iron was present in the system (SFeA and SFeF), similar phase assemblages were observed as for the baseline ash excepting the additional presence of Fe_2O_3 and zinc ferrite which formed at 1100 and 1250°C. When ceria was present in the ash (SCeA and SCeF) the same phases were observed as for the baseline ash excepting that CeO_2 was present at all temperatures and $CePO_4$ was present after heat treatment at 1250°C.

From all the observations, it is clear that accelerated oxidation of SiC takes place at higher temperatures but no catastrophic corrosion e.g. gross sintering / melting effects are observed.

Table I. Phase assemblages of heat treated SiC, SiC – ash pellets (SA, SFeA, SCeA) and treated SiC filter surfaces (SF, SFeF, SCeF) (phases listed in order of decreasing intensities)

	900°C	1100°C	1250°C
SiC	α-SiC, SiO_2 (crist.)	α-SiC, SiO_2 (crist.)	α-SiC, SiO_2 (crist.)
SA	α-SiC, SiO_2 (crist.+ quartz), $CaSO_4$	α-SiC, SiO_2 (crist.+ quartz), Glass phase	α-SiC, SiO_2 (crist.+ quartz), Glass phase
SF	α-SiC, SiO_2 (crist.), $CaSO_4$, $Ca_3(PO_4)_2$	α-SiC, SiO_2 (crist.), $CaSO_4$, $Ca_3(PO_4)_2$	α-SiC, SiO_2 (crist.), $Ca_3(PO_4)_2$, SiO_2 (quartz)
SFeA	α-SiC, SiO_2 (crist.+ quartz), Fe_2O_3, $CaSO_4$	α-SiC, SiO_2 (crist.+ quartz), $ZnFe_2O_4$, Glass phase	α-SiC, SiO_2 (crist.+ quartz), Glass phase
SFeF	α-SiC, $Ca_3(PO_4)_2$, $CaSO_4$, Fe_2O_3	α-SiC, SiO_2 (crist.), $Ca_3(PO_4)_2$, Fe_2O_3, $ZnFe_2O_4$	α-SiC, SiO_2 (crist.), $Ca_3(PO_4)_2$, Fe_2O_3, $ZnFe_2O_4$
SCeA	α-SiC, SiO_2 (crist.+ quartz), CeO_2, $CaSO_4$	α-SiC, SiO_2 (crist.+ quartz), CeO_2, $CePO_4$,	α-SiC, SiO_2 (crist.+ quartz), $CePO_4$, Glass ph.
SCeF	α-SiC, SiO_2 (crist.), $Ca_3(PO_4)_2$, CeO_2	α-SiC, SiO_2 (crist.), $Ca_3(PO_4)_2$, CeO_2	α-SiC, SiO_2 (crist.), $Ca_3(PO_4)_2$, CeO_2, $CePO_4$

Cordierite - Ash Interactions

No thermal events were observed in the DTA/TGA data for cordierite (figure 1a). For the CA pellets (figure 1b), a significant endothermic event occurs at about 1185°C. No radial shrinkage effects were observed below 1250°C where complete melting was noted. XRD analyses of the heat treated cordierite - ash pellets are given in Table II. For the CA pellets after heat treatment at 900 and 1000°C, calcium sulphate and tricalcium phosphate, arising from the ash constituents, and $ZnAl_2O_4$ are present. Above 1100°C, calcium sulphate is no longer present. As expected from the sintering behaviour, the pellet heat treated at 1250°C contains a glassy phase as well as cordierite and $ZnAl_2O_4$. Based upon these observations, it can be concluded that whilst liquefaction of portions of the ash constituents occur, liquid levels and compositions are insufficient to cause gross degradation of the cordierite, except at 1250°C. At temperatures of

418

900°C and above, ZnAl₂O₄ forms from degradation of cordierite. For the heat treated sprayed filters, ZnAl₂O₄ is the only phase present at 1100°C and above suggesting glass formation.

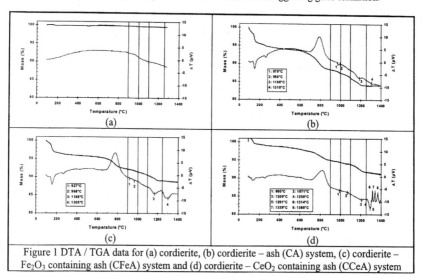

Figure 1 DTA / TGA data for (a) cordierite, (b) cordierite – ash (CA) system, (c) cordierite – Fe₂O₃ containing ash (CFeA) system and (d) cordierite – CeO₂ containing ash (CCeA) system

Table II.   Phase assemblages of heat treated cordierite and cordierite – ash pellets (phases listed in order of decreasing intensities)

	900°C	1000°C	1100°C	1250°C
Cordierite	Cordierite	Cordierite	Cordierite	Cordierite
CA	Cordierite, ZnAl₂O₄, CaSO₄, Ca₃(PO₄)₂	Cordierite, ZnAl₂O₄, Ca₃(PO₄)₂, CaSO₄	Cordierite, ZnAl₂O₄, Ca₃(PO₄)₂	Cordierite, ZnAl₂O₄, Glass Phase
CFeA	Cordierite, Fe₂O₃, ZnAl₂O₄, CaSO₄, Ca₃(PO₄)₂	Cordierite, Fe₂O₃, ZnAl₂O₄, Ca₃(PO₄)₂	Cordierite, Fe₂O₃, ZnAl₂O₄, Ca₃(PO₄)₂, FeAl₂O₄	Cordierite, Fe₂O₃, FeAl₂O₄, Glass Phase
CCeA	Cordierite, CeO₂, ZnAl₂O₄, CaSO₄, Ca₃(PO₄)₂	Cordierite, CeO₂, ZnAl₂O₄, Ca₃(PO₄)₂, CaSO₄	Cordierite, CeO₂, ZnAl₂O₄, Ca₃(PO₄)₂, CaSO₄, CePO₄	Cordierite, CePO₄, CeO₂, ZnAl₂O₄, Ca₃(PO₄)₂, Glass

When iron is present in the system, two major endothermic peaks were observed at 1168 and 1305°C (Figure 1c). Whilst no radial shrinkage occurs at temperatures lower than 1250°C, melting of the pressed pellets was observed at this temperature, indicating that Fe₂O₃ did not suppress gross liquefaction. Fe₂O₃ augments the phase assemblages observed for the CA system after heat treatment at 900 and 1000°C. At 1100°C, FeAl₂O₄ is observed, and at 1250°C Fe₂O₃, FeAl₂O₄, and a glassy phase are present. Thus, apart from the formation of the iron aluminate spinel, indicative of further cordierite degradation, Fe_2O_3 appears to have little effect on the degradation of cordierite at temperatures of 1100°C and below. For the heat treated sprayed filter surfaces, only cordierite and Fe_2O_3 were detected up to 1100°C and at 1250°C, only cordierite.

The addition of ceria to the ash appears to increase the temperature at which any major endothermic event occurs to 1290°C (Figure 1d), although an endothermic event of less significance appears at 1209°C. No measurable radial shrinkage effects were observed for heat treated pellets of any of the cordierite - ash pellets at 900, 1000 or 1100°C. However, after heat treatment at 1250°C, the CA and CFcA pellets were totally melted and the CCeA pellets only partially melted. After heat treatment at 1100 and 1250°C (Table II), cerium phosphate is present in addition to cordierite, ceria, tricalcium phosphate and $ZnAl_2O_4$. The CeO_2 containing ash seems to suppress gross liquefaction at 1250°C as a result of the formation of $CePO_4$ which removes phosphorous, one of the major liquefying constituents in the systems examined. For the heat treated sprayed filter surfaces, only cordierite and ceria were detected at all temperatures.

CONCLUSIONS

Silicon carbide suffers little oxidative or chemical degradation in the presence of any of the ashes tested at 900°C and for the CeO_2 containing ash this temperature limit may extend to 1100°C. Accelerated oxidation is however induced by Ca, Zn, and P. The only significant degradation mechanism which might be experienced by SiC DPF systems at temperatures in excess of 970°C is pore blockage due to the ash agglomerating and sintering in gas transfer pores, if neither Fe nor Ce is present in the ash. The behaviour of cordierite is similar to SiC in that no catastrophic degradation effects occur at temperatures of 1100°C and below. There is however evidence to show that cordierite undergoes reaction with lubricant constituents (e.g. Zn) which causes some chemical degradation.

REFERENCES

1. J.P.A. Neeft, *Catalytic Oxidation of Soot: Potential for the Reduction of Diesel Particulate Emissions*, PhD Thesis, Technische Universiteit Delft, The Netherlands, (1995).

2. P. Stobbe, H.G. Petersen, J.W. Høj, and S.C. Sorensen, "SiC as a Substrate for Diesel Particulate Filters" SAE Paper No. 932495 (1993).

3. W.A. Cutler and G.A. Merkel, "A New High Temperature Ceramic Material for Diesel Particulate Filter Applications" SAE Paper No. 2000-01-2844 (2000).

4. K. Ohno, K. Shimato, N. Taoka, H. Santee, T. Ninomiya, T. Komori, and O. Salvat, "Characterization of SiC-DPF for Passenger Car," SAE Paper No. 2000-01-0185 (2000).

5. D.M. Young, W.J. Warren, K.P. Gadkaree, and L. Johannesen, "Silicon Carbide for Diesel Particulate Filter Applications: Material Development and Thermal Design" SAE Paper No. 2002-01-0324 (2002).

6. J. Kracklauer, Econalytic Systems Inc., private communication, (1998).

7. E.M. Levin, C.R. Robbins, H.F. McMurdie, "Phase Diagrams for Ceramists", Ed. M.K. Reser, Pub. Am. Ceram. Soc., Columbus, Ohio, USA, 1964

8. D. O'Sullivan, S. Hampshire, M.J. Pomeroy, R.J. Fordham, M.J. Murtagh, "Resistance of Cordierite Diesel Particulate Filters to Corrosion by Combustion Products of Diesel Fuel Containing Ferrocene" Paper 00POW032, Proc. ISATA 2000, Dublin, Ireland, Vol. Powertrain Technology - Developments in Petrol/Diesel Engines and Transmissions, 545, (2000).

9. L. Ogbuji, E. J. Opila, "A Comparison of the Oxidation-Kinetics of SiC and Si_3N_4", *J. Electrochem. Soc.*, 142 (3): 925-930 1995.

OVERVIEW OF CERAMIC MATERIALS FOR DIESEL PARTICULATE FILTER APPLICATIONS

Willard A. Cutler
Corning GmbH
Abraham-Lincoln-Strasse 30
D-65189 Wiesbaden
Germany

ABSTRACT

Diesel particulate filters effectively trap the soot (elemental carbon and associated hydrocarbons) and ash present in diesel engine exhaust. This soot buildup causes an increase in the pressure drop across the filter, resulting in poorer engine performance. This soot buildup must be periodically eliminated by oxidizing the soot inside the filter. This soot oxidation/burning is exothermic and can create various temperature profiles inside the filter depending on the exhaust gas chemistry and flow rate, soot amount and composition, and filter material and configuration. Porous ceramic materials are ideal candidates for diesel particulate filters due to their refractory nature and the ability to control their porous microstructure. Several ceramic candidates including aluminosilicates, phosphates, carbides and nitrides will be discussed in relation to this application. In addition, the performance requirements for diesel filters as well as the trade-offs associated with diesel particulate filter performance, including filtration efficiency, clean and soot-loaded pressure drop, regenerability, and chemical-thermal-mechanical survivability, will also be discussed.

INTRODUCTION

The ceramic-based, wall flow, diesel particulate filter (DPF) was first invented in the early 1980s [1] as a novel offshoot of the porous, ceramic, honeycomb-based catalytic converter [2-3]. DPFs have been in use for heavy duty engine applications on a limited scale for over two decades. Diesel particulates can be composed of a variety of components including: solid carbon, adsorbed hydrocarbons, sulfur containing particulates and other inorganic oil-based or fuel based residuals. These particulates can cover a wide range from 5 nm to 50,000 nm (0.005 to 50 μm) depending on the fuel, engine and engine operating conditions. Diesel particulates have long been seen as a public nuisance, but more recent studies have shown that these particulates are a public health concern [4-6]. Due to the potential health implications of diesel particulates, entities such as the California Air Resource Board (CARB), United States Environmental Protection Agency (US-EPA), European Union (EU) and several major cities have enacted legislation that will require significant reductions in particulate matter in the near future (2005-2012). It is likely that nearly all diesel powered engines will have to be equipped with particulate reduction devices by 2015.

Another significant driver for the introduction of DPFs prior to these legislated limits is the "green" or "environmentally friendly" marketing value such filters provide. However, regardless of the motivation for introducing filters, it has been proven that ceramic-based DPFs are capable of reaching significant particulate reduction levels and are the leading candidates for mobile emission sources.

421

Diesel vehicles are the mainstay of the heavy duty trucking industry throughout the world due to their power, torque and fuel efficiency. In Europe, diesel engines have made significant in-roads into the passenger car market, with a >40% market share and rising [7]. This has been possible due to: (a) the technological advances that have made diesel drivability on par with or in some cases, better than gasoline engines, as well as (b) the excellent fuel economy and the lower price of diesel fuel relative to gasoline in Europe.

Ceramic-based DPFs will play an important role in the future of improving our air quality. A fair treatment of the topic would require a book rather than a short paper; however, this paper attempts to provide a short overview of how a filter works and the challenges associated with developing and implementing ceramic filters for mobile applications.

PARTICULATE FILTRATION AND REGENERATION

Ceramic-based diesel particulate filters trap particulates or "soot" on the outside of the porous ceramic walls (cake filtration), within the porous wall (deep bed filtration) or some combination of the two. As the soot layer builds up within the filter, the soot layer itself does much of the filtration. This soot layer, combined with the "sticky" nature of diesel soot, allows micrometer size porosity to trap nanometer size soot. However, the filtration efficiency of these filters is strongly linked to the pore size, pore size distribution, percent porosity and "pore connectivity," a measure of how well one pore is connected to the next pore.

As in the case of traditional liquid or air filters, as the filter collects more and more particulate matter the pressure drop across the filter increases. Unlike traditional filters, which must be removed or exchanged, the diesel particulate filter is cleaned or "regenerated" in-situ by heating the filter in an oxidizing atmosphere (O_2 or NO_2), either with or without the aid of a catalyst, to oxidize the soot. Oxidation using NO_2 takes place in the ~220 to 400°C temperature window while O_2 oxidation takes place above 550°C [8]. As exotherms created during regeneration can be severe, the filter must be refractory and capable to withstand substantial thermal gradients.

There are two main types of systems of DPF systems, active and passive. Passive systems rely on the conditions inherently available in the exhaust system to regenerate the filter. This often involves creating NO_2 to burn the soot at low temperatures [9] or promoting normal soot oxidation at lower temperatures [10]. Many retrofitted bus and truck fleets successfully use passive regeneration, as drive cycles and thus exhaust conditions are generally predictable. In addition these systems are fairly easy to retrofit. Passive systems, however, experience regeneration problems when the drive cycle moves out of the desired range (cool exhaust temperatures for example) such that the regeneration no longer occurs as desired.

Active systems are designed to promote the "controlled" regeneration of the DPF even when the conditions are not naturally available during the drive cycle to passively regenerate the filter. In addition to the DPF itself, the system components often include a flow-through oxidation catalyst substrate(s), catalysts (on the filter or in the fuel), sensors (pressure drop, temperature, oxygen, etc.), heat initiation source(s) (burner, heater, post injection, etc.) and the use of very sophisticated engine controls to calculate the filter performance before, during and after a regeneration event, and to control exhaust gas chemistry, multiple fuel injections, etc. These systems are often designed with multiple redundancies aimed at protecting the filter or the catalyst on the filter from seeing an excessively high temperature or stress. Such redundancies aimed at triggering a controlled regeneration to protect the filter from soot overloading may include: (a) time limit between regenerations to deal with applications where the engine does a lot of idling, (b) mileage limit between regenerations, (c) pressure drop limit between regenerations, and/or (d) engine soot production calculations based on fuel consumption and driving patterns between regenerations.

In spite of these system redundancies, system designers must still understand the DPF behavior in an "upset" or "uncontrolled" regeneration condition. The DPF must still survive the uncontrolled regeneration and continue to function as a filter. This places constraints on the thermal properties of the DPF material and on its design.

Uncontrolled regeneration is most often described as an unplanned event created by the following combination of worst case conditions: First, the filter is heavily loaded with soot. Second, the filter gets very hot either (a) as a result of the engine running at high loads and speeds (exhaust flow rate is high and the oxygen content low), or (b) as a result of the initiation of a controlled regeneration. Third, the filter begins to regenerate due to the high temperature of the exhaust gas. Fourth, due to a change in drive cycle (fast driving followed by a sudden idle) the engine load/speed is dramatically reduced, the exhaust gas flow declines precipitately, but the oxygen content in the exhaust increases dramatically. The combination of these four conditions leads to burning the soot in an "uncontrolled" fashion, such that the hot, regenerating filter suddenly sees a low exhaust flow, which no longer effectively cools the filter, and a high oxygen concentration, creating a significant exothermic reaction during the soot burnout. This condition is most often studied when developing ceramic materials and designs for DPFs.

FILTER AND FILTER MATERIAL ATTRIBUTES

There are five main important attributes in a filter: (a) good filtration efficiency, (b) low pressure drop (clean, soot loaded, soot & ash loaded), (c) good regenerability, (d) good survivability and (e) low cost. Unfortunately, all attributes are not easily achievable in a single filter. While most of the ceramic literature is dedicated to removing porosity to create dense materials, ceramic filters require the intentional, controlled addition of porosity. The starting inorganic materials (composition, particle size & distribution), the final material/structure formation method (solid state sintering, liquid phase sintering, and transient phases), the use of fugitive pore formers and the sintering temperature/time all determine the final microstructure of the porosity.

Filtration efficiency is most commonly measured in terms of the mass of soot removed, although efficiency can also be measured in terms of number of particles removed [11-12]. Current legislation is based on the former, while future legislation will likely be based on the latter. This change is likely to occur, as newer engines are thought to create fewer large particles, which contribute most to mass, but more small particles, which contribute most to health effects [13]. Fortunately ceramic-based DPFs can show high effiencies on both a mass and a number basis [14]. Filtration efficiency is determined primarily by the pore size [15] and the percent porosity [16]. Large mean pore (~25 µm) bodies with broad pore size distributions at moderate porosities (≤50%) show poor filtration efficiency initially (<70%), but as the soot layer builds, the filtration efficiency improves (>90%). However, after each regeneration, the filtration efficiency again declines and must build over time. It is uncommon for large pore bodies with high porosities (>60%) to reach high to reach high efficiencies. Pore connectivity is also thought to influence filtration efficiency, but at present the understanding has not matured. Currently standard pore sizes for commercial filters range from $8 \leq X \leq 20$ μm. There is some belief that large pore size bodies can be uniformly plated down to small pore size bodies using a catalytic washcoat to improve the efficiency and provide space to store catalyst, but this is very difficult to do in practice.

The pressure drop of a filter is determined by a host of material and design factors. The material factors that affect pressure drop, as before, include pore size, pore size distribution, percent porosity and "pore connectivity," pore to pore communication. From a strictly pressure drop point of view, large pore size, high porosity and good communication between pores lead to low pressure drop. However, these are often at odds with filtration efficiency, heat capacity, strength and other constraints. The permeability of the clean filter wall correlates well with modeled and experimental values. Thus composition screening for clean filter pressure drop can be accurately accomplished on small samples. Unfortunately, the DPF is only truly clean before being fitted to the vehicle. Once soot is introduced into the filter, pressure drop characteristics can change dramatically, depending on the location of the soot relative to the pores. In structures with poorly connected pores or low porosity the rise in pressure drop can be dramatic as porosity for exhaust gas passage becomes blocked or soot density increases [16-17]. More porosity or better connected pores lead to lower soot loaded pressure drop, but additional porosity compromises strength and heat capacity. The accumulation of ash from lube oils or fuel borne catalysts also leads to an increase in pressure drop. As ash permeability is orders of magnitude higher than soot permeability [18] this increase only occurs with time and must be dealt with by creating larger volume filters to store ash or with specially designed filters [19-20].

Filter regenerability relates to the ease or difficulty with which the filter is regenerated. Regenerability is influenced by the gas flow within the filter, the filter heat capacity, filter surface area, the use of regeneration aids (catalysts) and system controls. Fast filter regeneration initiation is aided by high surface area (good mass transfer), low heat capacity, uniform flow and regeneration aids. Non-uniform regeneration can lead to higher than desired soot loadings on subsequent filtration cycles. Regenerability is an active focus of filter study and product regenerability coupled with system controls is often viewed as proprietary.

Filter survivability encompasses thermal shock, fatigue, corrosion, etc. The DPF environment is widely variable and in addition to soot includes water vapor (byproduct of combustion), hydrocarbons, lean & rich atmospheres, gaseous acids (nitrogen and sulfur based), etc. Passenger car filters must typically survive for >250,000 km, while heavy duty truck filters must typically survive for >500,000 to 1,600,000 km. Current commercially available passenger car systems regenerate every 400 to 1000 km, depending on conditions. A filter should be able to survive hundreds (passenger car) or possibly thousands (heavy duty) of regenerations. Typically, crack initiation as a result of these regenerations is not acceptable, as the crack opening displacement will likely increase with time and lead to a compromise in the filtration efficiency. The material parameters that may be important to the survivability of the filter include its melting point, coefficient of thermal expansion, elastic modulus, strength, heat capacity, thermal conductivity, chemical reactivity and pore characteristics (size, percent porosity, pore connectivity).

The melting point of a candidate DPF material must be sufficiently high, such that when considered in combination with the bulk heat capacity of the filter, thermal conductivity and the targeted soot loading, the filter will not melt locally or globally. Even localized melting of the filter will, at the very least, compromise its filtration efficiency, and widespread melting raises the concern for blockage of the exhaust system. In addition, the melting point of the filter can be locally affected by the presence of impurities deposited on the walls of the filter during use. These deposits, which include ash and exhaust manifold debris can act as a flux by reacting with the filter to form a eutectic that has a lower melting point than that of the originally uncontaminated filter [21]. Such localized fluxing could eventually lead to degradation of strength, thermal shock resistance, filtration efficiency, or pressure drop. While melting point is an important parameter, a more important measure for non-oxide materials is the material's oxidation resistance. Oxidation and crystal growth, which changes filter properties and can interfere with filter flow, often occur at significantly lower temperatures than the melting point.

The coefficient of thermal expansion (CTE) of a body is a measure of how much it expands (or contracts) on heating, and is defined as $\Delta L / (L \Delta T)$, where L is the original length of the body; ΔL is the length change of the body, resulting from the thermal gradient, ΔT. A large positive CTE means that the body experiences a large length/volume expansion when heated. A material's CTE has serious implications, as filters are not uniformly heated, particularly during regeneration. This means that one part of the body is expanding more than another part, thus straining the material between the hotter and cooler regions, and creating thermal stresses. For a given thermal gradient, ΔT, the strain, ε is found simply from Equation 1:

$$\varepsilon = (CTE)\,(\Delta T) \qquad (1)$$

Thus, a high material CTE in combination with a large thermal gradient will create a high internal strain. It is desirable for filters to have as low a CTE as possible to minimize thermal strains and stresses. Some ceramic materials have anisotropic CTEs, which must also be considered. High CTE materials can be successfully employed as filters by using additional, costly engineering methods such as either (a) creating compliant, low density structures, or (b) segmenting the filter into smaller regions to limit strain build-up must be employed. While filters can be segmented in the radial direction, they can't be segmented in the direction of flow. Thus high CTE materials in long lengths can also pose concerns.

The elastic modulus, a measure of the degree of stiffness of a body, is also important in determining the filter's resistance to failure under an imposed thermal gradient. The amount of stress that is generated within the body due to differences in thermal expansion of different regions of a filter is related to the corresponding amount of deformation, or strain, by the elastic modulus through Equation 2,

$$\sigma = (E)\,(\varepsilon) \qquad (2)$$

where σ is the stress and E the elastic modulus. The stiffer the body (large E), the higher the stress for a given strain. To limit the stress within the body, the elastic modulus should be as low as possible.

The strength of a DPF is important both for canning and subsequent use in the vehicle exhaust stream. The type of intracrystalline and intercrystalline bonding within the material, and the porosity, pore size distribution, and flaw population control the strength of the ceramic. From a fracture mechanics point of view, pores are stress concentration sites, thus the pore size distribution must be carefully controlled. The strength of the filter itself is further dictated by its dimensions, cross sectional symmetry, cell density, channel geometry, wall thickness, etc. The minimum strength needed to withstand the mechanical stresses imposed during canning is generally fairly low (<1 MPa)[22] if performed correctly, but higher values can be required depending on the bulk density of the filter and the gap bulk density of the mat material required to produce the appropriate holding force in the can. The highest stresses generated within the filter are those imposed by the thermal loading accompanying sudden changes in temperature. If the entire filter could be heated and cooled uniformly then no thermal stresses would develop. Unfortunately, this is not possible. A material will crack when the stress to which it is exposed exceeds the strength of the material. The higher the CTE, the stronger the material must be to withstand thermal gradients and avoid cracking. The addition of a catalyst to the structure can modify its strength, CTE and elastic modulus, so care must be taken to evaluate the composite system (filter and coating) for performance.

In simplistic terms a thermal shock parameter (TSP) can be formulated as a relative measure of the ability of a material to be able to withstand steep temperature gradients (ΔTs) [23]. The TSP is a calculated parameter based on the strength, elastic modulus and the CTE according to Equation 3.

$$TSP = (\sigma)/[(E)(CTE)] \qquad (3)$$

High TSP values mean that the body can withstand large temperature gradients (ΔTs) without cracking. Low TSP values mean a body can withstand only small ΔTs without cracking. In some instances the thermal conductivity has been included in the TSP in the Biot modulus, but its effect is seen to be secondary [24]. Filters are subjected to wide ΔT range both within the driving cycle itself and within the filter during a regeneration cycle. For example, on a cold day, the filter may begin at -30°C before starting the vehicle and can see >900°C during a regeneration. The temperature gradients within a regeneration cycle can exceed 500°C/cm if not managed properly, thus the filter product should have a high resistance to thermal shock either due to intrinsic material properties or created through engineering design.

The bulk volumetric heat capacity of the DPF is a product of the specific heat capacity of the ceramic comprising the filter (material property) and the bulk density of the filter (the amount of material present in grams per liter of filter – a function of the wall porosity and cell structure of the DPF). The specific heat of most ceramic materials rises continually on heating. The heat capacity determines the amount of thermal energy the filter can absorb. The higher the heat capacity, the more heat that is required to raise the temperature of the filter. Thus, the filter's heat capacity determines how much heat is required to raise the temperature of the filter to the temperature at which regeneration initiates, the temperature at which the catalyst is damaged or its melting point. For DPFs which contain catalysts, the desire is that the filter can be heated rapidly such that the catalyst becomes active to fulfill its CO or hydrocarbon oxidation function. For normal controlled regeneration cycles, rapid heating of the filter is desired, thereby enabling quick regeneration and, thus, minimizing the fuel penalty to start regeneration. This is best accomplished with a low heat capacity filter. However, for uncontrolled regenerations, a high heat capacity filter may be desired so that the large exotherm produced does not heat the filter beyond its catalyst de-activation temperature (<900°C), melting or oxidation point (see Table 1) or beyond the point where chemical interactions or ash sintering (<1050°C) can occur [19]. Thus, the requirements of the application place opposing demands on the filter heat capacity. A DPF with a low heat capacity would enable a more rapid regeneration cycle with minimum fuel penalty, and may be best suited for a regeneration strategy based on more frequent regenerations (lower soot loadings). Conversely, a DPF with a high heat capacity will reach lower temperatures during regenerations (controlled or uncontrolled) providing a larger safety margin, but is more difficult to heat. The initial choice for active regenerations was high heat capacity filters, but as regeneration controls have become more sophisticated and as regeneration understanding has broadened, lower heat capacity filters are often being applied. The thermal conductivity of a filter is a measure of how quickly it can conduct heat from hotter to cooler regions of the body. The thermal conductivity of a material generally declines rapidly with increasing temperature. Thermal conductivity has much less of an influence at high exhaust flow rates, where the heat is primarily carried by the gas, than at low exhaust flow rates [25]. Some degree of thermal conductivity is needed to successfully employ high CTE materials, as thermal conductivity helps create more uniform strain fields. While there is little conductivity influence at high flow rates, the ideal material for low flow rates would have a low conductivity to heat the filter to initiate regeneration and a high conductivity during a severe regeneration to carry the heat away.

MATERIALS CONSIDERED

There are a variety of potential oxide and non-oxide ceramic material candidates for DPFs. Many of these could be technically successful DPF candidates, but each has its pros and cons. The design of the filter must accentuate the pros and compensate for the cons. Table 1 shows some of the material candidates and some relevant material properties. Some of these materials including: alumina [27], cordierite [15-17, 28], mullite [29], aluminum titanate [30], NZP [31], silicon carbide [32-34] and silicon nitride [35] have been proposed in various forms. Additional novel materials continue to be developed for this application [36-37].

Alumina's formability, refractory nature, chemical resistance and high heat capacity are attractive, but its very high CTE makes it a difficult choice to implement in a low pressure drop, thermal shock resistant form. Alumina filters are typically made by sintering alumina powders, but more novel fiber approaches have also been employed [27]. Cordierite's formability, very low CTE, experience in exhaust systems and low cost are attractive, but its low material heat capacity has required modifications in the filter form to increase bulk heat capacity or alternatively requires the use of adequate system controls. Cordierite filters are typically made using clay, talc and alumina sources to create cordierite through a variety of complex reactions. Like alumina, mullite's formability, refractory nature, chemical resistance are attractive, but its high CTE makes it a difficult choice to implement in low pressure drop, thermal shock resistant forms using conventional mullite starting powders. Novel processing routes [29, 38-39] using mullite precursors and a gas-solid phase reaction in a silicon tetra-flouride or similar atmosphere have produced more needle-like or "straw pile" microstructures which are more thermal shock resistant, although these structures have high porosities and low heat capacities. Aluminum titanate is refractory, chemical resistant, has high heat capacity and a very low CTE, but was initially not considered a viable candidate due to its thermodynamic instability. Recent work has overcome this limitation, making such composites viable candidates. Aluminum titanates are typically made using

427

alumina, titania, silica and stabilizers, which go through a variety of complex reactions to create the stabilized structure. NZPs are refractory and have very low CTEs, but also have very low heat capacities (lower than cordierite). These materials are usually made from a zirconia, zirconium phosphate, silica, and alkaline earth sources, which create the resultant NZP structure through a series of complex reactions. Silicon carbide's refractory nature, chemical resistance and high heat capacity are attractive, but its high CTE means that thermal shock resistant shapes can only be created using either (a) assemblies (gluing smaller sections together to build larger filters) or (b) highly compliant composite structures. The most common SiC products are made using; (a) silicon carbide powders and high temperature, inert gas sintering, (b) using silicon carbide/silicon carbide precursors mix and high temperature inert gas sintering [40], or (c) by using silicon carbide/silicon metal mix and high temperature inert gas sintering [33]. Silicon nitride should display similar properties to its carbide cousin (refractory nature, chemical resistance and high heat capacity), but be more thermal shock resistant due to its lower CTE. However, as silicon nitride starting powders are not an economically viable route to producing structures, nitriding silicon powders must be employed. The resultant porous microstructure does not generally hold the theoretical advantages over SiC that has been observed in practice in dense bodies.

Once the material and microstructural features have been selected, the functional filter form that the material is produced in is also important. Such filter forms may include extruded honeycombs with porous walls, porous foams or fiber wrapped filters. Some materials can be produced in a wide variety of forms, while others may only be viable in a single form due to property or processing constraints. For example, while alumina can be produced in all three forms, due to the high CTE of alumina, it really only has a chance to survive severe thermal shock as in the fiber wrapped form. Porous honeycombs are the most prevalent filter form due to their high surface area to volume ratio, low pressure drop, ease of microstructural control and manufacturing experience (low cost).

CONCLUSIONS

Ceramics in general are "perfect" candidate for DPFs applications, due to the refractory nature, tailorable porosity characteristics and formability. Unfortunately, there is not a single "perfect" ceramic material for all filter applications. Thus the choice of a filter material and filter form, must be matched with the regeneration strategy, application, and economically viability (cost vs. benefit). Each ceramic material has pros and cons which must be carefully balanced; however, ceramic filters have been and will continue to be successfully implement to improve the quality of the air we breathe. The large number or resources dedicated to the development and improvement of these novel ceramic materials and the products based on these materials is likely to provide continued improvements in filter performance over the coming years.

Table 1: Filter Material Candidates

Porous Material	T_{melt} (°C)	T_{max} (°C) Est. use in air	CTE_c $(x10^{-7}/°C)$	Intrinsic Density (g/cm^3)	Specific Heat @ 500°C[d] $(J/g°C)$	Thermal Conductivity @ 500°C[e] (W/m K)
Alumina (Al_2O_3)	2050	1900	88	3.97	0.88[f]	~8[g]
Cordierite ($Mg_2Al_4Si_5O_{18}$)	1460	1350	6	2.51	1.11	~1
Mullite ($Al_6Si_2O_{13}$)	1810	1600	53	2.50	1.15[26]	~2[26]
Aluminum Titanate (Al_2TiO_5)	1600	1500	10	3.40	1.06	~1
NZP ($XZr_4P_6O_{24}$, X =Sr, Ba, etc)	1900	1800	5	3.44	0.75	~1
α-Silicon Carbide (SiC)	2400[a]	1350	45	3.24	1.12	~20
β-bonded α-Silicon Carbide (SiC)	2400[a]	1350	45	3.24	1.11	~12
Si-bonded α-Silicon Carbide (Si-SiC)	1400[b]	1350	43	3.19[26]	1.12	~10
Silicon Nitride (Si_3N_4)	1900[a]	1350	30	3.0	1.15	~5

[a] sublimes, decomposes or second phase formation predominates

[b] silicon metal melts

[c] 25 to 1000°C

[d] measured at 500°C using a ~50% porosity ceramic wall (except where stated)

[e] computed from heat capacity and thermal diffusivity measured at 500°C

[f] 25°C value

[g] estimated from dense alumina values

REFERENCES
[1] R.J. Outland, "Ceramic Filters for Diesel Exhaust Particulates," US 4,276,071 (1981)
[2] R.D. Bagley, "Method of Forming an Extrusion Die," US3803951A (1974)
[3] I.M. Lachman and R.M. Lewis, "Anisotropic Cordierite Monolith," US3885977A (1975)
[4] A. Peters, H.E. Wichman, T. Tuch, J. Heinrich and J. Heyder, "Respiratory Effects are Associated with the Number of Ultrafine Particles," Am. J. of Respiratory Critical Care Medicine 155 1376-1383 (1997)
[5] N. Kunzli, R. Kaiser, S. Medina, M. Studnicka, G. Oberfeld and F. Horak, "Air Pollution Attributable Cases – Technical Report on Epidemiology," Prepared for the Third WHO Ministerial Conference of Environment and Health, London Institute for Social and Preventive Medicine, University, Basel Switzerland (1999)
[6] R.H. Harrison, "Source Apportionment of Airborne Particulate Matter in the United Kingdom," DETR London, ISBN 0-7058-1771-7 and www.environmental.detr.gov.uk/airq/ (1999)
[7] M. Love, S. Whelan and B.M. Cooper, "Diesel Passenger Car and Light Commercial Vehicle Markets in Western Europe," Ricardo Marketing Report to AECC (2003)
[8] C.F. Goersmann and A. Walker, "Catalytic Coatings for DPF Regeneration," FAD Conference, Dresden (2003)
[9] B.J. Cooper, H.J. Jung, and J.E. Thoss, US 4,902,487 (1990)

[10] K. Voss, et. al., "Engelhard's DPX Catalyzed Soot Filter Technology for Particulate Emissions Reduction from Heavy Duty Diesel Engines with Passive Regeneration," SAE TopTec, Gothenburg (2000)

[11] ACEA Programme on Emissions of Fine Particles from Passenger Car [2] Report (2002)

[12] Mayer et.al., "VERT Particulate Trap Verification," SAE 2002-01-0435

[13] A. Leipertz, "Combustion and Emissions Formation in Diesel Engines," SAE TopTec, Gothenburg (2003)

[14] S. Pischinger, "Exhaust Aftertreatment for Diesel Engines," Volkswagen Diesel Powertrains (2001)

[15] M. Murtagh et. al., "Development of a Diesel Particulate Filter Composition and its Effect on Thermal Durability and Filtration Performance," SAE 940235(1994)

[16] G.A. Merkel, T. Tao and W.A. Cutler, "New Cordierite Diesel Particulate Filters for Catalyzed and Non-Catalyzed Applications," Sixth International Congress on Catalysis and Automotive Pollution Control, (2003)

[17] T. Tao, W.A. Cutler, K. Voss and Q. Wei, "New Catalyzed Cordierite Diesel Particulate Filters for Heavy Duty Engine Applications," SAE 2003-01-3166 (2003)

[18] D.L. Hickman, Corning Incorporated, personal communication (2003)

[19] D. Beall et. al., US6696132 (2004)

[20] I. Gege, K. Ohno, S. Hong, H. Sato, "Diesel Particulate Filter: Filter Material, Innovation and Design," FAD Conference, Dresden (2003)

[21] G.A. Merkel, W.A. Cutler and C.J. Warren, "Thermal Durability of Wall-Flow Ceramic Diesel Particulate Filters," SAE 2001-01-0190 (2001)

[22] C. Washington, Corning Incorporated, personal communication (2004)

[23] W.D. Kingery, H.K. Bowen and D.R. Uhlmann, *Introduction to Ceramics* John Wiley & Sons, 823 (1976)

[24] K.T. Faber, M.D. Huang and A.G. Evans, "Quantitative Studies of Thermal Shock in Ceramics Based on a Novel Test Technique," J. Am. Ceram. Soc., 64 [5] P. 296 (1981)

[25] G. Bhatia and N. Gunasekaran, "Heat-up of Diesel Particulate Filters: 2D Continuum Modeling and Experimental Results," SAE 2003-01-0837 (2003)

[26] C. Warren, Corning Incorporated, personal communication (2004)

[27] 3M Diesel Filter Cartridges For Particulate Control, Technical Bulletin 9185-EMC J (1999)

[28] G.A. Merkel, D.M. Beall, D.L. Hickman and M.J. Vernacotola, "Effects of Microstructure and Cell Geometry on Performance of Cordierite Diesel Particulate Filters," SAE 2001-01-0193 (2001)

[29] S.A. Wallin, A.R. Prunier and J.R. Moyer, "Mullite Bodies and Methods of Forming Mullite Bodies, US 6,306,335 (2001)

[30] S. Ogunwumi and P. Tepesch, "Aluminum-Titanate Based Materials for Diesel Particulate Filters," 28th International Cocoa Beach Conference on Advanced Ceramics & Composites (2004)

[31] W.A. Cutler and G.A. Merkel, "A New High Temperature Ceramic Material for Diesel Particulate Filter Applications," SAE 2000-01-2844 (2000)

[32] K. Ohno et. al., "Characterization of SiC-DPF for Passenger Car," SAE 2000-01-0185 (2000)

[33] S. Miwa, et. al., "Diesel Particulate Filters Made of Newly Developed SiC," SAE 2001-01-0192 (2001)

[34] D.M. Young, C.J. Warren, K.P. Gadkaree and L. Johannesen, "Silicon Carbide for Diesel Particulate Applications: Material Development and Thermal Design," SAE 2002-01-0324 (2002)

[35] N. Miyakawa, H. Maeno and H. Takahashi, "Characteristics and Evaluation of Porous Silicon Nitride DPF," SAE 2003-01-0386 (2003)

[36] L. Pinckney, G.H. Beall and K. Chyung, "High Alumina, Low-Expansion Ceramics in the Li2O-Al2O3-SiO2 System, 28th International Cocoa Beach Conference on Advanced Ceramics & Composites (2004)

[37] E.M. Yorkgitis and L. Hornback, "Characterization of a Composite Diesel Particulate Filter and Substrate," 28th International Cocoa Beach Conference on Advanced Ceramics & Composites (2004)

[38] I. Talmy and D.A. Haught, US 4,910,172 (1990)

[39] C.L. Buhrmaster, "Extrusion, Joining, Microstructure and Properties of Mullite Whisker Honeycombs," Corning Incorporated internal report (1991)

[40] K.P. Gadkaree, et. al., US 6,555,031 (2003)

SOOT MASS LIMIT ANALYSIS OF SiC DPF

Hiroki Sato, Kazutake Ogyu, Kazunori Yamayose, Atsushi Kudo, and Kazushige Ohno
IBIDEN.Co.,LTD
1-1,Kitagata,Ibigawa-cho,Ibigun,Gifu Pref.,501-0695,Japan

ABSTRACT

In re-crystallized SiC segment filter made of coarse and fine grains, the amount of soot mass limit did not change when mean size of coarse SiC grains became larger. We found that the soot mass limit of the re-crystallized SiC filter segments which consisted of the coarse and fine SiC powder was not lowered even if the grain size of the coarse power was increased. The soot mass limit was kept constant at 12 g/L for the grains of 10 to 25 μm in size. During the investigation, we noticed that under the uncontrolled regeneration of the soot the maximum temperature, as well as the maximum temperature gradient, tended to be suppressed with increased grain size. The suppression of the maximum temperature can be considered desirable for a substrate as the catalyzed DPF (Diesel Particulate Filter).

To identify the factors affecting the soot mass limit, the physical properties, such as the bending strength, the thermal conductivity, the Young's modulus and others of the filter for the grain sizes of 10 to 30 μm were measured. The calculated results of the thermal shock fracture resistance parameter R', including the thermal conductivity, by using the measured values showed that R' remained constant for the grain sizes within the range given above, and that the thermal shock resistance of the filter during the regeneration accompanied with a gradual temperature change can be well represented by the thermal shock fracture resistance parameter R'.

1.INTRODUCTION

In recent years, the controls on the exhaust gas emissions of the diesel cars have become extremely rigorous. To clear this hurdle, the development of the engine management and the after-treatment technologies have become very important in recent years. In 2000, the first series production of DPF for passenger cars has been mounted on the vehicle by Peugeot-Citroen, which is one of the largest car manufacturers in Europe. It was spotlighted worldwide [1].The after-treatment technologies by DPF can be roughly classified into two categories; one is the type in which soot is accumulated and then burned and the other is the type in which the soot is burned continuously.

When the soot is accumulated and burned, the properties required for the filter include the high

heat resistance, the high heat conductivity, the high thermal shock resistance, and the hig
strength. The ceramic material which has long been known as a ceramic material we
possessing these properties is SiC. However, since SiC is the brittle material with
comparatively high coefficient of linear expansion, it is vulnerable to having a low thermal shoc
resistance, causing cracking problems when the thermal stress limit is exceeded. To solve thi
problem, the filter is formed of smaller segments each of which is reduced in size, and forme
into a structural body (filter) elastically supported by special the cement reinforced by cerami
fibers to increase the thermal shock resistance . The DPF actually mounted on PSA cars i
formed of segments of the re-crystallized SiC. The durability of the filter is demonstrated by th
use of it on 500,000 cars in actual use.

There are some properties including the strength required for the filter substrate to assure it
performance. Among them, the most important one is the thermal shock resistance. The DP
forms the substrate with comparatively large particles (10 μm or larger) to form a porous body
Accordingly, its strength is lowered due to an increase in the grain siz
of the SiC. However, the behavior of it against the thermal shock is not made clear.

Under the circumstances, we attempted to identify the relationship between the grain size c
the SiC and the thermal shock fracture resistance parameter R'[2]. Actually, the evaluation c
the segments is performed by the soot mass limit test to clarify the relationship between the grai
size of the SiC and the limit of cracking. In addition, the physical properties of the filter, such a
the bending strength, the Young's modulus, thermal conductivity and others, are measured whe
the grain size of the SiC is changed, then the thermal shock fracture resistance parameter R' i
calculated from the measured results. Based on the data thus obtained, an examination is mad
hereunder on the overall test results of the soot mass limit of the segmented filter.

2.EXPERIMENTAL

2.1 Sample preparation

1) the samples for soot mass limit test

Samples were manufactured using a pilot production line. The fine and coarse SiC powder
were mixed with organic binder, then extruded to honeycomb, plugged, binder removed, an
sintered. Table 1 shows the specifications of the evaluated samples. The four grain sizes of Si
used were; 10, 15, 20, and 25 μm.

Table.1 Samples for soot mass limit test

SiC grain size (μm)	Porosity (%)	Cell structure (mil/cpsi)	Segment length (mm)	Segment weight (g)
10	40	10/300	150	115
15	40	10/300	150	113
20	40	10/300	150	113
25	40	10/300	150	115

2)The samples for calculating the the thermal shock fracture resistance parameter R'

Samples were manufactured using pilot production line. The fine and course SiC powders were mixed with organic binder, then extruded, binder removed, and sintered.The sintered bodies obtained by burning were formed in specified sample shapes to evaluate for physical properties. The specifications of the evaluated samples are shown in Table 2.   The three grain sizes of SiC used were; 10, 20, and 30 μm.

Table 2 Test samples

SiC grain size (μm)	Porosity (%)	Pore diameter (μm)
10	40	10
20	40	10
30	40	10

2.2 Experimental procedure

1)Soot mass limit

The filter in which a certain amount of soot was accumulated was heated to 720°C in an inert atmosphere and then, the cool air at room temperature was fed into the filter. Here, burning of the soot accumulated in the filter was started. After the soot was completely burned, the filter was visually checked to see if any crack had generated. We refer this maximum amount of soot as the soot mass limit[1].

2) Physical property for calculating the thermal shock fracture resistance parameter R'

The grain sizes of the SiC were taken as a parameter, and each physical property value providing the thermal shock fracture resistance parameter R' was evaluated.   The evaluated items and evaluation methods are shown in Table 3.

Table 3 Evaluated items and evaluation methods

Evaluated items	Bending strength	Young's modulus	thermal conductivity	coefficient of linear expansion
Evaluated method	3 point bending	Resonace	Laser flash	Dilatmeter

3. RESULT AND DISCUSSION

3.1 Soot mass limit

1) Soot mass limit result

Figure 1 shows the soot mass limit values of the respective samples.   Figures 2 and 3 show the maximum temperatures and maximum temperature gradients of the segments vs. the amount of soot.

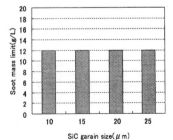

Fig.1 Soot mass limit v.s. SiC grain size

433

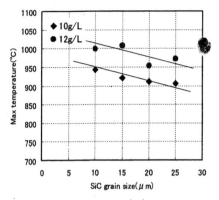

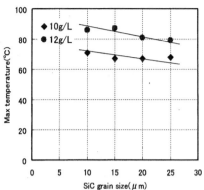

Fig.2 Max temperature during
the regeneration v.s. SiC grain size

Fig.3 Max temperature gradient during the
regeneration v.s. SiC grain size

As shown in Fig.1, the soot mass limit values remained unchanged even if the grain sizes c
the SiC were increased.

As shown in Fig.2 and 3, the maximum temperature and maximum temperature gradien
decrease slightly as the grain size of the SiC was increased. The slope does not change despite c
difference of soot amount. It is considered that the effect of smaller surface area by enlargeme
of grain size prevent the fast transferring of exothermal heat of soot. The delay of heat transfe
could suppress the severe thermal stress in the substrate. We estimate, however, that this eve
occurs only on a substance with a high thermal conductivity such as SiC. As mentioned late
the evaluation results for the physical properties actually show that the thermal conductivity wa
lowered slightly as the grain size was increased.

Anyway lowering the maximal temperature could be a grate advantage as a substrate fc
catalyzed diesel particulate filter. It provides more safety margin to prevent catalyst sintering

2) Summary

In the actual soot combustion test, it could be seen that, up to a grain size 25 μm, both th
maximum temperature and the maximum temperature gradient during the regeneration tended t
decrease.

It was clear that the regenerative limit value did not decrease irrespective of increases in the Si
grain size up to 25 μm.

3.2 The thermal shock fracture resistance parameter R'

1) Physical property

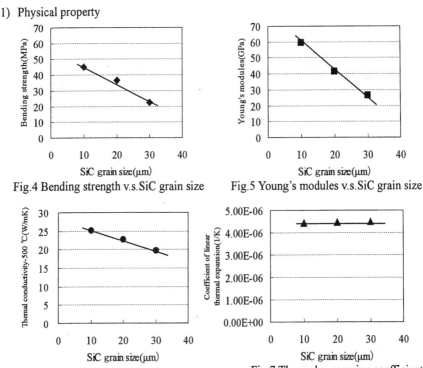

Fig.4 Bending strength v.s.SiC grain size

Fig.5 Young's modules v.s.SiC grain size

Fig.6 Thermal conductivity v.s.SiC grain size

Fig.7 Thermal expression coefficient
v.s.SiC grain size

Figures 4 to 7 show the dependencies of the bending strength, Young's modulus, thermal conductivity and coefficient of linear expansion on the grain size of the SiC, respectively. The coefficient of linear expansion hardly changed despite the changes in the grain size of the SiC. While, the thermal conductivity showed a tendency to decrease slightly as the grain size is increased. Also both the strength and Young's modulus had a tendency to decrease as the grain size of the SiC increased. In this evaluation for physical properties, it was verified that the lowering of the Young's modulus is slightly larger than the lowering of the strength.

As generally understood, the thermal shock fracture resistance parameter R' is represented in different ways depending on the heat transfer conditions. The heat transfer conditions can be roughly classified into two categories; one when an abrupt change in the temperature gradient occurs, as in the case of cooling in water, and the other when a gradual temperature gradient occurs. As can be seen from the temperature chart (Fig.8) of the DPF during the general regeneration, the combustion of the soot has a rather gradual temperature gradient. Accordingly, the thermal shock resistance of the DPF may be expressed by the following the thermal shock

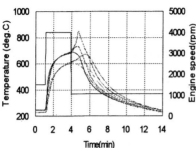

Fig.8 Temperature chart of the DPF during the general regeneration

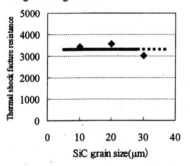

Fig.9 Relationship between the thermal shock fracture resistance parameter R' and the grain size of the SiC

fracture resistance parameter R' including th thermal conductivity (expression (1))[2].

$$R' = \frac{\sigma(1-v)k}{\alpha E} \qquad (1)$$

σ: bending strength α: Coefficient of linear
E: young modulus   thermal expansion
κ: thermal conductivity   V: poisson's ratio

The relationship between the thermal shock fractu resistance parameter R' and the grain size of the Si is shown in Fig.9 according to the basic physic property data obtained before. As shown in th figure, the grain size exerts almost no effect on th thermal shock fracture resistance parameter R', ar the thermal shock fracture resistance parameter I remains approximately constant within th measurement range of the properties.

2) Summary

As for the physical properties, the evaluation resul confirm that though the strength is remarkab lowered as the grain size is increased, the therm shock fracture resistance parameter R' is hard lowered. The reason is considered to be that, when the grain size is increased (10 to 30 μm), th decrease of the Young's modulus is larger than that of the strength irrespective of the slight decreasing of the thermal conductivity.

4.CONCLUSION

1. The regeneration limit of the soot accumulated in the filter, which consists of the coarse ar the fine SiC powder is nearly constant irrespective of the average grain size.

2. This tendency could be clearly explained by the thermal shock fracture resistance parameter I obtained from the physical properties of the filter.

REFERENCE

[1]K.Ohono, K.Shimato, et al.,"Characterization of SiC-DPF for Passenger can " SAE pap No.2000-01-0185

[2]R.W.Dravidge,"Mechanical behavior of ceramics"(translated to Japanese)p.134 Kyori publibishing company(1982)

A MECHANISTIC MODEL FOR PARTICLE DEPOSITION IN DIESEL PARTICLUATE FILTERS USING THE LATTICE-BOLTZMANN TECHNIQUE

Mark Stewart, David Rector, George Muntean*, and Gary Maupin
Pacific Northwest National Laboratory
PO Box 999, Richland, WA 99352
(* author to whom correspondence should be addressed)

ABSTRACT

Cordierite diesel particulate filters offer one of the most promising aftertreatment technologies to meet the quickly approaching Environmental Protection Agency 2007 heavy-duty emissions regulations. A critical yet poorly understood component of particulate filter modeling is the representation of soot deposition. The structure and distribution of soot deposits upon and within the ceramic substrate directly influence many of the macroscopic phenomena of interest, including filtration efficiency, back pressure, and filter regeneration. Intrinsic soot cake properties such as packing density and permeability coefficients remain inadequately characterized. The work reported in this paper involves subgrid modeling techniques that may prove useful in resolving these inadequacies. The technique uses a lattice-Boltzmann modeling approach. This approach resolves length scales that are orders of magnitude below those typical of a standard computational fluid dynamics representation of an aftertreatment device. Individual soot particles are introduced and tracked as they move through the flow field and are deposited on the filter substrate or previously deposited particles. Electron micrographs of actual soot deposits are compared with the model predictions.

Descriptions of the modeling technique and the development of the computational domain are provided. Preliminary results and comparisons with experimental observations are presented.

INTRODUCTION

The U.S. Environmental Protection Agency's 2007 heavy-duty diesel engine exhaust emissions standards mandate a 90% reduction in particulate matter from current levels. The majority opinion in the diesel industry is that these standards, by implication, mandate exhaust particulate filtration for diesel engines [1]. Unfortunately, there is no commercially viable filtration technology that can be applied universally to all on-highway, heavy-duty diesel engines without regard to duty cycle [2]. Several key technical hurdles must be overcome for diesel particulate filters to become practical, including filter plugging, thermal failures, size, cost, filtration performance, deactivation, and durability. Cordierite diesel particulate filters (DPFs) offer one of the most promising alternatives and have been the focus of much study (Figure 1).

Resolving these technical issues requires a detailed, system-level understanding of the interaction between the diesel engine's exhaust characteristics and the filtration device. Attention must be given to all possible application duty cycles and ambient conditions. Modeling can greatly expedite this process.

System models can be used to optimize engine control strategies, help determine the physical dimensions of the catalyst system, and simulate numerous real-world applications that may be prohibitively expensive to field test. Detailed component models can be used to investigate novel filtration techniques or optimize existing substrate geometries for soot oxidation or capacity.

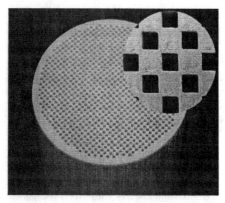

Figure 1. Cordierite diesel particulate filter

When exhaust first passes through a clean diesel particulate filter, a linear increase in pressure is observed for a time, and then an "elbow" in the pressure drop history is reached. After that the pressure drop continues to increase but at a much slower rate. Another transition is seen in the particulate removal efficiency, which is relatively low at first and improves dramatically as soot accumulates. These observations suggest that two separate modes of filtration are manifest. The first may involve accumulation of soot in the pores of the filter substrate (deep bed filtration), while the second may be characterized by the accumulation of a blanket of soot on top of the substrate (cake filtration).

The exact nature of these two filtration modes and the transition between them is not well understood, nor are the details of the physical mechanisms involved. For example, do particles bridge over the tops of the pores or fill some of the pores until they reach the surface? If so, how much of the pore volume is filled in this way? Are there features within the three-dimensional structure of the substrate that facilitate this process? What is the three-dimensional structure of the soot deposits, and how does this structure affect macroscopic phenomena such as pressure drop, filtration efficiency, and oxidation rate? How does the structure of the substrate affect the final structure of the soot cake layer? The answers to these questions could lead to better predictions of key performance parameters for existing filters and to the tailoring of more effective filter media. Micrographs can offer some information about soot deposit structure, but it is difficult to ensure that soot deposits are neither damaged nor dislodged as the samples are prepared. Mechanistic models could offer corroborating evidence for proposed deposition processes and a deeper understanding of the underlying phenomena involved.

Konstandopoulos and associates have modeled the growth of soot deposits on ceramic filters as individual soot particles are captured [3,4]. The soot particles in this approach moved in a ballistic trajectory and were either scattered or captured upon impact. This approach does not allow for deep bed filtration. Also, the diversion of flow around deposits and the resulting effect on subsequent deposition are not considered. The work presented herein derives the particle motion from the fluid flow field and continuously adjusts the flow field as soot is deposited.

This paper reports on a technique that may help develop a deeper understanding of soot deposition via detailed subgrid modeling. This technique uses a lattice-Boltzmann modeling approach that resolves length scales which are orders of magnitude below those typical of a standard computational fluid dynamics (CFD) representation of an aftertreatment device. {*Note:*

theoretically, CFD techniques can be used down to these smaller length scales. However, for reasons discussed in the next section, the geometry of the pore structure makes this approach impractical.} This improved resolution allows for the characterization of functionality not previously employed in the CFD analysis.

THE LATTICE-BOLTZMAN TECHNIQUE

The basic conservation equations (continuity, momentum, and energy) that govern fluid dynamics are typically formulated as a set of coupled, nonlinear, partial differential equations. These equations have no general closed-form analytical solution except in the case of very trivial initial and boundary conditions (e.g., the Blasius solution for a flat plate). These equations are often referred to collectively as the Navier-Stokes equations—given the additional assumptions that Newton's viscosity law and Fourier's conduction law are used [5]. Because these equations are intractable, analytical solutions rarely exist for the practical problems in engineering. Therefore, computational techniques must be employed.

Conventional CFD methods involve the discretization of the Navier-Stokes equations using a finite volume or finite element formulation. In other words, the partial differentials are approximated locally over an "element" as algebraic differences. Given the size and shape of the element and the form of the algebraic approximation, varying degrees of accuracy can be obtained. However, as a general rule, the solution is governed largely by the quality of the computational mesh describing the physical domain. Highly intricate and complex geometries such as the pores in a cordierite soot filter substrate are very difficult to grid accurately and efficiently.

In addition, the solution procedure typically involves solving for pressure throughout the entire domain at each time step to satisfy continuity. This pressure solution does not scale linearly on a parallel computer, limiting the maximum number of nodes that can be used to resolve a geometry. In addition, boundary conditions are difficult to implement for complex geometries such as porous media. For all these reasons, traditional CFD approaches have typically not resolved the fine-scale structures in diesel particulate filters.

Lattice-Boltzmann computation is a powerful alternative to classical CFD. The classical techniques use finite elements or finite volume formulations. The lattice-Boltzmann method takes a different approach to modeling fluid systems than conventional solvers do [6,7,8]. The spatial domain is discretized into a finite number of lattice sites, each of which has values for density, pressure, flow, etc. The single-particle distribution function, which describes the probability of a particle traveling along a particular direction and speed, is discretized to form a finite set of displacement vectors connecting each lattice site to adjacent sites (Figure 2). Similar distributions are also created for energy and each chemical species.

 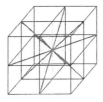

Figure 2. Lattice displacement vectors for two- and three-dimensional systems

The lattice-Boltzmann equation describes the evolution of the discretized particle distribution function, $f_i(x,t)$, along direction i as a function of time. The new time distribution function is given by the following equation:

$$f_i(x + \Delta te_i, t + \Delta t) - f_i(x,t) = -\frac{1}{\tau}\left[f_i(x,t) - f_i^{eq}(x,t)\right]$$

(1)

where τ is a linear relaxation parameter, and f^{eq} is the local equilibrium distribution. The local equilibrium is expressed in the form of a quadratic expansion of the Maxwellian distribution:

$$f_0^{eq} = w_0\left[A_0 + D_0 u^2\right]$$

(2)

$$f_i^{eq} = w_i\left[A + B(e_i \cdot u) + C(e_i \cdot u)^2 + Du^2\right]$$

(3)

where the coefficients are defined as

$$A_0 = \rho - 6A; \quad D_0 = -\frac{\rho}{c^2}$$

(4)

$$A = \frac{p}{3c^2}; \quad B = \frac{\rho}{3c^2}; \quad C = \frac{\rho}{2c^4}; \quad D = -\frac{\rho}{6c^2}$$

(5)

and c is the reference lattice speed, $c = \Delta x/\Delta t$. The weight coefficients, w_i, for a three-dimensional system are

$$w_i = \begin{cases} \frac{1}{3} & i = 0 \\ \frac{1}{18} & i = 1-6 \\ \frac{1}{36} & i = 7-18 \end{cases}$$

(6)

The linear relaxation parameter, τ, is related to the kinematic viscosity by the expression

$$v = (\tau - 0.5)\Delta tc/3$$

(7)

Solution methods consist of a streaming stage and a collision stage for each time step. During the streaming stage, each lattice site transmits distribution information to adjacent sites. The incoming information is then relaxed toward an equilibrium distribution that is determined by local conditions. New time pressure and velocity values are determined using these expressions:

$$p = \frac{c^2}{3}\sum_i f_i$$

(8)

$$\rho u = \sum_i f_i e_i$$

(9)

440

This procedure is repeated for the specified number of time steps until a steady-state solution is obtained. The result is a second-order solution to the Navier-Stokes and continuity equations:

$$\rho \frac{\partial u}{\partial t} + \rho u \cdot \nabla u = -\nabla p + \rho \nu \nabla^2 u \qquad (10)$$

$$\nabla \cdot (\rho u) = 0 \qquad (11)$$

A similar procedure is followed for solving the energy and chemical species transport equations.

The advantages of this procedure center on the fact that no global solution methods are required for determining field variables. Each lattice site only requires information from adjacent sites. This makes the lattice-Boltzmann method inherently parallelizable and makes it possible to simulate complex geometries in great detail. Pore-scale geometries consisting of millions of lattice nodes are typical. Lattice-Boltzmann is well suited to compute flows with elaborate boundaries, such as flows through the pores of a ceramic diesel particulate wall filter. Lattice-Boltzmann can also treat phenomena on boundaries, such as chemical reactions at the surfaces of catalysts or soot particles, in addition to chemical reactions in bulk.

CONSTRUCTION OF THE PHYSICAL DOMAIN

To obtain meaningful prototypic modeling results using lattice-Boltzmann, the physical domain used in simulations must be represented as accurately as possible. Therefore, a wall-flow cordierite diesel particulate filter was selected, representing the most common filter media currently used by the aftertreatment industry. Corning EX-80 soot filters measuring 14.3 cm in diameter and 15.2 cm long were procured and analyzed, and the geometry of the filter wall pore-structure was digitized for use in defining the physical domain and boundary conditions.

The development of the lattice-Boltzmann model was a non-trivial matter due to the intricate nature of the physical domain. The flow path through the porous substrate of the EX-80 filter is extremely tortuous. The substrate has a high fraction of void space and is three dimensional in its structure, with a wide pore size distribution. Because the accuracy of the modeling results depends on the quality and accuracy of the physical domain representation, detailed three-dimensional geometries of the porous soot filter material were required.

A procedure was developed for defining and digitizing the soot filter geometry from a series of micrographs. An EX-80 filter was sectioned to provide several 1 cm samples of the substrate wall, which were potted in an acrylic epoxy microscope mount. The samples were then ground to the filter wall surface and polished flat using a series of mechanical grinding and polishing techniques. The final polishing step used a water-based colloidal suspension of 1-micron diamond polishing compound.

Optical microscopy images were taken and digitized for each sample, as shown in Figure 3a. After the first round of images was taken, all of the samples were then repolished with a 3-micron diamond and water colloidal medium using a Buehler Automet® automatic polishing station. The subsequent repolishing step was performed for 60 seconds to remove approximately 6 microns of material from the filter surface. Additional images of each sample were then taken. This procedure of polishing and imaging was repeated 25 times and produced a series of 25 surface images of the porous filter wall structure spaced 6 microns apart.

Upon completion of the microscopy work, the 25 images were processed with image analysis software and converted into discrete digital representations of the solid filter material and void space, as shown in Figure 3b. Each image, or layer, was then combined with a custom interpolation code to fill the domain between each layer. Ultimately, this process created a continuous three-dimensional representation of the actual filter geometry seen in Figure 4.

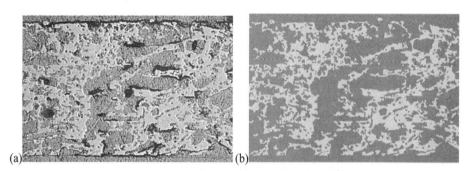

Figure 3. Micrograph of EX-80 substrate cross section (a) and corresponding digitized image (b)

Figure 4. Digitally reconstructed physical domain

Sample sets representing the three-dimensional structure were taken from multiple locations in the EX-80 filter for comparison of the homogeneity of the wall structure and pore-size distribution on a global scale and to provide multiple domain sets for flow field analysis. The digitized three-dimensional physical domain was then meshed and used in the lattice-Boltzmann model to define the filter geometry and boundary conditions.

The lattice-Boltzmann model of the filter material used in these simulations consisted of 240x240x52 lattice nodes with approximately 3-micron resolution. Each lattice node has an associated velocity vector, pressure, temperature, and molar concentration for each chemical species. This level of resolution allows the detailed three-dimensional representation of flow and field variables within the pore structure of the filter material.

MODELING OF SOOT PARTICLES

Soot particles are made up of elementary spherules with diameters on the order of 20 nm. Observed soot particles range from this size up to about 1 micron, with the bulk in the size range of hundreds of nm [9]. Although diesel soot is made up predominantly of submicron particles, soot particles were represented in this initial study by 1-micron spheres. Because the soot particle size is smaller than the lattice spacing in the lattice-Boltzmann model, it was not possible to resolve the irregular features of real soot particles or their rotational motion. Future work (at smaller length scales) will endeavor to examine the effects of soot particle size distribution and the irregular shapes of soot agglomerates on deposition and soot cake structure.

Motion is imparted to the particles by the exhaust flowing around them. Drag force can be calculated by Stoke's Law. Alternatively, the particle agglomerates can be considered porous structures, and drag force can be calculated by assigning a Darcy resistance to the flow of fluid through them. The local velocity at each particle was estimated from the eight surrounding lattice sites using tri-linear interpolation. Each particle was then displaced in the direction of this average velocity according to the particle time step. The model includes the ability to super-impose Brownian motion on the average motion of the particles due to the local velocity field. Brownian motion may be an important factor in moving small particles between fluid stream lines so they can interact with obstructions.

Particles may be captured by interacting with a substrate region or with previously deposited particles. Interactions between particles in motion are not considered in the present model. Particle capture probability in the model is currently set to 100%. In other words, whenever a particle touches the filter substrate or another particle, it becomes instantly fixed in place. Future refinements of the model will allow capture probability to be set as a parameter. A practical result of high capture probability is to artificially increase the initial efficiency of the clean soot filter and to shorten the time required in the model to accumulate soot deposits. While these effects are beneficial from the standpoint of computational requirements, a more accurate portrayal of capture probability would be necessary for qualitative predictions of initial filter performance.

Once particles are deposited, they impose a Darcy resistance to the flow of exhaust through and between them. The permeability of the soot deposit at a lattice point is scaled by the number of deposited particles present. Deposited particles in the model form rigid, unbreakable bonds. No effort is made at this point to model the deformation of the soot deposits due to drag from the flowing fluid.

The model presented here does not consider regeneration of the filter. The shapes, structures, and positions of soot deposits are an important consideration in the modeling of soot oxidation during regeneration. Previous modeling work has dealt with heat transfer and chemical transport phenomena involved with the oxidation of soot [10]. These models typically make simplifying assumptions about the three-dimensional geometry of soot deposits. In future work, this type of transport modeling may be coupled with the particle deposition model presented in this paper.

PRELIMINARY RESULTS

The model predicts some deep bed filtration in the early stages of filter operation as small deposits form throughout the substrate and relatively little soot is deposited near the surface. Soot deposition at key bottlenecks near the surface soon leads to the formation of large deposits at the mouths of pores. From this point on, most deposition occurs near the top surface of the substrate (see Figure 5). Most of the deep pores remain relatively clear. As has been observed in experiments, capture efficiency improves dramatically as soot accumulates.

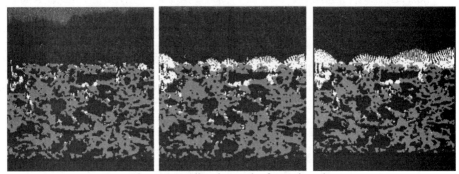

Figure 5. Predicted growth of soot deposits

Several features of the soot deposits predicted by the mechanistic model can be seen in observations of actual soot deposits. Mounds of soot forming around areas of low resistance to flow through the substrate are apparent both in the model predictions and in micrographs of partially loaded soot filters. Figures 6 and 7 include micrographs of actual soot deposits on a porous cordierite substrate. The soot deposits are present primarily as clumps near the top surface of the substrate and have a smooth or slightly feathered texture in contrast to the flakey appearance of the underlying substrate.

The effects of shadowing can be seen in the fingered structure of the predicted soot deposits. Similar "columnar" morphologies were predicted by Konstandopoulos [3] and are consistent

Figure 6. Micrographs showing mounds of soot and a feathery surface texture

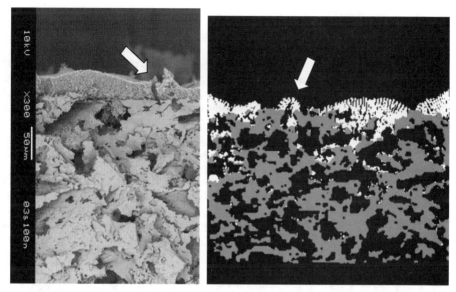

Figure 7. Pores through soot deposits connecting with pores in the substrate

with pictures of particulate deposits in the literature. A capture probability of less than 100% would likely decrease this effect by allowing rebounding particles to fill in gaps between the fingers of deposited soot.

The structure of the soot cake on the surface is markedly affected by the porous structure of the substrate beneath. The predicted soot deposit structures included pores, voids, and openings that connect with underlying pores in the substrate. These features can also be observed in micrographs of actual soot deposits (see Figures 6 and 7). Such openings in the soot cake may be an important factor in macroscopic phenomena such as pressure drop and capture efficiency.

CONCLUSIONS

The lattice-Boltzmann method is versatile, easy to use, and has proven effective for modeling flow through extruded ceramic structures with tortuous pores. With the application of some simplifying assumptions, a particle deposition model based on a lattice-Boltzmann flow solution was able to qualitatively reproduce several structural features of soot deposited on ceramic filters. Incremental improvements to this approach can be implemented easily, and the method shows promise for quantitative representation of several macroscopic phenomena such as particulate capture efficiency and pressure drop.

In previously published work, techniques were discussed for continuum modeling of the soot layer in diesel particulate filters [10]. The model presented in this paper could complement this approach by validating and improving the continuum representation of the complex soot deposit structure. These continuum representations are an important step in developing accurate device-scale models that can then be applied to the design and analysis of improved aftertreatment devices.

FUTURE DIRECTION

- Parametric studies will be performed for several key model parameters. Adjusting the Darcy resistance for deposited particle regions may affect the shapes of soot deposits and the macroscopic pressure drop across the domain. Adjustable capture probabilities of less than 100% will likely lead to some filling in of voids and pores of the deposit structure, also affecting the shapes of deposits and macroscopic pressure drops. These studies may lead to accurate quantitative predictions of important performance parameters such as pressure drop and capture efficiency.
- More detailed models at smaller length scales are planned. These studies will examine the effects of realistic particle size distributions, irregular particle shapes, and continuing agglomeration of moving particles, all of which could affect the shapes of soot deposits and macroscopic phenomena of interest.
- Future models could combine discrete particle deposition with studies of filter regeneration. This could involve simultaneous deposition of new particles while underlying layers in the soot deposits are removed by oxidation. Alternatively, deposition models could be used to impose irregular shapes and permeabilities for continuum representations of soot deposits.

ACKNOWLEGMENTS

We would like to thank Gurpreet Singh and Kevin Stork of the Office of FreedomCAR and Vehicle Technologies at the U.S. Department of Energy (DOE) for funding our research. We would also like to thank Pacific Northwest National Laboratory (PNNL) for providing the computational resources used in this effort. PNNL is operated by Battelle for DOE under Contract DE-AC06-76RL01830.

REFERENCES

[1]Leet, J. A., "2007-2009 USA Emission Solutions for Heavy-Duty Diesel Engines," U.S. DOE Diesel Engine Emissions Reduction (DEER) Conference, August 2002, San Diego, CA.

[2]Stover, T., "DOE/Cummins Heavy Truck Engine Program," U.S. DOE Diesel Engine Emissions Reduction (DEER) Conference, August 2001, Portsmouth, VA.

[3]Konstandopoulos, A. G., "Deposit growth dynamics: particle sticking and scattering phenomena," Powder Technology, 109, pp. 262-277, 2000.

[4] Konstandopoulos, A. G., Skaperdas, E., and Masoudi, M., "Microstructural Properties of Soot Deposits in Diesel Particulate Traps," SAE Technical Paper Series, 2002-01-1015.

[5]Panton, R. L. "Incompressible Flow," Wiley-Interscience.

[6]Chen, S., "Lattice Boltzmann Method for Fluid Flows," Ann. Rev. Fluid Mech. 1998, 30:329-64.

[7]Martys, N. S., "Simulation of multicomponent fluids in complex three-dimensional geometries by the lattice Boltzmann method," Phys. Rev. E53, pp. 743-750, 1996.

[8]Palmer, B. J., and Rector, D. R., "Lattice Boltzmann Algorithm for Simulating Thermal Two-Phase Flow," Phys. Rev. E., 61, 5 pp. 5295-5306, 2000.

[9]Park, K., Feng, C., Kittelson, D. B., and McMurray, P. H., "Relationship Between Particle Mass and Mobility for Diesel Exhaust Particles," Environ. Sci. Technol., 37, pp. 577-583, 2003.

[10]Muntean, G. G., Rector, D., Herling, D., Lessor, D., and Khaleel, M., "Lattice-Boltzmann Diesel Particulate Filter Sub-Grid Modeling - a progress report," SAE Technical Paper Series, 2003-FL-46.

DEVELOPMENT OF CATALYZED DIESEL PARTICULATE FILTER FOR THE CONTROL OF DIESEL ENGINE EMISSIONS

Yinyan Huang, Zhongyuan Dang, Amiram Bar-Ilan
Süd-Chemie Prototech, Inc.
32 Fremont Street
Needham, MA 02494

ABSTRACT

A catalyzed diesel particulate filter (CDPF) was prepared based on Corning cordierite wall flow filter substrate. A high surface area precious metal/base metal catalyst was deposited on the filter substrate. The deposition of catalyst generates negligible pressure drop. Lab testing shows that the CDPF has high activity for the oxidation of CO and hydrocarbon and low activity for the oxidation of SO_2. The catalyst is very active for the oxidation of diesel soot. The CPDF has balance point temperature of $350^{\circ}C$ as tested on 5.5KW Lister Petter LPA2 diesel engine genset with the use of federal diesel fuel. Presence of high surface area coating layer enhances the activity for CO, hydrocarbon and soot oxidation and the thermal stability of the catalyst.

INTRODUCTION

Diesel engines function by burning fuels (hydrocarbons) at high temperatures [1]. In theory, the products of the combustion process are CO_2 and water. But, it is not uncommon that the combustion process is incomplete resulting in the formation of undesirable byproducts such as carbon monoxide, hydrocarbons and soot. Other reactions occurring in internal combustion engines include the oxidation of nitrogen molecules to produce nitrogen oxides and the oxidation of sulfur to form SO_2 and small percentage of SO_3. Further, when the temperature decreases, the SO_3 can react with H_2O to form sulfuric acid. Other inorganic materials are formed as ash. The products of these reactions result in undesirable gaseous, liquid and solid emissions from internal combustion engine: gaseous emissions –carbon monoxide, hydrocarbons, nitrogen oxides, sulfur dioxide; liquid phase emissions – unburned fuel, lubricants, sulfuric acid; and, solid phase emissions – carbon (soot). The combination of liquid phase hydrocarbons, solid phase soot and sulfuric acid results in the formation of small size droplets often called total particulate matter.

Compared with gasoline fueled internal combustion engines, diesel engines emit more particulate matter and pose a greater threat to air quality and to the health of human beings [2]. To reduce these risks, tremendous efforts have been made for the control of diesel particulate emissions. One well-known approach is to use filters to trap exhaust particulate matter [3]. These filters are generally made of porous, solid materials having a plurality of pores extending therethrough and small cross-sectional sides, such that the filter is permeable to the exhaust gas which flows through the filter and yet capable of restraining most of all of the particulate materials. As the mass of collected particulate material increases in the filter, the flow of the exhaust gas through the filter is gradually impeded, resulting in an increased backpressure within the filter and reduced engine efficiency.

Conventionally, when the backpressure reaches a certain level, the filter is either discarded, if it is a replaceable filter, or removed and regenerated by burning the collected particulate materials so the filter can be reused. Regeneration of filters in situ can sometimes be accomplished by periodically enriching the air to fuel mixture. The enrichment produces a higher exhaust gas temperature. The high exhaust temperature burns off the particulate materials contained within the filter.

Thermal regeneration of diesel particulate filter at temperatures above 600°C is not generally desirable because it can lead to uncontrolled ignition of the soot, resulting in temperature overshoot and filter substrate damage. In addition, thermal regeneration consumes large amounts of energy. Rather, regeneration of diesel particulate filters at lower temperature is preferred. Such regeneration can be accomplished with the assistance of catalysts.

Precious metal catalysts are widely used as catalyst to reduce diesel engine emissions. They are highly active for the oxidation of CO, hydrocarbons, soot and SO_2 [4]. Oxidation of CO, hydrocarbons and soot is desired while oxidation of SO_2 into SO_3 is not [5, 6, 7]. Therefore, the desired catalyst should have high activity for CO, hydrocarbons and soot oxidation but low activity towards SO_2 oxidation. Some combinations of precious metal/base metal could suppress the activity of SO_2 oxidation, such those reported in literature. [8, 9]

Pressure drop across a filter substrate means fuel penalty of diesel engine operation. Therefore, it is desirable that a filter substrate has low pressure drop. Deposition of catalyst on a filter substrate usually increases pressure drop that leads to additional fuel penalty. Therefore it is critical that catalyst coating doesn't generate significant pressure drop increase across the substrate.

The present development deals with a catalyzed diesel particulate matter exhaust filter comprising a porous filter substrate coated with high surface area support and catalytic materials comprising both precious metal and base metals.

EXPERIMENTAL

Corning DuraTrap^{TM} CO diesel particulate Filter made of EX80 Cordierite ceramic is used as substrate for catalyst deposition. The substrate has size of 5.66" diameter by 6" length, cell density of 200cpsi and wall thickness of 12mil.

Corning DPF substrate is first washcoated with high surface area support followed by drying and calcinations to secure the washcoat layer. Then, the catalytic materials are deposited on the washcoated substrate. The catalyst contains both precious metals and base metals.

A core of 1.75" diameter by 6" length was taken from the full size catalyst and tested in lab reactor for pressure drop measurement and for gas phase reaction activity. Air was used for pressure drop measurement. Gas phase oxidation of carbon monoxide and propylene under was conducted under the conditions of 700ppm CO, 300ppm C_3H_6, 100ppm SO_2, 4% H_2O and 30,000/h GHSV. CO and hydrocarbon concentrations were measured with gas analyzers while SO_2 concentration was measured with HP 6890 GC-MS.

Balance point temperature of the diesel particulate was measured on 5.5KW Lister-Petter LPA2 diesel genset. The catalyzed DPF of 4.66" diameter by 6" length was taken from the full size core and used for the BPT testing. Federal diesel fuel from pump containing about 350ppm sulfur was used for the measurement.

RESULTS AND DISCUSSIONS

Pressure Drop Across the Catalyzed DPF

Figure 1 shows the pressure drop comparison between blank Corning filter substrate and the catalyzed filter substrate. Pressure drop across the catalyzed filter is almost identical to that across the blank substrate. The results clearly show that the catalyst coating doesn't add significant pressure drop across the substrate.

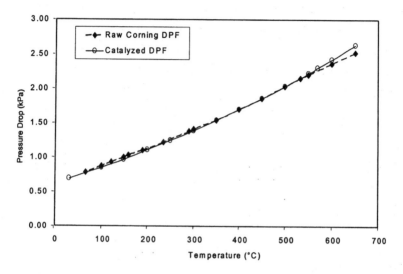

Figure 1. Pressure drop of blank and catalyzed filter (50,000/g GHSV, Air)

Gas Phase Oxidation of CO, Propylene and SO_2

Gas phase oxidation of carbon monoxide, propylene and SO_2 was conducted under the conditions of 700ppm CO, 300ppm C_3H_6, 100ppm SO_2, 4% H_2O and 30,000/h GHSV. As shown in the figure 2, catalyzed filter is highly active for the oxidation of CO and hydrocarbon. Under the testing conditions, complete conversions of CO and propylene are reached at about 250°C on the pristine catalyst.

The catalyst shows low SO_2 oxidation activity. Figure 2 shows that at 400°C and up to 500°C, the conversion of SO_2 is below 10%. Under the identical conditions, a pure Pt catalyst without base metals shows 40% SO_2 conversion at 400°C and 80% SO_2 conversion at 500°C. It is clear that the presence of base metals suppresses the oxidation of SO_2.

The catalyst has high thermal stability. After aging at 650°C for 48hours, the activities of the catalyst for CO and propylene show no deterioration.

Coating of high surface area support is critical for the high activity and high thermal stability of the catalyst. Catalyst prepared by direct deposition of PGM/base metals on filter substrate shows lower activity and thermal stability. Figure 3 shows the activity of the catalyst without washcoating layer for CO and propylene oxidation. The pristine catalyst reaches complete conversion of CO and maximum of 90% conversion of propylene at above 300°C. After aging at 650°C for 48hours, the activity of the catalyst drops significantly. The maximum conversion of CO is about 90% and the maximum conversion of propylene is about 85% at above 400°C.

449

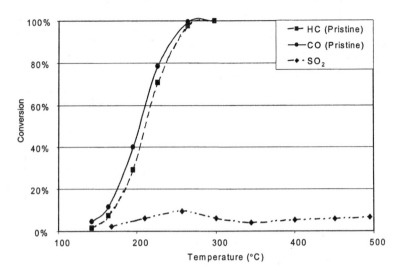

Figure 2. Activity of the catalyst with washcoat for CO, propylene and SO$_2$ oxidation

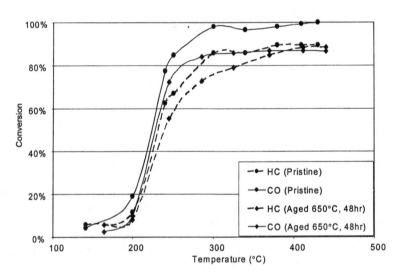

Figure 3. Activity of the catalyst without washcoat for CO and propylene oxidation

Balance Point Temperature Testing

Activity of the catalyst for the oxidation of soot was evaluated via balance point temperature measurement. Measurement was conducted on a 5.5KW Lister-Petter LPA2 diesel genset. The rated speed of the engine is 1800rpm. A 4.66" diameter by 6" length filter substrate was canned and placed in the passage of the engine exhaust. Before measurement, the filter substrate was loaded with soot under the condition of 3KW for 7 hour. The exhaust temperature is below 250°C during soot loading process. After soot loading process, the engine load was increased step-by-step and kept at each loading step for 15 minutes. The exhaust temperature and pressure drop across the filter substrate were recorded. Figure 4 show the temperature-pressure drop curves of the catalyzed filters. The exhaust temperature increases with load monotonously. The pressure drop across the filter increases with temperature initially and starts to drop at certain point. The point where the slope of the temperature-pressure drop curve changes direction from positive to negative is defined as balance point temperature at which the accumulation of particulate reaches equilibrium with its oxidation.

For a non-catalyzed filter substrate, the pressure drop increases with temperature up to the maximum engine load (about 450°C). It indicates that no regeneration takes place under normal engine operation conditions. For the catalyzed filter substrate, filter regeneration is achieved at high engine loading which provides just sufficient heat for the catalyst to react and ignite the soot. The present catalyst with washcoat layer shows a balance point temperature of about 350°C. In comparison, for catalyzed filter without washcoat layer the balance point temperature measured is about 375°C.

After aging at 650°C for 48 hours, the present catalyst has the same balance point temperature of 350°C.

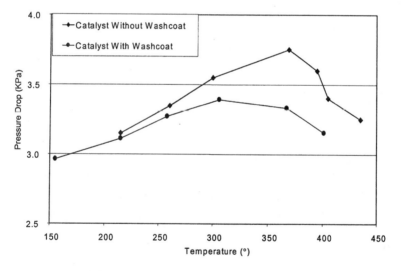

Figure 4. Balance point temperature measurement

Discussions

Precious metals are often used as catalysts for the reduction of diesel engine emissions due to their high activity for the oxidation of CO and hydrocarbon. In addition, precious metals are also active for the oxidation of SO_2 that is not desirable. Therefore, suppression of SO_2 oxidation is needed. A combination of precious metal/base metals shows high CO and HC activity and low SO_2 oxidation.

Combination of precious metal/base metals is also important for high activity for soot oxidation. Under the present conditions, a catalyst of pure precious metal doesn't show good diesel soot oxidation activity, no obvious regeneration activity of loaded filter is taking place under normal engine operation conditions.

Precious metals are usually supported on high surface materials to enhance their dispersion, to increase the number of active sites and to promote the thermal stability of a catalyst. High surface area support is conventionally applied to monolithic substrate via washcoating process. With the present catalyzed particulate filter, washcoating process results in negligible back pressure increase. It contributes to high activity and high thermal stability of the catalyst, and better fuel efficiency.

Diesel soot contains "hard" carbon molecules and has low mobility. It has been realized that good contact between the catalyst and soot "molecule" is critical for the oxidation of soot on catalyst [3]. The present catalyst is highly dispersed on the high surface support and therefore has better contact with "soot" particles. As results, the activity of the catalyst for soot oxidation is higher and the measured balance point temperature is lower than precious metal catalyst not supported on a washcoat layer.

CONCLUSIONS

A catalyzed diesel particulate filter was developed for reducing emissions of diesel engines. The catalyzed filter comprises of Corning's wall flow filter substrate and a precious metal/base metal catalyst supported on high surface area material. The present catalyst adds negligible pressure drop across the substrate. It has high activity for the oxidation of CO and hydrocarbons and low activity, i.e. high selectivity towards SO_2 oxidation. The catalyst shows also high activity towards soot oxidation. Deposition of washcoat layer leads to high activity and high thermal stability of the catalyst.

REFERENCES

[1] J.B. Heywood, Internal Combustion Engine Fundamentals, McGraw-Hill, 1988
[2] J.P.A. Neeft, M. Makkee, J.A. Moulijn, Fuel Processing Technology, 47 (1996) 1
[3] B.A.A.L. van Setten. M. Makkee, J.A. Moulijn, Catal. Rev. Sci. Eng., 43 (2002) 489
[4] B.J. Cooper & S.A. Roth, Platinum Metals Rev., 35 (1991) 178
[5] R.J. Farrauto, K.E. Voss, Appl. Catal. B: Env. 10 (1996) 29
[6] H.J. Stein, Appl. Catal. B: 10 (1996) 69
[7] P. Zelenka, W. Cartellieri, P. Herzog, Appl. Catal. B: Env. 10 (1996) 3
[8] M.Wyatt, W.A. Manning, S.A. Roth, M.J. D'Aniello, Jr., E.S. Andersson and S.C.G. Fregholm, SAE technical paper 930130, "The design of flow through diesel oxidation catalyst"
[9] H.Ueno, T. Furutani, T. Nagami, N. Aono, H. Goshima, K. Kasahara, SAE technical paper 980195, "Development of catalyst for diesel engine"

THE USE OF TRANSPARENT PLZT CERAMICS IN A BIOCHEMICAL THIN FILM INTERFEROMETRIC SENSOR

Thomas Nicolay

Department of RF and Microwave Engineering
Saarland University, Bld 22
66123 Saarbrücken, Germany

ABSTRACT

A key technology in characterizing chemical and biological processes is the determination of local thin films properties on substrate surface layers that are coated with biochemical active materials. With increasing demands for data logging, the need for cost efficient, simple and reproducible biochemical sensors steadily rises. The optical measurement device presented in this paper is based on the principle of thin film interferometry. Modulating the transparent substrate of the sensor allows us to extend the measurement principle to slowly varying or static biochemical processes. In order to produce cost-efficient transparent substrates with high optical qualities, PLZT 9/65/35 ceramics prepared by electrophoretic deposition have been used and characterized.

INTRODUCTION

While most detectors for physical and electrical measurands are available as mass market parts, measurements of chemical and/or biological signals are based on highly complicated measuring systems and accordant complex sensor systems. It is not possible to measure chemical or biochemical signals directly. Transducers have to be used to transform the measurand in an electrical output signal. State-of-the-art measurement devices consist of electrical or optical transducers. Because of their robustness and high selectivity, optical sensors gain more and more significance for biochemical sensors. Several different optical measurement principles are applied, e.g. extrinsic and intrinsic optodes, evanescent-field technologies, interferometry and ellipsometry [1]. With those devices it is possible to measure simultaneously the index of refraction and layer thickness of dynamic and static substrate surfaces. Disadvantages are found in complex adjustment procedures and ambiguous analysis methods as described in [2], e.g. the angle of incidence and of reflection must be strictly adhered to at least 0.01°. To increase the selectivity the angles have to be near the Brewster angle (~70° for transparent glass substrates). Another problem of ellipsometry is the necessary fast rotation of polarization planes which results in very high and difficult to handle data rates.

White light with a very short coherence wavelength was used in [3] for substrate thickness measurements. The device works with great precision but is too complex and too expensive for compact biochemical sensors. A robust and conventional thin layer interferometric sensor for dynamic processes is presented in [4],[5]. The sensor consists of a planar optical wave guide and integrated sensor elements. For static measurements, several sensors have to be installed and coupled which makes the data acquisition and analysis complex and costly. In order to measure changes in temperature of a substrate Donelly and Fang [6],[7] measured the thermal expansion of the device by thin film interferometry.

All state-of-the-art interferometric measurement devices with one light source and a photo diode as detector for the characterization of materials are only applicable in dynamic measurements. If they were used for static measurements, the periodic interference structure leads to ambiguous data analysis.

EXPERIMENTAL

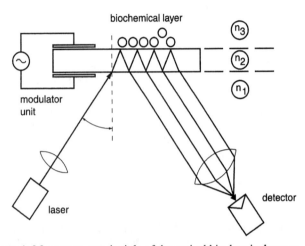

Figure 1: Measurement principle of the optical biochemical sensor

The biochemical thin film interferometer for static and dynamic measurements presented in this paper depends on the principle as described in [8]. Figure 1 illustrates the measurement system for chemical and/or biological processes at the boundary layer of a transparent substrate, which is able to transport optical radiation. Three media with different indexes of refraction (n_1, n_1 and n_3) are connected over two different boundary layers. If the index of refraction on both sides of the boundary differs, an incident ray becomes split into a refracted and a reflected part. The difference in the optical propagation path of reflected and refracted rays causes constructive or destructive interference. Lenses focus the reflected and transmitted parts of the light on a detector. The detector transforms the intensity into an electrical signal. In static measurements, the equivalent electrical signal is constant. If the index of refraction or the thickness of the substrate medium (n_2) varies, the electrical output signal shows a periodic structure in intensity. The shape is a very sensitive function of the ratio in refraction index on both sides of the optical boundary layers. Changes in the function pattern are based upon changes in the refraction index of the biochemical coated surface which are caused by external biochemical processes. Modulating the substrate allows us to apply the interferometric measurement principle to slowly varying or complex processes. Biochemical activities can be monitored while the biochemical layer itself remains in a more or less constant status.

The fundamental part of the sensor is the transparent substrate. With their electrostrictive behavior PLZT 9/65/35 ceramics are very suitable for the realization of the transparent substrate. The advantages of the electrostrictive material over conventional piezoelectrics are

454

the minimal or negligible hysteresis in the strain-field dependence and the voltage dependant dielectric and electrooptic properties [9][10].

In order to decrease production costs, a sub micron PLZT powder was produced by an improved mixed oxide route [11],[12]. By using nanosized zirconia- and titania powders, we have been able to synthesise a low cost PLZT powder suitable for the preparation of transparent specimen by sintering. Figure 2 shows the processing flowchart developed in this study. The abandonment of hot pressing processes denotes further cost reduction in the preparation process. By the use of electrophoretic deposition (EPD) and the membrane method [13], green bodies of high quality could be formed. Compared to the strongly varying optical quality of transparent ceramics by using cold pressed green bodies, the number and size of pores could significantly be reduced. An aftertreatment of the specimen by hot isostatic pressing was not necessary.

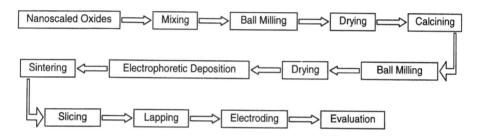

Figure 2: Processing flow chart for transparent PLZT ceramics prepared by EPD

The electrooptical behavior of the PLZT ceramic developed in this study is determined with the setup given in [14]. PLZT ceramics show optically uniaxial properties on a macroscopic scale when an electric field is applied. The optic axis features optical properties which differ from those of the other two orthogonal axis. Light traveling along the optic axis and vibrating in a direction perpendicular to it shows a refraction index n_0 that differs from the refraction index n_e of light traveling in an orthogonal direction and vibrating parallel to it. By entering the electrically energized ceramic, linear polarized light is split into two orthogonal components. Due to the different refraction indices n_e and n_0 the propagation velocity of the two components is different in the ceramic. This results in a phase shift known as optical retardation Γ in the output signal. The UV/VIS spectra were measured with a Perkin Elmer Lambda 35 UV/VIS spectrometer.

RESULTS AND DISCUSSION

Optical Transmission

The most outstanding feature of the PLZT ceramics is their very high optical translucency and transparency. The transmission of samples with different thickness at a wavelength of 1 μm is given in Figure 3. The transmission varies between 67.5 % for a 0.5 mm thick sample and 46 % for a sample with a thickness of 5.0 mm. Measured reflection losses of 31% (maximum transmission 69%) are very close to the theoretical value for reflection from two surfaces calculated with a refraction index of $n=2.5$ and assuming orthogonal incidence of light [15]. The picture of a sample in the insert of Figure 3 shows the high optical quality of the produced PLZT ceramics.

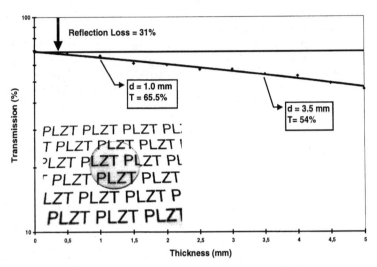

Figure 3: Optical transmission for PLZT 9/65/35

The electrooptic properties of PLZT ceramics are intimately connected to their ferroelectric properties. A variation of the ferroelectric polarization with an applied electric field results in a change of the optical properties.

Effective Birefringence
    As mentioned before the absolute value of the two different refraction indexes is defined as the effective birefringence Δn where n_e and n_o refer to the extraordinary and ordinary refraction indices of the ceramic.

$$\Delta n = n_e - n_0 = \frac{\Gamma}{l} \tag{1}$$

According to equation (1) the effective birefringence Δn was determined from measurements of the optical retardation Γ. The value l refers to the thickness of the PLZT ceramic plate along the optical path. In Figure 4 the effective birefringence is diagrammed as function of the applied electric field. The results show the typical memoryless behavior for PLZT 9/65/35. At room temperature the material is paraelectric. When the electric field is applied the ferroelectric anisotropy increases with the varying field. The resulting unique behavior is a classical quadratic electrooptic Kerr effect. The gap was first poled with a positive voltage up to an electric field of 6 kV/cm and then switched to an electric field of –6 kV/cm with a low frequency of 0.1 Hz. With an electric field of zero, the ceramic is optically isotropic. Since monochromatic laser light is used, an extension of the electric field beyond half wavelength retardation results in a progression of repeating light and dark bands as in an interferometer.

Phase shift

The resulting phase shift is a product of the electrically controlled birefringence Δn and the optical path length. As Figure 5 shows, with an applied electric field $|E|$=5.5 kV/cm a optical retardation of $\lambda/2$ is achieved which results in a phase shift of 90°. The phase shift shows a quadratic dependence to the electric field. With relaxor materials as PLZT 9/65/35 the phase shift is not permanent, but exists only as long as the electric field is present. Switching the ceramic from a state of zero phase shift (no voltage) to 90° phase shift (full voltage) allows the realization of an optical shutter in a setup with two cross polarized filters. With intermediate applied electric fields it is possible to create an analog modulator.

With a retardation of $n \cdot \lambda/2$ the transmitted ray is still linear polarized. A retardation of $n \cdot \lambda/4$ produces a circular polarized ray, a retardation of $n \cdot \lambda/8$ results in an elliptic polarization.

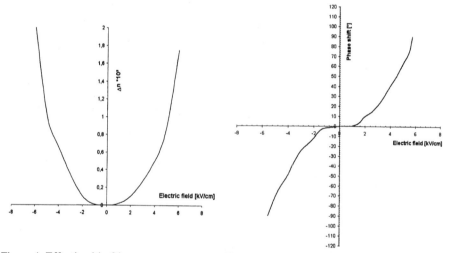

Figure 4: Effective birefringence vs electric field

Figure 5: Phase shift vs electric filed

CONCLUSIONS

With in the optical path length modulatable transparent substrates, thin film interferometry measurement systems can be extended to measure complex, slowly varying or static processes. PLZT ceramics prepared by the improved mixed oxide route, using nano sized oxide powders, are suitable for the production of high quality transparent ceramic substrates. The paper shows that expensive production steps such as hot pressing, vacuum sintering and hot isostatic pressing can be abandoned. The green body preparation by EPD instead of cold pressing or casting results in higher optical quality of the substrate. The electrical and optical properties match those known from literature. The preparation process of the ceramics shows a cost-efficient alternative to those used in today's industry.

REFERENCES

[1]   R. Berger et al., "Electrometer", U.S. patent No 940903, (1992).

[2]   T. Yamada et al., "Measuring method for ellipsometric parameter and ellipsometer ", U.S. patent No 5,335,066, (1991).

[3]   W. Riedel, "Measurement of layer thickness with white light interferometry," Annual rep. Fraunhofer Institut, Freiburg, 1995.

[4]   Caranto et al., "An optical fibre hin film thickness monitor," *Meas. Sci. Technol.*, **4** 865-869 (1993).

[5]   V. Lange, G. Higelin, "Static and dynamic test method for thin mono- and polycristalline Si-films," *Microsystem Technologies 1*, 185-195, Springer (1995).

[6]   Donelly, McCaulley, "Infrared-laser interferometric thermometry: A nonintrusive technique for measuring semiconductorwafer temperatures," *J. Vac. Sci. Technol.*, **A8** (1)84-92 (1990).

[7]   Fang et al., "A fiber-optic high temperature sensor," *Sensors and Actuators*, A44 19-24 (1994).

[8]   A.W. Koch, "Verfahren und Vorrichtung zur optischen Dünnschicht-Biosensorik," patend pend., DE19646770A1 (1998).

[9]   L.E. Cross, "Relaxor Ferroelectrics," *Ferroelectrics* **76**, 241-267, (1987)

[10] M.P. Harmer, D.M. Smyth, "Nanostructure, Defect Chemistry and Properties of Relaxor Ferroelectrics," *ONR Final Report* No. N00014-82-K-0190, Lehigh University Bethlehem, PA, Feb. 1992

[11] E. Bartscherer, A. Braun, M. Wolff and R. Clasen, "Improved Preparation of Transparent PLZT Ceramics by Electrophoretic Deposition and Hot Isostatic Pressing," *27th Annual Conference on Composites, Advanced Ceramics, Materials and Structures*, **23** 593-600 (2003).

[12] E. Bartscherer, K. Sahner and R. Clasen, "Preparation of PLZT powders from nano sized oxides," *26th Annual Conference on Composites, Advanced Ceramics, Materials and Structures*, **23** 601-607 (2002).

[13] R. Clasen, "Forming of compacts of submicron silica particles by electrophoretic deposition," *2nd Int. Conf. on Powder Processing Science*, 633-640 (1988).

[14] G. H. Haertling, "Ferroelectric Ceramics: History and Technology," *J. Am. Ceram. Soc.*, **82** [4] 797-818 (1999).

[15] C.E. Land, P.D. Thacher, The Physics of Opto-Electronic Materials, edited by W.A. Albers jr., Plenum Press, New York, 169-196 (1971).

Low Cost Synthesis of Alumina Reinforced Fe-Cr-Ni-Alloys

Thomas Selchert, Rolf Janssen and Nils Claussen
Advanced Ceramics Group
Technical University Hamburg-Harburg
Denickestr. 15 - 21073 Hamburg/ Germany

ABSTRACT

Refractory metal ceramic composites are formed by a novel synthesis route combining the advantages of powder metallurgical (PM) and pressure casting processing. For example, Fe-Cr-Ni alloys reinforced with alumina can be fabricated by a reactive infiltration. In the present paper, the essential reaction mechanism to form interpenetrating microstructures even at low ceramic loading are outlined.

INTRODUCTION

Various concepts have been developed in the past (e.g. ODS alloy) to improve the performance of HT-alloys. However, due to the high cost of the processing, applications of these materials are still limited /1-3/.

In the presented technique, first porous ceramic precursors are infiltrated with molten aluminium. In a second step, the specimens are heat treated to run the aluminothermic reaction and to form the desired alloy reinforced by alumina. This technique, developed recently by the Advanced Ceramics Group at the TUHH, offers a low cost fabrication route for ceramic reinforced metal composites /4-10/. The processing of starting powders, sintering cycle and mechanical properties of the synthesized material was described recently /11/.

In general, a porous sintered preform, consisting of Cr_2O_3 and Fe-Ni-alloy /8/ is infiltrated with molten aluminum at temperatures between 600°C and 700°C in less than one sec via squeeze or die pressure casting technique. Details about metal casting are given in /12,13,14/. During the infiltration a partial aluminide formation occurs, that coats the Fe-Ni-ligaments. After infiltration, the infiltrated compact is heat treated in a second processing step up to 1000°C. During this treatment, the final alumothermic reaction is initiated, resulting in the formation of Al_2O_3. Furthermore, the metal phase reacts with the admixed metals forming the desired alloy as a matrix.

This paper will focus on the reaction sequence where transient aluminide phases play an important rule. By careful controlling the processing, composites offering a composition close to a matrix commercial Fe-Ni-Cr-high temperature alloy can be formed.

EXPERIMENTAL

Precursor preparation:

As reference, a commercial HT-(Fe-Ni-Cr)alloy is used (Nicrofer 3127hMo, by Krupp). The corresponding notation is 1.4562 (SEW/VdTüv) or UNS N08031 (ASTM). The material composition is shown in Table 1. After sintering, the sintered preform consists of Cr_2O_3, Fe-Ni-alloy and Mo.

Table 1: Composition of the referenced Nicrofer and reactive precursor

Material [wt. %]	Fe	Cr$_2$O$_3$	Ni	Cr	Mo	Cu
Nicrofer 3127hMo	34	-	31	27	6	1
Composite	28.2	36.3	28.4	-	6.2	0.8

The precursor was sintered at temperatures between 1200 to 1300°C in order to achieve a open porosity of 10 to 20 vol.% and a median pore size of about 1.5 μm.

Aluminum infiltration:
The infiltration was performed in a laboratory squeeze casting device designed by TUHH. For homogeneous infiltration, the cavity and the samples were preheated to 400°C and 600°C, respectively. Subsequently, the melt (commercial Al-alloy, German code Al-Si231 (DIN 1725/2: 3.2982.05) melting range 530°C - 570°C) was heated up to 700°C – 800°C and poured in the cavity. Complete infiltration was achieved using an infiltration pressure of 65 MPa.

Heat treatment:
The infiltrated compacts were heat treated up to 1000°C in nitrogen atmosphere using a tube furnace. Furthermore, the heat treatment was performed also in a DSC. In order to control precipitation and grain size of the matrix, the compacts were subjected to a final heat treatment with temperature profile according to the heat treatment used also for the commercial Fe-Ni-Cr-alloy.

RESULTS AND DISCUSSION
Figure 1 shows the sintered metal-ceramic precursor. Additional XRD confirms that Fe-Ni-alloy and Cr$_2$O$_3$ are formed already by sintering at temperatures ~1350°C. Still, there is an open porosity (black areas) between the bright Fe-Ni-alloy ligaments and the grey Cr$_2$O$_3$ particles. Generally, the pores are located at the Cr$_2$O$_3$ particles.

Figure 1: Sintered porous precursor

After infiltration, the pore channels are infiltrated with aluminium (dark phase), see Figure 2 to 4. If the infiltration temperature is less then 560°C, no reaction occurred between molted aluminium and Fe-Ni-alloy ligaments or the Cr₂O₃ particles (Figure 2).

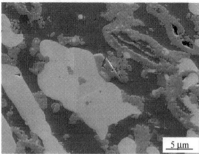

Figure 2: Precursor infiltrated at 560°C; no visible reaction

However, by an infiltration at 640°C a surface reaction on the Fe-Ni-ligaments appears. XRD analysis indicates the formation of aluminides. Sheasby /15/ observed a first exothermic reaction between Al and Fe to Fe-aluminide at about 640°C. Lee and German /16/ found approximately the same formation temperature for Ni-aluminides. They observed the formation of a first reaction layer on the metal particle that overlaid them. These observations confirm the present result, although here the Al reacts with the Fe-Ni-ligaments to Fe-,Ni-aluminides at the ligament surface, see Figure 3.

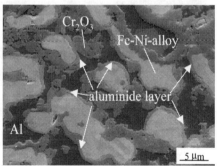

Figure 3: Precursor infiltrated at 640°C; partial surface reaction

These aluminides are formed by an exothermic reaction. DSC experiments show a first exothermic peak about 660°C. At this temperature, additional aluminide clusters are formed beside the aluminide layers on the alloy surface as shown in the figure 4. Further experiments confirm that the aluminide clusters are formed locally at the surface of the alloy ligaments. It is considered, that these clusters are formed where the aluminide layer formed first does not act as a diffusion barrier, i.e. on places where the layer is not continuous. This statement is confirmed by experimental observations by

461

Philpot /15/, Sheasby /16/ and Lee et al. /17/ showing that continuous Fe-respective Ni-aluminides surface layer prevent further reactions of Fe-Ni and Al to aluminides.

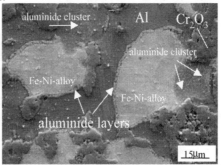

Figure 4: Precursor heat treated to first exothermic reaction at 660°C; partial aluminide cluster formation

The aluminide clusters are formed by a reaction between the Al and the Fe-Ni-alloy ligaments. The extent of this reaction increased by a further heat treatment at temperatures of about 1000°C. Finally, the product consists of aluminides and Cr_2O_3. No aluminothermic reaction occurs although this is thermodynamically expected. Detailed SEM investigations show that microcracks are present resulting in a partial spacing between Cr_2O_3 and aluminides in the product. It is likely that this spacing retards the start of the desired alumina formation, see Figure 5.

Figure 5: Precursor with high surface/volume ratio, heat treated up to 1000°C

However, detailed investigations demonstrated that specimens with almost continuous aluminide layer formed during infiltration can be reacted easily to the desired product phases, i.e. alumina and Fe-Ni-Cr-alloy. Figure 6 shows one of these products. Here, the aluminothermic reaction has taken place resulting in a composite of Al_2O_3 particles embedded in a Fe-Ni-Cr-alloy matrix. The composition of the metal matrix is close to the Krupp Nicrofer 3127hMo alloy used in this work as reference. Beside Al_2O_3 and Fe-Ni-Cr-alloy, also some isolated Mo precipitations (bright phase) are formed as it was intended to improve the HT-stability of the composite.

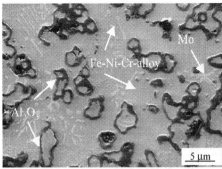

Figure 6: Precursor with low surface/volume ratio, heat treated in the furnace up to 1000°C, Al_2O_3 reinforcement in a alloy matrix

The formation of Cr_2O_3 and Fe-Ni-alloy as product phases depend essentially on the specimen size, i.e. on the surface to volume ratio and the resulting heat flow from the specimen to the furnace atmosphere during the exothermic aluminide formation. Obviously, the heat flow for small specimens with high surface/volume ration is high compared to large specimens. Due to this, the energy from the exothermic aluminide formation releases rapidly. This energy, however, is necessary for in situ heating of the specimen to a temperature level suitable to start the aluminothermic reactions. In the small specimen, however, all Al will be consumed before the ignition temperature of Al_2O_3 formation will be reached.

In the large specimens, the energy given by exothermic aluminide formation is higher and the outwards heat flow is smaller. Therefore, the start temperature for the aluminothermic reaction will be reached before all Al is consumed to aluminides.

During the aluminothermic reaction, the temperature in the specimen increases thereby melting partially the metal phase. The alloy incorporates all remaining phases, resulting in a final composition as desired. Figure 7 shows the microstructure of such a large specimen. Here a thin network of Al_2O_3 is embedded in a Fe-Ni-Cr-alloy matrix.

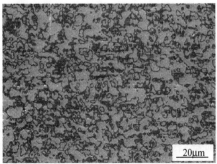

Figure 7: Alumina structure in a Fe-Ni-Cr-alloy matrix

SUMMARY

Fe-Ni-Cr-alloys reinforced with alumina particles can be fabricated by a reactive synthesis route using starting materials and processing techniques well established in the powder metallurgical and aluminum casting industry. During infiltration, the infiltration temperature should reach the level to obtain a nearly continuous aluminide layer on the metal ligaments. This layer protects the ligaments and retards an extensive aluminide formation.

A partial aluminide formation however, favors a aluminothermic reaction during annealing. As a result, Al_2O_3 is formed in a matrix of Fe-Ni-Cr-alloy. This process offers the potential for near net-shape manufacturing using low cost materials. Furthermore, promising HT-properties are expected.

LITERATURE:

/1/   Tien J.K., Caulfield T., "Superalloys, Supercomposites and Superceramics", Material Science Series, Academic Press, Inc. (London) LTD., (1989), ISBN 0-12-690845-1

/2/   Bradley E.F., Superalloys, 1988, ASM International Ohio

/3/   Siems D.T., Stotloff N.S., Hagel W.C., Superalloys II, 1987 John Wiley& Sons/New York

/4/   Claussen N., Beyer P., Janssen R., Kumar P., Travitzky N.A.,"Reactive Casting of Metal-Ceramic Composites", Polish Ceramic Bulletin 60 (2001) 14-20

/5/   Beyer P., Janssen R., Claussen N., "Synthesis of Alumina-Aluminide Composites by Reactive Squezze Casting", Adv. Eng. Mater. 2000 [2] (2000)734-736

/6/   Claussen N., German Patent Application No. 19957786, December 1999

/7/   Sanghage K.H., Claussen N., "Reaction Casting of Ceramic/Metal and Ceramic/Intermetallic Composites" Somiya-Handbook: Ceramic Materials And Their Applications, Vol 1, Chapt. 5, 2002

/8/   Kumar P., Travitzky N.A., Beyer P., Sandhage K.H., Janssen R., Claussen N., "Reactive Casting of Ceramic Composites (r-3C)", Scripta Mater. 44 (2001) 751-757

/9/   Claussen N., German Patent Application No. 10004471, February 2000

/10/   Travitzky N.A., Kumar P., Sandhage K.H., Janssen R., Claussen N., "Rapid Synthesis of Al_2O_3 reinforced Fe-Cr-Ni Composites", in press Mat. Sci. Eng. A

/11/   Selchert T., Janssen R., Claussen N., "Reaction Synthesis of Refractory Metall-Ceramic Composites", 27th International Cocoa Bech Conference on Advanced Cermics and Composites: B, Ceramic Engineering & Science Proceedings Volume 24, Issue 4, 2003, p.175-180

/12/   Smith W.E., Wallace J.F., "Gating of Die Castings", Modern Casting 44, 1936, 325 –357

/13/   Radtke S.F., "Pore-Free Die Castings" – A Progress Report, Die Casting Engineer, 1972

/14/   Brunhuber E., Gießereilexikon 15. Edition, 1991 Schiele&Söhn/ Berlin

/15/   Philpot K.A., Munir Z.A., Holt J.B., "An investigation of the synthesis of nickel aluminides through glassless combustion", J. Mater. Sci. 22, (1987), S.159-169

/16/   Sheasby J.S., "Powder Metallurgy of Iron-Aluminium", The International Journal of Powder Metallurgy & Powder Technology, Vol 15. No.4, (1979), p. 301-305

/17/   Lee D.J., German R.M., "Sintering Behavior of Iron-Aluminum Powder Mixes", Int. Jor. of. Powder Met. & Powder Techn., Vol. 21, No. 1, New York, (1985)

HIGH TEMPERATURE BEHAVIOUR OF CERAMIC FOAMS FROM SI/SIC-FILLED PRECERAMIC POLYMERS

Juergen Zeschky, Thomas Hoefner, Henning Dannheim, Michael Scheffler and Peter Greil
University of Erlangen-Nuremberg
Department of Materials Science, Glass and Ceramics
Martensstrasse 5
91058 Erlangen, Germany

Dieter Loidl, Stephan Puchegger and Herwig Peterlik
University of Vienna
Institute of Materials Physics
Boltzmanngasse 5
1090 Vienna, Austria

ABSTRACT

Foams of a preceramic polymer were manufactured using a self-foaming process of a Si/SiC filled poly siloxane resin at 270 °C. In a subsequent pyrolysis process at 1000 °C, Si-O-C micro composite foams with a high structural isotropy, spherical cells and dense struts were obtained. The ceramic foams exhibit a high compression strength exceeding 4 MPa at a total porosity of 70 – 80 %.

The Young's modulus, shear modulus and compression strength were measured at temperatures up to 1200 °C. The foams were oxidized up to 200 hours and the effect of the oxidation on the surface microstructure was examined by scanning electron microscopy. Formation of porosity in the struts resulted in a reduction of the compression strength, which, however, remained stable at prolonged oxidation treatment after 10 hours. The foams were subjected to thermal shock quenching with a temperature difference of 1100 °C and 1400 °C for up to 10 cycles. This treatment reduced the Young's modulus and compression strength less than 20 %.

INTRODUCTION

Ceramic foams are characterized by a linked network of irregularly shaped open or closed polyhedron pore cells. Highly porous ceramic foams offer non catastrophic fracture behaviour and a superior thermal shock resistance. Potential applications include catalyst carriers or heating media in porous gas burners.

Ceramic foams can be manufactured by the reticulation process[1], space holder methods[2], gel casting methods[3] and self foaming techniques[4]. Alternatively, low viscous preceramic polymer melts such as poly siloxanes were foamed using additional foaming agents such as poly urethane[5] or by release of gaseous polymerisation products that are formed during the curing process of the polymer[6].

In this study, a phenyl methyl poly (silsesquioxane) containing small amounts of ethoxy- and hydroxy-groups was foamed in situ. Heating in the temperature range of 220 – 300 °C gives rise for the cross linking of the polymer via condensation reactions forming water, ethanol and traces of benzene. The release of these gases leads to the formation of foam cells which get stabilized by the simultaneous increase of the polymer viscosity. Finally the thermoset foam is pyrolysed

resulting in a ceramic residue with an amorphous silicon oxycarbide matrix. The structural a
mechanical properties of the ceramic foam can be varied by adding filler particles to the polym
which can react with the polymer residue or the pyrolysis atmosphere to form carbides a
nitrides.

EXPERIMENTAL PROCEDURE

A methyl phenyl poly (silsesquioxane) ($[(C_6H_5)_{0.62}(CH_3)_{0.31}(OR)_{0.07}SiO_{1.5}]_n$ with n ~ 20, l
[–OH] and [–OC_2H_5], Silres H44, Wacker Chemie, Burghausen, Germany) was used
preceramic polymer. At room temperature the polymer is a solid white powder. It melts at 4C
60 °C and a density of 1.1 g/cm^3. A Si/SiC blend (Si: Silgrain, d_{50} = 7.6 μm, Elkem, Meerbusc
Germany; SiC: F500B, d_{50} = 16 μm, H.C. Starck, Selb, Germany) was added to the polymer ir
polymer:Si:SiC weight ratio of 35:12:53. Si and SiC were selected as inert filler powders in orc
to reduce the shrinkage and to increase the mechanical strength and structure stability of t
foam.

The polymer/filler blend was mixed in a ball mill for 2 hours. The mixture was foamed in
aluminium tray (100 mm wide, 140 mm long and 60 mm high) at 270 °C for 2 hours in air. T
thermoset foam was cut into rectangular bars (100 mm * 15 mm * 10 mm) and cylinders (10 m
in height and diameter). The foams were pyrolysed in nitrogen at 1000 °C for 4 hours. A heati
rate of 3 °C/min was applied. A temperature hold was applied at 500 °C for 2 hours.

The compression strength was determined at temperatures ranging from 25 °C to 1200
from samples with cylindrical geometry (10 mm in height and diameter) using a universal testi
machine (Instron 4202, Instron Corp., Canton, MA, USA) with an attached furnace (PMA-06/\
Maytec GmbH, Singen, Germany). The crosshead speed was set to 0.5 mm/min. The samp
were heated with 10 °C/min to the measurement temperature which was equilibrated for 30 n
prior to the measurement. The crushing strength was derived from the maximum applied fo
after the elastic buckling of the specimens. The elastic constants were measured at the Univers
of Vienna in the same temperature range using the Resonant Beam Technique (RBT) as describ
in detail elsewhere[7]. Ceramic foam bars of 100 mm * 15 mm * 10 mm were measured ir
graphite heated furnace in vacuum. The Young's modulus and the shear modulus were calcula
from the frequency spectrum.

Oxidation behavior was evaluated by heating pyrolysed cylindrical samples to 1200 °C fo
– 10000 minutes in air. The microstructure of oxidized samples was examined by scanni
electron microscopy (Quanta 200, Fey Deutschland GmbH, Kassel, Germany). Crystalline pha:
were identified by x-ray diffraction (Diffrac 500, Siemens AG, Munich, Germany).

The thermal shock resistance was measured after preheating the specimen in a furnace
1130 °C and 1430 °C for 10 minutes. The specimens were taken out of the furnace and quencl
in water of room temperature (25 °C). The specimens were stirred in the water until they cool
completely. The bending strength σ_n after n cycles was tested in a four point bending setup wit
crosshead speed of 0.5 mm/min. The Young's modulus E_n was measured by impulse excitati
method (ENV 843-2) using Buzz-o-Sonic software 4.05 (Buzzmac Software, Glendale, USA) a
a standard microphone and personal computer. n was 1 for ΔT = 1400 °C and n = 1, 2, 5 and
for ΔT = 1100 °C, respectively. Damage parameters D_σ and D_E were calculated according to e
1,

$$D_{\sigma,n} = 1 - \frac{\sigma_n}{\sigma_0} \qquad \text{and} \qquad D_{E,n} = 1 - \frac{E_n}{E_0} \qquad (1)$$

where σ_0 and E_0 are the bending strength and Young's modulus of the unquenched specimen.

RESULTS AND DISCUSSION

Figure 1 shows the compression strength and the elastic modules of the open cell ceramic foams in dependence of the temperature. The compression strength at room temperature is 4.3 MPa for a fractional density of 0.27. Increasing temperatures lead to higher values of the crushing strength up to 9 MPa at 900 – 1000 °C. At temperatures above 1100 °C the foams suffer from a noticeable decrease of the compression strength. The Young's modulus and the shear modulus showed similar temperature dependences.

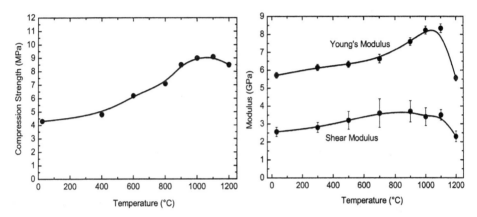

Figure 1. Compression strength and elastic moduli of Si/SiC filled ceramic foams as a function of temperature.

As no phase reactions took place during heating the samples up to the measurement temperature, it can be assumed that the increase of the elastic modulus and the compression strength is attributed to the high amount of free carbon (> 20 mole % in the polymer residue[8]) which is known for improved mechanical properties at high temperatures [12].

However, when heated above 1100 °C in air, the crushing strength σ_{cr} of the ceramic foams decreased. The time dependent reduction of σ_{cr} is given in Figure 2. An exponential decay was measured until, after 10 hours, a constant plateau at 2.7 MPa was reached.

Micrographs of polished strut cross sections of a pyrolysed foam before and after oxidation for 10 hours are shown in Figure 3. After pyrolysis in inert atmosphere an amorphous Si-O-C matrix surrounded the Si (light grey particles in Fig. 3a) and SiC powder (dark grey particles). No reactions between the filler particles and the polymer residue or the atmosphere could be observed.

Cracks of 10 – 50 µm were observed in the matrix, the total strut porosity was 8 – 10 %. The oxidation in air at 1200 °C lead to an increased porosity of V_P = 30 – 40 % due to the reaction of the matrix with the atmosphere according to:

$$Si + C + 1.5\ O_2(g) \rightarrow SiO_2 + CO(g) \tag{2}$$

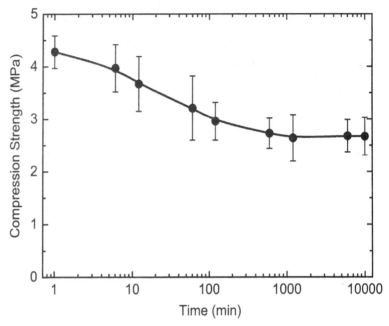

Figure 2. Compression strength as a function of the oxidation time at 1200 °C in air.

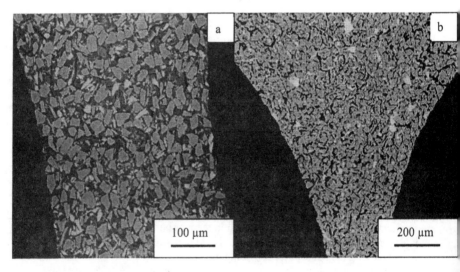

Figure 3. Cross section of a strut of non oxidized foam (a) and foam oxidized in air at 1200 °C for 600 min (b).

While carbon burnout in filler free Si-O-C is supposed to lead to spherical pores[9], the oxidation of the Si/SiC filled foams resulted in irregularly shaped pores, Fig. 3b. Cristobalite was identified in samples oxidized for 600 hours. The devitrification of the matrix phase was confirmed in filler free Si-O-C foams fired at 1200 °C in air [9]. The porosity increase is caused by the burnout of the free carbon. Due to the microcracks in the matrix phase the oxidation reactions proceeded throughout the whole strut volume.

The resistance of the Si/SiC foams to thermal shocks is displayed in Figure 4. Thermal shock with $\Delta T = 1100$ °C caused the most severe deterioration during the first shock cycle. The compression strength was reduced by 5 % and the Young's modulus by 9 %. SEM micrographs confirmed that micro cracks formed during quenching in the strut material. Additional thermal shock cycles did not cause noticeable additional crack growth. This thermal fatigue saturation behaviour is typical for ceramic foams due to decreased thermal stress formation[10]. The thermally induced degradation of the mechanical properties is in good agreement with data from filler free poly siloxane derived ceramic foams of comparable fractional density[11]. Thermal shocks with $\Delta T = 1400$ °C resulted in higher damage parameters D_σ and D_E of 10 - 15 %.

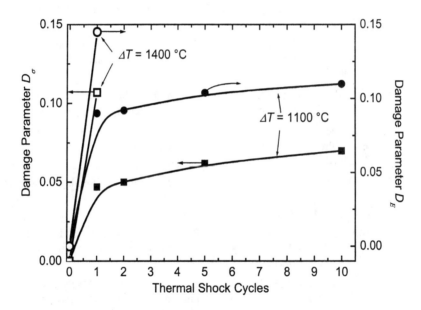

Figure 4. Influence of the number of thermal shocks on the compression strength and the Young's modulus.

CONCLUSIONS

Ceramic Si-O-C foams from Si and SiC loaded poly siloxanes are distinguished by a superior compression strength, Young's modulus and shear modulus at 1000 °C which are twice as high as compared to the corresponding values at room temperature. At 1200 °C, oxidation reactions result in a carbon burnout generating increased porosity in the struts which lead to a decreased

compression strength of 2.5 MPa. The crushing strength and Young's modulus are reduced b
less than 12 % after quenching from 1100 °C to room temperature. These properties make th
Si/SiC filled ceramic foam a promising candidate for applications in gas burners, catalyst carrie
or metal melt filters where high thermal shock resistance is require

ACKNOWLEDGEMENTS

Financial support from the Austrian Science Fund under P15670 and from the Deutsch
Forschungsgemeinschaft (DFG) under SCHE-628/1-2 is gratefully acknowledged.

REFERENCES

[1] K. Schwartzwalder and A.V. Somers, Method for Making Porous Ceramic Articles, U.S
Patent, USA, 1963.

[2] T. J. Fitzgerald, V. J. Michaud, and A. Mortensen, "Processing of Microcellular Si
Foams. Part II. Ceramic Foam Production," Journal of Material Science, 30 1037-104
(1995).

[3] Pilar Sepulveda, "Gelcasting Foams for Porous Ceramics," American Ceramic Societ
Bulletin, 76 [10] 61-65 (1997).

[4] X. Bao, M. R. Nangrejo, and M. J. Edirisinghe, "Sythesis of Silicon Carbide Foams fro
Polymeric Precursors and their Blends," Journal of Material Science, 34 [11] 2495-250
(1999).

[5] P. Colombo and M. Modesti, "Silicon Oxycarbide Ceramic Foams from a Preceram
Polymer," Journal of the American Ceramic Society, 82 [3] 573-578 (1999).

[6] T. Gambaryan-Roisman, M. Scheffler, T. Takahashi, P. Buhler, and P. Greil, "Formatio
and Properties of Poly(siloxane) derived Ceramic Foams"; pp. 247-251 in Euromat 9
Edited by G. Müller. DGM, Munich, Germany, 2000.

[7] D. Loidl, S. Puchegger, K. Kromp, J. Zeschky, P. Greil, M. Bourgeon, H. Peterli
"Elastic moduli of porous and anisotropic composites at high temperatures", Advance
Eng. Mat., in print.

[8] M. Scheffler, T. Gambaryan - Roisman, T. Takahashi, J. Kaschta, H. Muenstedt, I
Buhler, and P. Greil, "Pyrolytic Decomposition of Preceramic Organo Polysiloxanes
Ceramic Transactions, 115 239-250 (2000).

[9] P. Colombo, J. R. Hellmann, and D. L. Shelleman, "Mechanical Properties of Silico
Oxycarbide Ceramic Foams," Journal of the American Ceramic Society, 84 [10] 224
2251 (2001).

[10] V. R. Vedula, D. J. Green, and J. R. Hellmann, "Thermal Fatigue Resistance of Open Ce
Ceramic Foams," Journal of the European Ceramic Society, 18 [14] 2073-2080 (1998).

[11] P. Colombo, J. R. Hellmann, and D. L. Shelleman, "Thermal Shock Behavior of Silico
Oxycarbide Foams," Journal of the American Ceramic Society, 85 [9] 2306-2312 (2002).

[12] J. Haag, Heuer J., Kraemer M., Pischinger S., Wunderlich K., Arndt J., Stock M
Coelingh W., Reduction of hydrocarbon emissions from SI-engines by the use of carbo
pistons, SAE technical paper series, 952538 (1995).

STABILIZATION OF COUNTER ELECTRODE FOR NASICON BASED POTENTIOMETRIC CO₂ SENSOR

Yuji Miyachi
Department of Molecular and Material Sciences, Interdisciplinary Graduate School of Engineering Sciences, Kyushu University, 6-1 Kasuga Koen, Kasuga-shi, Fukuoka 816-8580, Japan

Go Sakai, Kengo Shimanoe, Noboru Yamazoe
Department of Material Sciences, Faculty of Engineering Sciences, Kyushu University, 6-1 Kasuga Koen, Kasuga-shi, Fukuoka 816-8580, Japan

ABSTRACT

Two kinds of materials, $Bi_2Cu_{0.1}V_{0.9}O_{5.35}$ (BICUVOX) and $NaCoO_2$, were investigated for their applicability as the counter electrode for a potentiometric CO_2 sensor using NASICON (Na^+ conductor, $Na_3Zr_2Si_2PO_{12}$). The BICUVOX counter electrode was more stable than an Au counter electrode. Thermal cycling between the operating temperature (450 °C) and room temperature under various conditions indicated the time needed for the BICUVOX electrode to reach the stationary potential after switching to the operating temperature was unacceptably long. $NaCoO_2$ was found unsuitable as a counter electrode material due to its reactivity with CO_2. This reactivity was suppressed completely when $NaCoO_2$ was coated with a layer of glass (SiO_2: Na_2O: B_2O_3: Al_2O_3 = 44: 20: 31: 5, in molar ratio). The glass-coated $NaCoO_2$ electrode worked well as a counter electrode at 450 °C. Thermal cycling indicated that the glass-coated $NaCoO_2$ electrode exhibited an initial ageing effect. Once aged, glass-coated $NaCoO_2$ showed the properties of a reliable counter electrode, i.e., rapid warm-up characteristics and a reproducible stationary potential at 450 °C.

1. INTRODUCTION

A potentiometric device combining NASICON (Na^+ conductor, $Na_3Zr_2Si_2PO_{12}$) with an auxiliary phase of metal carbonate is an attractive candidate for a CO_2 sensor in view of its simple structure, high selectivity and fast response [1-4]. In addition, adoption of a planar structure by using a NASICON thick film would make it possible to reduce fabrication costs as well as device size [5, 6]. A device of this type attached with an Au counter electrode develops a potential that depends not only on the partial pressure of O_2 (PO_2) but also on the activities of Na^+ and Na_2O of NASICON in the vicinity of the electrode. The potential is thus relevant to the surface chemical state of NASICON. The EMF response of this device to CO_2 is stable under continuous operation at elevated temperature (e.g. 400-500 °C), but once the device is cooled to room temperature the NASICON surface is vulnerable to attack by moisture and CO_2 to form surface contaminants. When sensor operation is restarted at 400-500 °C, it may take a long warm-up time for the EMF to recover the original value, depending on the degree of contamination, or it may be totally unable to recover if the contamination is too severe [7]. In order to overcome this problem, the counter electrode should be made independent of NASICON. For this purpose, we tried to introduce two kinds of materials, BICUVOX ($Bi_2Cu_{0.1}V_{0.9}O_{5.35}$, O^{2-} conductor) and Na_xCoO_2 (x = 1, Na bronze) [8, 9], into the counter electrode. BICUVOX is expected to function as an air reference electrode, the potential of which depends on PO_2 only, whereas Na_xCoO_2 may act as a solid reference electrode the potential of which is independent of PO_2.

471

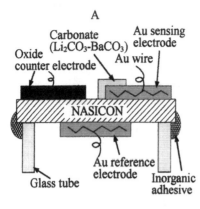

A

Carbonate (Li₂CO₃-BaCO₃) — rendered: Carbonate (Li_2CO_3-$BaCO_3$) Au sensing electrode / Au wire

Oxide counter electrode

NASICON

Au reference electrode

Glass tube / Inorganic adhesive

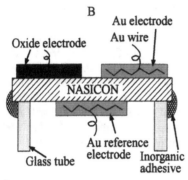

B

Au electrode / Au wire

Oxide electrode

NASICON

Au reference electrode

Glass tube / Inorganic adhesive

Fig. 1. Shematic drawings of three-electrode devices. A: Planar CO_2 sensing device attached with Au reference electrode. B: Device for investigating the warming-up characteristics of the upper electrodes.

2. EXPERIMENTAL

2.1 Preparation of materials

An amorphous precursor of NASICON w prepared by a sol-gel method starting from the alkoxid of constituent elements [10, 11], i.e. $Si(OC_2H_5$ $Zr(OC_4H_9)_4$, $PO(OC_4H_9)_3$ and $NaOC_2H_5$. A stoichiom ric mixture of the alkoxides dissolved in ethanol at °C was hydrolyzed with a designated amount of wat The resulting xerogel was dried at 120 °C for 24 h a calcined at 750 °C for 1 h. The calcined powder w compacted into a disk by mechanical pressing and si tered at 1200 °C for 5 h [12]. The NASICON disk c tained was polished into geometry of 8 mm diamet and 1 mm thick.

The powder of $Bi_2Cu_{0.1}V_{0.9}O_{5.35}$ was prepar from the constituent metal oxides (Bi_2O_3, V_2O_5 and Cu through mixing in an attrition ball-mill, preliminary c cination at 650 °C for 15 h, pulverization in an aga mortar, and final calcination at 750 °C for 5 h [13-1: The powder of $NaCoO_2$ was prepared from a stoichi metric mixture of Co_3O_4 and Na_2O_2 through mixing an agate mortar and calcination at 500 °C for 12 h und dry N_2 atmosphere [8]. The glass powder of the comp sition, SiO_2: Na_2O: B_2O_3: Al_2O_3 = 44: 20: 31: 5 in mo ratio, was prepared from a mixture of the componei by melting, cooling and crushing. In this glass comp sition, glass-transition and melting point were 560 and 800 °C, respectively.

2.2 Fabrication of devices

The two types of three-electrode devices (A and B) fa ricated in this study are shown in Fig. 1. Device A h both sensing and counter electrodes on the upper si of the NASICON disk, thus it is a planar CO_2 sensing device. The Au reference electrode attach on the backside, sealed off from the outside atmosphere by a glass tube and an inorganic adhesiv is always exposed to dry air. The potential changes of the counter and sensing electrodes, if any, c be monitored by referring to this electrode. Device B is the same as device A, except the carbona which partially covers the Au sensing electrode, is not present. The potentials of the oxide ele trode (BICUVOX or $NaCoO_2$) and the Au electrode at different gas conditions can be measu simultaneously by referring Au reference electrode. Both devices A and B were fabricated as ported elsewhere [16]: Au paste was applied and calcined at 850 °C for 15 min to deposit Au ele trodes. For the sensing electrode, a layer of carbonate (Li_2CO_3 + $BaCO_3$, 1:2 in molar ratio) w deposited on the Au electrode by a melting-and-quenching method. For preparing the oxide ele trodes, the powder of BICUVOX or $NaCoO_2$ was applied as a paste (using alpha-terpineol + 5w ethyl cellulose) and, after drying, calcined at 750 °C for 15 min or at 800 °C for 30 min, respectiv To prepare a glass-coated $NaCoO_2$ electrode, the $NaCoO_2$ paste was first deposited as above, a

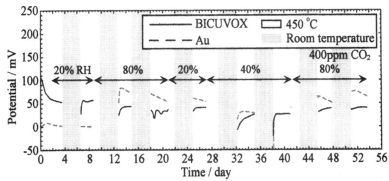

Fig. 2. Measured potentials of BICUVOX and Au electrodes vs. Au reference electrode at 450 °C during a series of heat-cycle tests (device B).

after drying, a paste of glass powder was applied on top of it. Then the whole assembly was calcined at 800 °C for 30 min.

2.3 CO_2 sensing experiments

All electrochemical measurements were carried out in a conventional gas flow apparatus equipped with a heating facility. By diluting a parent gas (2000 ppm CO_2 in dry synthetic air) with synthetic air, the concentration of CO_2 in the sample gas was varied in the ranges of 0-2000 ppm. Humid sample gas was prepared by allowing a part of the synthetic air to bubble through water. Relative humidity (RH) was controlled in the range of 10-80 % by controlling the volume ratio of wet air to dry air. When necessary, the device temperature was changed quickly between 450 °C and room temperature (25 °C) by removing the hot furnace off from the gas chamber the device was placed in or doing the reverse.

3. RESULTS AND DISCUSSION
3.1 Heat-cycle tests of BICUVOX electrode

As already reported [17], the device with an Au sensing electrode (covered with carbonate layer) and a BICUVOX counter electrode worked well under steady condition at elevated temperature (450 °C). The properties of the BICUVOX electrode and Au electrode were examined by fabricating the three-electrode device B (Fig. 1). This device was placed in the gas chamber and subjected to a series of heat cycle tests: The temperature of the device was switched between operating temperature (450 °C) and room temperature at an interval of 2-4 days, while humidity of the gas flowing over the device was changed up or down at an interval of 6-14 days, as shown in Fig. 2. The concentration of CO_2 was kept constant (400 ppm) throughout these experiments. As shown in this figure the potential of the Au electrode relative to the Au reference electrode was close to zero in the initial period, which is as expected. Once exposed to high humidity (e.g. 80% RH) at room temperature, however, the potential deviated significantly from zero on the next heating, though this deviation decreased very slowly with time. This instability of the electrode potential is believed to originate from the contamination of NASICON surface as described before [7]. The potential of the BICUVOX electrode, on the other hand, hardly showed such humidity-related instability and tended to converge to a constant EMF level as shown in Fig. 2. This appears to reflect that the electrode acts as an air-reference electrode. Unfortunately, however, it often took long times (0.5-3 days) for the BICUVOX electrode to reach a steady state after switching to the operat-

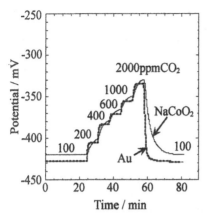

Fig. 3. Transients of potentials of NaCoO₂ electrode and Au sensing electrode vs. Au reference electrode on changes in CO₂ concentration at 450 °C under dry condition (device A).

ing temperature, as shown. With such slow warming-up characteristics, the BICUVOX electrode was judged to be not acceptable as a CO_2 sensing device.

3.2 Sensing device using NaCoO₂ counter electrode

A three-electrode device (A) was fabricated by using a NaCoO₂ counter electrode. The potentials of the sensing and counter electrodes were measured respectively at 450 °C relative to the Au reference electrode while the CO_2 concentration was changed stepwise under dry condition. As shown in Fig.3, not only the Au sensing electrode but also the NaCoO₂ electrode responded to changes in CO_2 concentration, though the latter electrode was more sluggish in response transients. This means that the latter electrode in fact acts as a CO_2 sensing electrode in this device. It is suspected that a carbonate auxiliary phase (e.g. Na₂CO₃) was formed spontaneously through a reaction between NaCoO₂ and CO_2 in gas phase. In order to protect the NaCoO₂ electrode from such a reaction, the electrode was coated with a layer of glass.

The CO_2 sensing Behavior of the resulting device is shown in Fig. 4. The potential of the glass-coated NaCoO₂ electrode was no longer responsive to changes in CO_2 concentration under dry (a) and humid (b) conditions. This confirms that the glass-coated electrode can now serve as a counter electrode. Since the Au sensing electrode potential relative to the reference electrode responded well to changes in CO_2 concentration, the same response behavior could also be obtained by referring to the glass-coated electrode. The EMF response observed between the Au sensing electrode and the glass-coated NaCoO₂ electrode was measured of different CO_2 concentration at

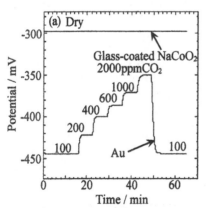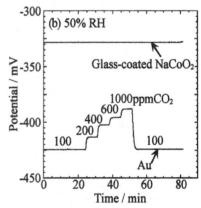

Fig. 4. Transients of potentials of glass-coated NaCoO₂ electrode and Au sensing electrode on changes in CO₂ concentration at 450 °C under dry (a) and 50% RH (b) conditions (device A with a glass coating on the oxide electrode).

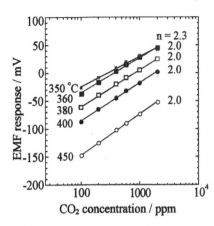

Fig. 5. The EMF response between Au sensing electrode and glass-coated $NaCoO_2$ counter electrode as correlated with CO_2 concentration at various temperatures under dry condition (device A with a glass coating on the oxide electrode).

operating temperatures ranging between 450 °C and 350 °C. As shown in Fig. 5, linear regressions of EMF vs. $\ln[CO_2]$ indicated (using the Nernst equation) that the number of electrons involved in the electrochemical reaction was 2. This is the number of reaction electrons for the electrochemical reduction of CO_2. This data indicates, the glass-coated $NaCoO_2$ counter worked well under steady state conditions. However, operation was only verified within a short term. Careful inspection of Fig. 4 reveals that the potential of the glass-coated $NaCoO_2$ electrode as well as that of the Au sensing electrode differ between dry condition (a) and humid condition (b). This suggests that the potentials were still drifting very slowly. In fact, the glass-coated $NaCoO_2$ electrode was found to show a significant ageing effect before being stabilized, as described in the next section.

3.3 Heat-cycle tests of glass-coated $NaCoO_2$ electrode

The glass-coated $NaCoO_2$ electrode was incorporated into the three-electrode device B to test its properties under heat-cycle test conditions similar to those for the BICUVOX electrode. The results are shown in Fig. 6. The Au electrode showed a large irregular shift in potential at each heated stage. In contrast, the potential of the glass-coated $NaCoO_2$ electrode increased monotonically for the initial period of about 15 days and converged to a constant level of about -200 mV after that. This indicates that the electrode shows an ageing effect in the initial period. Once aged, the electrode can reach the steady state fairly quickly on switching to the operating temperature. In this respect, this electrode is far superior to BICUVOX electrode. These results assure that the glass-coated $NaCoO_2$ electrode, once aged, can be used as a reliable counter electrode for the NASICON-based CO_2 sensor.

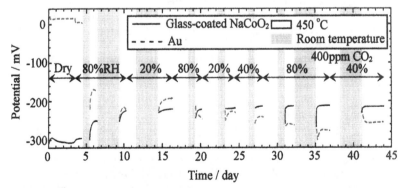

Fig. 6. Measured potentials of glass-coated $NaCoO_2$ electrode and Au electrode vs. Au reference electrode at 450 °C during a series of heat-cycle tests (device B with a glass coating on the oxide electrode).

4. CONCLUSION

BICUVOX provides an air-reference counter electrode, which works well under stead operating conditions, but warming-up transients are often rather too sluggish. On the other hand $NaCoO_2$, when coated with glass and sufficiently aged, provides a reliable solid-reference counte electrode with rapid warm-up time.

5. REFERENCES

[1] T. Maruyama, Y. Saito, Y. Yamamoto, Y. Yano, "Potentiometric sensor for sulfur oxide using NASICON as a solid electrolyte", *Solid State Ionics*, **17**, 281-286 (1985).

[2] T. Maruyama, S. Sasaki, Y. Saito, "Potentiometric gas sensor for carbon dioxide usin solid electrolytes", *Solid State Ionics*, **23**, 107-112 (1987).

[3] S. Yao, Y. Shimizu, N. Miura, N. Yamazoe, "Solid electrolyte CO_2 sensor using binar carbonate electrode", *Chem. Lett.*, **1990**, 2033-2036 (1990).

[4] S. Yao, S. Hosohara, Y. Shimizu, N. Miura, H. Futata, N. Yamazoe, "Solid electrolyte CC sensor using NASICON and Li^+-based binary carbonate electrode", *Chem. Lett.*, **1991**, 2069-207 (1991).

[5] T. Kida, Y. Miyachi, K. Shimanoe, N. Yamazoe, "NASICON thick film-based CO_2 sensc prepared by a sol-gel method", *Sensors and Actuators B*, **80**, 28-32 (2001).

[6] L. Wang, R.V. Kumar, "Thick film CO_2 sensors based on Nasicon solid electrolyte", *Soli State Ionics*, **158**, 309-315 (2003).

[7] T. Kida, K. Shimanoe, N. Miura, N. Yamazoe, "Stability of NASICON-based CO_2 sensc under humid conditions at low temperature", *Sensors and Actuators B*, **75**, 179-187 (2001).

[8] H. Schettler, J. Liu, W. Weppner, R.A. Huggins, "Investigation of solid sodium referenc electrodes for solid-state electrochemical gas sensors", *Applied Physics A*, **57**, 31-35 (1993).

[9] S. Bredikhin, J. Liu, W. Weppner, "Solid ionic conductor semiconductor functions fc chemical sensors", *Applied Physics A*, **57**, 37-43 (1993).

[10] Y.L. Huang, A. Caneiro, M. Attari, P. Fabry, "Preparation of NASICON thin films by dip coating on Si/SiO_2 wafers and corresponding C-V measurements", *Thin Solid Films*, **196**, 283-29 (1991).

[11] D.D. Lee, S.D. Choi, K.W. Lee, "Carbon dioxide sensor using NASICON prepared by th sol-gel method", *Sensors and Actuators B*, **24-25**, 607-609 (1995).

[12] K. Obata, S. Kumazawa, K. Shimanoe, N. Miura, N. Yamazoe, "Potentiometric sensc based on NASICON for detection of CO_2 at low temperature", Proceeding of the 30th Chemica Sensor Symposium, vol. 16, Suppl. A, 70-72 (2000).

[13] J.C. Boivin, C. Pirovano, G. Nowogrocki, G. Mairesse, Ph. Labrune, G. Lagrange, "Elec trode-electrolyte BIMEVOX system for moderate temperature oxygen separation", *Solid State Ionic.* **113-115**, 639-651 (1998).

[14] A. Boukamp, "Small signal response of the BiCuVOx/noble metal/oxygen electrode sys tem", *Solid State Ionics*, **136-137**, 75-82 (2000).

[15] M.H. Paydar, A.M. Hadian, K. Shimanoe, N. Yamazoe, "The effects of zirconia additio on sintering behavior, mechanical properties and ion conductivity of BICUVOX 1 material", *Jour nal of the European Ceramic Society*, **21**, 1825-1829 (2001).

[16] K. Obata, S. Kumazawa, K. Shimanoe, N. Miura, N. Yamazoe, "Potentiometric sensc based on NASICON and In_2O_3 for detection of CO_2 at room temperature - modification with fo eign substances", *Sensors and Acutuators B*, **76**, 639-643 (2001).

[17] Y. Miyachi, G. Sakai, K. Shimanoe, N. Yamazoe, "Fabrication of CO_2 sensor usin NASICON thick film", *Sensors and Actuators B*, **93**, 250-256 (2003).

MICROSTRUCTURAL CONTROL OF SnO_2 THIN FILMS BY USING POLYETHYLENE GLYCOL-MIXED SOLS

Go Sakai
Department of Applied Chemistry
Faculty of Engineering
Miyazaki University
Gakuen-Kibanadai
Miyazaki 889-2192, Japan

Chiaki Sato
Department of Molecular and Materials Sciences
Interdisciplinary Graduate School of Engineering Sciences
Kyushu University
Kasuga-shi, Fukuoka 816-8580, Japan

Kengo Shimanoe, and Noboru Yamazoe
Department of Materials Science
Faculty of Engineering Sciences
Kyushu University
Kasuga-shi, Fukuoka 816-8580, Japan

ABSTRACT

Microstructural control of SnO_2 thin films was attempted by introducing various amounts of polyethylene glycol (PEG) into an aqueous sol of SnO_2 to be spin-coated. As measured for the powder samples derived from the sols, the addition of PEG cut the specific surface area of SnO_2 almost in half, while keeping the crystallite size almost unchanged. Pore size distribution analysis revealed that the average pore diameter was about 6 nm for the powder derived from the neat SnO_2 sol, while it was enlarged to about 20 nm by the addition of PEG1000 by 18 wt%. SEM observation of spin-coated thin films revealed that morphology changed from a dense packing of very fine particles to a more porous packing of larger particles with the addition of PEG1000 or PEG6000. The thickness of spin-coated film could be increased by increasing amount of PEG1000 or PEG6000 added. The sensor response of these films to H_2 gas increased with increasing film thickness. This tendency is considered to result because the porous structure of the film becomes better developed as the amount of PEG increases.

INTRODUCTION

Semiconductor gas sensors using metal oxides have been studied extensively to detect reducing or oxidizing gases in air. Among the various metal oxides so far tested, SnO_2 is the most attractive sensing material from a viewpoint of sensing properties as well as chemical and physical stability. It has been reported that the gas sensing properties of tin oxide-based semiconductor gas sensors are influenced by extrinsic factors such as grain size and microstructure. As reported by Xu et al., the sensor response to reducing gases (H_2 and CO) begins to increase sharply as grain size decreases to be smaller than a critical value (6 nm), which corresponds to twice the thickness of space charge layer [1]. As reported by Sakai et al., on the other hand, the depth of target gas penetration inside the sensing body is determined by a competition between the rates of diffusion and surface reaction of the target gas so that a microstructural modification in favor of the gas diffusion leads to an improvement in sensor response [2].

Recently, it has been reported that the stannic acid gel derived from $SnCl_4$ by hydrolysis with NH_4HCO_3 can be converted, through a hydrothermal treatment in an ammonia water, into a colloidal suspension (sol) of SnO_2, in which SnO_2 grains of about 6 nm in mean size are dispersed monolithically [3]. Moreover, the SnO_2 grains obtained are far more resistant to thermal growth at elevated temperature than the conventionally prepared SnO_2; the grain size can be small as 7 or 13 nm after calcinations at 600 °C or 900 °C, respectively. Although the sols can be used for preparing uniform thin films of SnO_2 on an alumina substrate by a spin-coating method [4,5], microstructural control of thin films has remained yet to be investigated. The present study has aimed at disclosing the way to the microstructural control of thin films. Our strategy to this goal was to modify the properties of the SnO_2 sol prior to the spin-coating by mixing polyethylene glycol (PEG1000 and PEG6000) in it.

EXPERIMENTAL

The stannic acid gel was obtained by hydrolyzing an aqueous solution of tin chloride ($SnCl_4$) with ammonium hydrogen carbonate (NH_4HCO_3). The stannic acid gel obtained was washed with water by repeating the procedures of suspending the gel into water and collecting it back by filtration and centrifugation. The sol dispersing nanoparticles of SnO_2 was prepared by hydrothermal treatment of the stannic acid gel in ammonia water (pH 10.5) at 200 °C for 3 h. The content of SnO_2 was about 2 wt%, and the average particle size of SnO_2 was 6 nm with fairly narrow distribution, as evaluated by a laser particle distribution analyser (LPA, LPA-3000/3100, Ohtsuka Electronics Co., Ltd.)

For preparing spin-coating dispersions, the above SnO_2 sol was condensed up to 4.5 wt% of SnO_2 content by evaporating water and added with various amounts of polyethylene glycol (PEG) with an average molecular weight of 1000 (PEG1000) or 6000 (PEG6000). To fabricate thin film devices, these dispersions were spin-coated (1500 rpm) on an alumina substrate attached with comb-type Au electrodes (electrodes distance 70 mm), followed by sintering at 600 °C for h. The thickness and microstructure of thin films were observed by FE-SEM (JSM-6340F, JEOL). The size of SnO_2 grains was estimated from the width of X-ray diffraction (XRD; RINT 2100, Rigaku) peaks based on Scherrer's equation. Gas-sensing experiments were carried out in a conventional flow apparatus equipped with an external heating facility. The sample gas typically used was 200 ppm H_2 diluted in dry air. The gas flow over the thin films was switched between the sample gas and dry air to record their responses in electric resistance. The senso response (S) was defined as the ratio of resistance in air (Ra) to that in the sample gas (Rg); S=Ra/Rg.

RESULTS AND DISCUSSION

Influences of PEG on the microstructure of SnO₂ powder

It is possible that the microstructure of SnO_2 thin films derived from PEG-added sols of SnO_2 may be different from that derived from a neat SnO_2 sol. However, it was difficult to evaluate it experimentally for the thin films, so that the influences of PEG were investigated for the powder samples. Three kinds of SnO_2 sols, i.e., neat, 18 wt% PEG1000-added, and 18 wt% PEG6000-added, were evaporated to dryness and calcined at various temperatures in the range of 300-900 °C for 3 h. The resulting powder samples were examined for microstructural features, i.e., size of SnO_2 crystallites, specific surface area, pore size distribution and total pore volume. It was confirmed by separate experiments that the PEG added was decomposed (burnt) away from the SnO_2 powder after calcination at 300 °C for 1 h. Figure 1 shows the crystallite sizes of SnO_2 as a function of calcination temperature. As previously reported, the SnO_2 crystallites derived from the neat sol, which had been treated hydrothermally, showed remarkable resistance to thermal growth at elevated temperature: The crystallite size remained as small as 14 nm after calcination at 900 °C. As seen from the figure, the crystallite sizes as well as the thermal growth behavior of crystallites were almost unchanged for the SnO_2 powders derived from the PEG-added sols. This confirms that the addition of PEG hardly affects the crystallite size of SnO_2 and the thermal growth behavior of it.

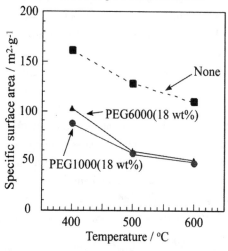 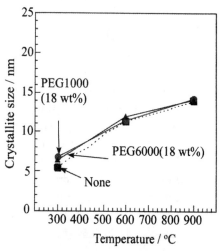

Fig. 1 Crystallite size of SnO_2 derived from neat or PEG-added sols as a function of calcination temperature (powder samples).

Fig. 2 Dependence of specific surface areas of SnO_2 derived from neat or PEG-added sols on calcination temperature (powder samples).

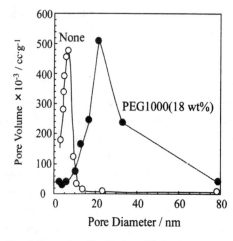

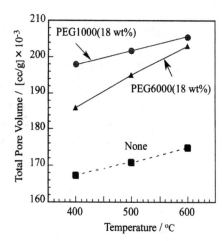

Fig. 3 Pore size distribution for SnO₂ powders derived from neat and PEG1000 (18 wt%)-added sols. (after calcination at 600 °C)

Fig. 4 Total pore volumes for SnO₂ powders derived from neat and PEG-added sols as correlated with calcinations temperature.

On the other hand, the specific surface area (SSA) of SnO₂ was found to be influenced significantly by the addition of PEG, as shown in Fig. 2. For the powder sample derived from the neat sol, SSA decreased monotonously with increasing calcination temperature, reflecting the thermal growth of SnO₂ crystallites. By the addition of PEG, however, SSA was reduced remarkably, being almost halved, as shown, despite that the crystallite size was almost the same as just mentioned. This indicates that several SnO₂ crystallites are bonded together into a fused particle when PEG has been added. When spherical particles are assumed, the SSA data after calcination at 600 °C correspond to particles of 7 and 17 nm in diameter for the powders derived from the neat and PEG (18 wt%) added SnO₂ sols, respectively. These sizes were comparable to those observed by SEM.

Pore size distribution analysis revealed that the average pore size was about 6 nm in diameter for the powder derived from the neat SnO₂ sol, while it was enlarged to about 20 nm when 18 wt% PEG1000 was added (Fig. 3). Moreover, the total pore volumes were made significantly larger by the addition of PEG, as shown in Fig. 4. These results indicate that, when the PEG-added SnO₂ sols dries up, SnO₂ crystallites are grouped into small clusters leaving pores among the clusters. In this way, clustering of crystallites and development of pores are brought into effect by the addition of PEG. These features are expected to be also common to the spin-coated thin films.

SEM observation of thin films

SEM images of the spin-coated thin films after calcinations at 600 °C are shown in Fig. 5.

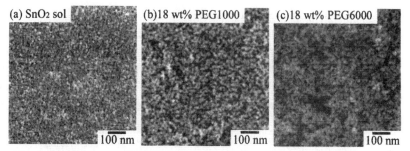

(a) SnO₂ sol   (b)18 wt% PEG1000   (c)18 wt% PEG6000

|100 nm |100 nm |100 nm

Fig. 5   SEM images of SnO₂ thin films derived from neat and PEG added sols.

Morphology changed from a dense packing of very fine particles for the neat SnO₂ sol-derived film to a more porous packing of larger particles for the films derived from 18 wt% PEG1000 or PEG6000-added sols, in coincidence with the microstructural changes just observed for the SnO₂ powders. The viscosity of the SnO₂ sols increased rather sharply with increasing content of PEG added, so that the film thickness attained by a single spin-coat could be varied between 100 nm and 1 mm by changing the PEG content.   As previously reported, the thin films derived from the neat SnO₂ sol suffered heavy cracks when thickness exceeded 300 nm.   However, the PEG-added SnO₂ sols gave heavy cracks-free films thicker than about 1 mm.

Sensing characteristics

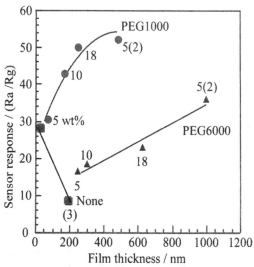

Fig. 6   Sensor response of three series of SnO₂ sol-derived thin films to 200 ppm H₂ at 350 °C as a function of film thickness.   The figure attached to each plot is the content of PEG 1000 or 6000 in the SnO₂ sols, while those in parentheses indicate the number of spin-coating times.

Figure 6 shows the sensor response of three series of thin films to 200 ppm H₂ at 350 °C as a function of film thickness. The sensor response of the films derived from the neat SnO$_2$ so decreased with increasing film thickness (by repeating spin coating), as already reported. In this case, the microstructure was kept essentially the same, and the decrease of response can be understood as a simple thickness-related effect on the basis of reaction-diffusion theory [2]. In contrast, the response of the films derived from PEG added SnO$_2$ sols increased with increasing film thickness. This tendency is considered to reflect that pore size and total pore volume increase with increasing PEG content. Although the increase of thickness is unfavorable for the response even in these cases, the development of porous microstructure gives an effect more than compensating the thickness effect. As also seen from the figure, PEG1000 tends to be more effective than PEG6000 in promoting the response. This is consistent with that the former gives larger total pore volume than the latter does (Fig. 4), though more careful investigations on the microstructure are needed to explain the observed differences in effectiveness in between.

CONCLUSIONS

The effects of the addition of polyethylene glycol (PEG) into an SnO$_2$ sol on the microstructure and gas sensing properties of SnO$_2$ powders or thin films were investigated. The followings could be concluded.

1) Although the addition of PEG hardly affect the size and thermal growth behavior of SnO crystallites, it facilitates clustering of SnO$_2$ crystallites and increases pore size and total pore volumes.

2) The morphology of the thin films changes from a dense packing of fine particles to a more porous packing of larger particles with the addition of PEG.

3) The sensor response increases with increasing PEG content, owing to the development of porous structure.

REFERENCES

[1] C. Xu, J. Tamaki, N. Miura, and N. Yamazoe, "Grain Size Effects on Gas Sensitivity of Porous SnO2-based Elements", Sensors and Actuators B, 3, 147-155 (1991).

[2] G. Sakai, N. Matsunaga, K. Shimanoe, and N. Yamazoe, "Theory of Gas-Diffusion Controlled Sensitivity for Thin Film Semiconductor Gas Sensor", Sensors and Actuators B, 80 125-13 (2001).

[3] N. -S. Baik, G. Sakai, N. Miura, and N. Yamazoe, "Preparation of Stabilized Nanosized Ti Oxide Particles by Hydrothermal Treatment", J. Am. Ceram. Soc., 83[12], 2983-2987 (2000).

[4] N. -S. Baik, G. Sakai, N. Miura, and N. Yamazoe, "Hydrothermally Treated Sol Solution of tin Oxide for Thin Film Gas Sensor", Sensors and Actuators B, 63, 74-79 (2000).

[5] N. -S. Baik, G. Sakai, K. Shimanoe, N. Miura, and N. Yamazoe, "Hydrothermal Treatment of Tin Oxide Sol Solution for Preparation of Thin Film Sensor with Enhanced Thermal Stability and Gas Sensitivity", Sensors and Actuators B, 65, 97-100 (2000).

MIXED-POTENTIAL TYPE CERAMIC SENSOR FOR NO$_x$ MONITORING

Balakrishnan Nair, Jesse Nachlas, Michael Middlemas, Charles Lewinsohn and Sai Bhavaraju
Ceramatec, Inc.
2425 South 900 West
Salt Lake City, Utah 84119

ABSTRACT

Ceramatec, Inc. is developing a NO$_x$ sensor through a program funded by the Environmental Protection Agency (EPA) to support the implementation of new emission control systems proposed for future diesel engine applications. In order to effectively implement these new emission control systems, a NO$_x$ sensor, capable of making real time measurements with fast enough response times, needs to be incorporated into the control system in order to provide the necessary feedback to the control system. Mixed potential type sensors offer the advantages of high-temperature operation and very good sensitivity to low NO$_x$ concentrations due to the logarithmic response to NO$_x$ concentration. Previously, mixed-potential type NO$_x$ sensor technologies have been limited by problems of cross sensitivity with other gas constituents commonly found in diesel exhaust, as well as an inability to provide a meaningful signal in varying NO/NO$_2$ mixtures. A novel sensor system design developed at Ceramatec has overcome many of the problems previously associated with such NO$_x$ sensors. Sensors have been fabricated and tested that have the following characteristics: (1) Operation temperature of 500-600°C; (2) excellent sensitivity in NO$_x$ levels of 1-1200 ppm; (3) response times as fast as 1-3 seconds; (4) insensitivity to various NO/NO$_2$ ratios in the exhaust stream; (5) very low cross-sensitivity to CO, CO$_2$, H$_2$O and low levels of SO$_2$; (6) ability to operate in oxygen-containing environments with no requirement for a pumping cell. The primary application targeted for the new sensor technology is heavy-duty diesel trucks. Other potential applications include advanced turbines, light-duty trucks, automobiles, heavy farm and construction machinery, off-road vehicles, and power generation as well as applications that utilize natural gas as the combustion fuel.

INTRODUCTION

Recently, a review paper was published that described the progress in utilizing mixed potential type ceramic sensors for use in combustion exhaust monitoring.[1] Mixed-potential based NO$_x$ sensors consist generally of an oxide sensing electrode on an oxygen-ion conducting electrolyte (e.g. zirconia), as shown schematically in Figure 1. The NO$_x$-sensing electrode makes electrical contact with a metal (usually platinum), that functions as a current collector. When the sensor is exposed to a NO$_x$ containing gas, a potential develops between the NO$_x$-sensing electrode and the reference electrode (E_2 in Figure 1). A part of this potential is due to the difference in oxygen concentration between the surfaces (E_1 in Figure 1), which can be measured by monitoring the potential difference between an O$_2$-sensing electrode and the air reference electrode. The difference, $\Delta E = E_2 - E_1$ represents the non-Nernstian mixed potential generated due to the presence of the NO$_x$ gas. The magnitude of ΔE is a measure of the concentration of the NO$_x$ species in the target gas.

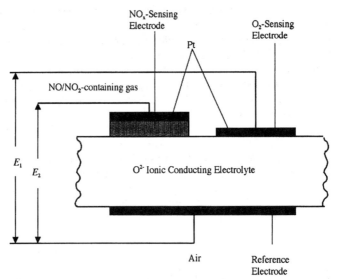

Figure 1   Basic schematic of a conventional NO_x sensor based on a mixed-potential based sensing electrode.

The mixed potential, also referred to as non-Nernstian potential, is generated due to the occurrence of two simultaneous electrochemical reactions at the NO_x-sensing electrode. A significant research effort has been carried out by a number of groups that has led to this understanding of mixed potential sensors.[2,3,4] The mixed potential generated is related to the following reactions occurring at the electrode/electrolyte interface:

$$\text{For NO:}$$
$$NO + O^{2-} \rightarrow NO_2 + 2e^- \tag{1}$$
$$1/2O_2 + 2e^- \rightarrow O^{2-} \tag{2}$$
$$\text{For } NO_2:$$
$$NO_2 + 2e^- \rightarrow NO + O^{2-} \tag{3}$$
$$O^{2-} \rightarrow 1/2O_2 + 2e^- \tag{4}$$

The NO_2 or NO in the gas mixture to be analyzed are partially (catalytically) reduced or oxidized to a mixture of NO and NO_2. Since the net current is zero under open circuit conditions, the mixed potential can be viewed as the potential required to drive the partial electrode reactions (1) and (2), or (3) and (4), at the same rates.

EXPERIMENTAL PROCEDURE

In the experiments reported here 8 mole % yttria stabilized zirconia was used as the electrolyte for the mixed potential sensor measurements. A variety of semi-conducting ceramic oxide materials were utilized as electrodes with varying degrees of success. The sensors were prepared by first coating a platinum air reference electrode on the inside of an $8Y-ZrO_2$ tube followed by coating the oxide material as the sensing electrode. This electrochemical cell was then installed in a sealed gas chamber that could be placed in a furnace for testing at a variety of temperatures. Experiments were performed by mixing oxygen, nitrogen and certified mixtures of NO/N_2, NO_2/N_2, CO/N_2 or SO_2/N_2 using mass flow controllers to measure and control the flow. In order to verify the gas composition we utilized a Eurotron Combustion Gas Analyzer to measure the gas after mixing. Water vapor was introduced into the gas stream by bubbling the gases through a temperature controlled water bath. As gas composition was varied to the sensor measurement chamber we measured the DC voltage signal using an Agilent 34970A data acquisition system. In addition, AC Impedance and DC polarization studies were performed using a Solartron 1260/1286 system.

RESULTS AND DISCUSSION

One of the main requirements for a NO_x sensor is that it be capable of operating in the temperature range typically found in an engine exhaust stream (i.e. 500 – 600 °C). The NO_x sensor response is shown in Figure 2 where the commonly observed phenomenon of mixed potential sensors is observed, i.e., as temperature increases the magnitude of the response decreases.

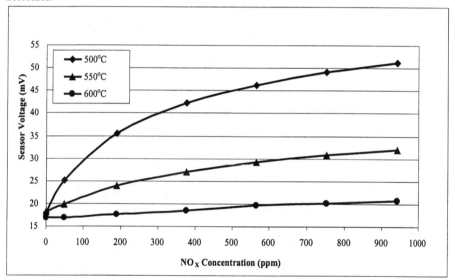

Figure 2 Sensor Voltage Response as a Function of Temperature

These results show that the ceramic based mixed potential sensor is suitable for operation in the temperature range required for automotive applications. The exact temperature of operation needs to be optimized while considering other system parameters such as response time and cross-sensitivity effects.

One of the biggest challenges with mixed potential sensors for measuring NO_x concentrations is the cross-sensitivity between NO and NO_2. It is commonly observed that the voltage signal for these two gases is of opposite sign, thereby making it impossible to establish a meaningful relationship between output voltage signal and total NO_x concentration. Typically, some means is employed to convert all of the gas to either NO or NO_2 to obtain a meaningful voltage relationship with NO_x concentration. We have experimented with a number of modifications to

485

our sensor system to achieve such a goal. As shown in Figure 3, four different types of modifications were performed to our sensor system. The total NO_x concentration was kept constant and the NO/NO_2 ratio was varied to determine which modification resulted in the most stable voltage signal. It can be seen that modification #1 resulted in the worst performance, with modification #2 showing slight improvement and modifications #3 and #4 showing acceptable performance. Essentially, modifications #3 and #4 demonstrate that regardless of the gas input NO/NO_2 ratio the sensor voltage maintains a constant voltage signal at a constant total NO_x concentration.

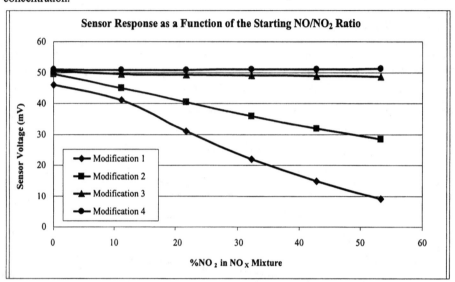

Figure 3   Demonstration of elimination of sensitivity to varying NO/NO_2 ratios in N2/5%O2/941 ppm NO_x mixtures through modifications to the sensor system

A series of tests were performed to test the effect on the sensor voltage response when exposed to other gases typically found in an exhaust gas stream. Figure 4 shows the effect of CO2 in the gas mixture. As can be seen there is no effect from the presence of CO_2 in the exhaust gas.

Initially, the sensor system showed substantial cross-sensitivity to CO leading us to modify the sensor system to eliminate this problem. In order to demonstrate this, we performed a test with CO prior to incorporating the modifications in the sensor system. The purpose of this test was to measure the sensor response to CO gas alone mixed into N_2 and O_2. Then, we incorporated the appropriate modifications and carried out the same test again with only CO mixed with N_2 and O_2. The results from these tests are shown in Figure 5. Figures 5(a) and 5(b) show that the voltage response to CO when no modification is present is a change of 62 mV for 1500 ppm CO.

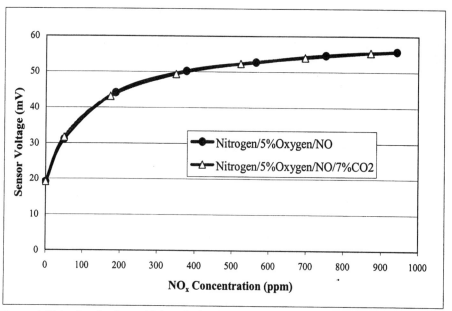

Figure 4 Data showing insensitivity of mixed potential sensor response to CO_2 concentration in the gas stream.

After incorporation of the appropriate modifications, the voltage signal generated at 1500 ppm CO is approximately 2 mV. These results clearly demonstrate the effectiveness of the modified sensor system in eliminating cross-sensitivity to CO.

The next gas constituent tested for cross-sensitivity was water vapor. The results from this test are shown in Figure 6. We chose 15% H_2O as this concentration represents the upper end of the concentration expected in an exhaust gas stream. As shown in Figure 6, this test demonstrated that the effect of water vapor on the sensor response is small enough to not be a problem for the successful development of a NO_x sensor.

Finally, we tested the cross-sensitivity effects of SO_2 on our sensor. SO_2 contamination is a serious problem in many solid state electrochemical devices and performance in SO_2 is a critical need for sensor development and commercialization. The EPA has mandated that by the year 2006, sulfur content in diesel fuel will be required to contain 15 ppm or less of sulfur while gasoline will be limited to 30 ppm. Therefore, we carried out experiments with SO_2 concentration of 15 ppm to test for cross-sensitivity. This test result was very encouraging, as we observed no measurable change in the sensor voltage in the presence of 15 ppm SO_2 as shown in Figure 7. This preliminary result is very promising as SO_2 is known to be a problem in many systems, particularly those where adsorption phenomena are involved; in previous work at Ceramatec it has been observed that ppm levels of SO_2 can rapidly degrade many solid state electrochemical devices. Long term tests need to be carried out in future research to study the effect of SO_2 further. While we had excellent results in our tests for SO_2 cross-sensitivity at the EPA-mandated limit of 15 ppm that diesel fuel manufacturers need to comply with by 2006, the tests carried out at the current emission limits of 300-500 ppm showed considerable cross-sensitivity and sensor degradation. Since the effects noticed at high SO_2 concentrations may be accelerated effects of processes that may occur during long-term testing at lower SO_2 concentrations, further testing will be performed on cross-sensitivity with SO_2 because of the importance of this issue.

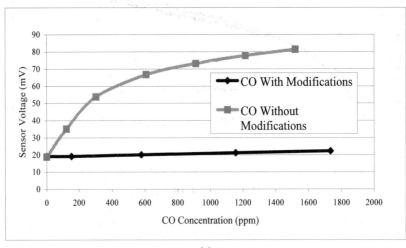

(a)

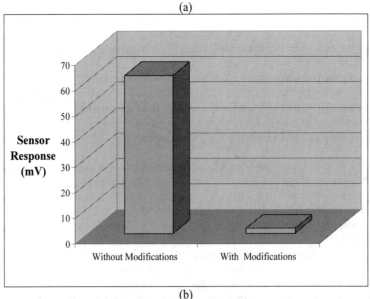

(b)

Figure 5 (a) Sensor Response to CO With and Without Modifications (b) Interpolated data from Figure 5(a) showing sensor response to 1500 ppm of CO with and without the modifications.

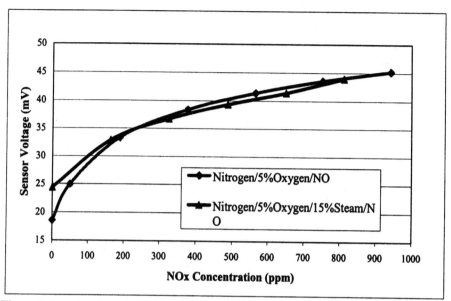

Figure 6 Data showing relative insensitivity of H_2O concentration on sensor mixed potential sensor response.

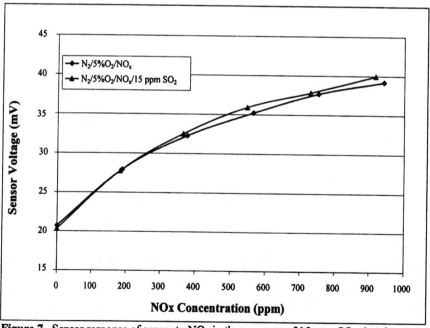

Figure 7 Sensor response of sensor to NO_x in the presence of 15 ppm SO_2 showing no significant cross-sensitivity or short-term degradation.

One additional objective in our sensor development program was to demonstrate a sensor detection limit of 15 ppm or less. Our experimental results showed that we were able to obtain a meaningful voltage signal at concentrations as low as 1 ppm total NO_x, thereby meeting the most stringent requirements of sensitivity for any of the current NO_x sensor applications. These test results are shown in Figure 8.

The final technical objective for our sensor development program was to demonstrate a sensor response time of approximately 1 second or less. This is an important requirement in order for the sensor to be useful as part of a control system. A typical response time curve for a sensor showing a 63% response time (time constant) of approximately 1.2 seconds is shown in Figure 9. The response is logarithmic and the 90% response time is approximately 2.4 seconds. This result was encouraging because it indicated that we are very near to the requirements necessary for a sensor to meet the necessary performance requirements. It is expected that with additional effort we will successfully meet the response time requirements necessary for our sensor to be utilized as part of a NO_x control and measurement system.

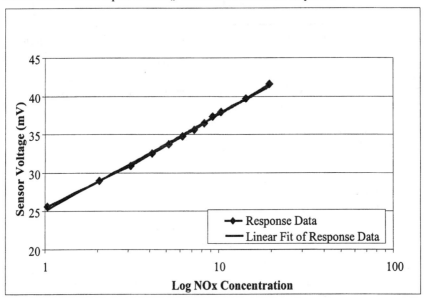

Figure 8 Sensor Response for Low NO_x Gas Composition

SUMMARY AND CONCLUSIONS
Preliminary results from Ceramatec's NO_x sensor development program indicate that some of the major problems associated with mixed potential sensors have been overcome. Specifically, the new sensor system eliminates the problem of cross-sensitivity between NO and NO_2 gases which has been a major stumbling block for mixed potential sensors previously. In addition, we have minimized the cross-sensitivity effects of CO and H_2O to the point where such effects do not present problems for sensor operation under targeted operating conditions.

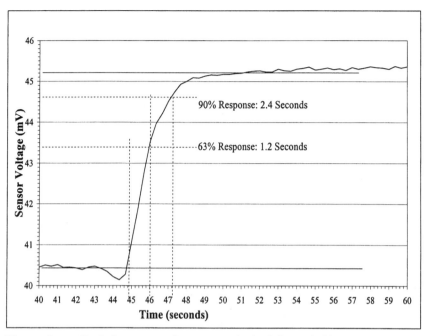

Figure 9 Typical response time curve obtained by switching the NO$_x$ concentration from 471 to 941 ppm (Flow-rate 500 cc/min).

We anticipate further improvements in performance in the presence of these gases by optimizing our sensor design. Initial experiments with SO$_2$ cross-sensitivity are promising if the requirement for tolerable limit is 15 ppm as mandated by the EPA. Finally, our preliminary response time testing has shown that we can achieve a response time close to one second. On-going work focuses on further optimization of sensor performance and long-term sensor testing.

REFERENCES

1   N. Miura, G. Lu and N. Yamazoe, "Progress in Mixed-Potential Type Devices Based on Solid
    Electrolyte for sensing Redox Gases," *Solid State Ionics*, **136-137** 533-542 (2000).
2   G. Lu, N. Miura and N. Yamazoe, "High Temperature Sensors for NO and NO$_2$ Based on Stabilized Zirconia
    and Spinel-Type Oxide Electrodes", *J. Mater. Chem.*, **7(8)**, 1445-1449, (1997)
3   S. Zhuiykov, T. Ono, N. Yamazoe and N. Miura, "High Temperature NO$_x$ Sensors Using Zirconia Solid
    Electrolyte and Zinc-Family Oxide Sensing Electrode", *Solid State Ionics*, **152-153**, 801-807, (2002)
4   W. Gopel, G. Reinhardt and M. Rosch, "Trends in the Development of Solid State Amperometric and
    Potentiometric High Temperature Sensors" *Solid State Ionics*, **136-137**, 519-531, (2000)

ELECTRODE MATERIALS FOR MIXED-POTENTIAL NO$_x$ SENSORS

D. L. West, F. C. Montgomery, and T. R. Armstrong
Oak Ridge National Laboratory
PO Box 2008, MS 6083
Oak Ridge, TN
37831-6083

ABSTRACT
The focus of this work is electrode materials for "mixed-potential" NO$_x$ sensing elements operating at temperatures near 650 °C. Several different metal oxides, including Cr-containing spinels and La-containing perovskites, were evaluated as sensing electrode materials and La-containing perovskites were used as conducting materials in the place of Pt. Evaluations included sensing response as a function of temperature, the differing response to NO and NO$_2$, and the effect of varying O$_2$ concentration. Strong responses to NO$_2$ were observed in several instances, and the NO responses were typically weaker and opposite in sign.

INTRODUCTION
The main pollutants from the combustion of low-sulfur fuels are carbon monoxide (CO), hydrocarbons (HC), and oxides of nitrogen (NO$_x$, a mixture of NO and NO$_2$). Combustion exhausts from spark-ignited, fuel-injected engines are currently passed over a three-way catalyst (TWC) that greatly reduces the levels of all three pollutants. This TWC loses its effectiveness for NO$_x$ removal at high O$_2$ concentrations,[1] so it cannot be employed for NO$_x$ remediation of the relatively O$_2$-rich exhausts from diesel and lean-burn gasoline engines.

NO$_x$ remediation of these exhausts may require techniques such as selective catalytic reduction (SCR) with reagent (HC and/or urea) injection. The amount of reagent injection during SCR is critical, as the reagents are themselves pollutants. Therefore it is essential to develop sensors that can rapidly and accurately assess the NO$_x$ levels in these exhausts. A suitable sensor for reagent injection would be operative at T ~ 700 °C and able to measure NO$_x$ concentrations in the range 100 ppm $\leq$ [NO$_x$] $\leq$1000 ppm.[2] At these temperatures the "NO$_x$" is predicted to be primarily nitrogen monoxide (NO), the dominant species above 500 °C.[3]

Two approaches to NO$_x$ sensing at "high temperature" (T ~700 °C) have appeared in the literature: "Amperometric"[4,5] and "mixed-potential"[6,7] methods. Amperometric techniques rely on measuring the oxygen-ion current generated by the electrochemical decomposition of NO$_x$. In mixed potential sensing, cathodic (reduction) and anodic (oxidation) reactions involving NO$_x$ and O$_2$ occur simultaneously on a sensing electrode, across which the *net* current (cathodic + anodic) is held at zero.[8,9] Since each of the reactions (cathodic and anodic) occurring on the sensing electrode should have a single-valued I-V characteristic, a unique voltage should be developed for a given [NO$_x$] and [O$_2$].

In this work we are investigating materials for use in high-T, mixed-potential NO$_x$ sensing elements. In particular, different transition metal oxides are investigated as sensing electrode materials and electronically conducting oxides are investigated as replacements for Pt.

EXPERIMENTAL METHODOLOGY
Figure 1a) shows the geometry of the sensing elements prepared for the present investigation.

The YSZ (8 mol% Y_2O_3-substituted ZrO_2, TZ-8Y, Tosoh, NJ) substrate was tape cast, laminated, and sintered at 1350 °C for 2 hr in air to produce disks about 16 mm in diameter and 1 mm in thickness. An electronically conducting layer, comprising the reference electrode (RE) and current collector (CC) was screen-printed onto one broad face of the YSZ disks, air-dried, and fired at 1100 °C for 0.3 hr in air. A second screen-printed layer, the sensing electrode (SE), was then patterned over a portion of the CC, air-dried, and fired at 1100 °C for 1 hr in air.

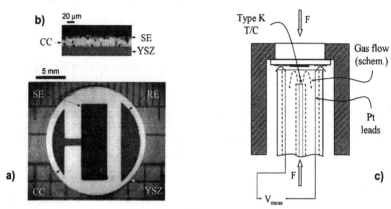

Figure 1: Specimen geometry and test fixture. a) shows the electroded surface of a prototype sensing element. The current collector (CC) extends underneath the sensing electrode (SE) as shown in b). A schematic of the fixture used for electrical connections and provision of the test atmosphere is shown in c).

The sensing elements were mounted in the fixture shown in Fig. 1c). A gas mixing unit (Environics (Tolland, CT) 4000) was used to mix N_2, O_2, and NO_x (NO or NO_2, 5000 ppm_V in N_2)) at room temperature. Mixture compositions were in the range 7 vol % $\leq [O_2] \leq$ 20 vol %, 300 $ppm_V \leq [NO_x] \leq$ 1500 ppm_V, with the balance being N_2. These gas mixtures were presented to the electroded side of the sensing elements as shown schematically in Fig. 1c), and the voltage developed between the SE/CC and RE (V_{meas} in Fig. 1c)) was measured with a Keithley (Cleveland, OH) 617 electrometer. To simulate elevated temperature service, the fixture shown in Fig 1c) was placed (centrally located) in a horizontal tube furnace.

The sensing performance of the elements was evaluated with two different techniques. Isothermal testing was conducted (at 600 and 700 °C) to characterize the sensing performance as a function of $[NO_x]$ (at fixed $[O_2]$) and $[O_2]$ (at fixed $[NO_x]$). For these isothermal measurements the "sensing response" (ΔV) was taken as the change in voltage with a given $[NO_x]$ and $[O_2]$ relative to that measured with 0 ppm_V input NO_x and 7 vol % O_2. This "baseline" voltage was small, typically less than 2 mV in magnitude. To characterize the sensing performance as a function of T the sensing elements were subjected to 5 minute "pulses" of 450 ppm_V NO_x (in 7 vol % O_2, balance N_2) while the temperature was ramped at 120 °C/hr.

Table I lists some representative compounds that have been investigated for use as sensing electrode materials. Inks for screen-printing were produced from the powders in Table I using proprietary methods and materials. All the materials in Table I were used as sensing electrodes with Pt (ElectroScience (King of Prussia, PA)) as the RE/CC (recall Fig. 1) material. Significantly more compounds than those listed in Table I have been investigated as SE materials, but the observations presented here about the materials in Table I hold generally for most of the SE compounds (and mixtures) evaluated with the sensing element geometry shown in Fig. 1a).

Table I: Compounds investigated for use as sensing electrode materials

Material	Supplier
NiO	J. T. Baker (Philipsburg, NJ)
Cr_2O_3	EM Science (Gibbstown, NJ)
$NiCr_2O_4$ (NC2)	Cerac (Milwaukee, WI)
$CoCr_2O_4$ (CC2)	" "
$La_{0.80}Ca_{0.21}Cr_{1.01}O_3$ (LCC)	Praxair (Woodinville, WA)
$La_{0.6}Sr_{0.4}Co_{0.2}Fe_{0.8}O_3$ (LSCF)	" "

In addition to investigating materials for use as sensing electrodes, we have also explored the use of the electronically conducting oxides $La_{0.85}Sr_{0.15}CrO_3$ (LSC) and $La_{0.80}Sr_{0.20}FeO_3$ (LSF) as substitutes for Pt. The SE oxides chosen for this were LSCF and NC2. As was the case with the different SE materials, inks for screen printing the LSC and LSF were produced from commercially available (Praxair) powders. Table II lists the components of sensing elements that were fabricated to investigate the substitution of electronically conducting oxides for Pt.

Table II: Specimens prepared to investigate the substitution of oxides for Pt.

SE material	RE/CC material	Sample ID
$La_{0.6}Sr_{0.4}Co_{0.2}Fe_{0.8}O_3$	Pt	LSCF/Pt
$NiCr_2O_4$	Pt	NC2/Pt
$La_{0.6}Sr_{0.4}Co_{0.2}Fe_{0.8}O_3$	$La_{0.85}Sr_{0.15}CrO_3$	LSCF/LSC
$NiCr_2O_4$	$La_{0.85}Sr_{0.15}CrO_3$	NC2/LSC
$La_{0.6}Sr_{0.4}Co_{0.2}Fe_{0.8}O_3$	$La_{0.80}Sr_{0.20}FeO_3$	LSCF/LSF
$NiCr_2O_4$	$La_{0.80}Sr_{0.20}FeO_3$	NC2/LSF

RESULTS AND DISCUSSION
Evaluation of Sensing Electrode Materials
Figure 2a) shows NO_2 response traces (at 7 vol % O_2) for a sensing element constructed with LSCF as the SE material and Pt as the RE/CC material. A rapid and reproducible response to NO_2 at 600 and 700 °C was possible with the LSCF SE. Fig. 2b) shows that both LSCF and NC2 were capable of a strong response to NO_2 at 600 °C in 7 vol % O_2. The responses of these sensing elements to varying $[NO_2]$ at fixed $[O_2]$, as well as to varying $[O_2]$ at fixed $[NO_2]$, were well-described by logarithmic expressions:

$$\Delta V_{fixed\,[O_2]} = V' + m_1 \log[NO_2], \quad \Delta V_{fixed\,[NO_2]} = V'' + m_2 \log[O_2], \tag{1}$$

where V', V'', m_1, and m_2 are constants. m_1 and m_2 are taken here to define the *sensitivity* of the sensing element to $[NO_2]$ (at fixed $[O_2]$) or $[O_2]$ (at fixed $[NO_2]$), respectively. Smaller absolute value(s) of m_1 and m_2 correspond to smaller changes in ΔV with gas composition over the concentration ranges investigated.

The NO sensing performance for LSCF and NC2 (and the other materials listed in Table I) was poor, with input NO producing only a small (in comparison to NO_2) signal that was opposite in sign to that produced by NO_2. This asymmetry in sensing response held over a wide temperature range for all the materials in Table I. Representative data are shown in Fig. 3, which shows the response of the sensing elements with Cr_2O_3 and $NiCr_2O_4$ SE's to 450 ppm$_V$ NO$_x$ (in 7 vol % O_2) over the temperature range 450 °C ≤ T ≤ 700°C. Always, the NO$_2$ response is larger in magnitude and opposite in sign than that for NO. These observations are not unique, and have been reported by other investigators for "mixed-potential" NO$_x$ sensing elements of various geometries.[10]

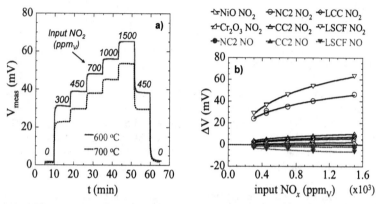

Figure 2: NO$_2$ response of a sensing element with a $La_{0.6}Sr_{0.4}Co_{0.2}Fe_{0.8}O_3$ (LSCF) SE and Pt RE/CC in 7 vol % O_2 (a)). b) shows the NO$_2$ response (at 600 °C and 7 vol % O_2) for the samples in Table I. The NO responses for selected materials in Table I are also shown. Lines drawn in b) are logarithmic fits for NO$_2$ and polynomial fits (2nd order) for NO.

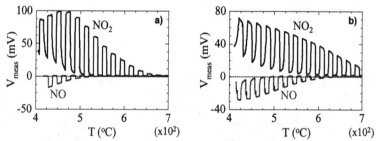

Figure 3: NO$_x$ response of sensing elements with Cr_2O_3 (a)) and $NiCr_2O_4$ (b)) sensing electrodes as a function of T. Data collected by pulsing NO$_x$ (5 min. 450 ppm$_V$ NO$_x$, 5 min. 0 ppm$_V$ NO$_x$, both in 7 vol % O_2) while ramping T at 120 °C/hr.

It is believed that the reduction of NO$_2$ to NO is responsible for the NO$_2$ sensing response, while the response to NO involves the oxidation of this species up to NO$_2$.[8] Given that NO is the dominant NO$_x$ species above 500 °C, it is not surprising that the response to NO is poor at higher temperatures if the sensing mechanism involves the oxidation of NO to NO$_2$. However, the weak NO response even at lower temperatures (~450 °C, Fig. 3) suggests that the oxidation of NO up to NO$_2$ is more difficult than the reduction of NO$_2$ to NO on these sensing electrode materials.

Substitution of Conducting Oxides for Pt

Figures 4a) and b) show the response (at 600 °C and 7 vol % O_2) of the sensing elements in Table II to input NO_2 and NO, respectively. The substitution of either LSC or LSF for Pt consistently resulted in a decreased sensitivity to NO_2 (i.e., $|m_I|$ in Eqn. (1) decreased), and an enhanced response to NO. Further, substituting LSC or LSF for Pt changed the algebraic sign of the NO_x responses for the sensing elements that had LSCF sensing electrodes.

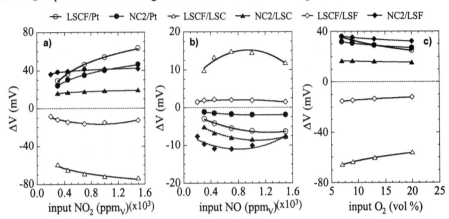

Figure 4: NO_x sensing performance (at 600 °C) of the sensing elements in Table II. The NO_x responses (7 vol % O_2) are shown in a) and b) while c) shows the $[O_2]$ dependence of the response to 450 ppmv NO_2. The lines drawn in a) (except for LSCF/LSF and NC2/LSF) and c) correspond to logarithmic fits, all other lines drawn are 2nd order polynomial fits.

Although use of LSC and LSF has enhanced the NO response at fixed $[O_2]$, Fig. 4b) indicates that the response is no longer a single valued function of [NO]. For example, for the LSCF/LSC sensing element, $(\partial V_{meas}/\partial[NO])$ is >0 for [NO] < 700 ppmv but <0 for [NO] > 1000 ppmv. Presently we believe that this reflects the sensing element responding to NO_2, as the two species can inter-convert via (where p stands for partial pressure)

$$NO(g) +1/2O_2(g) = NO_2(g), \quad K_p = p_{NO_2}/[p_{NO}\cdot(p_{O_2})^{1/2}]. \qquad (2)$$

The equilibrium (2) favors NO_2 at room temperature, and shifts to the left with increasing T.[3] It is difficult to state the exact position of the equilibrium (2) at the sensor surface, but formation of some NO_2 is predicted irrespective of the value of K_p. Further, the sensing elements with oxides as the conducting material had the smallest sensitivity for NO_2 (i.e., the voltage change induced by 300 ppmv NO_2 is nearly equal to that induced by 1500 ppmv NO_2 (Fig. 4a)) and thus these sensing elements might be predicted to respond strongly to small amounts of NO_2. Therefore, we tentatively attribute the observed "reversal" in the NO response (Fig. 4b)) to the formation of NO_2 in the mixture of NO, O_2, and N_2. The small "reversal" in the NO_2 sensitivity of the LSCF/LSF sample (Fig. 4a)), and the change in sign of the NO_x responses with the LSCF SE are more difficult to explain and remain under investigation.

Figure 4c) shows the $[O_2]$ dependence of the NO_2 response (450 ppmv NO_2, 600 °C) for the samples in Table II. The data are well fit by logarithmic expressions, but there is no systematic trend in the [O_2] sensitivity at fixed NO_2 (m_2 in Eqn. 1) as was found for the NO_2 sensitivity.

CONCLUSIONS

Investigation of compounds for sensing electrode materials has revealed two promising candidates for NO_2 sensing at 600 °C: $La_{0.6}Sr_{0.4}Co_{0.2}Fe_{0.8}O_3$ and $NiCr_2O_4$. The weak response of these and other materials to NO indicates that utilization of these "mixed-potential" sensing elements for NO_x sensing at T ~650 °C will be difficult, as NO is the dominant equilibrium NO_x species at these temperatures. It may be possible to substitute electronically conducting oxides for Pt in these types of sensing elements, and the substitution can affect the magnitude and sign of the NO_x and O_2 sensitivities.

ACKNOWLEDGEMENTS

The authors would like to thank B. L. Armstrong and C. A. Walls for producing screen-printing inks and the YSZ substrates, and T. Geer for metallography and microscopy. D. Kubinski and R. Soltis of Ford Research Laboratories provided consultation in exhaust sensing requirements.

Research sponsored by the Heavy Vehicle Propulsion System Materials Program, DOE Office of FreedomCAR and Vehicle Technology Program, under contract DE-AC05-00OR22725 with UT-Battelle, LLC.

REFERENCES

[1]J. T. Woestman and E. M. Logothetis, "Controlling automotive emissions," *The Industrial Physicist*, **1**, pp. 21-4, 1995.

[2]F. Menil, V. Coillard, and C. Lucat, "Critical review of nitrogen monoxide sensors for exhaust gases of lean burn gasoline engines," *Sensors and Actuators B*, **67**, pp. 1-23, 2000.

[3]K. B. J. Schnelle and C. A. Brown, "NO_x Control," in *Air Pollution Control Technology Handbook*. CRC Press, Boca Raton. 2001, pp. 241-55.

[4]W. Gopel, R. Gotz, and M. Rosch, "Trends in the development of solid state amperometric and potentiometric high temperature sensors," *Solid State Ionics*, **136-7**, pp. 519-31, 2000.

[5]N. Docquier and S. Candel, "Combustion control and sensors: A review," *Progress in Energy and Combustion Science*, **28**, pp. 107-50, 2002.

[6]N. Miura, G. Lu, and N. Yamazoe, "Progress in mixed-potential type devices based on solid electrolyte for sensing redox gases," *Solid State Ionics*, **136-7**, pp. 533-42, 2000.

[7]F. H. Garzon, R. Mukundan, and E. L. Brosha, "Solid-state mixed potential gas sensors: Theory, experiments, and challenges," *Solid State Ionics*, **136-7**, pp. 633-8, 2000.

[8]N. Miura, H. Kurosawa, M. Hasei, G. Lu, and N. Yamazoe, "Stabilized zirconia-based sensor using oxide electrode for detection of NO_x in high-temperature combustion-exhausts," *Solid State Ionics*, **86-8**, pp. 1069-73, 1996.

[9]M. N. Mahmood and N. Bonanos, "Application of the mixed potential model to the oxidation of methane on silver and nickel-zirconia catalysts," *Solid State Ionics*, **53-6**, pp. 142-8, 1992.

[10]N. Miura, M. Nakatou, and S. Zhuiykov, "Impedancemetric gas sensor based on solid electrolyte and oxide sensing electrode for detecting total NO_x at high temperature," *Sensors and Actuators B*, **93**, pp. 221-8, 2003.

STUDY OF HIGH SURFACE AREA ALUMINA AND GA-ALUMINA MATERIALS FOR DENOX CATALYST APPLICATIONS

Svetlana M. Zemskova, Julie M. Faas, Carrie L. Boyer, Paul W. Park
Caterpillar, Inc. Technical Center E/854, Peoria IL 61656

J. Wen and I. Petrov
Center for Microanalysis of Materials, University of Illinois at Urbana- Champaign, 104 South
Goodwin Ave., Urbana-Champaign, IL 61801-2902

ABSTRACT
    Various alumina and Ga-alumina samples were synthesized by sol-gel and mesoporous
material synthesis techniques employing Lauric acid and Pluronic P123. The pore properties of
the aluminas were thoroughly characterized using various analytical techniques including BET,
XRD, SEM and TEM. The surface area of the materials varied between 117-460 m^2/g, pore
diameter between 2.1-18.2 nm and pore volume between 0.27-1.24 cm^3/g. The Lean-NO$_x$
performance of the synthesized aluminas was evaluated on a powder test bench system using
propene as a reductant. The materials' deNOx performance was sensitive to synthesis conditions
and varied from 3-68% for NO conversion and 7-82% for NO$_2$ conversion.

INTRODUCTION
    Porous alumina materials have been widely used as a catalyst support because the materials
provide high surface area to disperse catalytic metal or metal oxide particles and enhance the
catalyst hydrothermal durability. In addition, the alumina actively participates in the reaction
mechanism to reduce NOx with hydrocarbon reductants in the presence of excess oxygen (so-
called Lean-NO$_x$ catalysis) [1]. Previous studies showed that a pure alumina (non-catalyzed)
demonstrated high NO$_x$ reduction performance especially when NO$_2$ or oxygenated
hydrocarbons were present in the gas stream [1]. As a result, the alumina has been considered to
be a strong candidate material for Lean-NO$_x$ and Plasma Assisted Catalysis technologies. On the
other hand, previous studies indicated that gallium doped alumina exhibited both high catalytic
activity for lean-NOx catalysis and high tolerance to sulfur poisoning [2-5]. The aim of this study
was to prepare alumina and Ga-alumina catalyst supports by different methods (precipitation,
sol-gel and sol-gel with structure directing agent) and compare their deNOx performance.

EXPERIMENTAL
    Sol-gel alumina and Ga-alumina were prepared in accordance with [2] using different
complexing agents: 2-methyl-2,4-pentanediol (MPD), 2,4-dimethyl-2,4-pentanediol (DMPD),
ethylene glycol (EG) and 2-methyl-2-propyl-1,3-propanediol (MPP). Ga(NO$_3$)$_3$, Ga(Acac)$_3$
(Acac = C$_5$H$_7$O$_2$) and Ga(OPr-i)$_3$ (OPr-i = i-C$_3$H$_7$O) were used as Ga-precursors. Ga-aluminas
were prepared by impegnation (IW) of sol-gel alumina with Ga(NO$_3$)$_3$, as well as by single-step
sol-gel (SS) using a mixture of Al(OPr-i)$_3$ and different Ga-precursors. Mesoporous alumina and
Ga-alumina were templated with Lauric Acid (LA) and Pluronic P123 following procedures
described in [6, 7]. Surface area (SA), pore volume (PV) and pore diameter (PD) distribution
measurements (Brunauer-Emmett-Teller analysis, BET) were performed using a Micromeritics
ASAP 2000 system. The powder samples were out-gassed under vacuum at 450°C overnight
prior to taking measurements. The surface area was determined by multi-point measurements

using several relative pressures of N_2 to He (N_2 surface area 0.162 nm^2) at 77 K. X-ray analysis (XRD) was performed on a Bruker AXS instrument with Cu Kα radiation in range from 10 to 90 degree 2θ. SEM (Scanning Electron Microscopy) analysis was performed on a LEO 1550 instrument at different magnifications. Transmission electron micrographs (TEM) were taken on a JEOL 2010LaB6 TEM equipped with a lanthanum hexaboride (LaB$_6$) gun operating at an accelerating voltage of 200kV. The samples were prepared by dispersing the powder product as a slurry in ethanol, which was then deposited and dried on a holey carbon film on a Cu grid. Selective Catalytic Reduction tests at temperatures of 350-625 °C were performed on a catalyst powder bench system at a space velocity of 30,000 hr^{-1} with a gas feed composition of 0.1%NO (or 0.1%NO$_2$), 0.1%propene, 7%H$_2$O, 9%O$_2$. Catalyst performance in the temperature range of 350-600°C was calculated as % conversion of NOx into nitrogen detected by Agilent 6890 Series GC System.

RESULTS AND DISCUSSION
Structural Characterization

BET data for aluminas and Ga-aluminas prepared by different methods are summarized in Table I. N_2 isotherms for mesoporous aluminas with pore diameter less than 2.5 nm did not

Table I. BET characterization and deNOx performance of different materials.

Sample ID	Ga Precursor	SA, m^2/g	PD, nm	PV, cm^3/g	Conversion, % NO	NO$_2$
Various Commercial Aluminas						
Alumina 1	none	130-350	3-27	0.2-1.1	18-52	25-65
Precipitation Method Alumina						
Alumina 2	none	170	10.8	0.56	25	46
Sol-Gel Alumina						
Alumina 3, MPD	none	225	14.5	1.2	47	67
Aumina 3a, DMPD	none	267	19.5	1.3	49	66
Alumina 3b, EG	none	216	8.1	0.6	14	20
Alumina 3c, MPP	none	206	15.6	0.8	43	65
Incipient Wetness Ga-alumina						
2.5%Ga-Alumina 4	Ga(NO$_3$)$_3$	215	11.0	0.78	52	78
6%Ga-Alumina 5	Ga(NO$_3$)$_3$	202	9.6	0.65	55	79
24%Ga-Alumina 6	Ga(NO$_3$)$_3$	148	5.3	0.27	66	82
50%Ga-Alumina 7	Ga(NO$_3$)$_3$	117	7.1	0.28	68	73
Single-Step Sol-Gel Ga-alumina						
2.5%Ga-Alumina 8	Ga(NO$_3$)$_3$	201	14.1	0.71	42	56
2.5%Ga-Alumina 9	Ga(acac)$_3$	232	10.1	0.59	43	60
2.5%Ga-Alumina 10	Ga(OPr-i)$_3$	292	17.0	1.24	49	66
5%Ga-Alumina 11	Ga(OPr-i)$_3$	276	14.8	1.02	52	65
25%Ga-Alumina 12	Ga(OPr-i)$_3$	186	18.2	0.85	68	74
Mesoporous Alumina and Ga-alumina						
Alumina 13	none	427	2.5	0.27	34	38
Alumina 14	none	418	4.2	0.45	34	58
Alumina 15	none	460	9.9	1.13	20	33
2.5%Ga-Alumina 16	Ga(OPr-i)$_3$	292	2.1	0.15	14	24

exhibit hysteresis loops, while isotherms for aluminas with larger pore diameters had a pronounced hysteresis loop at relative pressures of $p/p_o\sim0.4\text{-}0.9$ (Fig. 1). It can be seen that pure alumina and 2.5%Ga-alumina prepared with a structure-directing agent have the highest surface areas. Employment of Pluronic P123 results in the formation of a mesophase with 9.9 nm pore

Figure 1. Typical N_2 Isotherm and Pore Size Distribution curve for mesoporous Alumina 14.

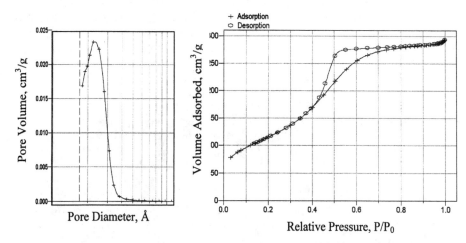

size (Alumina 16) while the mesophase prepared with LA had 2.1-4.2 nm pore size and low PV (Alumina 13, 14 and 16). Ga-loading onto alumina resulted in a decrease of SA, PD and PV. Different Ga-precursors resulted in the formation of different sol-gel 2.5%Ga-aluminas: Ga-alumina 10 prepared from Ga(OPr-i)$_3$ had the highest SA, PD and PV, while Ga-alumina 9 prepared from Ga(acac)$_3$ had quite low PV and PD. The comparison of the Ga-aluminas prepared by incipient wetness (IW) and single-step (SS) methods shows that pore diameter for IW samples gradually decreases as Ga content increases, while pore diameter for SS samples pass the minimum at 5%Ga load. Probably, there are Ga layers on pore surfaces in case of IW samples, while in the

Figure 2. XRD patterns of different Ga-aluminas

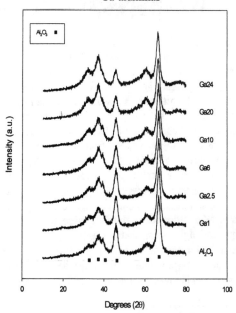

case of SS samples some amount of Ga is built into bulk material.

X-ray analysis of Ga-aluminas prepared by IW and SS methods showed that no crystalline phases were observed except γ-alumina (Fig. 2). Minor shifts of the peaks in the XRD patterns indicated that Ga coordinated into the alumina lattice and changed the alumina crystal structure. XRD patterns of Ga-alumina were identical before and after deNOx tests.

XRD patterns of mesoporous alumina and 2.5%Ga-alumina exhibited a single peak at ~1.2-1.7 and 0.7 degrees 2θ for mesophases templated with LA and P123 respectively. This leads to the conclusion that mesophases had a fully disordered pore structure. The d spacings calculated from peak positions were 5.3-7.7 and 13 nm (for LA and P123) and mesophase wall thickness was between 3.2-4.0 nm. HRTEM examination (Fig. 2 b) confirmed that the pore structure of mesophases is fully disordered and can be best described as sponge-like. Complimentary SAXS analysis is in progress to verify these conclusions.

Figure 2. SEM (a) and HRTEM (b) images of mesoporous alumina.

 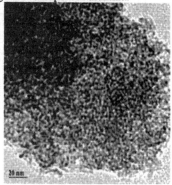

Catalytic performance

Selective catalytic reduction (SCR) of NO/NO$_2$ with propene over different aluminas are summarized in Table I. The deNOx performance of sol-gel alumina prepared with small size EG complexing agent is much lower than for aluminas prepared with bulky DMPD or MPD. The deNOx performance of sol-gel 2.5%Ga-aluminas prepared by single-step sol-gel with Ga(NO$_3$)$_3$ and Ga(acac)$_3$ were similar while employment of Ga(OPr-i)$_3$ resulted in higher NO/NO$_2$ conversion. The increase of Ga loading from 2.5% to 25% increases the NO conversion from 52 to 68% and NO$_2$ conversion from 56 to 82%. Ga-aluminas with low Ga content (<5%) prepared by IW showed better deNOx performance, however 25%Ga-alumina prepared by single-step sol-gel had the same performance as 50%Ga-alumina prepared by IW. In addition, it exhibited lower conversion temperatures of 525°C (NO) and 500°C (NO$_2$).

The maximum NO conversion over mesoporous aluminas is 3-34% and NO$_2$ conversion is 7-58%. NOx reduction activity of mesoporous aluminas prepared with LA is sensitive to solvent type and aging temperatures (Fig. 3, 4). Synthesis with sec-butanol (s-BuOH) and N, N'-Dimethylformamide (DMFA) as well as a lower aging temperature of 110° C results in formation of aluminas with better deNOx performance. The temperature for maximum NO conversion is lower for mesoporous aluminas than for sol-gel γ-alumina (525°C versus 550°C).

CONCLUSIONS

Various synthetic techniques (sol-gel, incipient wetness, structure directing agents) were employed to prepare alumina and Ga-alumina catalyst support materials. A single step process for Ga-alumina materials was less time consuming than the two step process of sol-gel alumina preparation with subsequent Ga loading by incipient wetness (IW) technique. The comparison of Ga-aluminas prepared by SS and IW methods showed that in both cases the surface area and

Figure 3. NO conversion with propene over mesoporous aluminas templated with LA at different temperatures and solvents.

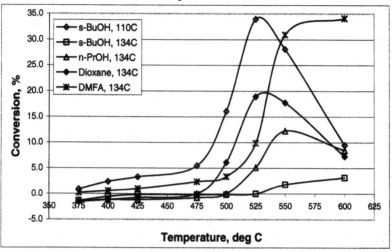

Figure 4. NO$_2$ conversion with propene over mesoporous aluminas templated with LA at different temperatures and solvents.

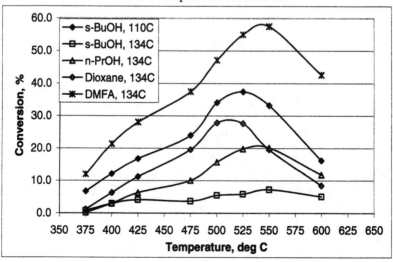

pore volume decrease as the Ga content increases, but SS samples have higher surface areas and pore volumes than the IW samples. Pore diameters for IW samples gradually decrease as the Ga content increases, while pore diameters for SS samples pass through a minimum at 5%Ga content. It was proposed that there were Ga layers on the pore surfaces in the case of IW samples, while in case of SS samples some amount of Ga was built into the bulk material. Therefore, SS samples may or may not have a continuous Ga layer on the pore surfaces. This resulted in different deNOx performance of the catalysts prepared by different methods.

XRD analysis indicated that no crystalline phases were present except γ-alumina for both IW and SS Ga-aluminas. Minor shifts of the peaks in the XRD patterns indicated that Ga coordinated into the alumina lattice and changed the alumina crystal structure parameters. XRD analysis of the mesophases performed at low angle showed that they were formed of amorphous alumina with a fully disordered pore structure.

In the case of the single-step sol-gel process the increase of Ga loading from 2.5% to 25% increases the NO conversion by 21% and NO_2 conversion by 15% over the parent alumina. 25%Ga-alumina prepared by SS method outperformed the 50%Ga-alumina prepared by IW. NOx reduction activity of mesoporous aluminas with propene is sensitive to synthesis conditions (solvent type, aging temperatures). There is not enough data at this point to identify correlations between pore structure and deNOx performance of mesoporous aluminas. However, it should be emphasized that mesoporous alumina is formed of amorphous alumina and thus differs from sol-gel γ-alumina which is crystalline.

ACKNOWLEDGEMENT

A part of the study was financially supported by the Office of Heavy Vehicle Technologies in U.S. Department of Energy under the contract #DE-FC05-97OR22579 and DE-AC05-00OR22725. TEM study was carried out in the Center for Microanalysis of Materials, University of Illinois, which is partially supported by the U.S. Department of Energy under grant DEFG02-91-ER45439.

REFERENCES

[1] P. W. Park, "Lean NOx Catalysis Research and Development," Diesel Engine Emission Reduction (DEER) Conference , Newport RI., August 24-28, 2003.

[2] T. Miyadera, K. Yoshida, "Alumina-supported catalysts for the selective reduction of nitric oxide by propene", *Chem. Lett.*, 1483-1486 (1993).

[3] T. Maunula, Y. Kintaichi, M. Inaba, et. al. "Enhanced activity of In and Ga-supported sol-gel alumina catalysts for NO reduction by hydrocarbons in lean conditions", *Appl. Catal. B.*, **15** (1998) 291-304.

[4] T. Haneda, Y. Kintaichi, H. Shimada, H. Hamada, "Ga_2O_3/Al_2O_3 prepared by Sol-Gel method as a highly active metal oxide based catalyst for NO reduction by propene in the presence of oxygen, H_2O and SO_2", *Chem Lett.*, 181-182, 1998.

[5] K. Shimizu, M. Takamatsu, K. Nishi, et. al. "Influence of local structure on the catalytic activity of gallium oxide for the selective reduction of NO by CH_4", *Chem. Commun.*, 1827-1828 (1996).

[6] F. Vaudry, S. Khodabandeh, and M. Davis, "Synthesis of pure alumina mesoporous materials", *Chem. Mater.*, 8 (1996) 1451-1464.

[7] W. Zhang, T. J. Pinnavaia, "Rare earth stabilization of mesoporous alumina molecular sieves assembled through an N°I° pathway", *Chem. Comm.*, 1998, 1185-1186.

DEVELOPMENT OF STRONG PHOTOCATALYTIC FIBER AND ENVIRONMENTAL PURIFICATION

Hiroyuki Yamaoka, Yoshikatsu Harada, Teruaki Fujii, Shinichirou Otani and Toshihiro Ishikawa

Ube Industries,Ltd.
1978-5 Kogushi, Ube City, Yamaguchi prefecture
755-8633, Japan

ABSTRACT

We developed an new *in-situ* process for production functional ceramic fibers with gradient surface layer by means of precursor methods using polycarbosilane. Using this new technology, high-strength photocatalytic fiber, which is composed of anatase-TiO_2 and amorphous SiO_2 phase, was developed. This fiber could effectively decompose harmful organic chemicals by irradiation with UV light.

INTRODUCTION

The photocatalytic activity of anatase-TiO_2 has attracted a great deal of attention. At present, most research has been performed using a film or powder material. Of these, powder photocatalysts have some difficulty in practical use, e.g., they have to be filtrated from treated water. Film photocatalysts cannot provide sufficient contact arca with harmful substances. In order to avoid these problems, other types of research concerning fibrous photocatalysts have been conducted. However, up to the present, they have not succeeded in combining excellent photocatalytic activity and high fiber strength as long as a sol-gel method was adopted. We have developed a new *in-situ* process by which functional surface layers with a nanometer-scale compositional gradient can be readily formed during the production of bulk ceramic components [1]. The basis of this process is to incorporate selected low-molecular-mass additives into either the precursor polymer from which the ceramic forms, or the binder polymer used to prepare bulk components from ceramic powders. Thermal treatment of the resulting bodies leads to controlled phase separation ('bleed out') of the additives, analogous to the normally undesirable outward loss of low-molecular-mass components from some plastics [2 –6]; subsequent calcination stabilizes the compositionally changed surface region, generating a functional surface layer. By using this new process, we have developed high strength photocatalytic fiber (TiO_2/SiO_2 fiber). In this paper, we describe the production process and properties of this photocatalytic fiber, and also introduce the environmental purification using this photocatalytic fiber.

EXPERIMENTAL PROCEDURE

Polycarbosilane containing an excess amount of titanium alkoxide was synthesized by the mild reaction of polycarbosilane $(-SiH(CH_3)-CH_2-)_n$ (20kg) with titanium (IV) tetra-n-butoxide (20kg) at 220°C in nitrogen atmosphere. The obtained precursor polymer was melt-spun at 150°C continuously using melt-blow type spinning equipment. The spun fiber, which contained excess amount of nonreacted titanium alkoxide, was heat-treated at 100°C and subsequently fired up to 1200°C in air to obtain photocatalytic fiber. In the initial stage of the heat-treatment at 100°C, effective bleeding of the excess amount of nonreacted titanium compound from the spun fiber occurred to form the surface layer containing a large amount of titanium compound. During the next firing process, the heat-treated precursor fiber was converted into a titania-dispersed, silica-based fiber with a sintered titania layer on the surface. The concept for production process of photocatalytic fiber is shown in Fig.1.

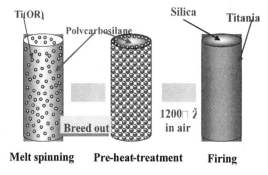

Fig.1   A new process for producing high strength photocatalytic fiber with gradient structure.

The microstructure of the fiber was investigated by SEM (Hitachi S-5000) and TEM (JEM 2010F). X-ray diffraction pattern of the fiber was measured by Rigaku X-ray diffractometer with CuKα radiation with a Ni filter. Chemical analysis of the fiber was made for three elements: Si (by a gravimetric method); Ti (by an inductivity coupled plasma method); O (by a combustion volumetric method).

Tensile strength of the fiber was measured according to the ASTM D3379-75 standard with a 25-mm gauge length.

The photocatalytic activity of the fiber was investigated by coliform-sterilization ability as follows. The fiber (0.2g) was placed in water (20ml) containing coliform (Escherichia coli IFO 3972) of 2×10^5/ml. Irradiation of UV light (352nm, $2mW/cm^2$) was performed at room temperature, and then a small amount of the water was extracted every 1 hour. After cultivation using the extracted water, the amount of active coliform was calculated from the number of formed colonies.

The purification effect for industrial wastewater was investigated using the purification equipment with a module composed of the cone-shaped felt material made of the fiber which was shown in Fig.2

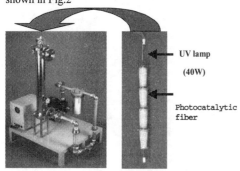

UV lamp

(40W)

Photocatalytic
fiber

Fig.2 The purification equipment
using photocatalytic fiber.

RESULTS AND DISCUSSION

This fiber contained 9.4% titanium, 43.6% silicon and 47% oxygen. And as can be seen from Fig. 3, this fiber was mainly composed of anatase-TiO_2 along with amorphous SiO_2. Although this fiber was fired at 1200°C in the production process, no obvious rutile phase could be observed. It is well known that anatase-TiO_2 converts to rutile at the temperature ranging from 700°C to 1000°C [7]. In particular, pure nonocrystalline anatase easily converts to rutile at lower temperature [8]. In the case of this fiber, it is thought that the surrounding SiO_2 phase caused the stabilization of the anatase phase. At the interface between TiO_2 and SiO_2, atoms constructing TiO_2 are substituted into the tetrahedral SiO_2 lattice forming tetrahedral Ti sites [9]. The interaction between the tetrahedral Ti species and the octahedral Ti sites in the anatase is thought to prevent the transformation to rutile.

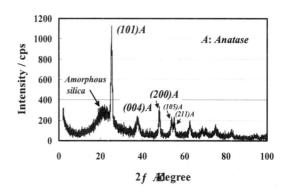

Fig.3 X-ray diffraction pattern of photocatalytic fiber.

Investigation by SEM showed that the outer surface of the fiber appeared to be smooth (Fig. 4a) and was covered with titania particles (Fig. 4b). Furthermore, the TEM image of the cross-section near the surface showed a TiO_2-sintered surface and particle-dispersed bulk structures (Fig. 4c). Most of the surface TiO_2 crystals (crystalline size: 8 nm, Fig. 4d) were directly sintered, whereas the internal TiO_2 crystals were bound with the amorphous SiO_2 phase (liquid phase sintered structure). The gradient-like structure resulted in strong adhesion between the surface TiO_2 layer and the bulk material, which is different from the behavior of other coating layers formed on substrates by means of conventional methods. Furthermore, a strengthening effect of the fine particles (<10 nm) dispersed in the bulk ceramics can be achieved at the same time. Tensile testing of the monofilaments according to the ASTM D3379-75 standard with a 25-mm gauge length demonstrated that the strength was markedly higher (>2.5 GPa) than that of ordinary sol-gel TiO_2/SiO_2 fibers (<1 GPa) [10].

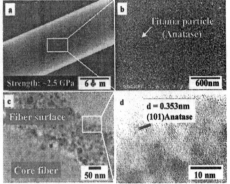

Fig.4 SEM micrographs and TEM image of photocatalytic fiber. (a, b: SEM micrographs of fiber surface; c, d: TEM image of cross section near fiber surface)

Figure 5 shows the results of extinction activity of coliform. In this experiment, using the desirable fiber covered with very fine TiO_2 crystals (8 nm), all of the coliform in the water was completely sterilized within 5 hours. In the comparative study, using undesirable fibers covered with large TiO_2 crystals (9 – 11 nm), sterilization of the coliform was markedly slow. From this results, the size of TiO_2 crystal is found to be closely related to the photocatalytic activity.

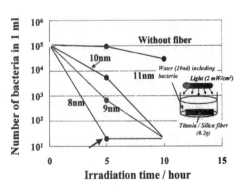

Fig.5 The test results of extinction activity of coliform using photocatalytic fiber.

Figure 6 shows the micrographs of water containing coliform before and after UV irradiation.

Coliform was no observed in the water after UV irradiation. In addition, we observed the generation of CO_2 gas from the water during the UV irradiation. This result means that coliform are not only sterilized, but also decomposed to CO_2 by photocatalytic activity.

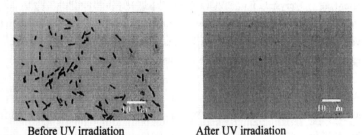

Before UV irradiation   After UV irradiation

Fig.6 micrographs of water containing coliform before and after UV irradiation.

Figure 7 shows the results regarding a decomposition of dioxin contained in a wastewater using the purification equipment with the photocatalytic fiber by circulation for 10 hours. Dioxins are well known as the very harmful organic chemicals which is very difficult to be decomposed. In this case, 95.1% of the dioxin was found to be decomposed after only 2 hours.

Figure 8 shows the results regarding a decomposition of PCB (polychlorinated biphenyl) contained in a wastewater using the purification equipment with the photocatalytic fiber. PCB is also well known as the very harmful organic chemicals which is very difficult to be decomposed. PCB was completely decomposed by passage through the purification equipment only one time.

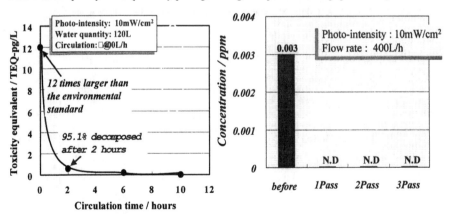

Fig.7 Decomposition of dioxin.   Fig.8 Decomposition of PCB.

CONCLUSION

A photocatalytic fiber was developed by using new in-situ process. This fiber was composed of anatase-TiO_2 and amorphous SiO_2 phase, and the surface of the fiber was covered with nanometer-size TiO_2 crystals. Furthermore, this fiber had very high tensile strength of over 2.5 GPa. This strength was sufficient to use this fiber on water purification.

The water purification equipment, which uses this photocatalytic fiber, could effectively decompose harmful organic chemicals like dioxins and PCB.

REFERENCES

[1]Ishikawa, T., Yamaoka, H., Harada, Y., Fujii, T., Nagasawa, T., Nature, 416(6876), 64-67 (2002).

[2]Kerk, S. L., Tay, S. C. & Hu, S. J., J. Electron. Mater.,18 Part 1, 117-121 (1989).

[3]Perovic, A., J. Mater. Sci., 20, 1370-1374 (1985).

[4]Perovic, A. & Murti, D. K., J. Appl. Polym. Sci., 29 Part 2, 4321-4327 (1984).

[5]Needles, H. L., Berns, R. S., Lu, W. C., Alger, K. & Varma, D. S., J. Appl. Polym. Sci., 25, 1737-1744 (1980).

[6]Perovic, A. & Sundararajan, P. R., Polym. Bull., 6, 277-283 (1982).

[7]C. K. Chan, J. F. Porter, Y. G Li, W. Guo, C. M. Chan, J. Am. Ceramic. Soc., 82 (3), 566-572 (1999).

[8]P. I. Gouma, P. K. Dutta, M. J. Mills, NanoStructured Materials, 11 (8), 1231-1237 (1999).

[9]C. Anderson, A. J. Bard, J. Phys. Chem. B, 101, 2661-2616 (1997).

[10]Abe, Y., Gunji, T. & Hikita, M., Yogyo_kyokai-shi, 94 (12), 1243-1245 (1986).

Processing of Biomorphous SiC Ceramics from Paper Preforms by Chemical Vapor Infiltration and Reaction (CVI-R) Technique

Daniela Almeida Streitwieser, Nadja Popovska, Helmut Gerhard, Gerhard Emig
Institute of Chemical Reaction Engineering, University Erlangen-Nuremberg
Egerlandstrasse 3, D-91058 Erlangen, Germany

Abstract

Chemical Vapor Infiltration and Reaction (CVI-R) technique is used to produce biomorphic high porous SiC ceramics derived from paper preforms. The first step is the carbonization of the biological structure to obtain a biocarbon (C_B) template. The C_B – template is then infiltrated with Si/SiC by CVI technique using methyltrichlorosilane in excess of hydrogen as a precursor. The reaction between carbon and silicon to silicon carbide is completed by a following thermal treatment (1250°C – 1600°C) of the samples as an additional processing step. The resulting SiC ceramics have extreme low density and show excellent mechanical properties, such as high strength and damage tolerance. Such properties make them suitable as catalysts supports for automotive applications, filters, foams or heat and noise insulation materials. The main advantage of the CVI-R technique compared to the other siliconization methods is the possibility to define the properties of the resulting ceramics by varying the infiltration and reaction conditions. Due to the relative mild infiltration temperatures (800°C – 900°C) it is possible to obtain ceramics retaining the exact micro as well as macro structure of the initial carbon preform.

Introduction

The development of advanced composites and fibrous materials has been at the frontier of materials science and technology for the last 3 decades [1]. Since the high demands in material performance cannot be fulfilled by conventional manufacturing technologies new methods have been developed to produce high porous ceramics [1]. Biomorphic ceramics are high porous ceramics obtained by carbonization and subsequent ceramization of biological preforms. Up to now the widest investigated methods are liquid Si infiltration [2, 3, 4], Si gas infiltration [2, 5, 6], SiO vapor infiltration [5, 7], polymer infiltration [8] and, the method used in this paper, CVI – R technique [9, 10, 11, 12]. SiC ceramics and its processing routes are so important, because of their excellent mechanical and physical properties and its chemical inertness even at high temperatures [13]. Porous SiC ceramics with low density and high strength are suitable as catalysts supports for automotive applications, filters, foams or heat and noise insulation materials.

Chemical Vapor Infiltration (CVI) has been used in the last decade for protection and densification of fiber reinforced ceramic matrix composites [14, 15] and for the surface modification of porous preforms [16]. CVI has the advantage, compared to the other processing methods, of depositing SiC at a relative low temperature range between 800°C and 1100°C [17]. For the ceramization process of biological structures into SiC ceramics, first a pyrolysis step is applied to convert the paper preform into a carbon lattice, obtaining thus the C_B – template. The next step is the deposition of a Si/SiC layer around the carbon fibers of the preform by Chemical Vapor Infiltration. In the subsequent high temperature reaction step the Si reacts with the carbon of the template forming a SiC ceramic [9]. By this conversion the highly specified structure of paper remains unchanged down to the micrometer level. The ceramics obtained from such biological preforms show high strength at very low density [9]. Strength, density and degree of conversion to SiC depend mainly on the infiltration and reaction conditions.

The isothermal and isobaric Chemical Vapor Infiltration process (ICVI) has been used in this work for the infiltration of C_B – templates derived from paper with the precursor gase Methyltrichlorosilane (MTS) in excess of hydrogen at low temperature (800°C – 900°C). MT deposits under such conditions SiC with excess of Si. This film composition is ideal for th processing of biomorphic ceramics from C_B – templates, since the free Si can react with th carbon of the preform during the thermal treatment. Additionally, the deposited SiC reinforce the structure in the micrometer level, since it deposits around the individual fibers and fortifi them. In this paper the degree of conversion of the biocarbon templates into a biomorphic Si ceramic will be analyzed. The final characteristics of the obtained ceramic can be affected b varying the MTS molar fraction during the infiltration process.

EXPERIMENTAL RESULTS

Experimental Equipment and Process Conditions

The experimental equipment used for the conversion of paper preforms by CVI – R to Si ceramics is a conventional tubular hot wall CVD reactor described previously [12]. The pap substrates are first pyrolyzed at 850°C using a low heating rate. The C_B – templates ca afterwards be infiltrated in the isothermal and isobaric CVI equipment. The infiltration performed at 850°C for 3 hours. The total pressure is kept constant at atmospheric pressure. Th MTS molar fraction is varied between 0.01 and 0.05 with excess of hydrogen. At this depositic conditions the films consist of Si rich SiC deposits, with an average Si/C ratio of 3. The therm treatment of the infiltrated preforms is set for 1 hour at 1400°C. During this treatment the solid solid reaction between the silicon and the carbon from the template takes place. In Fig. 1 a flo chart of the ceramization process with the individual steps is presented.

Fig. 1: Flow chart of the ceramization process of paper preforms to SiC ceramics

Ceramization of the Paper Preforms

Pyrolysis: The carbonization reactions have been studied by various authors extensively [2, 3, € Aim of the pyrolysis is the transformation of the biological molecules of the fibers, basical cellulose, hemi cellulose and lignin, to a carbon lattice. Therefore the samples are exposed to high temperature treatment in inert atmosphere. The heating rate is set very low to 1K/min up 350°C and 2K/min up to 850°C. This guaranties that the initial microstructure of the fibers is n destroyed during the carbonization process. The mass loss during the pyrolysis depends on th paper type and is around 80%. Shrinkage in all directions is also observed. The shrinkage length and height is 26% and lies somewhat higher in width at 33%. The major mass loss observed between 220°C and 350°C where volatile species like H_2O, CO, CO_2, aliphatic acid carbonyls and alcohols are released from the substrates [3]. At high temperatures polyaromat carbon compounds are formed. These species react further to build the carbon lattice structure they are transported partially out of the reactor as high viscous bitumen-like by-products. Th geometrical density of the C_B – templates is 0.125 g/cm³ with a void volume fraction of 0.91.

Fig. 5 a) and b) SEM micrographs of the paper fibers and the C_B – templates are shown. The fibrous structure of the paper preform is retained with high fidelity.

Chemical Vapor Infiltration: In the second step of the ceramization process the obtained C_B – templates are infiltrated with methyltrichlorosilane (MTS) in excess of hydrogen. A solid Si/SiC layer is deposited around each fiber in the template. The infiltration conditions are set to 850°C, 3 h infiltration time, residence time of the gases in the reactor of 1.5 s and MTS molar fraction between 0.01 and 0.05. In Fig. 2 the weight gain in percent is represented as a function of the MTS molar fraction. A linear dependence of the deposition rate expressed as a weight gain on the molar fraction is observed.

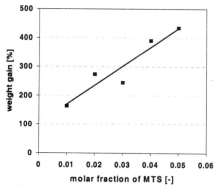

Fig. 2: Samples infiltrated at 850°C for 3 hours infiltration time

Reaction: The solid – solid reaction between the excess of silicon and the carbon from the C_B – template is performed at a temperature of 1400°C. At this temperature Si diffusion through the imperfections of the SiC layer and reaction with the C of the template take place. The degree of conversion of this reaction depends mainly on the available amount of Si in the deposited layer.

CHARACTERIZATION OF THE BIOMORPHIC CERAMICS

Composition of the Ceramics

The composition of the ceramized paper is analyzed by Raman spectroscopy. Using this method the different phases such as Si, SiC and C on the surface of the fibers can be detected. The intensity of the peaks increases with the crystallinity of the phases. As shown on Fig. 3 a), at low MTS molar fractions of 0.01 and 0.02 carbon is still detected on the fibers surfaces. The samples infiltrated with 0.03 molar fraction of MTS show only SiC peaks. And the samples infiltrated at 0.04 and 0.05 have an excess of Si. These results indicate that Si has not diffused completely through the layer and reacted to SiC. The total molar Si/C ratio of the ceramized samples, including the carbon from the template, can be calculated since the ratio of the deposited layers is known. The results are shown in Table 1. It can be seen that, including the carbon from the template, the samples infiltrated with 0.01 and 0.02 molar fraction MTS have less silicon than carbon, so not all carbon was able to react to SiC. The samples with 0.03 to 0.05 molar fraction MTS have an excess of silicon. This result confirms the observation done by Raman spectroscopy. When analyzing the samples thermogravimetrically (TGA) in air atmosphere a weight loss due to carbon oxidation is observed for all samples, as shown in Fig. 3 b). The weight loss, expressed in weight percent with regard to the initial mass, decreases with increasing molar fraction of MTS. So the samples infiltrated at 0.05 molar fraction MTS have a weight loss of only 5% after the TGA, whereas the samples infiltrated at 0.01 MTS molar fraction have 25% weight loss. The SEM micrograph in Fig. 4 shows the sample infiltrated at 0.03 MTS molar fraction. Residual carbon in the inner core of the fibers is detected. The shell consists of the infiltrated SiC with excess of Si.

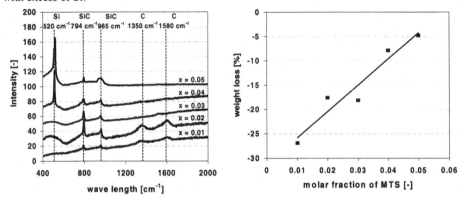

Fig. 3: a) Raman spectra and b) TGA of the samples infiltrated at different MTS molar fractions

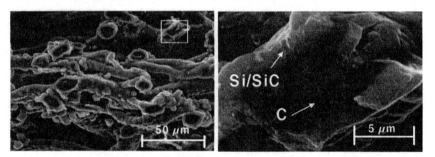

Fig. 4: SEM micrograph of biomorphic ceramic: a) 500 x magnification, b) 2000 x magnification.

Table 1: Analysis of the composition of the deposited layers

Infiltration parameters			total mass gain Δm	mass of infitrated Si [g]	mass of infitrated C [g]	total mass of C* [g]	moles Si, n_{Si}	moles C, n_C	ratio n_{Si} / n_C
T [°C]	x_{MTS} [-]	α [-]	[%]	Si [g]	[g]	C* [g]	n_{Si}	n_C	n_C
850	0.01	90	163.6	0.15634	0.02229	0.09408	0.00557	0.00783	0.73
850	0.02	45	274.0	0.21881	0.03119	0.10192	0.00779	0.00849	0.94
850	0.03	30	318.3	0.31596	0.03974	0.12606	0.01125	0.01050	1.08
850	0.04	24	390.6	0.28475	0.03635	0.10645	0.01014	0.00886	1.17
850	0.05	19	433.9	0.30743	0.04200	0.11104	0.01095	0.00925	1.21

* total mass of carbon is obtained by summing up the mass of the C_B- template and the mass of infiltrated C

Morphology of the SiC Ceramics

SEM micrographs of the C_B – template and the ceramized samples after oxidation by TGA are shown in Fig. 5. As it can be seen, the C_B – template retains more or less the fibrous characteristics of the paper preform after pyrolysis. The samples infiltrated at 0.01 up to 0.03 MTS molar fraction consist of relative thin ceramic shells. They are hollow, because of the oxidation of the residual carbon by TGA. These samples show also low strength values of 5 to 8

MPa. The samples infiltrated with 0.04 and 0.05 MTS molar fraction have much thicker layers that form clusters on the contact surfaces. The inner holes where the unreacted C remained before TGA are very small in both cases. This confirms the results of Raman and TGA that the reaction between silicon and carbon has taken place to a wide extend. The double ring bending strength measurements show strength values up to 15 MPa.

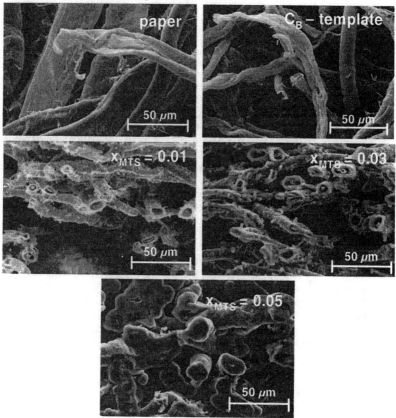

Fig. 5: SEM micrographs of the C_B – template and the ceramized samples after TGA

CONCLUSIONS

Paper preforms have been converted successfully into SiC biomorphic ceramics by CVI – R technique. After a pyrolysis step at 850°C with slow heating rates, the samples are infiltrated at 850°C with MTS in excess of hydrogen for 3 hours depositing Si/SiC. The reaction between the excess Si and the carbon from the template takes place during the subsequent thermal treatment. The resulting ceramics are analyzed by Raman spectroscopy, TGA and SEM micrographs. Depending on the infiltration conditions the samples show different properties. At low MTS molar fractions of 0.01 to 0.03 carbon is still detected by Raman and TGA. At higher molar fractions of 0.04 to 0.05 free Si is still observed on the surfaces by Raman. TGA shows in this case that only little carbon is left unreacted in the core of the fibers. These ceramics show also good mechanical properties at still high porosity.

ACKNOWLEDGEMENTS

The authors thank the German Ministry of Education and Research (BMBF) for the financial support as part of the program: Development of Biomorphic Ceramics for the Exhaust Gas Treatment (03N8018E).

REFERENCES

1 Zhou, B., Biomimetic Design and Test of Composite Materials, J. Mater. Sci. Technol, Vol. 9 (1993), pp. 9 – 19

2 Sieber, H., Hoffmann, C., Kaindl, A., Greil, P., Biomorphic Cellular Ceramics, Advances Engineering Materials, Vol. 2, No. 3 (2000), pp. 105 – 109

3 Herzog, A., Vogt, U., Graule, T., Zimmermann, T., Sell, J., Characterization of the Pore Structure of Biomorphic Cellular Silicon Carbide derived from Wood by Mercury Porosimetry, Ceramic Materials and Components for Engines, [Int. Symposium], 7th, Goslar, Germany (2001), pp. 505 – 512

4 Sieber, H., Friedrich, H, Kaindl, A., Greil, P., Crystallization of SiC on Biological Carbon Precoursers, Bioceramics: Materials and Applications III (Ceramics Transactions 110), The American Ceramic Society (2000), pp. 81 – 92

5 Sieber, H., Vogli, E., Greil, P., Biomorphic SiC – Ceramic Manufactured by Gas Phase Infiltration of Pine Wood, Ceramic Engineering and Science Proceedings (2001), 22 (25th Annual Conference on Composites, Advanced Ceramics, Materials, and Structures: B), pp. 109 – 116

6 Vogli, E., Sieber, H., Greil, P., Biomorphic SiC – ceramics prepared by Si – vapor phase infiltration of wood, J. of the European Ceramic Society 22 (2002), pp. 2663 – 2668

7 Vogli, E., Mukerji, J., Hoffmann, C., Kladny, R., Sieber, H., Greil, P., Conversion of Oak to Cellular Silicon Carbide by Gas – Phase Reaction with Silicon Monoxide, J. of American Ceramic Society, Vol. 86 6 (2001) pp. 1236 – 1240

8 Sieber, H., Friedrich, H., Zeschky, J., Greil, P., Lightweight ceramic composites from laminated paper structures, Ceramic Engineering and Science Proceedings (2000), 21(4, 24th Annual Conference on Composites, Advanced Ceramics, Materials, and Structures: B, 2000), pp. 129 – 134

9 Sieber H., Vogli, E., Müller, F., Greil, P., Popovska, N., Gerhard, H., CVI – R gas phase processing of porous biomorphic SiC – ceramics, Key Engineering Mat., 206 – 213 (2001), pp. 2013 – 2016

10 Ohzawa, Y., Sadanaka, A., Sugiyama, K., Preparation of gas – permeable SiC shape by pressure – pulsed chemical vapor infiltration into carbonized cotton cloth preforms, J. of Materials Science 33 (1998), pp. 1211 – 1216

11 Ohzawa, Y., Nakane, K., Gupta, V., Nakajima T., Preparation of SiC based cellular substrate by preesure – pulsed chemical vapor infiltration into honeycomb – shaped paper preforms, J. of Materials Science 37 (2002), pp. 2413 – 2419

12 Almeida Streitwieser, D., Popovska, N., Gerhard, H., Emig, G., Application of the Chemical Vapor Infiltration and Reaction Technique (CVI – R) for the Preparation of High Porous Biomorphic SiC Ceramics Derived from Paper, J. of European Ceramic Society, submitted for publication, Dec. 2003

13 Papasouliotis, G. D., Sotirchos, S., On the homogeneous Chemistry if the thermal decomposition of Methyltrichlorosilane, J. of the Electrochemical Society 141, 6 (1994), pp. 1599 – 1611

14 Kowbel, W., Rashed, A., Novel Ceramic Filter for Hot Gas Cleanup, Proceedings of 13th International Conference on CVD, Electroechemical Society, New Jersey (1996), pp. 607 – 611

15 Golecki, I., Recent Advances in Rapid Vapor Phase Densification of Refractory Composites, in Proceedings - Electrochemical Society (1996), 96-5 (Chemical Vapor Deposition), pp. 547 – 554

16 Dekker, J.P., Moene, R., Schoonman, J., The influence of surface kinetics in modeling CD processes in porous preforms, J. of Materials Science 31 (1996), pp. 3021 – 3033

17 Suzuki, K., Nakano, K., Kume, S., Chou, T.W., Fabrication and Characterization of 3D carbon fiber reinforced SiC matrix composites via slurry and pulsed CVI joint process, Ceramic Engineering and Science Proceedings 19, No. 3 (1998), pp. 259 – 266

FORMATION OF POROUS STRUCTURES BY DIRECTIONAL SOLIDIFICATION OF THE EUTECTIC

F. W. Dynys[*] and A. Sayir[*§]
NASA - Glenn Research[*] Center and Case Western Reserve University[§]
21000 Brookpark Rd.
Cleveland, OH 44135

ABSTRACT

Directional solidification of eutectics has self assembly characteristics that can fabricate two dimensional periodic arrays of two or more phases. The ordering of the phases can be utilized as a "lithographic technique" to produce porous structures for fabrication of miniature devices. Directional solidification by the Bridgman technique was applied to the eutectic system Si-TiSi$_2$. Patterned growth of TiSi$_2$ rods occur in the silicon matrix during solidification. Micro-channel structure of pillar arrays of TiSi$_2$ was produced by silicon dissolution by KOH. Average rod diameter was 2.5 μm with 99% of the population falling in the 2-3 μm range. The inter-rod spacing showed small dependence upon growth rate, varying from 4.3 μm to 5.7 μm. Colony formation could not be suppressed for the composition at and around the eutectic invariant point due to low thermal gradient to growth rate ratio and contamination from the crucible.

INTRODUCTION

Chemical reactions have long been the mainstay thermal energy source for aerospace propulsion and power. Micro-channel devices for in-situ and on-demand chemical production is an emerging technology that offers enhancements over conventional macro-processes.[1] Micro-channel devices are usually defined as miniaturized reaction systems with channel dimensions ranging from the sub-micrometer to the sub-millimeter range. Miniaturization enhances heat and mass transfer, offers portability, compactness and weight reduction. Various components are now reported in the literature, including injectors, pumps, mixers, electrodes, filters, heat exchangers and valves.

Micro-channel devices are practical for aeronautics application because of the reduced physical dimensions, weight reduction and potential efficiency enhancement. Aerospace applications include micro-reformers for electric propulsion of aircraft, micro-thrusters for micro-satellites to maintain precise orbit and sustain life support for planetary missions.[2,3]

Silicon has been a natural candidate material for testing micro-channel devices because of the matured micro-fabrication technology. Aeronautic operating environments will require materials capable to perform at higher temperatures under harsh chemical environments than silicon can sustain. A mature technology for metals and lesser degree for ceramics is directional solidification of the eutectic (DSE). It offers a potential method to fabricate micro-components. DSE has self-assembly characteristics that can fabricate two-dimensional periodic arrays of two or more phases by crystallization from the melt. The size and periodic spacing of rods or lamella normally range from 100 nm to 100 μm. DSE as a "lithographic technique" has been investigated for metallic and ceramic systems by Cline,[4] Angers et al.,[5] and Orera et al.[6] This paper reports work on DSE as a lithographic technique for the eutectic system Si-TiSi$_2$. The Si-TiSi$_2$ system was selected because it exhibits patterned growth of TiSi$_2$ rods. Relationship of microstructure with processing parameters will be discussed along preliminary results on etching of the phases.

EXPERIMENTAL PROCEDURE

Starting materials for ingot preparation were silicon granules and titanium powder of commercial purity, ≥99.9%. The silicon alloy compositions investigated are shown in Table 1. Boron nitride crucibles were used containing 30-40 g of the silicon alloy composition. Compositions were melted in an argon atmosphere in a graphite resistance heated furnace. Temperature was controlled by an optical pyrometer. Table 1 shows the melt temperature for each composition. Alloy compositions were soaked at the melt temperatures for 4-5 hours prior to directional solidification to achieve homogenization.

The directional growth of the alloy compositions was accomplished using the vertical Bridgman method. Pull rates for each alloy composition is given in Table 1. The measured temperature gradient was 8.5 °C/mm. Ingots were approximately 20 mm in length and 22 mm in diameter.

Table 1.
Solidification Conditions

Alloy (Wt. %)	Temp. (°C)	Pull Rate (mm/h)
25%Ti-75% Si eutectic	1525	130, 30 & 6
30%Ti-70% Si	1525	30 & 6
20%Ti-80% Si	1525	30 & 6

The BN crucible is reactive to titanium, this has been noted by Crossman and Yue for Ti-Ti_5Si_3 system.[7] Attempts using Al_2O_3 crucibles were not successful because they mechanically failed during solidification. The BN crucibles were sufficient in containing the melt under these process conditions.

Precipitates of TiB_x were observed at the crucible interface, as shown Figure 1. The precipitates range in size from 10-100 μm. Eutectic structure was not observed where the density of precipitates was high, as shown in Figure 1a. When the precipitate density was low, the eutectic structure co-existed as shown in Figure 1b. There appears to be no apparent interaction between the TiB_x precipitates and the eutectic structure. No effort was made to identify the exact chemical composition of the precipitates. Elemental analysis on the precipitates by EDAX show that the presence of B, Ti and N. Boron was not detected in the Si matrix or in the $TiSi_2$ rods. It is probable that the melt is contaminated with a small level of

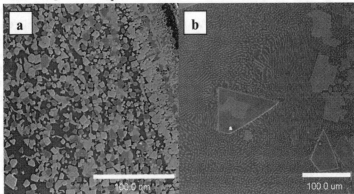

Figure 1. Polished cross-sections showing TiB_x precipitates near the crucible interface.

boron. Further analytical techniques were not utilized for this work. The crystal shapes of the TiB$_x$ precipitates were revealed by etching away the silicon matrix using KOH solution. Octahedrons, dodecahedrons and hexagonal shapes were observed.

RESULTS AND DISCUSSION

The understanding of the microstructural development of the eutectic systems is critical to control the pore or channel dimensions. The microstructural development of the Si-TiSi$_2$ system is initially examined under isothermal cooling (similar to casting) and structures are compared with those produced by unidirectional solidification.

The Si-TiSi$_2$ eutectic forms a rod structure under isothermal cooling condition. Figure 2 shows typical transverse microstructures of a solidified Si-TiSi$_2$ eutectic composition following 1 h soak at 1575 °C in an Al$_2$O$_3$ crucible. The dark and light gray regions are Si and TiSi$_2$, respectively. Colonies containing TiSi$_2$ rods and Si matrix have formed along with large TiSi$_x$ precipitates. Composition of the large precipitates (TiSi$_x$) is currently not known but incomplete homogenization of the melt could be the cause for their formation. It should be noted that TiSi$_2$ rods formed in the absence of boron contamination.

Directional solidification was employed along the thermal gradient to promote homogeneous rods or lamellae structure. The size of the rods or lamellae is dependent upon rate of solidification (R) and thermal gradient (G). The eutectic composition containing 75 wt. % Si was solidified at rates of 130, 30 and 6 mm/h. Phase analysis of the specimens by x-ray diffraction showed the expected phases, Si and TiSi$_2$. Figure 3 shows a transverse micrograph of the Si-TiSi$_2$ eutectic solidified at 6 mm/h. Large colonies, >100 μm, containing TiSi$_2$ rods are observed. Examination of the microstructure in the longitudinal direction showed no preferred alignment of the TiSi$_2$ rods with the thermal gradient. The boundary structure between colonies is irregular and contains coarser precipitates. Increasing the solidification rates to 30 and 130 mm/h does not significantly change the microstructure. The higher solidification rate produces larger size colonies. Similar microstructures are observed for off-eutectic compositions containing 70 and 80 wt.% Si.

The observed microstructures may be examined using Hunt and Jackson criteria. Hunt and Jackson[8] developed a method to predict eutectic microstructures based upon entropy of fusion.

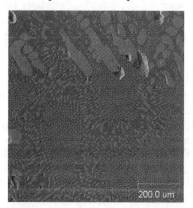
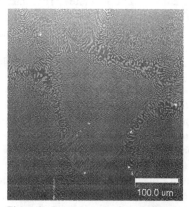

Figure 2. Transverse section of non-directional solidified Si-TiSi$_2$ eutectic

Figure 3. Transverse section of directionally solidified Si-TiSi$_2$ eutectic composition at 6

The eutectic structure will exhibit a regular morphology (rod-like or lamellar) if both phases possess low entropy of fusion, typically, $\Delta S/R < 2$, where ΔS is the entropy of fusion and R is the gas constant. For the Si-TiSi$_2$ system, both the Si phase and TiSi$_2$ have high entropies of fusion; ΔS_{Si} is 30 J/K mol[11] and ΔS_{TiSi_2} is estimated to be 54 J/K mol.[9] Both phases have high value of $\Delta S/R > 2$. The expected microstructure based on Hunt and Jackson

Table 2.
TiSi$_2$ Inter-rod Spacing

Solidification Rate (mm/h)	Spacing (μm)
6	5.7
30	5.6
130	4.3

criteria would be the irregular (faceted for both faces) structure. In contrary to this expectation, Figures 2 through 4 depicts a "complex regular" microstructure which has many of the features of fibrous eutectic microstructure within the colonies.

This apparent contradiction could be explained through the stabilization of isothermal liquid-solid interface through kinetic undercooling. In high entropy melting Si-TiSi$_2$ system, the kinetic undercooling of TiSi$_2$ phase is large. In this case, the kinetic undercooling can balance the undercooling due to composition undercooling providing local isothermal interface and consequently leading to rod-like structure formation. This has been also observed for eutectic (75 wt.% Si) and off-eutectic compositions (70 and 80 wt. % Si). At eutectic invariant point, the starting composition is expected to yield volume fraction of minor phase of 31 v% Si phase. However, the eutectic composition became Si rich by depletion of Ti through the reaction of BN crucible with the melt. The amount and spatial distribution of boron and its effect on the stabilization of the liquid-solid isotherm needs further study. Constitutional undercooling due to boron buildup can stabilize cellular interface producing cellular microstructure, Figure 3. Current efforts are focusing to attain homogeneous TiSi$_2$ rod-like structures using controlled amount dopant additions under high G/R ratio and inhibiting colony formation.

A significant aspect of eutectic growth under directional conditions is the relationship between solidification rate (R) and inter-rod spacing (λ). A common observation is that $\lambda^2 R =$ constant. Table 2 shows the dependence of inter-rod spacing on solidification rate, suggesting a linear relationship of $\lambda R =$ constant. Theoretically, however, for a given lamella spacing, a range of growth rates is possible[10] and we did not made any attempt to assign specific relation ship between the solidification rate (R) and inter-rod spacing (λ). Normally, the spacing (λ) for other material systems shows a larger dependence upon solidification rate. The mean TiSi$_2$ rod diameter is 2.5 μm as shown in Figure 5. The size ranges from 2-3 μm, 99% of the rods fall into this size range.

Chemical etching was performed on polished specimens using KOH solution. A 5 minute soak at 80 °C in a 10M KOH solution was sufficient to etch the silicon. Figure 6 shows that a pillar array of TiSi$_2$ can be readily fabricated. It was observed that there was a minor population where there was deep undercutting at the TiSi$_2$-Si interface. It was not determined if the undercutting was caused by cracks, impurity segregation or some other unknown cause. A selective wet chemical etch for TiSi$_2$ was not found in the literature[13]. Plasma etching was performed to evaluate its effectiveness using a CF$_4$/O$_2$ mixture. Figure 7 shows the etched microstructure. The sample was polished prior to the dry plasma etch. Selective etching is observed around the Si-TiSi$_2$ interface. Craters form around the TiSi$_2$ rods. Some of the TiSi$_2$ rods exhibit deep craters. The TiSi$_2$ is pyramidal in shape, ideal geometry for electrodes for ion discharges. Changing the O$_2$ content in the plasma changes the etch chemistry. The sample was etched with an O_2 content <5% which enhances F ion concentration and etch rates. Experiments have not been tried by using a passivating film to control the etching, this can be attained using O_2 content >7.5%.

The premise of this work was to produce ordered porous architectures that may be applicable to micro-device fabrication or template for part fabrication. Figures 6 and 7 depict structures that

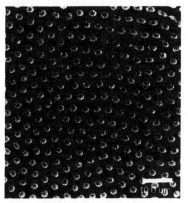

Figure 4. TiSi₂ rods in a Si matrix

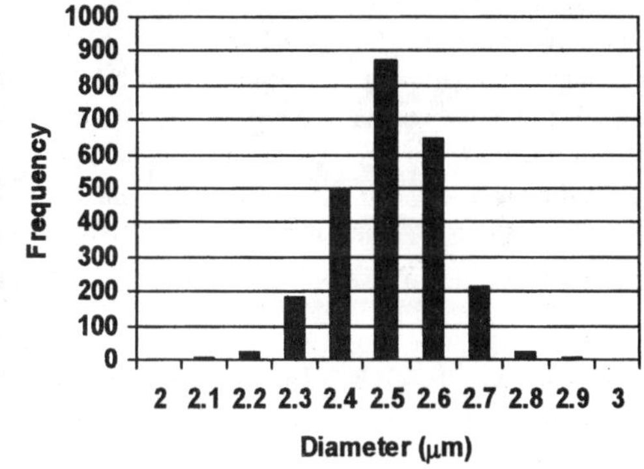

Figure 5. Diameter distribution of TiSi₂ rods.

can be fabricated using cost efficient directional solidification technology and environmentally benign etching procedure. This technique is a "bottom-up" method that exploits self-assembly characteristic of DSE. DSE offers a wide selection of materials where their unique properties can be utilized for operation in harsh environments. Anisotropic etch behavior between different phases provides the opportunity to fabricate structures with high aspect ratios (height to width on the order of 1000:1). Currently, aspect ratios are limited in semiconducting materials because the anisotropic etch properties are controlled by crystallography. Other benefits include fabrication simplicity, process control of microstructure and stability of the structure at high temperatures. Like other bottom-up methods, DSE is not suitable to make complex or interconnected patterns.

Nanotechnology has created a demand for new fabrication methods with an emphasis on simple and low cost techniques. DSE is an unconventional approach compared to the low temperature biomimetic approaches. Technical challenge for DSE is producing microstructural architectures on the nanometer scale. In both processes the driving force is the minimization of Gibb's free energy. Self assembly by biomimetic approaches depends upon weak interaction forces between organic molecules to define the architectural structure. The architectural structure for solidification is dependent upon strong chemical bonding between atoms. Constituents partition into atomic level arrangements at the liquid-solid interface to form polyphase structures. This atomic level arrangement at the liquid-solid interface is controlled by the atomic diffusion and total undercooling due to composition (diffusion), kinetics and curvature of the boundary phases. Judicious selection of the materials system and control of the total undercooling are the key to produce structures in nanometer scales.[10]

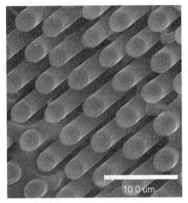

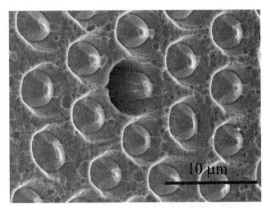

Figure 6. KOH etched DSE Si-TiSi₂.   Figure 7. Plasma etched Si-TiSi₂ microstructure.

SUMMARY

Directional solidification by the Bridgman technique was applied to the eutectic system Si-TiSi$_2$. Patterned growth of TiSi$_2$ rods occurred at compositions of 70, 75 and 80 wt. % Si. Titanium reaction with BN crucible caused possible boron contamination Thermal gradient in process was insufficient to align the TiSi$_2$ rods unidirectional. Large colonies with uniform rods were observed, average rod diameter was 2.5 μm. Inter-rod spacing showed small dependence upon pull rate. Pillar arrays of TiSi$_2$ were fabricated by selective etching of Si using KOH. Dry chemical etch using a CF$_4$/O$_2$ plasma exhibit selective etching at the Si/TiSi$_2$ interface. The dry etch transforms the TiSi$_2$ rods into a pyramidal shape. Further etch process mapping is needed to selective etch components to produce uniform porous architectures.

ACKNOWLEDGEMENT

This work was supported partly by NASA Aerospace & Power Base Nanotechnology Project, NASA Cooperative agreement NCC3-850 and Air Force Office of Scientific Research.

REFERENCES
[1] W. Ehrfeld, V. Hessel & H. Löwe , *Microreactors: New Technology for Modern Chemistry*, Wiley-VCH, NY, (2000).
[2] A. V. Pattekar et al., "A Microreactor for in-situ Hydrogen Production by Catalytic Methanol Reforming", *Proc. 5th Int. Conf. on Microreaction Technology*, Strasbourg, France. (2001).
[3] C. Rossi et al., Smart Materials & Structures, 10, (2001).
[4] H.E. Cline, In Situ Composites IV, Eds. P.D. Lemkey, H.E. Cline & M. McLean, Elsevier Science Pub. Co., Inc., pp.217 (1982).
[5] L.M. Angers et al., Ibid., pp. 205 (1982).
[6] V. M. Orera et al., Acta Materialia, **48**,4683 (2000).
[7] F. W. Crossman & A. S. Yue, Met. Trans., **2**,1545 (1971).
[8] R. Elliott , *Eutectic Solidification Processing,* Butterworths & Co., London, (1983).
[9] S. Meschel & O. Kleppa, Journal of Alloys and Compounds **267**,128 (1998).
[10] K. A. Jackson & J. D. Hunt, Trans. Met. Soc. AIME, **236**,1129 (1966).
[11] B. Vinet et al., J. Colloid & Interface Sci., **255**, 363 (2002).
[12] B. Lee, J. Mat. Res., **14**, 1002 (1999).
[13] P. Walker, W. Tarn, S. Smolinske, E. Previato & E. Marchisotto, *CRC Handbook of Etchants for Metals*, CRC Press, Boca Raton,FL, (1990).

HIGH SURFACE AREA CARBON SUBSTRATES FOR ENVIRONMENTAL APPLICATIONS

Kishor. P. Gadkaree
Corning Incorporated
SP-FR-05-1
Corning, New York 14831
USA

Tinghong Tao
Corning Incorporated
SP-DV-0201
Corning, New York 14831
USA

Willard A. Cutler
Corning GmbH
Abraham-Lincoln-Strasse 30
D-65189 Wiesbaden
Germany

ABSTRACT

A family of high surface area carbon honeycomb substrates has been developed for a variety of environmental applications. This technology is based on synthetic carbon precursors combined with a ceramic backbone to produce strong, abrasion resistant and durable carbon structures. These structures do not contain any binders and as a result are highly durable to solvent attack and are also highly electrically conductive. These structures contain from 5-95% carbon, which can then be activated to obtain surface areas in excess of 1000 m^2/g. In addition, catalyst precursors can be added to the carbon precursors to achieve catalyst dispersed in-situ within the carbon structure. These substrates are ideal candidates for industrial air pollution control, water purification, catalyst support, capacitive deionization and other applications. The materials and processing for these high surface area carbon substrates will be discussed. In addition, two applications will be discussed in more detail- their use as catalyst supports and in capacitive deionization of water.

INTRODUCTION

Activated carbons are widely utilized for removal of contaminants from fluid streams in many industrial applications. These applications include water purification, food processing, and VOC removal from paint booth exhausts, heterogeneous catalysis and others. Typically activated carbon powders or granules which are synthesized from natural precursors such as coal, wood, nutshells are often used. Although these materials are very successful, there are drawbacks associated with their use. For instance, the natural raw materials, properties of which vary depending on various factors such as the origin and the batch of material used, make it difficult to control the properties of the carbons from batch to batch. In addition, the powdered or pellet form necessitates the use of packed beds with associated high pressure drop, channeling and other disadvantages.

Monolithic honeycomb structures may be fabricated from synthetic resins and pyrolyzed to create activated carbon structures that address all the issues mentioned above. These structures have a very high surface area to volume ratio to allow efficient utilization of the material, very low pressure drop even at high flow rates and highly reproducible adsorption characteristics because of the synthetic precursors used. Two methods to fabricate such structures have been reported earlier [1-3]. One involves coating of porous ceramic honeycombs with a high carbon yields phenolic precursor resin followed by curing, carbonization and activation of the resin at high temperatures to yield a strong, controlled porosity carbon honeycomb structure. This method yield carbon honeycombs with a maximum carbon content of about 20wt%. For simplicity we will refer to monoliths produced by this method as carbon impregnated honeycombs (CIH). Another method involves extrusion of the phenolic resin with various additives to yield a resin honeycomb which is then cured, carbonized and activated to obtain a carbon honeycomb. The carbon percentage in the honeycombs obtained by this method may be controlled anywhere between 5-95wt%. For simplicity we will refer to monoliths produced by this method as extruded resin carbon honeycombs (ERCH). The honeycombs fabricated by both the methods are characterized by the fact that these honeycombs contain no binders and have a continuous carbon structure which is electrically conductive. To use these honeycombs as catalyst supports, either metal salts of noble or transition metals may be incorporated in the phenolic resin solution followed by the curing, carbonization and activation steps or the catalysts may be impregnated onto carbon after activation. Such structures have been successfully demonstrated in water purification/filtration [4], volatile organic compound (VOC) adsorption/desorption [5-6], mercury removal [7], catalysis [8] and capacitive deionization [9] applications. In this paper we describe two applications of the honeycombs structures related to capacitive deionization of water and as catalyst supports for certain industrial applications.

PROCESSING

CIH products start with a highly porous, commercially available, ceramic honeycomb (Corning Incorporated, Corning, NY). These honeycombs, which are most frequently cordierite ($Mg_2Al_4Si_5O_{18}$) in their fired composition may be fabricated with a wall thickness of anywhere from 0.003 to 0.05 cm and with cell densities from ~15 to 140 cells/cm^2. The cells may be square, rectangular, triangular, hexagonal or other shapes. Wall porosity may be adjusted from ~10% to 70%. The honeycomb is impregnated with high carbon yield, low viscosity polymeric resin. The resin chosen for this work was a liquid phenolic resole (Occidental Chemical Co. Niagara Falls. NY) due to its inexpensive nature (plywood glue) and very high carbon yield (50% of the cured weight). This aqueous resole has a viscosity of ≤ 100 cP and contains about 65% solids. This low viscosity allows the resin to soak into the ceramic honeycomb structure. The structure is drained and excess resin is removed by blowing air through the cells. The excess resin is then drained and the resin-coated honeycomb is dried (95°C) and cured (150°C) to crosslink the resin. The resin forms an interpenetrating network within the ceramic. The cured honeycomb is then pyrolyzed or carbonized in nitrogen at moderate temperatures (~900°C) to convert the polymer to carbon. Most of the cured resin remains in the cell wall porosity with a thin layer of resin on the surface. The coated honeycomb is then subjected to an activation process, in a steam or CO_2 environment. The activation is carried out at the same temperature to obtain a burn-off of ~25-30 wt.% carbon. The resulting structure has a continuous, high surface area carbon layer which is inseparable from the ceramic backbone.

ERCH products start with the same phenolic resin sources, combined with organic and inorganic additives [3]. The resultant viscous paste is extruded in through a die into the honeycomb shape. The extruded resin honeycomb is dried (95°C) and cured (150°C) to crosslink the resin. The resin forms the backbone of the structure and the resultant body is very strong. The cured resin honeycomb is then carbonized in nitrogen at moderate temperatures (~900°C) to convert the polymer to carbon. Some shrinkage is associated with the body at this point, but the resultant honeycomb is extremely strong and abrasion resistant (unlike extruding activated carbon powders) and is largely carbon (up to 95%). The extruded honeycomb is then subjected to an activation process, in a steam or CO_2 environment. The activation is carried out at the same temperature to obtain a burn-off of ~25-30 wt% carbon (larger carbon mass to start with than for CIH). The resultant structure has carbon surface areas of 1000 m^2/g.

For catalyst applications the catalyst can be applied in one of two ways. First, the catalyst can be introduced as a catalyst precursor and mixed intimately with the carbon precursor resin. Lower temperatures are then used to carbonize and activate the catalyst so that the catalyst dispersion is maintained. Alternately, activated carbon honeycombs (CIH or ERCH) can be impregnated with catalysts using insipient wetness method.

CARBON CATALYST SUPPORT EXAMPLE

Activated carbon has been widely used as a catalyst support due to its high surface area and unique catalyst supporting functionality. Due to versatility in raw material and variability of process, the activated carbon normally shows certain levels of variability in physical properties, especially, surface area and pore size. Using the novel carbon monolith making technology described previously, the activated carbon physical properties including surface area and pore size/distribution can be tailored for catalytic applications to provide superior performance as compared to conventional granular activated carbons.

Activated carbon honeycombs were fabricated using high carbon yield polymeric resin with low viscosity. Ceramic honeycombs coated with polymer resin were subjected to various carbonization and activation conditions to modify pore size and its distribution. Widely different pore size distributions have been obtained through tailoring the processing conditions. The pore properties were measured by nitrogen adsorption at 77K. Table 1 shows the physical properties of these honeycomb carbon catalysts. Surface area of these carbon catalysts ranges from 700 to 1000 m^2/g and pore volume ranges from 0.6 to as large as 1.55 cc/g carbon. At least 50 vol% of the pores are in the mesopore range. The fraction of meso-pores can be increased to about 80 to 90 vol%. In addition, pores centered at various sizes from 38 to 100 angstroms with the capability of engineering the bi-modal and tri-modal distributions in the mesopore range have been obtained.

Table 1. Physical Properties of Activated Carbon Supported Honeycomb Catalysts

sample	S.A., m2/g	Pore volume			total pore volume cc/g	frac. micro	frac. meso	frac. macro
		t-micro	meso	macro				
1	698	0.27	0.32	0.02	0.61	0.45	0.52	0.04
2	1041	0.29	0.81	0.03	1.12	0.26	0.72	0.02
3	863	0.17	0.91	0.07	1.16	0.15	0.79	0.06
4	828	0.04	1.30	0.08	1.41	0.03	0.92	0.06
5	939	0.32	0.59	0.03	0.95	0.34	0.62	0.03
6	922	0.09	1.28	0.17	1.55	0.06	0.83	0.11
7	853	0.01	1.35	0.61	1.97	0.01	0.68	0.31

Activated carbon supported honeycomb catalyst was produced by incipient wetness impregnation method, one of the most popular commercial approaches to make carbon-supported catalysts. About 1wt% of Pt was loaded onto the carbon honeycomb. The sample was dried at 120°C overnight and then calcined at 400°C for 2 hours in inert gas. The catalyst was reduced at 400°C in hydrogen for 2 hours. Toluene hydrogenation was selected to evaluate catalyst performance because there is a well-established relationship between catalytic activity and catalyst properties [10]. The reaction was carried out at 100°C and 50 mL/min H_2 and 0.01 mL/min liquid toluene. The space velocity is 864 (1/hr). For commercial catalysts, the equal amount of Pt in catalysts was charged into the reactor. The obtained samples were diluted with cordierite granules.

The catalytic activity (turnover frequency: s^{-1}) of the carbon honeycomb supported Pt catalysts in gas-phase toluene hydrogenation is shown in Figure 1 with comparison to commercial beaded catalysts of platinum on carbon (Pt/C) and platinum on alumina (Pt/ Al_2O_3). It is clear that novel carbon honeycomb Pt/C catalysts perform at least an order of magnitude better than commercial Pt/C and Pt/Al$_2$O$_3$ catalysts in terms of metal use efficiency. Furthermore, the four carbon honeycombs formed at different conditions (as listed in Table 1), which possess different pore size distributions have significant impact on their corresponding catalytic activity.

The data also indicate that the pore size distribution is a dominating factor in optimizing the hydrogenation activity of the novel catalysts. More importantly, the unique surface chemistry of this carbon creation process could also contribute to its superior catalyst performance.

CAPACITIVE DEIONIZATION EXAMPLE

Removal of ions from water is very important in a number of applications such as water softening, removal of heavy metals from industrial wastewater, ultra pure process water for electronics industry as well as desalination of brackish water and sea water. There are many industrial processes for removing ions, though each has some associated drawbacks, for example high-energy usage, durability, reliability, waste generation etc. A very important concept was proposed in 1960s [11-12] to develop such a process. The concept involves passing water-containing ions between a pair of oppositely charged electrodes. The ions are attracted to the appropriate electrode and removed, thus producing deionized water. For such a system the electrode material must be chemically inert, durable, electrically conductive and of high surface area. Although the concept showed promise, electrodes problems plagued early efforts. More recent efforts using carbon-coated titanium plates [13] resulted in significant improvements over the earlier work, but still may not be attractive from a commercial point of view.

Novel electrodes for a deionization device fabricated from CIH or ERCH structures due to their high surface area, chemical inertness and electrical conductivity. Conductive patches were intimately adhered to the surface of the carbon honeycombs and electrical contacts were attached to create honeycomb electrodes. The electrodes are arranged in an alternate manner where the flow of water is perpendicular to the honeycomb and water flows through the honeycomb cells, instead of between the electrodes as is traditionally done. Such a flow pattern allows for simple construction and a very low pressure drop (energy) penalty. By controlling the cell density and wall thickness, the effective flow area may be controlled to obtain very efficient contact between the ions in water and the electrode.

The electro adsorption experiments were done in a set up with the electrodes of 50mm x 50 mm x 5 mm thick immersed in the salt solution to a depth of 40mm. The solution was circulated through the electrodes via a peristaltic pump. Measurements of salt concentrations were done with an Orion model 1 15-conductivity device, which was first calibrated with standard solutions. All the experiments were done at 1.2 V DC potential applied across the electrodes. The salt concentrations were measured continuously as a function of time for various electrodes as well as the various salt solutions as noted. Salt concentrations were easily removed by washing or reversing the polarity of the electrodes.

Deionization experiments were done with calcium chloride and sodium chloride solutions at 500 ppm, 5000 ppm and 10,000 ppm concentrations. These particular salts were chosen to understand the effect of carbon structure on removal of both mono and divalent salts. These two compounds are also important to study from the point of view of applications in desalination and water softening. The salt concentration levels were chosen based on the fact that hard water has around 500 ppm salt concentration and brackish water has salt concentration of around 5000ppm. For these experiments the solutions were made by dissolving the appropriate amount of salt in distilled water. An ultrasonic preconditioning step, to remove air from the porosity was required for the electrodes before the start of experiments. Figure 2a shows sodium chloride removal as a function of time for an electrode pair at 10,000 ppm sodium chloride concentration as a function of surface area of electrode carbon. The initial slope of the curve gives the removal rate and the maximum amount adsorbed gives the capacity of the electrodes. The surface area measured on the electrodes ranged from 500 m^2/g to 1769 m^2/g. It may be expected that as the surface area increases the removal rate and capacity will also increase, since more carbon surface is available for ion removal, but experimentally there appeared to be an optimum surface area, where higher surface areas did not improve performance. This is thought to be due to the likelihood that increased surface area is likely more associated with increased pore depth rather than increased pore width.

Calcium chloride removal experiments were also done at the same concentration levels as the sodium chloride levels. The trends observed are identical to those seen in sodium chloride experiments, but calcium chloride is removed at slower rates and has a lower peak capacity (see Figure 2b).

To help separate the "structural" performance advantage, both plate and honeycomb electrodes were prepared in the same manner, with the same composition, carbon structure and carbon amount. Experiments were done under the same conditions of solution concentration, applied potential etc. The obtained results show the honeycomb electrode performance is substantially better than the plate electrodes because of the high surface area to volume ratio of the honeycomb structure. The total ion removal capacity is fourfold higher and the high removal rate is maintained fourfold longer than the plate electrode.

CONCLUSIONS

Novel, durable, high surface area honeycombs have been created which can be applied to a variety of environmental applications. Capabilities of tailoring pore size and its distribution of the polymer based honeycomb catalysts have been demonstrated. Two examples of their use have been demonstrated. As catalyst supports, these structures offer important efficiency benefits. As flow through honeycomb electrodes, high ion removal efficiencies and capacities are possible and the structures appear easy to clean.

REFERENCES
1. K.P. Gadkaree. US patent 5,820,967 (1998)
2. K.P. Gadkaree, Carbon 1998 ;36:981
3. K.P. Gadkaree, M. Jaroniec, Carbon 2000; 38: 983
4. W.A. Cutler, K.P. Gadkaree, T. Tao, US 6,227,382 (2001)
5. K.P. Gadkaree, WO9925449A1 (1999)
6. D.L. Hickman, T.V. Johnson, D.S. Weiss, US5759496A (1998)
7. K.P.Gadkaree, T.Tao, US 6,258334 (2001)
8. S. Dawes, K.P. Gadkaree and T. Tao, WO9917874A1 (1999)
9. K.P. Gadkaree and P. Marques, WO9826439A1 (1998)
10. J.L. Garnett, Catal. Rev. Vol. 5, 229(1971).
11. B.B.Arnold. and G.W.Murphy, J.Phy. Chem, 65,135(1961)
12. D.D. Caudle, J.H. Tucker, J.L.Cooper, B.B.Arnold -Res & Dev Progress report No 188, U.S. Dept of Interior (1966)
13. J.C. Farmer, J. Electrochem. Soc., Vol. 143, No., 159 (1996)

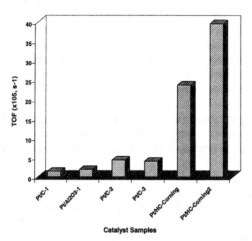

Figure 1. Turn-over Frequency of Carbon Supported Honeycomb Catalysts and Three Commercial Beaded Catalysts in Toluene Hydrogenation Reaction.

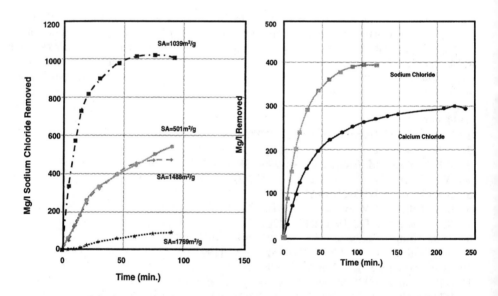

Figure 2: Graphs of: (a) NaCl removed as a function of time for various honeycomb electrodes with different surface areas (10,000 ppm concentration); and (b) NaCl and $CaCl_2$ removal as a removed as a function of time for identical honeycomb electrodes (5,000 ppm concentration).

DEVELOPMENT OF HIGH SURFACE AREA MONOLITHS FOR SULFUR REMOVAL

L. He, L. K. Owens, W. A. Cutler and C. M. Sorensen
Corning Incorporated
SP-DV-2-1
Corning, NY 14831

ABSTRACT
Removal of trace sulfur compounds from fuels is important in a wide variety of applications such as transportation and stationary fuel processors for fuel cells, diesel engines, and gasoline direct injection engine NOx traps. Several materials have been found useful for sulfur removal such as nickel metal on oxidic supports, zinc-oxide adsorbents, molecular sieves, and metal-sulfur sorbents. This paper discusses structured monolith development and processing of commercially available sulfur trap materials, including methods to make various sizes of monolith substrates with honeycomb cell geometry. Challenges to develop crack-free large samples from high surface area materials will also be discussed. The physical properties of monolithic parts are compared with that of pellets.

INTRODUCTION
Sulfur trap materials have a wide variety of applications [1-6]. These include transportation and stationary fuel processors, where sulfur poisons the downstream catalysts and the PEM fuel cell, and diesel engines and gasoline direct injection engines, where sulfur competes with NOx for adsorption sites on NOx traps. In these and other applications sulfur removal would extend the lifetime of catalysts and bring the technology to the forefront in a timely manner. Depending on the application, a monolithic sulfur trap could be used to remove the organic sulfur from either liquid (e.g. gasoline, diesel) or gaseous fuels (e.g. natural gas, vaporized gasoline). Two types of products are commonly used: (a) adsorbents to adsorb sulfur compounds by physi- or chemisorption mechanisms or (b) catalysts on which these impurities are chemically reacted and sulfur is retained.

Fuel reforming systems for generating hydrogen fuel from liquid or gaseous hydrocarbon fuel feed streams are well known. Published patent applications [7,8] disclose integrated fuel reformer systems incorporating multiple interconnecting reactor sections for continuous fuel processing, while another application [9] describes miniaturized fuel reforming systems offering shortened starting times and improved energy efficiency. Typical fuel reforming systems for processing hydrocarbon feed streams into hydrogen will comprise multiple stages or reactors to carry out the various steps of the hydrogen generation process. A common system design includes an initial reforming stage to produce carbon monoxide and water from an air-hydrocarbon-water feed, followed by a water-gas shift stage to generate hydrogen and carbon dioxide, followed by a preferential oxidation stage to oxidize residual carbon monoxide present in the feed prior to delivery to a fuel cell module. Conventional reformer and water-gas-shift (WGS) catalysts are sensitive to sulfur poisoning. Removal of trace sulfur will extend the lifetime of these catalysts and improve the performance of the fuel processor.

Fuel processing, including sulfur removal, can be accomplished with adsorbents or catalysts in several forms including monolith honeycomb structures, foams or beads. A monolithic sulfur trap provides advantages [10] including (a) lower pressure drop, (b) higher surface to volume ratio, (c) controllable thermal mass, and (d) thin walls to minimize transport by pore diffusion processes. Figure 1 illustrates pressure drop for various geometric forms including honeycomb monoliths, foams, and pellet packed bed systems. Monoliths clearly show the lowest pressure drop per unit length for the various structures tested[10]. Moreover, increasing the honeycomb cell density to increase the geometric surface area of the beds does not change bed pressure drop as significantly as design changes in the other media. In general, given constant honeycomb length, open frontal area, gas viscosity and volumetric flow rate, the pressure drop of these monolith adsorbents increases with increasing cell density, and decreases with decreasing wall thickness if cell density is kept constant. Thus monoliths are one of the most efficient methods available to pack high adsorbent surface area into a fixed volume while still maintaining low pressure drop.

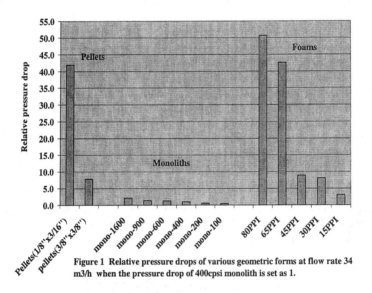

Figure 1 Relative pressure drops of various geometric forms at flow rate 34 m3/h when the pressure drop of 400cpsi monolith is set as 1.

This paper describes the development of high surface area monoliths for sulfur removal. Monoliths are extruded materials made from the same sulfur removal formulations currently available in pellet form. Batch compositions and processes to make cellular monolithic materials with various cell densities and web thicknesses will be discussed. The challenges of scaling up high surface area monoliths are also highlighted, including methods to solve drying- and firing-crack problems.

EXPERIMENTAL

Honeycomb sulfur adsorbents were made by the extrusion method. The honeycomb sulfur traps were composed of commercially available sulfur removal catalysts (high surface area nickel or zinc based materials). The commercially available sulfur removal pellets or tablets were ground to fine powder with the resulting powder having an average particle size of about 20 μm. The sulfur trap catalyst powder and solid additives such as organic and inorganic binders, surfactants, and plasticizers were mixed thoroughly in a turbula mixer. The powder mixture was placed in a muller and the liquid constituents of the batch were added as the mixture mulled to obtain a paste with the appropriate rheological properties. The mulled batch was forced through a desired substrate die to obtain monolithic substrates. The cell density and web thickness were controlled by the geometries of the dies. The green monoliths were then dried and fired. A variety of physical properties were measured including the BET surface area, and the pore size distribution measured using mercury porosimetry with pressures to 15,000 psi or 60,000 psi.

RESULTS AND DISCUSSION

Examples of representative supported nickel-based sulfur removal materials are reported in Table 1 below. All batch compositions are reported in parts by weight of the final batch mixture. The main ingredients of each batch consist of catalyst powder, water, and a temporary extrusion binder. In some cases additional extrusion aides such as oleic acid, lubricants such as metal stearate soaps, and/or permanent binders such as colloidal alumina are also employed. Extrudates of various outer diameters (OD) are formed, dried and fired, Table 1.

Table 1 - Extrusion Batches

Extrusion Batch Number	Extrudate OD cm - inch	Batch Composition (parts by weight)
1	1.9 - 0.75	100 parts catalyst/adsorbent powder; 5 parts methyl cellulose binder; 5 parts alumina binder; 30 parts oleic acid emulsion; 42.5 parts deionized water.
2	1.9 - 0.75	70 parts catalyst/adsorbent powder; 14 parts oleic acid emulsion; 3.5 parts cellulose binder; 37 parts water
3	14.4 - 5.66	80 parts catalyst/adsorbent powder; 5.6 parts cellulose binder; 0.8 parts stearate lubricant; 54.5 parts water

In their original pelleted form, the two types of Ni-based sulfur removal precursor materials had surface areas of ~ $300 m^2/g$. These high surface area materials generally require large liquid additions (60-80%) to reach appropriate rheological conditions. The large amount of liquid required in the extrusion composition makes the drying process more complicated and extremely difficult to obtain crack-free dried large samples with conventional drying methods. Normally, controlled humidity drying is needed to make crack-free large parts. Firing speeds, atmospheres and maximum temperature are also critical to control the decomposition of non-permanent binders and to ensure good part quality.

Following extrusion the wet honeycombs are slowly dried to avoid cracking, and then consolidated by heating to remove temporary binders and activate the permanent binders. A suitably slow drying method comprises gradual drying in a heated, controlled humidity enclosure at a rate sufficiently low to accommodate the volume change and avoid crack formation in the honeycombs. For example, small honeycombs can be dried in a humid atmosphere by heating at 90°C over a drying interval of 96 hours. Larger honeycomb may be dried over drying intervals of several days at temperatures in the 55-60°C range and at high relative humidity. The actual drying treatment in each case depends on the water content as well as the concentration and composition of volatile organics in the extruded honeycombs, but can readily be determined by routine experiment.

Obtaining crack-free fired adsorbent honeycombs with the correct physical and catalytic properties requires some attention to heating rates and heating temperatures, as well as to the firing atmospheres used, and the compositions and concentrations of extrusion aides and binders present in the dried honeycombs that are not removed in the course of the drying process. The maximum temperature and holding time were limited by the need to achieve good permanent binding and adequate strength in the honeycomb products, while not exceeding temperatures at which the material will lose its capability to adsorb and hold sulfur compounds due to agglomeration, sintering, or other mechanisms that destroy active sites.

Heating rates on the order of 20°C/hour with intermediate holding intervals of several hours at several soak points, are normally sufficient to reduce any honeycomb cracking in greenware to negligible levels. The best firing treatments for each batch example shown in Table 1 can readily be determined by routine experiment. For the smaller extrudates in Table 1, a firing schedule based on heating and cooling rates of 25°C/hour and a 10-hour hold at 350°C produces good results.

Typical properties for selected honeycomb monoliths produced from the batch compositions in Table 1 are compared with the properties of the commercial pelletized adsorbent starting material as shown in Table 2 & 3. The tables include the honeycomb cell density in channels per cm^2 and channels per in^2 and honeycomb channel wall thicknesses in mm and inches. Pore properties were measured at pressure range of 0-15,000 in Table 2 and 0-60,000 psi in Table 3. The physical property data show that extruded honeycomb sulfur traps have similar pore properties to pellet materials regardless of honeycomb geometry. Surface areas of honeycombs are lower than that of pellets probably because they have been subjected to two calcination steps, the first to make commercial pellets, and the second after converting these materials to honeycomb monoliths.

Table 2: Adsorbent Pellet and Honeycomb Properties

Adsorbent Structure	Cell Density cells/cm² - cells/in²	Cell Wall Thickness mm – inch	Surface Area (m2/g)	% Wall Porosity	Mean Pore Diameter (μm)
Pellets	-	-	293	38.9	0.0475
Honeycomb	31 - 200	0.56 – 0.022	247	27.3	0.0470
Honeycomb	15.5 – 100	0.46 – 0.018	247	30.2	0.0444
Honeycomb	279 – 1800	0.20 – 0.008	260	30.5	0.0454
Honeycomb	15.5 – 100	0.46 – 0.018	213	32.3	0.0465
Honeycomb	279 – 1800	0.20 – 0.008	252	38.8	0.0541
Honeycomb	31 – 200	0.56 – 0.022	244	31.4	0.0313

Table 3: Adsorbent Pellet and Honeycomb Properties

Adsorbent Structure	Cell Density cells/cm² - cells/in²	Extruded Cell Wall Thickness	Surface Area (m2/g)	% Wall Porosity	Mean Pore Diameter (um)
Pellets	-	-	293	51.8	0.0259
Honeycomb	31 - 200	0.56 – 0.022	244	55.8	0.0214

In addition to Ni-based reactive adsorbents, other metal and metal oxide sulfur adsorbents may adapted for use as honeycombs in desulfurizing reactors. For example, there are a number of high-surface-area materials known to be effective for trapping sulfur and/or sulfur compounds that can offer particularly efficient desulfurizing activity for gas-phase fuel feed processing. Representative examples of such materials include metal-loaded activated carbon, various zeolitic or molecular sieve materials, and certain high-surface-area metal oxides both with and without added reactive metal phases.

Conventional metal oxide sulfur adsorbents include pellets containing zinc oxide. Typical properties for pelletized zinc oxide adsorbents include porosities of 50%, mean pore diameters of 0.03 um, a pore intrusion volume of 0.25 ml/g and a surface area of 52 m^2/g. Sulfur adsorbents combining zinc oxide with metallic copper additions are also known. Extrusion batches for each of these adsorbent types are reported in Table 4 below. The compounding and extrusion of these batches to form adsorbent honeycomb structures may be carried out following the same processes as employed for the production of the Ni-based monoliths described earlier. The alumina and methyl cellulose binders, oleic acid emulsion and metal stearate lubricants employed in these batches are the same as or functionally equivalent to those employed in Table 1.

Table 4 - Honeycomb Adsorbent Batch Compositions

Adsorbent Type	Batch Composition
ZnO	100 parts of ZnO powder; 10 parts alumina binder; 33 Parts oleic acid emulsion; 5.5 parts cellulose binder; 8.86 parts water
Cu-ZnO	100 parts of CuZnO powder; 10 parts alumina binder; 30 parts oleic acid emulsion; 5.5 parts cellulose binder; 10 parts water

Honeycomb adsorbents are extruded from the batch composition described above, and the green honeycombs are slowly dried and then fired to achieve structural consolidation. Peak firing temperatures below 400°C are employed in these cases to consolidate the honeycombs into unitary, crack-free adsorbent structures.

Pore properties of the zinc and zinc-copper oxide honeycomb adsorbents are reported in Table 5. Included is data for the parent pellet starting materials.

Table 5: Extruded Honeycomb Adsorbent Properties

Adsorbent Structure	Cell Density cells/cm^2 - cells/in^2	Cell Wall Thickness mm - inch	Porosity (%)	Mean Pore Diameter (μm)	Pore Volume (ml/g)	Surface area (m^2/g)
ZnO Pellets	-	-	49.7	0.028	0.245	52
ZnO Honeycomb	124 - 800	0.30 - 0.012	50.9	0.031	0.253	54
ZnO Honeycomb	31 - 200	0.56 – 0.022	55.6	0.032	0.266	54
ZnO Honeycomb	279 - 1800	0.20 – 0.008	51.3	0.032	0.246	55
Cu-ZnO Pellets	-	-	43.8	0.03	0.20	64
Cu-ZnO Honeycomb	31 - 200	0.56 – 0.022	46.9	0.03	0.23	61

As is evident from the data in Tables 2, 3 and 5, converting conventional sulfur adsorbent materials to honeycomb shapes can be achieved while preserving most of the pore properties and surface areas of the starting materials. This strongly suggests that comparable monolith shapes can be made from the same type of powder raw materials used to make commercial pellets. Monolithic honeycomb geometry should translate to improved sulfur removal technology for the reasons outlined in the introduction.

CONCLUSIONS

Several commercially available sulfur removal catalysts were developed successfully into monolithic honeycombs. Batch compositions and process parameters were defined and optimized so that high quality crack-free parts were obtained. Slow drying under controlled humidity conditions is an important method to obtain good parts at diameters up to 14 cm. Calcination parameters such as ramp rate, maximum temperature, binder compositions and concentration, and firing atmosphere are also important parameters. Larger part sizes are more sensitive to process conditions than small diameter parts. Methods to increase the yield of large parts include reducing the amount of binder, selecting organic additives with different burn-off temperatures, changing the burn-off path by modifying the firing schedule, and controlling the burn-off rate.

The geometric structure of the substrate should play an important role in the performance of sulfur adsorbents by influencing both pressure drop, contact between fluid and solid, and pore diffusion within the solid matrix. The geometric forms shown here have straight-through rectangular channels. Other designs such as wall-flow systems may demonstrate additional benefit by forcing greater contact between fluid and solid, albeit at the expense of additional pressure drop. Future investigation needs to include studying mechanisms for specific sulfur removal reactions.

REFERENCES

[1] "Process reduces sulfur from gasoline" R&D magazine, September 2000, P123

[2] S. Blankenship, et al., "An adsorbent for a hydrocarbon stream and process", WO 99/34912

[3] E. Lox, et al., "Catalyst for the purification of exhaust gases of internal combustion engines with reduction of hydrogen sulfide emission", US 5,045,521

[4] L. J. Bonville, et al., "Method for desulfurizing a fuel for use in a fuel cell power plant", US 6,159,256

[5] L. J. Bonville, et al., "System for desulfurizing a fuel for use in a fuel cell power plant", US 6,159,084

[6] D. A. Zornes, "Magneto absorbent", WO 200038831

[7] P. Chintawar, et al., "Process for converting carbon monoxide and water in a reformate stream and apparatus therefore", WO 00/66486

[8] F. Abe, et al., "Reformer", EP 967174

[9] "Reformer and fuel cell power generating apparatus using the same reformer" JP 2000-159502

[10] W. A. Cutler and L. He, "The Influence of catalyst supports in fuel processor applications", paper99f, the AIChE 2000 Spring National Meeting, Atlanta, Georgia

Processing of Porous Biomorphic TiC Ceramics by Chemical Vapor Infiltration and Reaction (CVI-R) Technique

Nadja Popovska, Daniela Almeida Streitwieser, Chen Xu, Helmut Gerhard
Department of Chemical Reaction Engineering
University of Erlangen-Nuremberg
Egerlandstrasse 3
D-91058 Erlangen, Germany

Heino Sieber
Department of Material Science, Glass and Ceramics
University of Erlangen-Nuremberg
Martensstrasse 5
D-91058 Erlangen

Abstract

Chemical vapor infiltration and reaction (CVI-R) is used for producing biomorphic porous TiC ceramics derived from wood and paper. The samples are first pyrolysed in inert atmosphere to yield biocarbon template structures (C_b-template). Subsequently, three routes for converting the C_b-templates into TiC ceramics are studied. The first route includes CVI-R with $TiCl_4$-H_2 and the carbon biotemplate as a carbon source. The effect of methane as additional carbon source is investigated on the second route ($TiCl_4$-H_2-CH_4). Finally, a two step CVI process (route 3), first $TiCl_4$ - H_2 and subsequent $TiCl_4$-H_2-CH_4, is performed in order to improve the infiltration depth into the C_b-template as well as to strengthen the mechanical properties of the resulting porous TiC ceramics by depositing of additional TiC. It is found, that good infiltration of wood derived C_b-templates with subsequent conversion into TiC can only be achieved with $TiCl_4$-H_2 as a precursor (route 1). Paper derived TiC ceramics with good mechanical properties, however are obtained by the two step procedure (route 3).

Introduction

Biomorphic cellular ceramics are a new class of ceramic materials with a naturally designed porous structure. They are obtained by carbonization and consecutive conversion of biological preforms into ceramics, retaining the initial microstructure of the biotemplates. Various biotemplate processing technologies have been developed for manufacturing of biomorphic SiC-based ceramics. Among these methods the most investigated are the Si liquid infiltration [1], the Si gas infiltration [2], the SiO vapor infiltration [3] and, the method used in this paper, the CVI – R method [4,5]. Compared to the other infiltration methods the chemical vapour infiltration (CVI) has the advantage of using relatively low processing temperatures and retaining exactly the initial structure of the biotemplate. Biomorphic SiC porous ceramics could be applied as high temperature filters or catalytic support structures due to their high thermal conductivity, good oxidation and corrosion resistance as well as high strength at elevated temperatures.

Porous TiC ceramics show improved corrosion resistance in phosphoric acid, high electrical conductivity as well as very good wettability by metal melts. These properties make these materials interesting candidates for specific applications such as ceramic-metal composites [6] or catalyst supports for chemical and biochemical reactions [7,8].

There exist only a few reports about production of porous TiC ceramics. One of them describes the preparation of porous TiC ceramics by liquid vacuum infiltration of wood derived charcoal with tetrabutyl-titanate. First, the tetrabutyl-titanate decomposes to TiO_2, and then reacts with the carbon biotemplate at temperatures of about 1400°C forming TiC [9]. In [10] we reported the first results about the processing of TiC porous ceramic from wood by chemical vapour infiltration (CVI-R).

In this paper, chemical vapour infiltration and reaction (CVI-R) process is investigated to produce biomorphic cellular TiC ceramics by converting carbon biotemplates derived from wood or paper into TiC according to the following scheme (Fig. 1).

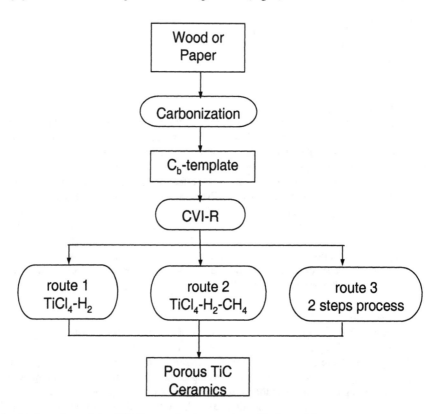

Fig. 1: Flow chart of the CVI-R ceramization process with three different routes

Three routes for converting the carbon biotemplates into TiC ceramics are investigated. The first route includes chemical vapour infiltration (CVI-R) with $TiCl_4$-H_2. The effect of methane as additional carbon source is investigated on route 2. Finally, on the third route, two step CVI-R process is performed with $TiCl_4$-H_2 and subsequently with the $TiCl_4$-H_2-CH_4 system. The grade of infiltration of the C_b-template as well as the mechanical properties of the resulting porous TiC ceramics should be improved with the two step process.

EXPERIMENTAL

Wood and paper are used as substrates to be converted into biomorphic TiC ceramics. Three types of wood - beech, pine and rattan with different pore structure (Fig. 2) are investigated. Also, flat paper preforms of 0.80 mm thickness with a geometrical density of 0.22 g/cm³ and initial porosity of 82 % are used.

The pyrolysis of the wood and paper preforms and the chemical vapour infiltration of the resulting carbon biotemplates is performed in a horizontal hot-wall tubular flow reactor operated at atmospheric pressure.

The samples are pyrolysed in inert atmosphere (He, 5 cm/s) at the following conditions: 1 K/min ramp, up to 350°C, 1h dwell time, followed by 2 K/min ramp up to 850°C with also 1 h dwell time.

$TiCl_4$ is used as a Ti source. It is vaporized in a bubbler and carried into the reactor by a carrier gas H_2 or He.

RESULTS AND DISCUSSION

1. CVI-R of TiC in Carbon Biotemplates using $TiCl_4$-H_2 (route 1)

The carbon biotemplates derived from wood and paper are converted into TiC ceramics according to the following equation:

$$TiCl_4 + 2H_2 + C_b \rightarrow TiC + 4HCl \qquad (1)$$

$TiCl_4$ in excess of hydrogen is supplied as titanium source and the carbon biotemplate acts as a carbon source. The gaseous precursors diffuse in the pores of the carbon biotemplate, where decomposition of $TiCl_4$ and chemical reaction with the carbon to TiC takes place simultaneously.

The SEM micrographs show a good infiltration depth in the pores of the wood derived C_b-templates (Fig. 3) as well as a homogeneous infiltration of the paper performs (Fig. 2). However, a complete conversion of the carbon into TiC could not be achieved due to the slow carbon diffusion through the grown TiC layer. The maximum bending strength achieved in the case of paper derived C_b-templates is only 4-5 MPa, which is too low to ensure stability, so that the samples break easily even by handling them.

2. CVI-R of TiC in Carbon Biotemplates using $TiCl_4$-H_2 with Addition of Methane (route 2)

In order to improve the strength of the porous TiC ceramics, methane (CH_4) in under-stoichiometric amount is added to the precursor and supplied as carbon source additional to those from the C_b-biotemplate (eq. 2).

$$(x+y)\,TiCl_4 + xC_b + yCH_4 \xrightarrow{\;H_2\;} (x + y)TiC + 4(x + y)HCl \qquad (2)$$

Because of the higher deposition rate the infiltration into the pores is not as homogeneous as in the case of the $TiCl_4$-H_2 system as can be seen in the SEM micrographs of wood derived TiC ceramics (Fig. 3). In the case of paper, however, the methane as additional carbon source contributes to an improved mechanical stability of the obtained porous TiC ceramics.

3. CVI-R of TiC in Carbon Biotemplates in a Two Steps Process (route 3)

In order to combine the advantages of the both processing routes described above, a two step procedure is investigated. Two steps infiltration of the C_b-templates derived from paper, first with $TiCl_4$-H_2 (eq. 1) followed by $TiCl_4$-H_2-CH_4 (eq. 2) is applied to produce biomorphic cellular TiC ceramics achieving high infiltration depth and good mechanical properties.

During the first step with the $TiCl_4$-H_2 system, the carbon fibres of the biotemplate are converted into TiC by $TiCl_4$ decomposition and a subsequent reaction between Ti and C_b (eq. 1). In the second step (eq. 2), CH_4 is introduced as additional carbon source and reacts with the decomposed Ti to TiC, reinforcing in this way the porous ceramic structure. The maximum bending strength achieved in the case of paper derived C_b-templates is around 10 MPa. This value is high enough to ensure a mechanical stability of the samples, while they are remaining high porous.

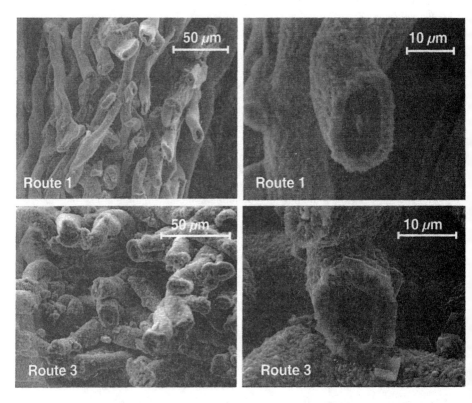

Fig. 2: SEM micrographs of paper derived TiC ceramics processed by route 1 and route 3

For the wood derived C_b-templates only routes 1 and 2 have been performed. The results on Fig. 3 show that due to its higher density compared to the paper preforms, it is more difficult to infiltrate the precursors into the centre of the preform. And since the deposition rate on route 2 under the addition of methane is much faster than with the $TiCl_4$-H_2 system, no TiC diffuses into the pores centre and only the surface is being coated with TiC. For this reason, only with route 1 a homogeneous infiltration can be achieved. The different wood types also show, that depending on the pore size and the porosity, a high or low infiltration is possible.

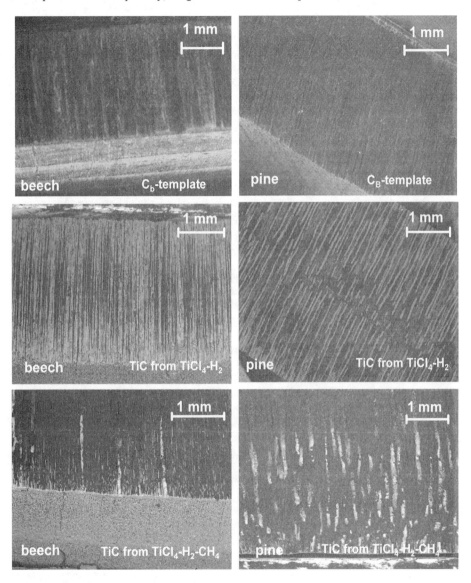

Fig. 3: SEM micrographs of wood derived C_b-templates before and after CVI-R with $TiCl_4$-H_2 and $TiCl_4$-H_2-CH_4 system

CONCLUSIONS

Cellular TiC ceramics derived from three different wood types- beech, pine and rattan - can be produced by CVI-R technique. Uniform infiltration along the pores can be achieved using $TiCl_4$-H_2 as a precursor.

Porous TiC ceramics with good mechanical properties, derived from carbonized paper preforms (C_b-template) can be produced by CVI-R in a two steps process. In a first step, the pores of the carbon biotemplate are infiltrated with $TiCl_4$ in excess of hydrogen, where thermal decomposition and chemical reaction to TiC take place simultaneously. A complete conversion of the C_b-template to TiC is limited by the slow solid-solid diffusion of the reactants - Ti and C, leading to poor mechanical properties of the resulting ceramics. For this reason, in a second processing step, methane in understoichiometric amount is added to the reaction gas as an additional carbon source. The additional deposited TiC strengthens the porous structure, resulting in improved mechanical properties of the obtained ceramics.

REFERENCES

1. Sieber, H., Hoffmann, C., Kaindl, A., Greil, P., Biomorphic Cellular Ceramics, Advances Engineering Materials, Vol. 2, No. 3, 2000, pp. 105 – 109
2. Vogli, E., Sieber, H., Greil, P., Biomorphic SiC – ceramics prepared by Si – vapor phase infiltration of wood, J. of the European Ceramic Society 22, 2002, pp. 2663 – 2668
3. Vogli, E., Mukerji, J., Hoffmann, C., Kladny, R., Sieber, H., Greil, P., Conversion of Oak to Cellular Silicon Carbide by Gas – Phase Reaction with Silicon Monoxide, J. of American Ceramic Society, Vol. 86 6, 2001, pp. 1236 – 1240
4. Sieber H., Vogli, E., Müller, F., Greil, P., Popovska, N., Gerhard, H., CVI – R gas phase processing of porous biomorphic SiC – ceramics, Key Engineering Materials 206 – 213, 2001, pp. 2013 – 2016
5. Greil, P., Vogli, E., Fey, T., Bezold A., Popovska, N., Gerhard, H., Sieber, H., Effect of Microstructure on the fracture of biomorphous silicon carbide ceramics, J. of the European Ceramic Society 22, 2002, pp. 2697 – 2707
6. US 6451385, Pressure infiltration for production of composites, publ. date 2002-09-17
7. JP2001233682, Method for producing ceramics carrier capable of proliferating bacteria degrading organic substance and inorganic substance at high speed, publ. date 2001-08-28
8. JP2003160385, Method of manufacturing ceramic carrier improved in carrying ability of bacteria decomposing organic and inorganic substance and made capable of quick proliferation of bacteria, publ. date 2003-06-03
9. Sun B., T. Fan, D. Zhang, Porous TiC Ceramics Derived from Wood Template, J. Porous Mater. 9 (2002) 275
10. Sieber H., C. Zollfrank, N. Popovska, D. Almeida, H. Gerhard, Gas phase processing of porous, biomorphic TiC-ceramics, Key Eng. Mater. (in print)

CHARGE TRANSPORT MODEL IN GAS-SOLID INTERFACE FOR GAS SENSORS

Sung Pil Lee and Yer-Kyung Yoon

Depart of Electrical and Electronic Engineering, Kyungnam University

449 Wolyoung-dong, Masan, Kyungnam 631-701, Korea

ABSTRACT

An analytical model for gas adsorption on a solid surface is studied qualitatively. The electric field and potential distribution in the depletion region of a gas-solid interface depend on the barrier height, the adsorbed gas concentration and image force. These dependences are obtained from the solution of the one-dimensional Poisson equation and Coulomb's force. If an electron from the adsorbed gases in the interface is at a distance x from the solid surface, an electric field exists perpendicular to the interface. This field may be calculated by assuming a hypothetical positive image charge q located at the distance (-x) from the gas. This field lowers the energy barrier of the interface, and the electrons from the adsorbed gases come to solid more easily.

INTRODUCTION

Many types of gas sensors have been developed to detect chemical species in the gas phase. Chemical sensors show an electrical response, such as a resistance change, a capacitance change, a change of electro-motive force, to a chemical change. FET(field effect transistors), MIS(metal insulator semiconductor) diodes, metal semiconductor capacitors and ferro-electric based devices have been studied for gas sensing with surface conduction mechanism[1].

A surface conduction-based sensor uses the change in the concentration of the conduction band electrons (or valence band holes) resulting from the chemical reactions with adsorbed gas species. Morrison[2] reported that chemoresistive oxide sensors often change their resistances by more than a factor of 100 upon exposure to a trace of reducing gases such as H_2, CH_4, ethanol, CO and propane. Oxygen adsorbed on the surface of the sensor extracts an electron from the bulk to ionize the oxygen absorbed into O^- or O_2^-, which increases the resistance of the sensors. However, as Weisz[3] pointed out, the concentration of charged oxygen molecules or atoms are limited to less than 1% of the total number of surface states. It is impossible that a change less than 1% of the surface coverage changes the total resistance by a factor of 100. Windischmann et al.[4] have attributed changes in the characteristics of SnO_{2-x} gas sensor to the decrease in the work function upon exposure to CO gas. Strassler et al.[5] explained this process in detail as the open Neck and Schottky contact.

In this study, a new model for the charge transport in a gas-solid interface is proposed for chemical gas sensors. The image force theory, which is applied in metal-semiconductor contact, is included in space charge model[6], and the current density is calculated by computer simulation.

SPACE CHARGE LAYER BY GAS ADSORPTION

When the diameter of ceramic particles or grain size is larger than 2 times the Debye length (L_D)[7], the Schottky barrier limited system can be considered similar to the mechanism of metal-semiconductors contact. Upon exposure of the sensor surface to a reducing gas such as CO, the CO reacts with the adsorbed O⁻, and the trapped electrons in the surface released to the conduction band and subsequently, these electrons lower the sensor resistance. The amount of resistance change is proportional to the concentration of the reducing gas in ambience.

The number of chemisorbed atom per unit area is

$$N_g = x_0 n_q(x) \tag{1}$$

where x_0 is the depth of the space charge region and $n_q(x)$ is the number of electrons per unit volume. A quantitative derivation of the depletive space charge layer, upon transfer of electrons from the oxide to the surface caused by the adsorption of gas species, can be easily obtained from the one dimensional Poisson's equation for a semiconductor containing donor and acceptor impurities per unit volume:

$$\frac{d^2 E(x)}{dx^2} = \frac{q n_q(x)}{\varepsilon_s} = E_m (1 - \frac{x}{x_0}) \tag{2}$$

$$E_m = -\frac{q}{\varepsilon_s} n_q(x) x_0$$

where $E(x)$, ε_s and $\rho(x)$ are the electric field, the dielectric constant of the semiconductor and the net charge density consisting of electrons (n), holes (p), ionized donor (N_D^+) and ionized acceptor (N_A^-) in the material, respectively.

The potential barrier $q\phi_b$ and the depth of the space charge region x_0 are expressed as

$$q\phi_b = -\frac{q x_0^2 n_q(x_0)}{2\varepsilon_s}, \quad x_0 = [\frac{2\varepsilon_s}{q n_q(x_0)} \phi_b]^{\frac{1}{2}} \tag{3}$$

Therefore,

$$N_g = (\frac{2\varepsilon_s n_q(x_0)\phi_b}{q})^{\frac{1}{2}} \tag{4}$$

If every donor defect level in the n-type SnO_2 has yielded its electron to the conduction band, $n_q(x_0)$ becomes the donor concentration. Therefore N_g is about $1.42 \times 10^{12}/cm^2$. The number of atoms on the surface of metal oxide is about 10^{15}, so that the coverage for depletive adsorption is less than 1%.

When gases are adsorbed on the solid surface, an electron in a dielectric at a distance x from the surface will create an electric field. The field lines must be perpendicular to the surface and will be the same as if an image charge, q^+, is located at the same distance from the surface. This image effect is shown in Fig. 1. The potential, due to the Coulomb attraction with the image charge, can be found as follows:

$$E_1(x) = \int_x^\infty F_1(x)dx = \frac{-q}{16\pi\varepsilon_d x} \tag{5}$$

Fig. 1 shows a plot of the potential energy assuming that no other electric field exists. With an electric field present in the depletion layer, the potential is modified and can be written as follows:

$$PE(x) = \frac{q}{16\pi\varepsilon_d x} + E(x) \tag{6}$$

The barrier height lowering $\Delta\phi_b$ by an image force is written by the condition d[PE(x)]/dx=0 or,

$$\Delta\phi_b = [\frac{q^3 n_q(x_0)}{8\pi^2\varepsilon_s\varepsilon_d^2}\phi_b]^{\frac{1}{4}} \tag{7}$$

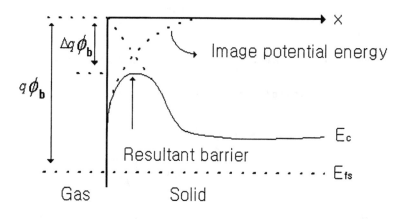

Fig. 1. Barrier height lowering by image force

CARRIER TRANSPORT AND CURRENT DENSITY

The current transport in Schottky barrier-limited contacts is mainly due to majority carriers. Fig. 2 shows the basic transport process by TFE (Thermionic Field Emission) under the image force effect.

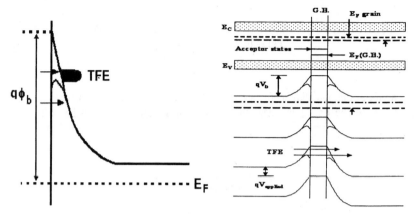

Fig. 2. Carrier transport process by image force effect

The shape of the barrier profile is immaterial and the current flow depends on the barrier height. The current density J_{TFE} is given by the concentration of electrons with energies sufficient to overcome the potential barrier and traversing in the x direction,

$$J_{TFE} = \int_{E_c}^{\infty} qV_x dn \qquad (8)$$

where, E_c is the minimum energy required for TFE and V_x is the carrier velocity in the direction of transport.

The electron density in an incremental energy range is given by

$$dn = N(E)F(E)dE$$
$$= \frac{4\pi(2m^*)^{\frac{3}{2}}}{h^3}\sqrt{E-E_c}\exp(-\frac{E-E_F}{kT})dE \qquad (9)$$

where, $N(E)$ and $F(E)$ are the density of states and the distribution function, respectively. Therefore, the charge current density is expressed as follows:

$$J_{TFE} = 2q(\frac{m^*}{h})^3 \exp(-\frac{qV_{applied}}{kT})\int_{v_{ox}}^{\infty} v_x \exp(-\frac{m^*v_x^2}{2kT})dv_x \times \int_{-\infty}^{\infty} v_y \exp(-\frac{m^*v_y^2}{2kT})dv_y \int_{-\infty}^{\infty} v_z \exp(-\frac{m^*v_z^2}{2kT})dv_z$$

$$= (\frac{4\pi qm^*k^2}{h^3})T^2 \exp(-\frac{qV_{applied}}{kT})\exp(-\frac{m^*v_{ox}^2}{2kT}) \qquad (10)$$

544

where, v_{ox} is the minimum velocity required in the x direction to surmount the barrier. And a normalized current density by CO gas adsorption is written by

$$\frac{J_{TFE(CO)}}{J_{TFE(air)}} = \frac{\exp(-\frac{q\phi_1}{kT})}{\exp[-\frac{q(\phi_b - \Delta\phi_b)}{kT}]} = \exp[\frac{q\Delta\phi_b(N_g)}{kT}]$$ (11)

where, $\phi_1 = \phi_b - \Delta\phi_b - \Delta\phi(N_g)$ and $\Delta\phi(N_g)$ is a potential barrier lowering by gas adsorption.

Fig. 3 shows a current density change in response to the CO concentration in SnO_2 based gas sensors. An increase of CO concentration gives rise to a decrease of ϕ_b and subsequently an increase of the sensor conductance. The carriers from thermionic field emission and/or gas adsorption easily come over the lowered barrier height by the image force effect. When we calculate the ratio of the current density considering the image force effect, the slope of the current density in air and in CO gas ambient appears to be 1.9 to 10,000 ppm CO gas change. It is a higher value than 1.5, which is the value calculated considering no image force effect.

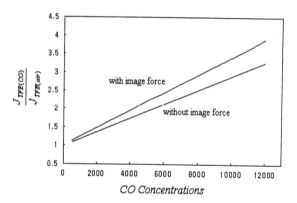

Fig. 3. Current density change to CO gas concentrations

CONCLUSION

A charge transport model for chemical sensors, which includes the image force effect in a gas-solid interface, is proposed. The number of chemisorbed atom per unit area is about $1.42 \times 10^{12}/cm^2$ and the number of atoms on the surface of metal oxide is about 10^{15}, so that the coverage for depletive adsorption is less than 1%. The carrier from thermal field emission or gas adsorption easily goes over the lowered barrier height by the image force effect. With consideration of the image force effect, the slope of the current density in air and in CO gas ambient appears to be 1.9 to 10,000 ppm CO gas change. It is a higher than that of the case without the image force effect.

ACKNOWLEDGEMENT

The authors wish to acknowledge the financial support of Kyungnam University made in the program year of 2003.

REFERENCES

[1] N. Yamazoe, "New approaches for improving semiconductor gas sensors," Sensors and Actuators B, 5, 7- 25 (1991).

[2] S.R. Morrison, "Selectivity in semiconducting gas sensors," Sensors and Actuators, 12, 425-440 (1987).

[3] P. B. Weisz, J. Chem. Phys. 20, 1483 (1952).

[4] H. Windischmann and P. Mark, J. Electrochem. Soc., 126, 62 (1979).

[5] S. Strassler and A. Reis, Sensors and Actuators, 4, 465 (1983).

[6] K.S. Kang and S.P. Lee, "CO gas sensors operating at room temperature," J. of Materials Sci., 38[21] 4319-4323 (2003).

[7] H.J. Engell and K. Hauffe, Ztschr. Electrochem., 56, 336-44 (1952).

CORROSION RESISTANT REFRACTORY CERAMICS FOR SLAGGING GASIFIER ENVIRONMENT

Eugene Medvedovski*
Ceramic Protection Corporation
3905 – 32nd St. N.E.
Calgary, AB, T1Y 7C1, Canada

Richard E. Chinn
Albany Research Center - USDOE
1450 Queen Ave. S.W.
Albany, OR 97321-2198, USA

ABSTRACT

Integrated gasification combined cycle power systems are the most efficient and economical power generation systems with a relatively low environmental impact. The gasification process requires the optimal design of gasifiers with extremely corrosion resistant refractory lining. The majority of the refractory materials tested for gasifier lining applications cannot resist the action of slagging corrosive environment combined with high operation temperatures as high as 1600°C and possible thermal shocks and thermal expansion mismatch between the lining and the slag.

Silicon carbide-based ceramics and some zirconia- and zircon-based ceramics manufactured by Ceramic Protection Corporation (CPC) have been tested in a simulated coal-fired slagging gasifier environment at a temperature of 1500°C. Crucible ceramic samples have been examined after exposure to the slag at high temperature. Microstructure studies of the ceramic zone contacted with the slag have been carried out. The highest performance, i.e. the absence of corrosion damage and thermal cracking after testing, was observed for silicon carbide-based ceramics ABSC formed by silicon carbide grains with an optimized particle size distribution bonded by the aluminosilicate crystalline-glassy matrix. Dense zirconia and alumina-zirconia and slightly porous zircon ceramics demonstrated comparatively lower performance due to their lower corrosion resistance and greater thermal cracking. ABSC ceramics can be manufactured as thick-walled large components and may be considered as a promising material for gasifier refractory applications. Similar ceramics, but with finer grain sizes, may also be recommended for thermocouple protection.

INTRODUCTION

Integrated gasification combined cycle power systems are the most efficient and economical power generation systems with a relatively low environmental impact. In these systems the fuel (coal, petroleum or biomass) slurry is burnt in the gasifier at high temperatures using high pressure steam and oxygen that results in formation of CO, CO_2, H_2 and byproducts[1-3]. The gasification process requires an optimal design of gasifiers with an extremely corrosion resistant refractory lining. In addition, the thermocouples providing the temperature monitoring in the gasifier also require protection from corrosion through such resistant materials. Generally, the materials used for gasifier lining and thermocouple protection should work at high temperatures, should withstand thermal shocks and harsh corrosive environments such as high-temperature slags and high-temperature corrosive gases. The performance of the gasifier refractory materials depends on the composition and structure of these materials and on the service conditions. The slag composition and the operating temperature have a significant role in refractory performance. For example, the coal ash slags are silicate-based, and they typically consist of (wt%) 40-55 SiO_2, 20-25 Al_2O_3, 15-22 FeO, 5-10 alkali and earth-alkali oxides, i.e. the gasifier slags have low viscosity and they are very aggressive at the operating temperatures, which are generally rather high (1500-1600°C). Additionally, refractory lining should be wear-resistant to the sliding and spalling actions of the molten slags.

The majority of refractory materials tested for gasifier lining applications cannot resist the action of a slagging corrosive environment combined with temperatures of 1500-1600°C, thermal shocks and the thermal expansion mismatch between the lining and the slag. Presently, chromium oxide-based refractories are most commonly used for the gasifier lining. Recent studies carried out by the Albany Research Center of DOE showed that the use of phosphate-based additives (AlPO4, CrPO4, etc.) to these refractories promotes a significant decrease of the slag penetration into the lining and an increase of the general lining service cycle[3]. The phosphate-based materials are widely used as additives for different refractories, concretes and structural ceramics for improvement of their mechanical properties, thermal shock- and corrosion resistance[4, 5]. However, the development of the refractory materials for gasifier applications is under ongoing consideration in order to improve their performance and manufacturing ability and to extend the service life cycle. The refractory lining and thermocouple protection replacements cost several million dollars in the USA per annum due to the materials and installation costs and the system shutdowns. This maintenance occurs typically 3-4 weeks every 10-18 months depending on refractory performance, so the extension of service cycle is vital in the reduction of operation cost.

MATERIALS SELECTION

In general, the performance of refractory ceramics for gasifier applications may be improved by achieving the following:

- Increasing high-temperature corrosion resistance, i.e. optimization of the compositions to decrease high-temperature interaction between constituents of the refractory and constituents of the slag;
- Reduction of the wettability of the refractory with the slag, including reduction of the surface porosity and pore sizes;
- Increased bonding between refractory grains;
- Reduction of the interconnected porosity of the refractory;
- Increased thermal shock resistance of the refractories by optimizing their granulometric compositions and by using constituents with a lower coefficient of thermal expansion.

Several high-temperature ceramic materials, such as silicon carbide-based ceramics (ABSC), zircon ceramics, and dense fully-stabilized zirconia and alumina-zirconia (AZ) ceramics, developed and manufactured by Ceramic Protection Corporation (CPC) have been selected for the testing in the slagging gasifier environment. Although these types of ceramics have not commonly been cited in the literature related to the slagging gasifier environments, they may respond well in meeting the above-mentioned requirements for such refractory materials. All these ceramics were selected due to their high corrosion resistance, previously proven in some high-temperature acidic environments, molten metals and alloys, as well as due to their high wear resistance. This is especially so for alumina-zirconia and ABSC ceramics. ABSC ceramics demonstrate excellent thermal shock resistance as a result of the features of their macrostructure formed by silicon carbide grains with a specially selected particle size distribution bonded by a crystalline-glassy aluminosilicate matrix, of a presence of small porosity and micro-cracks appeared during a bonding phase formation, and also due to relatively high thermal conductivity and lower coefficient of thermal expansion inherent to silicon carbide. Zircon ceramics also have a high level of thermal shock resistance due to their lower coefficient of thermal expansion and small uniform porosity. By comparison, zirconia and alumina-zirconia ceramics are fully dense, and they do not possess high thermal shock resistance. All the selected materials are able to work at high temperatures with no, or minimal, creep, especially ABSC and zircon ceramics.

ABSC ceramic products may be produced with a thickness of 4-5" (100-120 mm) due to the features of the composition, manufacturing and structure of these ceramics. Zircon ceramic products also may be manufactured with a relatively high thickness of up to 2" or 50 mm. Dense zirconia and alumina-zirconia ceramics, as fine-grade materials, may be generally produced only with a small thickness (0.5" or 10-12 mm and 1" or 20-25 mm, respectively). Some of the selected ceramics were described previously[6-8]. In accordance with the ability to manufacture the products with the particular shape and thickness, ABSC and zircon ceramics were initially considered as candidates for the gasifier lining applications, whereas zirconia, alumina-zirconia and zircon ceramics were initially considered as the candidates for the thermocouple protection.

EXPERIMENTAL
Raw Materials and Manufacturing
Silicon carbide powders (black, α-SiC) commercially produced by Saint-Gobain Ceramics and Plastics, Inc. and by Electro Abrasives, Inc., were used as the major constituents for manufacturing of ABSC ceramics. These powders were of different grades, i.e. with specially selected different particle sizes ranging from 3-5 μm to 2 mm and particle size distributions. As the Al_2O_3-constituents of ABSC ceramics, superground alumina powders commercially produced by Alcoa World Chemicals were used. For the zirconia and alumina-zirconia ceramic compositions zirconia powders doped with Y_2O_3 (8 and 3 mol.%, respectively) as the stabilizing agent commercially produced by Tosoh Corp. were used. In the case of alumina-zirconia ceramics, superground alumina powders commercially produced by Pechiney-Altech were used as the Al_2O_3-constituent. Zircon powders with different particle sizes (from 400 to 200 mesh) supplied by Continental Mineral Processing Corp. and Atofina Corp. were used as the major constituents for manufacturing of zircon ceramics.

All selected ceramics have been manufactured by the slip casting technology. Slurries were prepared selecting powders with proper particle sizes for each kind of ceramics with compositions optimized to achieve a high level of material densification. In the case of ABSC ceramics, vibration during casting was utilized. The ceramic samples prepared for testing were dried and then fired at temperatures of 1520-1550°C in natural gas-fired kilns.

Corrosion Resistance Testing
Ceramic crucibles (cups) were used as samples for the corrosion resistance testing. The test cups manufactured by a slip casting technique have diameters of 40-60 mm and heights of 50-70 mm with a wall thickness of 5-10 mm. They were filled about halfway with a slag prior to testing. The slag taken from an actual coal-fired slagging gasifier had an approximate composition (wt.%): 47 SiO_2, 20 Al_2O_3, 17 FeO, 6 CaO, 2 K_2O, 0.5 C, 1 organics and volatiles. The remainder is predominantly transition-metal oxides. The cups with the slag were placed into the furnace with the $MoSi_2$ heating elements and heated up to 1500°C with a soak of 1 hr. and then cooled. The heating-cooling rate was relatively fast (200°K/hr). In order to minimize the oxidative or reducing effects of the atmosphere during heating (i.e. in order to prevent the oxidizing of metastable FeO to more stable Fe_2O_3), the furnace chamber was purged with argon. After cooling, the test cups were cut in half such that the plane of the cut contains the cylindrical

axis. The cross-sections of the samples were ground and polished for ceramographic examination by methods described in [9, 10].

RESULTS AND DISCUSSIONS

Dense fully stabilized zirconia ceramics had a very uniform microstructure, consisting of zirconia crystals with an average size of 0.5-0.6 µm and with no glassy phase. Dense alumina-zirconia ceramics based on an optimized ratio between alumina and zirconia also did not have a glassy phase; the zirconia tetragonal crystals with a size of less than 1 µm were uniformly distributed between corundum grains with a size of 1-3 µm. The test cups from these dense materials permitted a little slag penetration (Fig. 1 and 2, respectively) with a depth of less than 1 mm. It should be noted that zirconia ceramics demonstrated higher corrosion resistance (less slag penetration) than alumina-zirconia ceramics. Both crucibles suffered from the thermal expansion mismatch between the ceramic body, slag and the formed product of corrosion. Cracks formed after cooling were greater for zirconia ceramics than for alumina-zirconia ceramics.

Zircon ceramics consisted of zircon grains with sizes from 5-20 to 50-75 µm depending on selected composition with a presence of insufficient content of zirconia and silica. Due to their porous structure, the wetting by the slag and the slag penetration to these ceramics was greater than for dense zirconia and alumina-zirconia ceramics, and this penetration achieved 2-2.5 mm (Fig. 3). Zircon ceramics demonstrated lower corrosion resistance. However, due to high thermal shock resistance of zircon ceramics, the cups did not crack away from the penetrated layer. Corrosion of zircon ceramics by the slag occurred, probably, due to the interaction between free SiO_2 and ZrO_2 of the ceramics and CaO, FeO and other oxides from the slag.

ABSC ceramics consisted of silicon carbide grains with different sizes ranged from a few microns to 2-2.5 mm bonded by the aluminosilicate crystalline-glassy phase. The specially selected ratio between silicon carbide grains provided a relatively high level of grain compaction and densification, although the ceramics had some intergranular porosity. Due to the features of high-temperature processes occurring in these ceramics during firing, the surface of ABSC ceramics was significantly denser than the middle, and had a higher content of the bonding phase. ABSC ceramics demonstrated a remarkably high level of corrosion resistance to the high-temperature gasification slag. Slag penetration did not occur at the testing conditions (Fig. 4). This may be explained by the general chemical stability of silicon carbide, by the similarity of the nature of aluminosilicate constituent of the bonding phase and the slag (that also has an aluminosilicate base) and the absence of the oxides in the bonding phase promoting the wetting by the slag. A relatively low surface porosity of ABSC ceramics may promote a decrease of the wettability of the surface with the slag and a decrease of the slag penetration. Although ABSC ceramics had some level of intergranular porosity, the SiC grains were coated by the aluminosilicate crystalline-glassy phase, which could not be separated due to a nature of the reaction-bonding mechanism occurring during firing. The presence of the mullite crystals reinforced the bonding phase promoted the corrosion- and thermal shock resistance of the ceramics. Among the possible high-temperature reactions, which may occur between the refractory material and the slag during longer exposure, the reaction between CaO and, especially, FeO contained in the slag with SiO_2 and Al_2O_3 from ABSC ceramics have the highest probability. These reactions and the formation of lower-temperature melting compounds (such as $CaSiO_3$, $FeAl_2O_4$ and some others) may increase damage of the ceramics. However, in the case of high-temperature interactions between the slag and the ceramic surface, the contact layer appear to be enriched by the Si-O and Al-O large complex ions forming at the ceramic surface and migrating to the contact layer. This may result in an increase of the viscosity of this contact layer and, hence, in a decrease of activity of the slag in the contact zone. In fact, the formation of products of the interaction between the ceramics and the slag was not found after the test. This may be explained by a relatively short action of the slag, by static testing conditions and by lower wettability and high corrosion and thermal resistance of ABSC ceramics. Due to the visual absence of the reaction zone and high thermal shock resistance of ABSC ceramics, the surface of the ABSC cup did not crack during cooling after the test. The exposure test results may be considered as preliminary to select promising materials.

CONCLUSIONS AND SUGGESTED FURTHER DIRECTIONS

Silicon carbide-based ceramics ABSC demonstrate the most promising test results, and this refractory material may be considered as a potential candidate for slagging gasifier lining applications. This is due to its high corrosion- and thermal shock resistance achieved by its optimal phase and granulometric compositions and a "self-formed" denser protective layer on the ceramic surface. Similar ceramics made by using silicon carbide grains with lower sizes (e.g. with the greatest particle size of 0.5-0.8 mm) may also be considered for manufacturing of thermocouple protection tubes. However, these results must be considered as preliminary, and further work should be continued to optimize the compositions by increasing the bonding of the major SiC-phase with more chemically inert constituents. Further testing of the modeling or actual ceramic components of the gasifier lining and thermocouple protection made from the existing and improved compositions should also be carried out also in dynamic conditions with a longer duration of exposure to the actual slagging environments.

ACKNOWLEDGEMENTS

The authors are grateful to Dr. Cynthia Dogan (Albany Research Center) for the discussion of this work.

REFERENCES

[1]C.P. Dogan, K.-S. Kwong, J.P. Bennett, et al., "Improved Refractory Materials for Slagging Coal Gasifiers"; in *Proc. 27th Int'l Conf. On Coal Utilization and Fuel Systems*, Clearwater, FL, 2002

[2]C.P. Dogan, K.-S. Kwong, J.P. Bennett, et al., "Refractory Loss in Slagging Gasifiers"; in *Proc. 7th Biennial Unified Int'l Technical Conf. on Refractories*, Cancun, Mexico, 2001

[3]C.P. Dogan, K.-S. Kwong, J.P. Bennett, et al., "Improved Refractories for IGCC Power Systems"; *Industrial Heating*, [9], 57-61 (2002)

[4]P.P. Budnikov, L.B. Khoroshavin, "Refractory Concretes with Phosphate Binders", Moscow, 1971 (in Russian)

[5]V.A. Kopeykin, A.P. Petrova, I.L. Rashkovan, "Materials on the Base of Metallophosphates", Moscow, 1976 (in Russian)

[6]E. Medvedovski, "Silicon Carbide-Based Wear-Resistant Ceramics"; *Interceram*, 50 [2], 104-108 (2001)

[7]E. Medvedovski, R.J. Llewellyn, "Oxide Ceramics for Abrasion- and Erosion Resistance Applications"; *Interceram*, 51 [2], 120-126 (2002)

[8]E. Medvedovski, "Wear-Resistant Engineering Ceramics"; *Wear*, 249, 821-828 (2001)

[9]A.H. Hunt, R.E. Chinn, "A Ceramographic Evaluation of Chromia Refractories Corroded by Slag"; *Structure*, 37, 6-10 (2001)

[10]R.E. Chinn, "Ceramography", ASM International and The American Ceramic Society, 19-44 (2002)

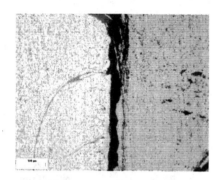

Fig 1: YSZ slag-refractory interface. The refractory is the left layer

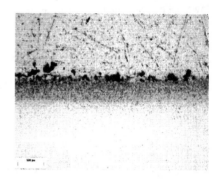

Fig 2: AZ slag-refractory interface, showing the slag-penetrated layer (dark band) and anorthite aciculae in the slag

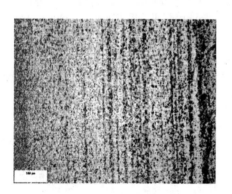

Fig 3: The slag-penetrated layer to the left of the left of the Z4 slag-refractory interface

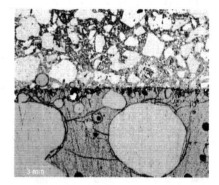

Fig 4: No penetration at the ABSC slag-refractory interface. The refractory is the upper layer

INFLUENCE OF THE DOPANTS AND THE METAL ELECTRODES ON THE ELECTRICAL RESPONSE OF HEMATITE BASED HUMIDITY SENSORS

Jean-Marc Tulliani and Paola Palmero
Polytechnic of Turin
Dipartimento di Scienza dei Materiali
Corso Duca degli Abruzzi, 24
10129 Turin - Italy

Pierre Bonville
Atomic Energy Commission
Centre d'Etudes de Saclay
Service de Physique de l'Etat Condensé
91191 Gif-sur-Yvette - France

ABSTRACT

Humidity sensing properties of hematite have been studied after doping with alkali and alkaline earth oxides precursors. Sensors were screen-printed and three different firing temperatures were investigated: 850°, 900° and 950°C. Secondary phases formation was evidenced by means of ^{57}Fe Mössbauer spectroscopy, for K-doped hematite samples, of XRD and dilatometry, for all the compositions. Sensors with Pt electrodes exhibited a strong decrease in resistance as a function of relative humidity, except those barium-doped, while with Pd/Ag ones only a limited number of thick-films gave an electrical signal.

KEYWORDS

Hematite, Thick films, Electrical response, Humidity sensors

INTRODUCTION

Demand for humidity sensors has increased over the last three decades for industrial, medical and home applications such as, for example, paper manufacturing, electronic industry, drugs preparation, respiratory equipment, home dehumidifiers and air conditioners[1, 2]. Many oxide ceramic materials have been successfully tested, based for example on alpha-Fe$_2$O$_3$, as such[3] or doped with silica[4], lithium[5] and other metals[6]. Sensors based on iron oxides belong to the protonic sensors family and their resistance decreases with adsorption of water molecules on their surface[2]. The electrical response of ceramic sensors is dependent on the microstructure[2] and on the electrodes geometry[7] and materials[8]. As there is no systematic study on the influence of the dopants and the nature of metals electrodes, the aim of this work is to investigate, by means of ^{57}Fe Mössbauer spectroscopy, X-ray diffraction, SEM observations, mercury porosimetry and dilatometry measurements the influence of several alkali (Li, Na, K and Rb) and alkaline earth (Mg, Ca, Sr and Ba) oxides onto the hematite (α-Fe$_2$O$_3$) microstructure, and to correlate these results with impedance measurements in function of relative humidity.

MATERIALS AND METHODS

Alpha-Fe$_2$O$_3$ powder (Aldrich > 99 %, finer than 2 µm) was mixed in ethanol with carbonates precursors of the alkali and alkaline earth dopants (Fluka > 99 %), except for strontium and barium where nitrates were used, in ethanol and in agate jars with agate balls by means of a planetary mill for 1 hour, followed by a drying overnight at 80°C. These doped mixtures were uniaxially pressed at 300 MPa into bars for dilatometry measurements and into pellets for mercury porosimetry measurements, and heat treated up to 900°C with a 10°C/min heating ramp followed by a 1 hour isothermal step at the maximum temperature. XRD measurements were performed in the 5-70° in 2θ, on crushed doped pellets heat treated at 900°C for 1 hour. In the case of Mössbauer spectroscopy measurements, pure and potassium-doped hematite powders were uniaxially pressed at 100 MPa, prior to heat-treating at 900°C for 1 hour. To check the influence on the hematite structure of some commercial metallic screen-printing inks (Ag, Pd/Ag and Pt/Ag) used for electrodes deposition, these inks were heat-treated at 850°C for 1 hour and then milled. Pellets of pure and doped α-Fe$_2$O$_3$ were then crushed and separated into 2 batches: the first one to be mixed (again in a planetary mill for 1 hour and in ethanol) with the heat-treated inks (50 wt. %) and the other one to remain as such. All the powders (Table 1) were again uniaxially pressed at 100 MPa and heat-treated at 850°C for 1 hour. The ^{57}Fe Mössbauer absorption measurements were made using a ^{57}Co:Rh source at room temperature. Additions of K$_2$CO$_3$ to α-Fe$_2$O$_3$ were investigated in details because it is known that potassium has beneficial effect on the humidity response of hematite[1].

Table 1. Investigated compositions by means of Mössbauer spectroscopy

Composition	Equivalent K$_2$O content (wt. %)	Denomination
α-Fe$_2$O$_3$	0	ND
α-Fe$_2$O$_3$ + K$_2$O	1, 2, 5, 10	K$_1$, K$_2$, K$_5$, K$_{10}$
α-Fe$_2$O$_3$ + 50 wt. % Ag ink	0	ND/A
α-Fe$_2$O$_3$ + 50 wt. % Pd/Ag ink	0	ND/PDA
α-Fe$_2$O$_3$ + 50 wt. % Pt/Ag ink	0	ND/PTA
α-Fe$_2$O$_3$ + K$_2$O + 50 wt. % Ag ink	5	K$_5$/A
α-Fe$_2$O$_3$ + K$_2$O + 50 wt. % Pd/Ag ink	5	K$_5$/PDA
α-Fe$_2$O$_3$ + K$_2$O + 50 wt. % Pt/Ag ink	5	K$_5$/PTA

Sensors were screen-printed through a 325 mesh stainless steel screen by means of an automatically controlled machine. Screen-printing inks were made of doped hematite powders plus an organic vehicle made of a volatile organic solvent and a polymer, which acted as a temporary binder for unfired films and conferred the appropriate rheological properties to the paste. All the inks were initially prepared with the same composition; however, corrections were necessary to obtain a screen-printable paste in function of the nature of the dopant. After deposition onto commercial α-Al$_2$O$_3$ substrates, sensors (5 × 8 mm^2) were dried in air for one night and then fired at 850, 900 or 950°C for 1 hour, with a 2°C/min heating ramp (Table 2). Pt or Pd/Ag (1:3) interdigitated metallic electrodes were then screen-printed on fired thick-films (4 digits on each electrode with a 0.4 mm gap), followed by a second thermal treatment at 850, 900 or 950°C for 1 hour, again with a 2°C/min heating ramp. In the case of magnesium, strontium and barium additions, thick-films presented a poor adhesion onto alpha-alumina substrates, thus metallic electrodes were directly screen-printed onto unfired sensing materials and these sensors underwent only a single step thermal treatment at 900°C, for 1 hour. Finally, humidity sensors were mounted in a laboratory apparatus in which relative humidity could be varied between 10 % and 98 %, with a total flow rate of 0.1 l/s at 1 m/s. All the system was immersed in a thermostatic bath and the temperature was set to 30°C during all the tests. Each sensor was alimented by an external circuit (V_{AC} = 3.6 V @ 1 kHz) where it constituted an unknown variable resistance. Sensitivity is defined as R_d/R_h ratio, where R_d is the resistance under dry conditions and R_h is the resistance at a specified humidity. Two sensors of each composition and thermal treatment temperature were tested at each time. More details on sensors preparation and testing are available elsewhere[9]. Thick-film adhesion onto alumina substrates was very poor for samples F/900 and K_1 and K_2, whatever the temperature treatment, thus metallic electrodes could not be screen-printed onto these compositions.

Table 2. Screen-printed compositions

Compositions	Ionic radius (pm)	Designation after thermal treatment		
		850°C, 1h	900°C, 1h	950°C, 1h
α-Fe_2O_3	---	---	F/900	---
α-Fe_2O_3 + 1 wt.% eq. K_2O	K^+ : 133	K_1/850	K_1/900	K_1/950
α-Fe_2O_3 + 2 wt.% eq. K_2O	---	K_2/850	K_2/900	K_2/950
α-Fe_2O_3 + 5 wt.% eq. K_2O	---	K_5/850	K_5/900	K_5/950
α-Fe_2O_3 + 10 wt.% eq. K_2O	---	K_{10}/850	K_{10}/900	K_{10}/950
α-Fe_2O_3 + 5 wt.% eq. Li_2O	Li^+ : 60	Li_5/850	Li_5/900	Li_5/950
α-Fe_2O_3 + 5 wt.% eq. Na_2O	Na^+ : 95	Na_5/850	Na_5/900	Na_5/950
α-Fe_2O_3 + 5 wt.% eq. Rb_2O	Rb^+ : 148	---	Rb_5/900	Rb_5/950
α-Fe_2O_3 + 5 wt.% eq. MgO	Mg^{2+}: 65	---	Mg_5/900	---
α-Fe_2O_3 + 5 wt.% eq. CaO	Ca^{2+}: 99	Ca_5/850	Ca_5/900	Ca_5/950
α-Fe_2O_3 + 5 wt.% eq. SrO	Sr^{2+}: 113	---	Sr_5/900	---
α-Fe_2O_3 + 5 wt.% eq. BaO	Ba^{2+}: 135	---	Ba_5/900	---

RESULTS AND DISCUSSION
Potassium additions

Mössbauer spectroscopy results showed that in all the samples, a Fe^{3+} charge state was identified and that for all non-doped samples (ND, ND/A, ND/PDA & ND/PTA) the same spectrum was observed (hematite sextet with a hyperfine field of 521 kOe), indicating that the thermal treatment had no effect on hematite. In the presence of K_2CO_3 dopants, the samples showed two extra sextets with hyperfine fields of 490 kOe and 441 kOe, corresponding to the $KFe_{11}O_{17}$ ferrite. The relative $KFe_{11}O_{17}$ content in the doped samples increased with dopant content and about 10 % of ferrite was formed for each K_2CO_3 % added, leading to 100 % ferrite for an addition of 10 wt. % of potassium carbonate. K-doped samples mixed with precious metals (K_5/A, K_5/PDA & K_5/PTA) also showed lower or quite equal ferrite contents with respect to samples doped with potassium and no metals additions, confirming the first results on non-doped samples, where additions of precious metals tended to favour the hematite structure. The presence of potassium ferrite was confirmed by XRD spectra on the K_5/900 and K_{10}/900 ground pellets (JCPDS card 25-0651).

Table 3. Total porosity (%) and mean pore radius (μm) by Hg porosimetry
(pellets heat treated at 900°C for 1 hour, average of 3 measurements)

Pellet	K_2	K_5	K_{10}	Li_5	Na_5	Rb_5	Mg_5	Ca_5	Sr_5	Ba_5
Total	39.83	41.37	40.46	46.68	20.63	41.24	44.24	39.91	44.83	42.81
porosity	±2.76	±0.38	±0.04	±2.10	±1.65	±1.96	±1.15	±0.98	±3.31	±0.31
Pore	0.10	0.13	0.13	0.19	0.10	0.13	0.13	0.13	0.11	0.10
radius	±0.001	±0.001	±0.001	±0.04	±0.001	±0.001	±0.001	±0.001	±0.02	±0.001

(For F samples: total porosity = 30.57 % ± 2.72 % and mean radius = 0.10 μm ± 0.001 μm)

Dilatometric curves of K_5 and K_{10} bars showed a continuous, quite regular, expansion of 1. % up to 810°-820°C, followed by a limited shrinkage of 0.05 % until 850°C and a second expansion step up to around 2 % at the end of the isothermal step at 900°C. Finally, at ambient temperature, bars showed only a slight expansion of ca. 1 %. Potassium additions to hematite significantly increased the total porosity and the mean pore diameter in the pellets (Table 3) however, porosity distribution remained nearly unchanged (Figure 1).

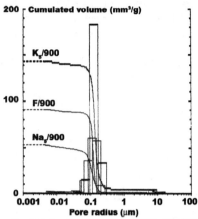

Figure 1. Porosity distribution in pellets F/900, K_5/900 and Na_5/900

Figure 2 reports K_{10}/850 sensors sensitivity for Pd/Ag electrodes with respect to Pt ones sensitivity curve presented a more regular negative slope with Pd/Ag electrodes from 40 % RH However, K_5 sensors did not gave any response with Pd/Ag electrodes.

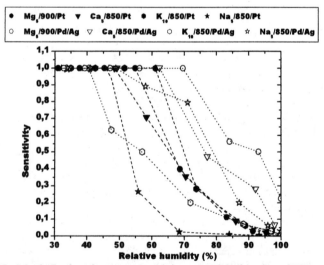

Figure 2. K, Na, Mg & Ca-doped sensors sensitivity at 30°C in function of RH and electrodes

Lithium additions

XRD spectrum, on the ground pellet heat treated at 900°C for 1 hour revealed the presence of α-LiFe$_5$O$_8$ (JCPDS card 38-0259) and the disappearance of hematite peaks. Dilatometric curve showed a continuous and rather limited expansion up to 500°C and a more important dilation above this temperature to reach 3.7 % at 850°C. A shrinkage then started and continued until the end of the isothermal step at 900°C (0.5 %); at ambient temperature bars shrank slightly (-0.8 %). The lithium ferrite formation strongly modified the total porosity and the mean pore diameter (Table 3), but not the pore size distribution. Thick-films resistances decreased continuously from 60 % RH[9], with platinum electrodes, while no response was recorded with Pd/Ag electrodes.

Sodium additions

Only hematite was identified on XRD spectrum and dilatometric curve showed a similar trend to lithium-doped α-Fe$_2$O$_3$: a continuous expansion of 0.7 % up to 780°C followed by an important shrinkage until the end of the isothermal step at 900°C (-8.7 %). At ambient temperature bars shrank significantly (-10 %). This strong shrinkage was confirmed by mercury porosimetry measurements which evidenced a total porosity reduction (Table 3) associated with a limited variation of pore radii distribution (Figure 1). Sensors resistances decreased rapidly from 45 % to 70 % RH however, the thick-films decrease in resistance with Pd/Ag electrodes was more progressive respect to sensors with platinum ones (Figure 2).

Rubidium additions

XRD spectrum allowed to identify the presence of hematite and β"-Fe$_2$O$_3$ (JCPDS card 40-1139), after 1 hour at 900°C. Dilatometric curve of Rb-doped hematite presented a trend close to that of lithium-doped one, with a maximum expansion of 2.2 % at 800°C, followed by a limited shrinkage up to 1.9 % after 1 hour at 900°C. Bars were slightly expanded (0.8 %) at ambient temperature. Rb-additions produced the same effect as potassium ones onto porosity evolution (Table 3). Sensors resistances quickly decreased from 40 % to 70 % RH[9] with platinum electrodes, similarly as for sodium additions, while no signal was recorded with Pd/Ag ones.

Magnesium additions

No new phase was detected by XRD and the dilatometric curve evidenced a continuous expansion of 0.4% up to 760°C, followed by a shrinkage of 3.4 % after 1 hour isothermal step at 900°C which reached 4.6 % at ambient temperature. However, Hg-porosimetry measurements didn't show any reduction in the total porosity (Table 3), as for sodium additions. The sensors presented a continuous decrease in the resistance from 60 % RH, with Pt metallic electrodes, and only from 70 % RH with Pd/Ag ones (Figure 2).

Calcium additions

XRD evidenced the formation of CaFe$_2$O$_4$ (JCPDS card 32-0168) and the dilatometric curve showed a behaviour close to that of magnesium-doped hematite: a continuous expansion of 0.7 % up to 850°C, followed by a 3.8 % shrinkage after 1 hour isothermal step at 900°C and 5% at ambient temperature. Ca-additions produced the same effect as potassium ones onto porosity evolution (Table 3). Sensors resistance showed a regular negative slope from 50 % to 95 % RH, with Pt electrodes and an irregular decrease only from 65 % RH with Pd/Ag ones (Figure 2).

Strontium additions

XRD spectrum on a heat treated pellet showed that strontium additions led to the formation of Sr-ferrite (SrFe$_{12}$O$_{19}$, JCPDS card 84-1531). The dilatometric curve presented a first continuous regular expansion step of 1.2 % up to 880°C, followed by a second one up to the end of the isothermal plateau at 900°C (1.6 %). At ambient temperature the bar showed an expansion of 0.6 %. The total porosity was also increased by Sr-doping (Table 3). Screen-printed sensors exhibited a decrease in their resistance in a rather limited range, from 50 % to 70 % RH, with platinum electrodes[9], while no response was recorded with PdAg ones.

Barium additions

BaFe$_{12}$O$_{19}$ was formed after thermal treatment (JCPDS card 39-1433) and the dilatometric curve presented only a reversible expansion step due to increasing temperature. Surprisingly, Ba additions led to an increase in the total porosity comparable to all the other dopants, except sodium (Table 3), but sensors gave no response, whatever the materials used for electrode deposition[9]. Barium ferrite did not seem to be a suitable material for humidity sensors.

CONCLUSIONS

Microstructural characterisations have evidenced that alkali and alkaline-earth doping of hematite could lead to new phases formation and that total porosity was significantly increased except with sodium additions. All the sensors, except barium ones, evidenced strong resistance decreases in function of relative humidity, but also a scarce sensitivity below 40-50 % RH probably because of a pore size distribution not significantly modified by alkali or alkaline-earth additions and rather centred on the mean pore radius. The effect of the dopant cation was not clear at all, except for barium: the residual porosity of the films, due to the solvent evaporation and the temporary binder calcination during drying and thermal treatment stages, hindered the influence of the doping. Then, sensitivity curves in function of RH seemed to be more dependent on thick-films porosity than on ionic radius of the dopant. Finally, sensors resistances died down in all the investigated compositions with Pt electrodes and only for K, Na, Mg and Ca addition with Pd/Ag ones. Only with K$_{10}$/850 and Na$_5$/850 sensors, Pd/Ag electrodes gave a better result in terms of a wider RH range over where the variation of resistance was observed. Therefore, Pt inks seemed to be more reliable for humidity sensors preparation, probably because of a lack of porosity of Pd/Ag pastes[8].

REFERENCES

[1] H. Arai and T. Seiyama, "Humidity sensors"; pp.982-1012 in *Sensors: A comprehensive survey* Vol.3 Chemical and biochemical sensors part II, Edited by W. Göpel, T.A. Jones, M. Kleitz, Lunsdtröm, and T. Seiyama, VCH, Weinheim, 1992

[2] N. Yamazoe and Y. Shimizu, "Humidity sensors: principles and applications", *Sensors and Actuators*, **10** 379-398 (1986)

[3] C. Cantalini and M. Pelino, "Microstructure and humidity-sensitive characteristics of α-Fe$_2$O$_3$ ceramic sensor", *Journal of the American Ceramic Society*, **75** [3] 546-551 (1992)

[4] M. Pelino, C. Cantalini, H.T. Sun and M. Faccio, "Silica effect on α-Fe$_2$O$_3$ humidity sensor" *Sensors and Actuators B*, **46** 186-193 (1998)

[5] G. Neri, A. Bonavita, S. Galvagno, C. Pace, S. Patanè and A. Arena, "Humidity sensing properties of Li-iron oxide based thin films", *Sensors and Actuators B*, **73** 89-94 (2001)

[6] M. Zucco, A. Negro, L. Montanaro, Italian Patent IT-MI2001A001910 (2001)

[7] M. Pelino, C. Colella, C. Cantalini, M. Faccio, G. Ferri and A. D'Amico, "Microstructure and electrical properties of an a-hematite ceramic humidity sensor", *Sensors and Actuators B*, **7** 464-469 (1992)

[8] W. Qu, "Effect of electrode materials on the sensitive properties of the thick-film ceramic humidity sensor", *Solid State Ionics*, **83** 257-262 (1996)

[9] J.M. Tulliani and P. Bonville, "Influence of the dopants on the electrical resistance of hematite based humidity sensor", submitted to *Ceramics International*

LIGHT WEIGHT CERAMIC SANDWICH STRUCTURE FROM PRECERAMIC POLYMERS

Thomas Hoefner, Juergen Zeschky, Michael Scheffler and Peter Greil
University of Erlangen-Nuremberg
Department of Materials Science, Glass and Ceramics
Martensstrasse 5
91058 Erlangen, Germany

ABSTRACT

Light weight ceramic sandwich structures with a foam core and surface cover tapes were manufactured from filler loaded preceramic polymer systems.

A poly (silsesquioxane) loaded with 58 wt.-% of a mixture of Si and SiC filler powders was tape cast. A polymer-filler blend with 65 wt.-% of the same filler composition was *in situ* foamed between two green tapes by a controlled heat treatment (blowing, curing and stabilization) at 270 °C. The sandwich element was subsequently pyrolyzed at 1000 °C in nitrogen atmosphere to form a Si-O-C micro composite material.

The microstructure and the mechanical properties of the sandwich structure were characterized. The interface strength between the foam core and the cover tape was controlled by varying the filler load of the foam, which effects the dimensional change misfit upon pyrolysis and thermal contraction. The foam of the sandwich structure exhibits a Weibull modulus $m = 13$ (tension) and $m = 8$ (compression) and a crushing strength exceeding 4 MPa at a fractional density of 0.27.

INTRODUCTION

Ceramic tapes and foams can be manufactured from silicon resins. Silicon resins have been used for coatings, and for near-net-shape manufacturing of ceramic components by active-filler controlled polymer pyrolysis [1-3]. Ceramic sandwich panels of low weight can be used for many applications including thermal insulation, high temperature design elements and kiln furniture [4]. Foamed in appropriate shape, the sandwich structure can be applied as a catalyst carrier, or as a catalytic trap for diesel particulate removal under harsh environmental conditions (acid, elevated temperature) [5,6]. Electrically conductive sandwich material can be used as a lightweight heating element as well as a regenerable absorber component [7]. Furthermore, the panels give benefit as acoustic insulation, as a carrier of silicon based photovoltaic cells and as a carrier of large size telescope mirrors [8].

In this study a *ceramic sandwich panel* was fabricated from filler (Si, SiC) loaded preceramic polysiloxane in one thermal processing step. Bonding between the cover tapes and the core material was achieved by the same material composition e.g. no adhesive interface was applied.

EXPERIMENTAL PROCEDURE

For the tape fabrication MK–resin ($[CH_3SiO_{1.5}]_n$) was dissolved in MTMS (($CH_3Si(OCH_3$ and subsequently MSE 100 ($CH_3SiO_{1.1}(OCH_3)_{0.8}$) (all Wacker Chemie, Burghausen, German and the curing agents, oleic acid ($C_{18}H_{34}O_2$) (Riedel de Haen AG, Seelze, Germany) a aluminum acetylacetonate ($C_{15}H_{21}AlO_6$) (Merck, Darmstadt, Germany), were added. The soluti was stirred for 10 minutes. Silicon powder (d_{50} = 7.6 µm; Elkem, Meerbusch, Germany) and S Grade 05 (d_{50} = 1.6 µm, H.C. Stark, Selb, Germany) were added as fillers. The filler content the cover tapes was kept constant at 58 wt.-%. The slurry was homogenized for 24 hours in a ball mill, containing silicon nitride (Si_3N_4) balls. The slip was cast on a silicone coated PET f by two doctor blades. The as-cast tape was dried in a saturated solvent atmosphere on the PE foil in the tape casting facility at room temperature for 72 hours. After drying the tape with average thickness of 600 µm was removed from the polyethylene support and cut to 100 x 1 mm² and 90 x 90 mm² sheets.

The foam core was fabricated from a poly (silsesquioxane) (Silres H44, Wacker Chem Burghausen, Germany) which has the general formula $[(C_6H_5)_{0.62}(CH_3)_{0.31}(OR)_{0.07}SiO_{1.5}]_n$ (n 20), where R is both a hydroxyl- (-OH), and an alcoxy-group (-OC_2H_5). The polymer was load with silicon powder as used for the tape fabrication and SiC F500B (d_{50} = 16 µm) (H.C. Sta Selb, Germany). The Si:SiC weight ratio in the filler blend was 1:4 and the filler content was s to 65 wt.-%. The filler powders and the silicon resin H44 were homogenized for 2 h hours using 5 l ball mill. The filler/resin powder mixture was filled into an aluminum tray and cross–linked a preheated furnace at 270 °C for 2 hours. Subsequently, the cured foams were cut to rectangu bars of 6 x 6 x (8–16) mm³ for dilatometer measurements, 30 mm in diameter and 8.5–30.4 m in height for compression tests and µCT measurements, 20 x 20 x 100 mm³ for flexural tests a 100 x 75 x 25 mm³ for shear tests. The filler grain size distribution was measured by las granulometry (Mastersizer2000, Malvern Inc. USA, version 1.2).

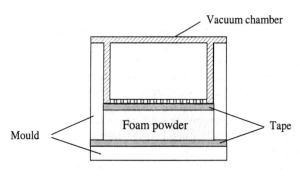

Figure 1. Schematic illustration of the set-up used for the sandwich fabrication.

For the sandwich fabrication the set-up as illustrated in Figure 1 was used. The sandwi preform was placed in a preheated furnace at 270 °C and cured for 2 h. The samples we pyrolyzed at 1000 °C for 4 hours under flowing nitrogen. A heating rate of 3 °C/min was appli up to 500 °C, hold for 2 hours and finally heated up to 1000 °C. The processing scheme of t sandwich fabrication is shown in Figure 2.

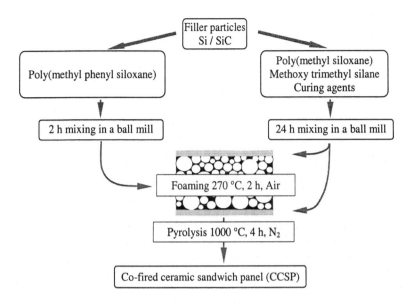

Figure 2. Flow chart of the sandwich fabrication.

The microstructure was examined by SEM (Quanta 200, FEI Company, Hillsboro, USA). The pore structure was characterized by micro computer tomography (μCT40, Scanco Medical, Bassersdorf, CH). The system uses a fan beam at $\lambda = 0.024$ nm from a micro focus X–ray tube (50 kV, 80 μA), detected by a CCD line array. From a selected volume of interest (VOI) the morphometric structure model index (SMI), the strut thickness and the cell size distribution, the degree of anisotropy (DA), the interconnectivity and the fractional density (ρ_{fd}) of the measured sample were calculated [9,10]. The coefficient of thermal expansion (CTE) as well as the expansion and shrinkage of foams and tapes were measured by dilatometry (Model 402E/7, Netzsch, Selb, Germany). Strength measurements were carried out in a universal testing machine (Instron Model 4204, Instron Co., Canton, USA). The samples were loaded in compression applying a cross–head speed of 1 mm/min. The Young's modulus was measured using a resonance beam technology [11]. Rectangular bars were used for a four–point loading method (40/20 mm). Double ring tests were performed according to DIN 52292-1 [12]. The samples were loaded applying a cross–head speed of 0.5 mm/s.

RESULTS AND DISCUSSION

Figure 3 shows a pyrolyzed sandwich fabricated with the test set-up displayed in Figure 1. The foam core exhibits a fractional density of $\rho_{fd} = 0.27$ and a linear shrinkage upon pyrolysis of 4.2 %, whereas the shrinkage of the cover tapes equals 3.8 %. Foam core and cover tapes have comparable CTE (foam: $\alpha_{200\text{-}1000} = 7.8 \times 10^{-6}$ K^{-1}; tape: $\alpha_{200\text{-}1000} = 7.9 \times 10^{-6}$ K^{-1}) and exhibit the same shrinkage and expansion behaviour in the temperature regime. Samples pyrolyzed at 1600 °C showed no delamination.

The core attained a crushing strength σ_{cr} of 4.25 ± 0.6 MPa and a flexural strength σ_{fl} of 4.3 ± 0.4 MPa. The Young's modulus was 5.35 ± 0.2 GPa, and the shear modulus

$G_c = 2.7 \pm 0.55$ GPa. The mechanical reliability of the foam core exhibits a Weibull modulus of $m_{cr} = 8$ under compression with a crushing strength $\sigma_0 = 4.5$ MPa and of $m_{fl} = 13$ for flexural loading with $\sigma_0 = 4.4$ MPa . The Weibull plots of the flexural and crushing strength are depicted in Figure 4.

Figure 3. Pyrolyzed ceramic sandwich.

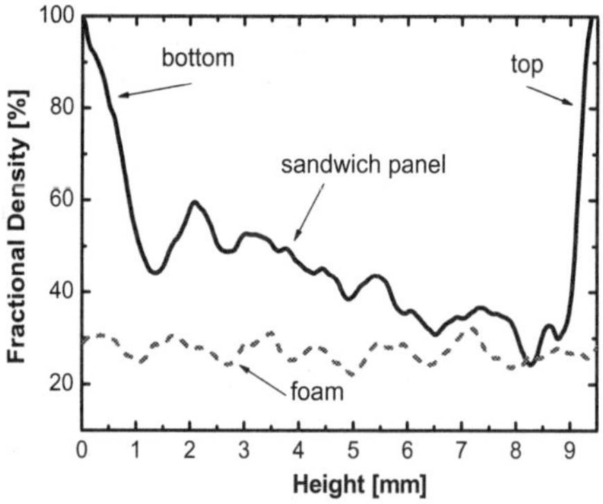

Figure 4. Density distribution in dependence on the sample height.

The pores have a plate like structure which is indicated by a SMI value of 1.0. However, the DA of 1.05 indicates a moderate isotropy of the structure. The struts have a thickness of 0.44 ± 0.19 mm, whereas the pores show a diameter of 1.3 ± 0.36 mm. Although the struts are relatively thick the interconnectivity density is characterized by a value of 0.9 1/mm³. Tapes pyrolyzed at 1000 °C exhibit a flexural strength σ_f of 72 ± 18 MPa. The contribution of the facesheet to the total strength of the sandwich can be disregarded. A flexural strength of the sandwich of $\sigma_{fls} = 3.2 \pm 0.9$ MPa was measured. The sandwich failed due to crack initiation in the cover tapes as schematically shown in Figure 5. A delamination crack propagated in the first raw of pore cells in the foam parallel to the facesheet until the whole assembly failed. Figure 4 shows the density distribution of the sandwich panel and the foam in dependence on the sample height. The cover tape prevents the release of the gaseous polycondensation products from the polymer melt. Due to pore ascent and the coalescence of the pores beneath the cover tape, the porosity increased from the bottom to the top of the core material. The gradient of the fractional density in the sandwich panel can be clearly seen, whereas the fractional density of the freely foamed ceramic core material (e.g. without cover tapes) remained constant throughout the total cross section . At a fractional density of $\rho_{fd} \approx 0.55$ a flexural strength of $\sigma_{fls} = 3.9$ MPa was determined. According to[13], the flexural strength σ_{fls} in foams is proportional to the fractional density ρ_{fd}:

$$\sigma_{fls} \propto (\rho_{fd})^{1.5} \qquad (1)$$

Due to the lower fractional density of $\rho_{fd} \approx 0.3$ at the top of the foam core, a reduced flexural strength of $\sigma_{fls} = 1.6$ MPa can be expected (eqn. (1)), when the sample is stressed bottom up.

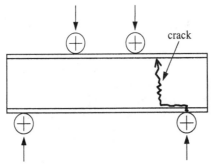

Figure 5. Crack propagation during bending.

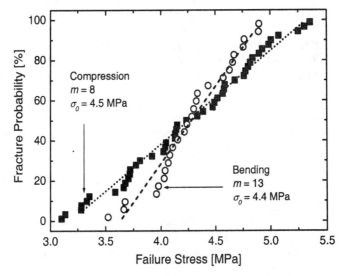

Figure 6. Weibull plot for the crushing and flexural strength of foam core.

CONCLUSIONS

In the present study a light weight ceramic sandwich structure was manufactured from a filler loaded polysiloxane in one processing step. The dimensional change misfit upon pyrolysis and thermal contraction between facesheets and foam core was minimized by different filler particle sizes. The foam core shows a high mechanical reliability, a high crushing and flexural strength at low fractional density.

ACKNOWLEDGEMENT

Financial support from the Deutsche Forschungsgemeinschaft (DFG) under SCHE-628/1-2 is gratefully acknowledged.

REFERENCES

[1] Peter Greil, "Polymer Derived Engineering Ceramics," Advanced Engineering Materi 2 [6] 339-348 (2000).

[2] P. Greil, "Active Filler Controlled Pyrolysis of Preceramic Polymers," Journal of American Ceramic Society, [78] 835-848 (1995).

[3] Peter Greil, "Near Net Shape Manufacturing of Polymer Derived Ceramics," Journa the European Ceramic Society, 18 [13] 1905-1914 (1998).

[4] Andrew J. Sherman, Robert H. Tuffias, and Richard B. Kaplan, "Refractory Cera Foams: A Novel, New High-Temperature Structure," Ceramic Bulletin, 70 [6] 1025-1 (1991).

[5] G. Saracco, N. Russo, M. Ambrogio, C. Badini, and V. Specchia, "Diesel particu abatement via catalytic traps," Catalysis Today, 60 [1-2] 33-41 (2000).

[6] J. - M. Tulliani, L. Montanaro, O. Morey, and J. P. Lecompte, "Caractérisation Mécani de Mousses Céramiques pour la filtration des émissions Diesel," Matériaux & Techniq (in French) [11-12] 55-60 (1996).

[7] P. Colombo and J. R. Hellmann, "Ceramic foams from preceramic polymers," Mater Research Innovations, 6 [5-6] 260-272 (2002).

[8] Y. V. Danchenko, V. N. Antsiferov, and N. A. Nazarenko, "Attempt to fabricate lig weight mirrors from a composite ceramic material," Journal of Optical Technology, 65 169-171 (1998).

[9] T. Hildebrand and R. Rüegsegger, "Quantification of Bone Microstructure with Struct Model Index," Computer Methods in Biomechanical Engineering [1] 15-23 (1997).

[10] T. Hildebrand and R. Rüegsegger, "A new Method for the Model-independ Assessment of Thickness in three-dimensional Images," Journal of Microscopy, [185] 75 (1997).

[11] D. Loidl, S. Puchegger, K. Kromp, J. Zeschky, P. Greil, M. Bourgeon, and H. Peter "Elastic moduli of porous and anisotropic composites at high temperatures"; Advan Eng. Mat., in print.

[12] Normenausschuss Materialprüfung (NMP) Im DIN Deutsches Institut Für Normung e Testing of glass and glass ceramic; determination of bending strength; coaxial double r bending test on a flat test piece with small test area. Beuth Verlag, Berlin (German 1984, p. 6.

[13] Lorna J. Gibson and Michael F. Ashby, "Cellular Solids: Structure and Properties"; 510 in Cambridge Solid State Science Series, 2nd Edition. Edited by D. R. Clarke Suresh, I. M. Ward FRS. Cambridge University Press, Cambridge (U.K.), 1997.

SELECTIVE CATALYTIC REDUCTION AND NO$_x$ STORAGE IN VEHICLE EMISSION CONTROL

Eric N. Coker, Sonia Hammache, Donovan A. Peña and James E. Miller
Sandia National Laboratories, PO Box 5800, Albuquerque, NM 87185-1349, USA

ABSTRACT

Selective Catalytic Reduction with non-vanadia catalyst formulations has been studied for NO$_x$ abatement using ammonia as the reductant. Laboratory testing has been carried out on both powder and monolithic catalysts using simulated vehicle exhaust. Performance as a function of temperature, NO:NO$_2$ ratio and space velocity has been monitored. Catalysts comprising transition metals supported on a titania-based substrate, in both powder and monolithic form, exhibit high NO$_x$ conversion with minimal N$_2$O production over a wide range of conditions, and have shown resistance to hydrothermal ageing, low-level SO$_x$ ageing and residual hydrocarbons in the feed stream.

In separate studies, the NO$_x$ storage-reduction properties of two representative formulations based on Pt/BaO/Al$_2$O$_3$, and the influence thereon of CO and CO$_2$ are discussed in detail, with particular attention to the uptake and release dynamics of NO$_x$.

INTRODUCTION

Control of vehicle emissions is a topic of ever-increasing concern to regulatory groups as well as environmentalists. Of particular concern are the tailpipe emissions of NO$_x$, SO$_x$, CO, un-burnt hydrocarbons and particulate matter (PM). Our research focuses on removal of NO$_x$ from lean burn engine exhausts, particularly light and medium duty Compression Ignition Direct Injection (CIDI, or diesel engines) using exhaust aftertreatment – identified as one of the key enabling technologies for CIDI engine success.[1] Two of the most promising technologies to achieve acceptable NO$_x$ reductions are Selective Catalytic Reduction (SCR) using ammonia, and NO$_x$ adsorption, known as Lean NO$_x$ Trap (LNT) or NO$_x$ Storage Reduction (NSR).[2] In NH$_3$-SCR, aqueous urea is injected into the exhaust upstream of the SCR catalyst; urea decomposes to yield NH$_3$ and CO$_2$.[3] While urea is a relatively cheap commodity chemical, integration of the technology into an existing vehicle platform (e.g., additional urea storage vessel, urea injection hardware and control, method of co-fueling a vehicle) are barriers to market acceptance.[4] In NSR, the catalyst oxidizes NO to NO$_2$, which adsorbs principally onto BaO to form Ba nitrates during lean operation (normal mode). During a subsequent brief rich operation mode, the stored nitrates decompose, re-releasing NO$_x$ which is reduced by a hydrocarbon-SCR type process.[5,6] While integration of NSR technology into an existing vehicle would be easier than SCR, an NSR catalyst requires periodic switching between lean and rich conditions, which can impact vehicle performance.

Infrastructure issues notwithstanding, the SCR process has the greatest potential to attain the >90% NO$_x$ reduction required for CIDI engines to meet the new EPA Tier II emission standards phasing in 2004, while NSR technologies are of interest because they could be implemented with only minimal impact on existing infrastructure.

EXPERIMENTAL SECTION

I) Ammonia SCR

The catalysts consist of various base metals (not vanadium) supported on an ion-exchangeable hydrous metal oxide (HMO) substrate, typically hydrous titanium oxide stabilized with some silica (HTO:Si).[7] The powder HTO:Si substrates were prepared from a mixture of tetra-isopropyl titanium and tetra-ethyl orthosilicate by first metathesizing with NaOH/MeOH, then hydrolyzing by slow addition to a water/acetone solution to precipitate the raw HMO. Next, the material was acidified, and particular metals[7] were introduced via ion exchange. After calcination at 600°C in air for 2 hours, other transition metals[7] were added by incipient wetness impregnation before repeating the calcination as before.

To prepare a monolithic catalyst, the final powder (above) was formed into a slurry using boehmite (Disperal, Sasol) and γ-alumina (Puralox SCFa-100, Condea) as binders. The cordierite monolith core (300 – 400csi) was then dip-coated with slurry, allowed to dry in ambient air and calcined in air at 425°C for 4 hours prior to testing.

Typical experimental conditions are shown in Table 1. The $NH_3:NO_x$ ratio was 1:1 for all runs, while the majority of experiments were performed at $NO:NO_2$ ratios of 1:1 and 4:1. The temperature was ramped from 450 to 125°C with 30 minute isothermal holds during each experiment, and water was included in all runs. The performance goals were 50% NO_x conversion at 200°C with $NO:NO_2$ = 4:1, and 90% NO_x conversion from 200-400°C with $NO:NO_2$ = 1:1.

In a typical exhaust aftertreatment set up,[8] an oxidation catalyst is positioned upstream of the emissions control (SCR) catalyst; besides removing unwanted hydrocarbons, the oxidation catalyst oxidizes some of the engine-out NO to NO_2.[9] The two performance targets above relate to start-up conditions (engine and oxidation catalyst cold, hence low $[NO_2]$) and warm engine conditions (catalysts warm and $[NO] \sim [NO_2]$). From mechanistic studies, it is known that the NH_3-SCR reaction involving an equimolar mixture of NO and NO_2 is fastest.

Table 1. Experimental conditions for the SCR of NO_x with NH_3 for simulated diesel exhaust

Temperature (°C)	NO (ppm)	NO_2 (ppm)	NH_3 (ppm)	O_2 (%)	CO_2 (%)	H_2O (%)	Space Velocity (h^{-1})
450-125	280-175	70-175	350	14	5	4.6	30,000-140,000

II) NO_x Storage Reduction

Two Pt-BaO/γ-Al_2O_3 catalysts (1wt.% Pt; 20wt.% BaO) were prepared (Pt-BaO/Al_2O_3) by incipient wetness impregnation.[10] To one, 2wt.% Cu was added (Pt-Cu-BaO/Al_2O_3).

The NSR reaction was carried out in a tubular fixed bed reactor loaded with 0.25g of catalyst. A 4-port valve was used to switch between lean and rich cycles without disturbing the reaction. Reactions were studied isothermally at 380°C at a SV of 40,000 h^{-1} after 2 hours conditioning under reaction conditions. Typical test gas compositions were as in Table 2.

Table 2: Test gas compositions for NSR catalyst evaluation

Cycle	NO (ppm)	O_2 (vol. %)	Propene/propane (HC) (vol. %)
Lean (for 30 min)	420	8	0.1
Rich (for 10 min)	420	0.5	0.1

RESULTS

I) Ammonia SCR

The catalytic performance of a typical NH₃-SCR powder catalyst is shown in Figure 1. Under "cold engine" conditions (20% of NO_x input is NO_2), activity is low at low temperature, but rises steadily, reaching 90% NO_x conversion at 300°C. With a "warm engine" (50% NO_2 input), activity is high across a broad temperature range, dropping to a low value only below 175°C. Note that in both cases, NO_x conversion increases slightly on dropping from 450 to 400°C, due to a reduction in competitive oxidation of ammonia.

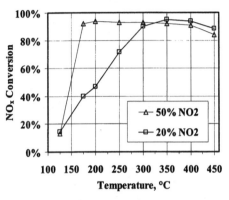

Figure 1: NH₃-SCR activity profile for typical powder catalyst at 20% and 50% NO_2:NO_x input. Space velocity 140,000 h⁻¹.

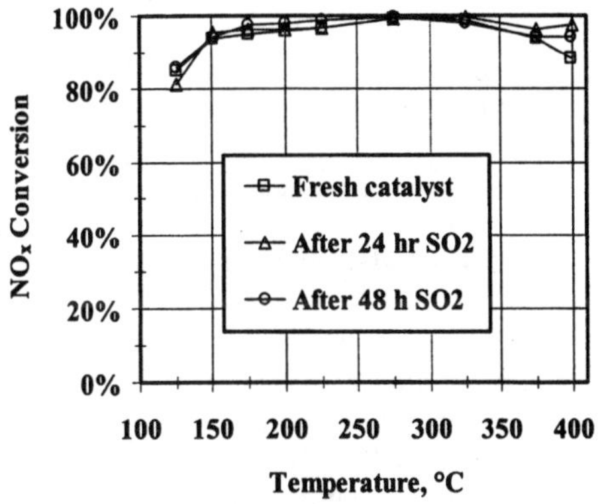

Figure 2: Effect of accelerated SO_2 ageing. For details see text.

In our research protocol, once successful catalysts have been identified they are subjected to durability testing, both hydrothermally, and in the presence of sulfur. A catalyst similar to the one represented in Figure 1 has shown excellent stability to 16 hours treatment in full test gas (Table 1) at 600°C, while only mild deactivation was seen at 700°C (results not shown). Sulfur tolerance is shown in Figure 2, where the catalyst was treated for 24 and 48 hours in the presence of 20ppm SO_2 at 350°C (full test gas without NH₃, SV 30,000 h⁻¹) prior to re-evaluating the catalyst's performance. The catalyst is extremely resilient to SO_2 under these conditions. Further durability tests are underway using elevated levels of SO_3, which is a more potent poison.

Figure 3 shows that increasing the loading of catalyst on a monolith leads to an improvement in performance. Further increases above 16wt.% lead to enhanced performance. Operation under "cold start" conditions (Figure 3b) amplifies the difference in performance between the high-loaded monoliths. The powder data shown in Figure 1 was recorded at a SV of 140,000 h⁻¹, which was calculated based upon the effective SV of a monolith loaded with 25 wt.-% catalyst operating at a bulk SV of 30,000 h⁻¹. Comparison of Figure 3 with Figure 1 indicates that high-loaded monoliths approach powder catalyst performance. A monolith loaded with 28 wt.-% catalyst (not shown) exhibited activity intermediate between the powder data and the 16 wt.-% loaded monolith. This result indicates that interaction between the catalyst and monolith (cordierite) does not adversely affect catalytic activity, and that powder data can be used to predict monolith performance.

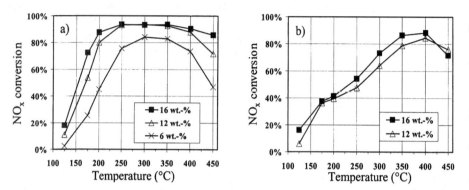

Figure 3. Catalytic performance as a function of catalyst loading on monolith; a) 50% NO_2:NO_x; b) 20% NO_2:NO_x.

II) NO_x Storage Reduction

The responses (reactor-out concentrations) of NO, NO_2, N_2O, and N total over Pt-BaO/Al_2O_3 and Pt-Cu-BaO/Al_2O_3 at 380°C are shown in Figures 4a and 4b. The inlet NO concentration is also shown for comparison. In both cases, NO is oxidized to NO_2 and stored as barium nitrates during the lean phase. Then, during the rich phase (regeneration), the stored NO_x is released. The hydrocarbons initially reduce the noble metal and then reduce NO and the released NO_x to N_2O and N_2. The results for the lean cycle indicate that Pt-BaO/Al_2O_3 is a better storage material for NO than Pt-Cu-BaO/Al_2O_3, with capacities of 108 and 64 µmol/gcat, respectively. Also, during this cycle, more NO_2 was detected over Pt-BaO/Al_2O_3 than over Pt-Cu-BaO/Al_2O_3. We can attribute this difference to the higher oxidation activity of Pt-BaO/Al_2O_3.

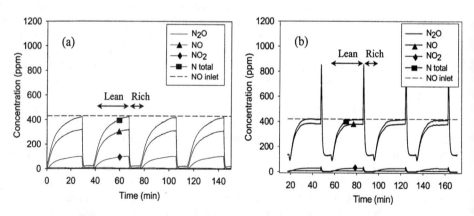

Figure 4: (a) NSR of Pt-BaO/Al_2O_3 at 380°C, (b) NSR of Pt-Cu-BaO/Al_2O_3 at 380°C

A spike in the NO concentration can be seen for Pt-Cu-BaO/Al$_2$O$_3$ after switching from lean to rich conditions. It has been reported in the literature that the origin of the NO breakthrough for Pt-BaO/Al$_2$O$_3$ could be either the slow reduction of the noble metal or the sweeping of NO$_x$ on the surface of the catalyst by the hydrocarbons.[11] Breakthrough due to the sweeping of NO$_x$ should be observed on both catalysts, however in our case no breakthrough was observed for Pt-BaO/Al$_2$O$_3$. Hydrogen chemisorption results (not shown) indicate that the dispersion of Pt is much higher on Pt-BaO/Al$_2$O$_3$ than on Pt-Cu-BaO/Al$_2$O$_3$. This suggests that breakthrough may be caused by insufficient active sites on Pt-Cu-BaO/Al$_2$O$_3$.

The effect of CO and CO$_2$ (CO$_x$) on NSR was evaluated at 380°C, in the absence and presence of hydrocarbons (HC) (Figure 5). CO$_x$ were added to the gas compositions in Table 2; lean cycle: CO (0.17 vol%), CO$_2$ (9.2 vol%); rich cycle: CO (1.3 vol%), CO$_2$ (14.3 vol%).

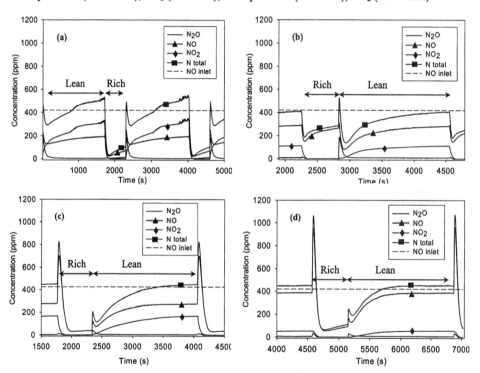

Figure 5: NSR profile of: Pt-BaO/Al$_2$O$_3$ in the presence of (a) CO and CO$_2$, (b) CO, CO$_2$ and HC; and of NSR profile of Pt-Cu-BaO/Al$_2$O$_3$ in the presence of (c) CO and CO$_2$, (d) CO, CO$_2$ and HC.

Addition of CO$_x$ had a negative impact on the performance of Pt-BaO/Al$_2$O$_3$, decreasing its storage capacity from 108 to 80μmol/g$_{cat}$. Moreover, in the absence of hydrocarbons, the net capacity dropped to only 1μmol/g$_{cat}$ (Figure 5a). The CO$_x$ may have caused desorption of stored NO during the lean cycle itself, thus decreasing the total amount of NO$_x$ stored. The presence of hydrocarbons (Figure 5b) decreased the damaging impact of CO$_x$. In contrast, CO$_x$ had a positive impact on Pt-Cu-BaO/Al$_2$O$_3$ (Figures 5c and 5d). The storage capacity of the catalyst increased from 64μmol/g_{cat} (no CO_x) to 78 and 74μmol/g_{cat} in the absence and in the presence of HC, respectively. Similarly NO_x conversion increased from 70% (no CO_x) to 90 and 86% in the absence and in the presence of HC, respectively (conversion = (NO_{inlet}-N_{total})/NO_{inlet}). This difference might be related to the stability of the nitrates on this catalyst.

CONCLUSIONS

The development of diesel exhaust catalysts will play an important role in meeting future energy and environmental needs. In the realm of urea-SCR catalysis, there is still a need for increased intrinsic activity along with increased hydrothermal and sulfur stability to successfully implement this emissions control strategy in commercial applications.

Addition of CO and CO_2 to the process gas resulted in a decrease in NO_x storage capacity of the benchmark NSR catalyst Pt-BaO/Al_2O_3. In addition, in the absence of hydrocarbons, CO_x caused desorption of the stored NO_x during the lean cycle itself. Despite a lower initial storage capacity, Pt-Cu-BaO/Al_2O_3 exhibited enhanced NO_x trapping when CO_x were present in the gas stream, and out-performed the benchmark catalyst.

ACKNOWLEDGEMENTS

This work was supported by the United States Department of Energy under Contract DE-AC04-94AL85000. Sandia is a multiprogram laboratory operated by Sandia Corporation, a Lockheed Martin Company, for the United States Department of Energy. Part of this work (SCR) was the result of a Cooperative Research and Development Agreement between Sandia National Laboratories and the Low Emission Technologies Research and Development Partnership (DaimlerChrysler, Ford and General Motors), and funded by the DOE Office of Advanced Automotive Technologies.

REFERENCES

[1] H. L. Fang and H. F. M. DaCosta, *Appl. Cat. B:Environ.*, **46**, 17-34, (2003).
[2] R. M. Heck and R. J. Farrauto with S. T. Gulati, "Catalytic air pollution control: commercial technology" John Wiley & Sons (2002).
[3] H. Bosch and F. Janssen, *Catal. Today*, **2**, 369, (1988).
[4] M. Koebel, M. Elsener and M. Kleemann, *Catal. Today*, **59**, 335-345, (2000).
[5] Y. Ikeda, K. Sobue, S. Tsuji and S. Matsumoto, *SAE paper* 1999-01-1279 (1999).
[6] E. Fridell, H. Persson, B. Westerberg, L. Olsson and M. Skoglundh, *Catal. Letters*, **66**, 71, (2000).
[7] "Method for Selective Catalytic Reduction of Nitrogen Oxides"; US patent application #10601255 assigned to Sandia National Laboratories, filed June 19th, 2003.
[8] B. Carberry, R. Hammerle, P. Laing, C. Lambert, C. Montreuil, R. Soltis, D. Upadhyay and S. Williams, *8th Diesel Engine Emissions Reduction Conference*, San Diego CA, Aug. 25-29, 2002.
[9] T. J. Gardner, L. I. McLaughlin, D. L. Mowery and R. S. Sandoval, *SAE Technical Paper Series* 2002-01-1880, (2002).
[10] A. Matsumoto, Y. Ikeda, H. Suzuki, M. Ogai and N. Miyoshi, *Appl. Catal., B. Environ.*, **25**, 115, (2000).
[11] E. Fridell, M. Skoglundh, S. Johansson, B. Westerberg, A. Törncrona and G. Smedler, in N. Kruse, A. Frennet, and J.-M. Bastin (eds), *Stud. Surf. Sci. Catal.*, **116**, 537, (1998).

Ceramic Armor

BALLISTIC IMPACT OF SILICON CARBIDE WITH TUNGSTEN CARBIDE SPHERES

M. J. Normandia and B. Leavy
U. S. Army Research Laboratory
Armor Mechanics Branch
Aberdeen Proving Ground, MD 21005

ABSTRACT

Ballistic impact experiments up to 1800 m/s were conducted on six different confined silicon carbides using tungsten carbide spheres. Partial results of normalized areal density penetrated versus impact velocity response curves are used to suggest comparative ballistic performance potential as a possible material screening technique and to provide validation data for numerical simulations. Complex response curve features are illustrated using an analytic model that captures critical data features. Data suggest an onset velocity above which the damage is no longer contained, but is inertially confined by the surrounding ceramic. Further increases in velocity increase penetration depths, while the impactor remains elastic until a critical value is reached where the impactor deforms inelastically (or shatters). The response curve then transitions to a reduced penetration rate, comparable to those measured in typical depth-of-penetration experiments and most ballistic applications. The response curves are then influenced by both target and penetrator dynamic properties. The data below the penetrator deformation velocity has potentially isolated an impact regime to enable direct measurement of target resistance for fully confined comminuted ceramics. Analysis of this using the conceptual framework of cavity expansion is discussed in a forthcoming paper.

INTRODUCTION

Extensive efforts to understand correlations between material properties and ballistic performance are summarized in US Army and ACER sponsored symposia proceedings, Sagamore[1] and PacRim[2]; containing extensive references. Normandia and Gooch[3] reviewed earlier attempts to screen ceramics using sphere impact tests by Donaldson[4, 5] and critiqued by Sternberg[6]. Low velocity impacts using tungsten carbide (WC) spheres to generate contained damage progression were presented by Shockey[7, 8] for silicon nitride and Kim[9, 10] for other ceramics by generating indentation stress-strain material response curves. Historical references of high-velocity penetration using WC data appear in Herrmann and Jones[11], Goldsmith[12] and Martineau[13]. Refer to reviews by Lawn[14], Stronge[15] and Fischer-Cripps[16] for contained damage analyses, and to the overview of ceramic impacts by Tanabe[17].

EXPERIMENTAL PROCEDURES

The test facility depicted in Figure 1 consists of a compressed gas gun launch tube, laser velocimeters, and dual x-rays. Also shown is a cutaway of the Ti6Al4V fixture confining the target. The target components depicted in Figure 2, show an impacted Cercom hot-pressed SiC-N at 800 m/s with the back of the front plate capturing an image of the ceramic damage. The crater is well defined crater as is the radial extent of the damage region. The rear surface of the front plate captures the damage features in the ceramic showing radial and circumferential cracks. All metal components are Ti6Al4V circular stock. The 3.125 mm thick front plate has an

entry hole to accommodate the 2 g, 6.35mm diameter, WC spherical penetrator purchased from
Machining Technologies Inc, and were Grade 25 (.000025" roundness tolerance), Class C-2 material (Rockwell A 92) and contained 6% Co binder with a 14.989 g/cc

Fig.1. Photographs of air-gun test facility (left) and target within a cut-away of the fixture (right).

measured density. The 9.525 mm thick back plate confines a 19.05 mm thick, 76.2 mm diameter ceramic, held to tight tolerances and enclosed in a 101.6 mm diameter ring. Three hot-pressed ceramic materials were purchased from Cercom Inc., a pressure-less sintered formulation under development from Superior Graphite, and two sintered versions from Saint-Go Bain. Numerous others are being tested in different thicknesses.

The test procedure consists of firing the spheres embedded in a flared sabot from the launch tube, measuring the time between two lasers at the end of the gun barrel to determine velocity, discarding the sabot prior to impact, and observing (when desired) the penetrator in flight, prior to impact, with two 150keV x-radiographs, which also served as a velocity measurement backup. A significant benefit of this technique is material recovery for microscopic analyses.

Fig. 2. Target components: 101.6 mm diameter Titanium 6Al4V cover with hole (inside shown), backing plate and confining ring containing 76.2 mm ceramic (typically 19.05 mm thick). This SiC-N tile was impacted at 800 m/s and exhibits well defined crater and radial damage extent.

RESULTS & DISCUSSION

The experimental procedure generates a final penetration depths used for model validation in numerical or analytic design tools, but does not provide time resolved data or load histories. The test firings conducted below 300 m/s produced no visible damage to the ceramics impacted, other than radial cracks. For SiC-N, impact at 393 m/s produced the first visible signs of damage, depicted in Figure 3. At the point of impact, the WC material excavated a small ring of damaged material, which was ejected from the area (see close up). The analysis is sensitive to this onset velocity, but for now, we utilize 393 m/s as a cutoff velocity until a more accurate determination can be made from ongoing experiments. For the hot-pressed SiC-N impacted at various impact velocities and shown in Figure 4, the damage extent can be identified as a function of impact velocity. This is carefully measured but not shown. There was no indentation into the rear backing plates, which depicted radial cracks, until the highest velocities, when circumferential cracks clearly appeared. The rear surface of the ceramic remained smooth.

Figs. 3 and 4. Photographs of confined Cercom SiC-N ceramic disk impacted by 6.25 mm WC sphere at impact velocities shown. Close up of 393 m/s impact area at lower left right (Fig 3). Note progression of damage.

After careful removal from the fixture, the front plate was removed, photographs taken, and careful excavation of the surface revealed the deepest point of penetration and the penetration directly under the impactor, which were not always identical at the low velocities. Well defined craters were observed and indentation/penetration depths and crater diameters were measured, as were the radial extent of the circumferential damage. We have preserved all targets, and plan to analyze them microscopically prior to, and after, sectioning.

Measurements for SiC-N are shown in Figure 5 as normalized penetration vs. velocity, with scatter evident at the highest velocities. An algorithm was used to predict the penetration if the sphere remained rigid using the measured hardness values to estimate dynamic yield strength of 10.08 GPa. Using the observed, and estimated, onset velocity of 393 m/s, we stopped the computation when this velocity was reached, and plotted the resulting prediction. It follows the data fairly closely, although it does not capture the details that are shown in Figures 6 and 7.

The CTH code with Adaptive Mesh Refinement was used to simulate the test results, five of which are shown in Figure 5. The constitutive model used for the ceramic was based on a modification of the Johnson and Holmquist[18] model for SiC-B that has a separate constitutive behavior for intact and fully-damaged material, and results are very sensitive to both a strain-based transition criterion and the pressure-shear dependence of the damaged ceramic. Ballistic data is required to determine these parameters.

The model constants were varied for each silicon carbide based on measured quantities of density, moduli, acoustic and shear wave speeds, Poisson's ratio, etc. Using these measured values, adjustments were made to two sensitive parameters in the model until good agreement was obtained with the data. These values were then utilized against a broader set of experimental data, available only for SiC-N, which were accurately captured. The addition of this sphere impact data to the suite of ballistic data permitted better model refinement. More notable, however, is the ability to calibrate two sensitive model parameters for different variants of SiC, from this data.

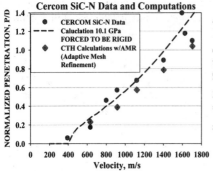

Fig. 4. SiC-N penetration/diameter vs. velocity, CTH simulations with modified JH-1 SiC-N model and rigid sphere analysis with 393 m/s onset velocity and 10.1 GPa dynamic flow stress.

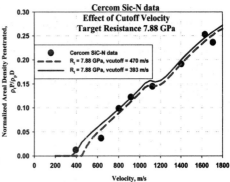

Fig. 5. Areal density normalized penetration for SiC-N and analytic model using target strength of 7.88 GPa (best matches data). Cutoff velocities of 393 m/s and 470 m/s compared.

An analytic tool based on the modified work of McDonough[4] was used to model a sphere impact, as shown in Figure 5. The dynamic yield strength for the WC was chosen as 4.9 +/- .25 GPa, based on recent Split-Hopkinson Pressure Bar experimental data by Wseesoryria. To verify this value, we examined data generated at LANL by Martineau[13] for High Strength Low Alloy (HSLA) 100 steel. A compressible cavity expansion pressure of 4.24 and 4.6 GPa were used based on the reported BHN hardness range of 280-310 for the HSLA steel. As seen in

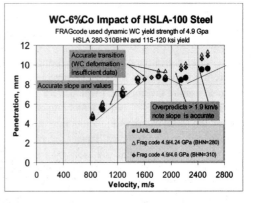

Figure 6, the many features of the data are captured up until 2 km/s. Material work hardening or WC shattering are possible explanations for the over-predictions. The velocity at which penetrator deformation occurs can be used to estimate the penetrator dynamic yield strength, if the target strength is known. The slope of the penetration data changes above this velocity. The slope of the penetration data below this velocity is determined solely on target material strength and elastic penetrator properties, while above the critical velocity, the slope is determined by a relative strength difference between the target and the deforming penetrator strengths, hence is lower. The deformation increases the contact area with the target and the penetration can decrease. Failure strain limits the maximal extent of penetrator deformation.

For ceramic computations, we used an onset velocity of 393 m/s, and 4.9GPa for the WC dynamic yield strength. A target strength of 7.88 GPa best matched the SiC-N data (see Fig. 5) showing normalized areal density penetrated. We also used a 470 m/s cutoff velocity, observed for SiC-1RN. In Figure 7, computations for SiC-B, SiC-N, and SiC-1RN using 7.88 GPa plus or minus 10% are shown as normalized penetrations using 393 m/s cutoff for SiC-B and SiC-N, and 470 m/s for SiC-1RN. In Figure 8, data for Saint-Go-Bain Hexaloy, Enhanced Hexaloy and Superior Graphite pressureless sintered SiC are also shown as normalized areal density penetrated. All response curves are contained within a band of about 10% in strength, still significant. Most sintered SiC's (not shown) produced significantly greater penetration depths.

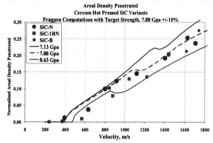

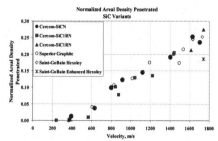

Fig. 7. Normalized penetration vs. impact velocity, for Cercom hot-pressed SiC-N, SiC-1RN and SiC-B. Analytic model computations used onset velocity of 393 m/s (470m/s in red), and 7.88 GPa (dashed) +/-10% illustrating subtle differences.

Fig. 8. Normalized areal density penetrated: Cercom hot-pressed SiC-N, SiC-1RN and SiC-B, Saint-Go Bain sintered Hexaloy and Enhanced Hexaloy, and Superior Graphite pressure-less sintered.

Measured target resistance stresses may not be sufficient to predict target penetration. The nature of the fracture or comminution for different ceramics may be much more significant for deep target penetration. Notice in Figure 9, the quite different behavior for three ceramics, hot-pressed SiC-N and Sic 1RN and the Superior Graphite material. Impacted at similar velocities the amount of damaged fragmented ceramic is quite different, which would show up in a cavity expansion analysis where the pressure at the penetrator target interface is determined from integration over the entire target area. Much larger comminuted and damaged regions, also imply less confinement, which would result in a lower target resistance pressure when the integrated. Direct measurement of cavity damage zones may be critical and two SiC-N tests have been conducted by SRI under contract to ARL to quantify the damage extent. Properties that control these damage regions are tensile or fracture strengths and yield strengths.

Fig. 9. Damage differences: Cercom SiC-1RN (left) vs. Superior Graphite pressureless sintered SiC (SG), (Left Photo) Cercom SiC-N (left) vs. SG (Right Photo). Greater damage zone would decrease target resistance.

SUMMARY

Ballistic response curves generated for six silicon carbide ceramic variants illustrate an onset velocity of an uncontained damage region, but inertially confined, a region of penetration with an elastic WC impactor, a velocity of impactor deformation above which the response curve transitions to a different penetration slope. We used analytic and numerical models to capture the response curve features and compared ceramics. Key observations are a velocity window to possibly directly measure strength of confined, comminuted ceramic, and the extent of damaged regions that may prove critical in predicting penetration in deep targets.

ACKNOWLEDGEMENTS
Gratitude is extended to: Dave MacKenzie for development of the low velocity test range and testing, and Dave Schall and technical crew for conduct of the high velocity experiments at ARL ballistic test facilities, and to Bruce Rickter for analytic model computations.

This research was supported by mission funding at the U.S. Army Research Laboratory.

REFERENCES
[1] E. S. C. Chin and J. W. McCauley *Proceedings: 45th Sagamore Conference: Armor Materials by Design;* 25 - 28 June 2001, St. Michaels, MD; Army Research Laboratory, p 773, 2001.

[2] W. McCauley, et al. "Ceramic Armor Materials by Design," *Proceedings of the Symposium on Ceramic Armor Materials by Design*, PAC RIM 4, Wailea, Maui, HI, November 4-8, 2001. Ceramic Transactions, Vol. 134, The American Ceramics Society, 2002.

[3] M. J. Normandia and W. Gooch, "An Overview of Ballistic Testing Methods of Ceramic Materials," *Proceedings of the Symposium on Ceramic Armor Materials by Design*, PAC RIM 4, pp. 113-138, Wailea, Maui, HI, November 4-8, 2001.

[4] C. duP. Donaldson and T. B. McDonough, " A Simple Integral Theory for Impact Cratering by High Speed Particles," *Aeronautical Research Associates of Princeton, Inc. A. R. A. P. Report No 201*, Dec. 1973.

[5] C. duP. Donaldson, R. M. Contiliano, and C. V. Swanson, "The Qualification of Target Materials using the Integral Theory of Impact," *A. R. A. P. Report No 295*, Nov. 1977.

[6] J. Sternberg, "Material Properties Determining the Resistance of Ceramics to High Velocity Penetration," *J. Appl. Phys.*, **65** [9] 3417-424 (1989).

[7] D. A. Shockey, D. J. Rowcliffe, K. C. Dao, and L. Seaman, "Particle Impact Damage in Silicon Nitride," *Journal of the American Ceramic Society*, 73 1613-19, 1990.

[8] D. A. Shockey, A.H. Marchand, S.R. Skaggs, G.E. Cort, M.W. Burkett, and R. Parker, "Failure Phenomenology of Confined Ceramic Targets and Impacting Rods," *Int. J. Impact Eng.*, 9 [3] 263-75, 1990.

[9] D. K. Kim and C-S Lee, "Indentation Damage Behavior of Armor Ceramics," *Proceedings of the Symposium on Ceramic Armor Materials by Design*, PAC RIM 4, pp. 429-440, Wailea, Maui, HI, November 4-8, 2001.

[10] D. K. Kim, C-S Lee, and Y-G Kim, "Dynamic Indentation Damage of Ceramics," *Proceedings of the Symposium on Ceramic Armor Materials by Design*, PAC RIM 4, pp. 261-268, Wailea, Maui, HI, November 4-8, 2001.

[11] W. Herrmann and A. H. Jones, "Survey of Hypervelocity Impact Information," *M. I. T. Aeroelastic and Structures Research Laboratory*, Report 99-1, Sept. 1961.

[12] W. Goldsmith, "Impact, The Theory and Physical Behavior of Colliding Solids," Arnold, London, 1960.

[13] R. L. Martineau, M. B. Prime, and T. Duffey, "Penetration into HSLA-100 steel with Tungsten Carbide Spheres at Striking Velocities Between 0.8 and 2.5 km/s," *Int. J. Impact Eng.*, to appear.

[14] B. R. Lawn, "Indentation of Ceramics with Spheres: A Century after Hertz," *Journal of the American Ceramic Society*, **81** 1977-94 (1998).

[15] W. Stronge, "Impact Mechanics," Cambridge Univ. Press 2000.

[16] A. C. Fischer-Cripps, "Introduction to Contact Mechanics," *Mech. Eng. Series*, Springer, 2000.

[17] Y. Tanabe, T. Saitoh, O. Wada, H. Tamura, and A.B. Sawaoka, "An Overview of Impact Damages in Ceramic Materials – For Impact Velocity Below 2 km/s," *Report of the Research Laboratory of Engineering Materials, Tokyo Institute of Technology*, 19, 1994.

[18] T. J. Holmquist and G. R. Johnson, "Modeling Ceramic Dwell and Interface Defeat," *Proceedings of the Symposium on Ceramic Armor Materials by Design*, PAC RIM 4, pps. 309-316, Wailea, Maui, HI, November 4-8, 2001.

[19] T. Weerasooriya, P. Moy, and W. Chen, "Effect of Strain-Rate on the Deformation and Failure of WC-Co Cermets under Compression," *Proceedings of ACER Cocoa Beach*, to be published, 2004.

TOUGHNESS AND HARDNESS OF LPS-SiC AND LPS-SiC BASED COMPOSITES

Karl A. Schwetz and Thomas Kempf
ESK Ceramics GmbH & Co. KG
P.O. Box 1530
D-87405 Kempten
Germany

Dirk Saldsieder and Rainer Telle
INSTITUT FUER GESTEINSHUETTENKUNDE
RWTH Aachen
Mauerstrasse 5
D-52064 Aachen
Germany

ABSTRACT

Liquid-phase-sintered silicon carbide with minimized residual porosity and very fine grain size (≤ 1 µm) is attractive as a high performance armor material due to its high hardness, high toughness and its potential for lower cost, especially when fabricated by gas pressure sintering. This study provides hardness (Vickers, HV-10) and fracture toughness (SEPB) data of YAG/AlN-doped LPS-SiC as a function of

(i)   the volume fraction of the YAG-based binder phase ,
(ii)   the content of free carbon,
(iii)   the toughening by in-situ formed SiC platelets and
(iv)   the toughening by dispersion of SiC, TiB_2 and Si_3N_4 particles.

The results indicate that higher amounts of the YAG based binder phase as well as additions of ceramic particles (>10 wt-%) reduce the hardness of LPS-SiC, whereas toughness may vary from 3.3 to 6.2 MPa $\cdot$ m$^{1/2}$. Toughness is controlled by (i) free carbon (chemical composition of the grain boundary), (ii) microstructure (grain size/shape of grains/interface strength), and (iii) amount of dispersed ceramic particles. Hardness is controlled primarily by the microstructural grain size and the grain boundary chemistry.

INTRODUCTION

EKasic®T is a liquid phase sintered SiC developed by WACKER CERAMICS in the early 1990's using an oxynitridic liquid phase based on AlN-YAG-(SiO_2) sintering additives [1], [2], [3], [4]. Today, EKasic®T is mainly used for rings in highly stressed gas seals and for dewatering elements in paper making machines. The extremely low wear required in such applications is realized by its mechanical properties [3] such as high hardness, high Young's modulus, and strength, the good thermal shock resistance, relatively high fracture toughness, and low density (3230 kg/m^3, TD). This combination of properties allows to apply the material also as a structural ceramic armor plate for protection against high impact projectiles. EKasic®T gave most promising results in preliminary ballistic tests performed at FHG Ernst-Mach-Institute, Freiburg/Germany.

The ballistic behaviour of EKasic®T was investigated in a depth of penetration (DOP) test using a AlCuMg1F40 backing, SiC-plates with a lateral dimension of 100x100 mm^2 and with different thickness (2.5/3.5/4.5 mm) and steel-core-projectiles 7.62x51 mm APFNB93 at impact velocities of 850 ±10 m/s.

Figure 1 shows the rankings of the ballistic resistances of different grades of SiC and HIP-treated sintered B_4C. EKasic®T (two qualities, T and T-O) performance was better than HIP-SB$_4$C (TETRABOR®) and the solid state sintered SiC materials (EKasic®F and F+, F+ designating post-HIPed -SSiC).

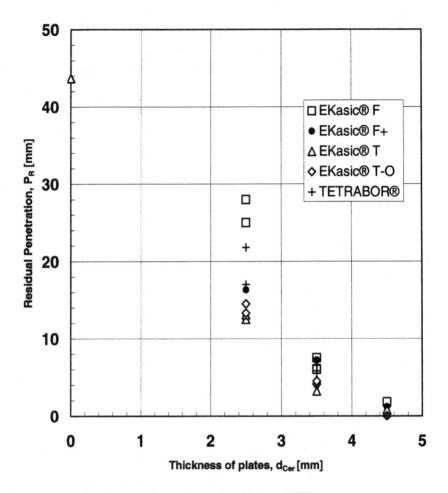

Figure 1. Ballistic resistance of ceramics produced by WACKER.

According to LaSalvia [5] the armor ceramics should have both high hardness and high toughness to prevent penetration of the projectile. In order to enable a high ballistic resistance, the technological efforts of the present work therefore focused on (i) raising the fracture toughness of LPS-SiC ceramics and (ii) preserving hardness of SiC as high as possible.

EXPERIMENTAL PROCEDURES

The LPS-SiC samples were prepared from a raw materials mix consisting of green α-SiC powder (specific surface area, 12.1 m^2/g; mean particle size, 0.7 µm; free carbon 0.3 wt-% C; residual oxygen, 0.45 wt-%), yttrium-aluminium-garnet (YAG) and aluminium nitride (AlN) powders having a mean particle size ≤ 1.5 µm [2], [6], [7]. Powder mixtures were wet-dispersed in deionized water and doped with 5 wt-% organic pressing aids and the slurries wet-sieved to – 20 µm mesh size. After spray drying, the granulated powders were further processed to shaped bodies (dimensions 30x30x60 mm) via cold isostatic pressing in rubber liners at 2000 bar. A two-stage gas pressure cycle (5 and 95 bar argon pressure, 1 h hold at 1960°C final temperature) with 5 vol-% YAG and AlN additives (composition 68 wt-% YAG – 32 wt-% AlN) was used for full densification of standard LPS SiC (black EKasic®T), and particle reinforced SiC-SiC/SiC-TiB$_2$ composites with 5, 10 and 20 wt-% particles. TiB$_2$ Grade F powder from H.C.Starck (specific surface area 4.6 m^2/g) and SiC Grade 1500 F from ESK-SIC GmbH were used. To correlate mechanical properties of LPS-SiC with increasing YAG-based binder phase contents, the quantity of sintering additives was varied within the limits of 1.5 to 12.0 vol-%. Green colored LPS-SiC materials [8] were obtained when before the sintering process the pressed bodies were decarburized in air at temperatures up to 500°C yielding a product with free carbon in the pressed body of less than 0.1 wt-%. In-situ platelet reinforcement was achieved by subsequent annealing (2050°C, 20 bar argon, 4 h hold) of the LPS-SiC samples with varying binder phase content. The SiC- Si$_3$N$_4$ particle reinforced composites were produced by axial hot-pressing of powder mixtures with 5, 10 and 20 wt-% nitride content (1900°C, 20 mins hold, 250 bar) into discs of d 70x7 mm size. The SiC- Si$_3$N$_4$ batches were prepared with 5 vol-% sintering additives on basis of the EKasic®T standard mix. α–Si$_3$N$_4$ powder used was Grade E10 from UBE, Japan (mean grain size 0.5 μm). Finally a standard EKasic®T composition was hot-pressed to evaluate properties compared to the hot-pressed composites and monolithic gas pressure sintered LPS-SiC samples, respectively. After gas pressure sintering or axial hot-pressing the bulk density was determined by water immersion method.

Before hardness and toughness were measured, samples were cut and ground to their final dimensions of 3x4x45 mm^3 with diamond tools and subsequently diamond-polished to a maximum surface roughness of 0.5 μm. To obtain Vickers hardness data [9], [10] a Dia-Tester 2RC-machine from Fa. Wolpert GmbH, Germany, with micrometer eyepiece was used. Load for Vickers indentations was 98.1 N (HV-10). Indent sizes were up to 100 μm, i.e. in all cases much greater than the gain size of the polycrystalline SiC materials. A minimum of 8 indentations were made for each sample and the results averaged. To determine the fracture toughness, the Single-Edge-Precracked Beam Method (SEPB) with a controlled flaw in the region of maximum tensile stress was used [11]. Sets of 3 – 5 specimens were used for the measurements, the data points and error bars representing the mean ± one standard deviation. The test bars were precracked on the center of the tensile surface with Knoop diamond indentation (98.1 N load) and the crack was then driven across the surface and halfway down into the sample with a hard metal bridge. Then the test bar was dipped for 5 mins into a penetrant dye (fluorescent ink), dried, and finally broken in four-point flexure using an inner span of 20 mm, an outer span of 40 mm and a Zwick 1474 universal testing machine. After fracture, the crack depth was measured to ± 0.01 mm at 25-fold magnification in a stereo microscope under ultraviolet light. The mean variation of K$_{IC}$ about the average was 0.2 MPa $\cdot$ m$^{1/2}$. For microstructural investigations polished samples were plasma-edged for around 4 mins with a 50 CF$_4$-50 O$_2$ (vol-%) gas mixture at a pressure of 10^{-2} mbar. Grain size was determined by image analysis using SEM micrographs under 2500 – 5000 fold magnification. X-ray diffraction analyses was used to determine the phase compositions of the samples.

RESULTS AND DISCUSSION

The properties of the liquid phase sintered SiC materials – relative density, hardness, toughness, microstructural data – are presented in Table 1 and compared with those of a solid-state-sintered SiC (EKasic®F, B/C-doped). The materials designated "C+S" were prepared by means of debindering without oxidation (carburization) followed by gas pressure sintering. Materials designated "D+S" were debindered by oxidation (decaburization) before gas pressure sintering.

Materials "HP" were axially hot-pressed. "D+S+A" denotes materials, which were decarburized and gas pressure sintered followed by annealing at 2050°C. X-ray diffraction revealed the presence of crystalline YAG in all LPS-SiC samples and additionally the foreign-particulate phase in case of the SiC matrix composite systems.

LPS-SiC Materials Prepared by Gas Pressure Sintering

For sintered densities of > 99.5 % theoretical density, $\geq$ 3 vol-% sintering additions in the starting mixture were necessary which corresponds to ~ 3 wt-% binder phase in the sintered body. Using only 1.5 vol-% sintering additives full densification is inhibited, when the pressed material is debindered in argon (C+S), which usually results in ~ 0.5 wt-% free carbon in the pressed body. The microstructure of dense LPS-SiC is characterized by globular SiC grains having a core/shell structure and a partially crystalline binder phase. According to recent HRAES analyses [8] the continuous binder network contains a segregation film (composition corresponding to x SiO_2 y AlN) and crystalline $Y_3Al_5O_{12}$ at the multiple grain junctions of the SiC grains. The SiC grains consist of a pure SiC core enclosed by a (Si, Al) (C, N, O) solid solution shell. In the samples sintered with 1.5 to 12 vol-% binder, the mean grain size decreases from 1.9 µm to 1.0 µm, but hardness and toughness are not much affected by the amount of sintering additives (22 – 23 GPa, ~ 3.5 MPa $\cdot$ m$^{1/2}$). Sintered bodies have a black color which has been attributed to the presence of free carbon [8]. With use of an oxidation treatment prior to densification the color of the sintered materials turns to green and the toughness of sintered bodies significantly increases up to 4.5 MPa $\cdot$ m$^{1/2}$ without loss in hardness. The operating toughening mechanism in the green LPS-SiC material is still not fully understood.

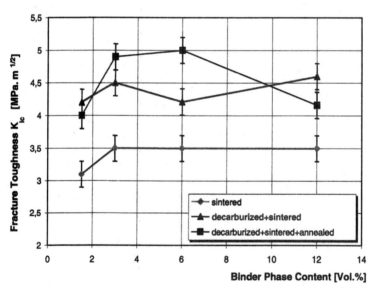

Figure 2. Toughness K_{IC} plotted versus the volume fraction of the binder phase in differently prepared LPS-SiC materials (sintering additions: 68 wt-% YAG – 32 wt-% AlN).

As shown by Figure 2 further improvement in the fracture toughness (up to 5.0 MPa · m$^{1/2}$) was achieved by modification of the microstructure using so-called "in situ platelet reinforcement" [2]. In this case a high temperature annealing treatment of the sintered bodies following the sintering resulted in the formation of platelet SiC grains with a mean grain size in the range of 2.5 to 4.0 μm and a mean aspect ratio of around 3 (see Figure 3).

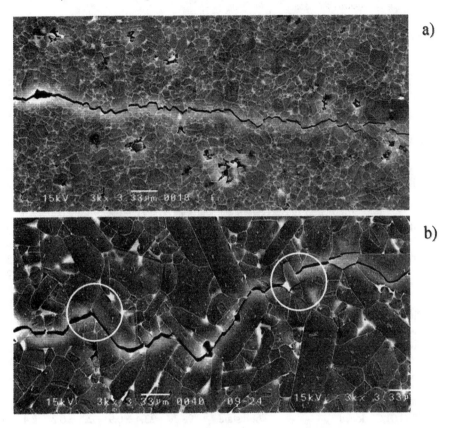

Figure 3. Intergranular crack propagation paths in LPS-SiC materials with globular (a) and platelet (b) microstructure; bar length = 3.3. μm; SEM after Vickers indentation and etching.

(a) K_{IC}= 3.3 MPa · m$^{1/2}$,   Left circle : crack deflection,
(b) K_{IC}= 5.0 MPa · m$^{1/2}$.   Right circle : bridging + transgranular fracture

Table 1

Comparison of LPS SiC and LPS SiC Based Composites with Solid Stated Sintered SiC

Designation (Composition, wt-%)	Vol-% YAG Based Binder	Densification	Density % TD	Hardness HV-10	Toughness MPa · m$^{1/2}$	Grain Size μm	Grain Shape
EKasic®T Standard	5	C+S	99.9	22.8 ± 0.8	3.3 ± 0.2	1.2	globular
SiC	1.5	C+S	91.2	12.2 ± 0.7	3.1 ± 0.2	1.9	globular
	1.5	D + S	98.9	21.8 ± 1.0	4.2 ± 0.2	1.9	globular
	1.5	D + S + A	98.7	20.5 ± 0.9	4.0 ± 0.2	4.0	platelet
SiC	3.0	C+ S	99.9	21.0 ± 0.8	3.5 ± 0.2	1.5	globular
	3.0	D + S	100.0	22.5 ± 1.0	4.5 ± 0.2	1.6	globular
	3.0	D + S + A	99.5	21.7 ± 0.7	4.9 ± 0.2	3.1	platelet
SiC	6.0	C + S	99.8	22.9 ± 0.8	3.5 ± 0.2	1.2	globular
	6.0	D + S	99.9	23.5 ± 0.9	4.2 ± 0.2	1.2	globular
	6.0	D + S + A	99.1	21.0 ± 0.8	5.0 ± 0.2	2.8	platelet
SiC	12.0	C + S	98.7	21.2 ± 0.9	3.5 ± 0.2	1.0	globular
	12.0	D + S	99.5	21.7 ± 0.8	4.6 ± 0.2	1.2	globular
	12.0	D + S + A	99.1	19.0 ± 0.7	4.2 ± 0.2	2.5	platelet
SiC - 5 SiC	5.0	C + S	99.8	23.9 ± 0.6	3.5 ± 0.2	1.4	globular
	5.0	S + A	99.7	23.8 ± 0.7	4.1 ± 0.2	3.2	platelet
SiC - 10 SiC	5.0	C + S	99.5	23.7 ± 0.9	3.7 ± 0.2	1.5	globular
	5.0	S + A	99.3	23.5 ± 1.0	4.2 ± 0.2	4.0	platelet
SiC - 20 SiC	5.0	C + S	98.7	20.8 ± 0.8	4.0 ± 0.2	1.7	globular
	5.0	S + A	98.5	19.7 ± 0.7	5.0 ± 0.2	4.2	platelet
SiC - 5 TiB$_2$	5.0	C + S	99.7	25.0 ± 0.6	4.2 ± 0.2	2.0	globular
SiC - 10 TiB$_2$	5.0	C + S	99.8	24.4 ± 0.7	4.3 ± 0.2	2.1	globular
SiC - 20 TiB$_2$	5.0	C + S	97.0	22.3 ± 0.7	4.8 ± 0.2	2.2	globular
SiC - 5 Si$_3$N$_4$	5.0	HP	99.9	26.2 ± 0.6	5.5 ± 0.2	0.6	whisker
SiC - 10 Si$_3$N$_4$	5.0	HP	99.7	26.1 ± 0.7	6.0 ± 0.2	0.5	whisker
SiC - 20 Si$_3$N$_4$	5.0	HP	99.8	22.8 ± 0.6	6.2 ± 0.2	0.5	whisker
EKasic®T	5.0	HP	100.0	26.5 ± 0.7	5.0 ± 0.2	0.6	globular
SSiC (B,C)-doped (EKasic®F)	-	solid state sintered	98.7	23.4 ± 0.9	2.5 ± 0.2	2.4	globular

Table 1 shows that the platelet reinforced microstructures lose hardness (1 – 2 GPa) due to a grain-size effect. As shown by Figure 3, the enlargement of the crack surface by crack deflection as well as load transmission by crack bridging with occurrence of the platelet crystals are regarded as mechanisms for microstructure toughening.

LPS-SiC Based Composite Materials Prepared by Particle Addition and Gas Pressure Sintering
Starting from SiC-based mixtures with up to 20 wt-% SiC or TiB$_2$ particles ($d_{50} \leq 2$ μm) it was possible to sinter SiC-SiC and SiC-TiB$_2$ composites to high densities using the standard gas pressure sintering cycle and 5 vol-% sintering aids on basis of YAG-AlN. The size distribution of the SiC grains in the microstructures of the SiC-SiC composites was bimodal, i.e. it consists of a matrix of fine grains with a mean grain size of ~ 1 μm and a second population of SiC crystallites of

around 3 – 4 μm, which was formed by dissolution and precipitation on the 2 μm seed grains. Substitution of SiC by TiB_2 second-phase-particles gave also microstructures with bimodal grain distributions where 2 – 5 μm globular TiB_2 grains were embedded in a fine grained SiC matrix (see Figure 4). Toughness was well correlated with increasing TiB_2 contents in the SiC matrix composites in accordance with [12]. The reinforcement due to enhanced crack deflection around larger particles lead to K_{IC} values of ~ 5 MPa $\cdot$ m$^{1/2}$ (for 20 wt-% particles added). However, hardness was unaffected only in materials up to 10 wt-% particle addition (~ 24 GPa, HV-10). With 20 wt-% particle addition hardness was ~ 3 GPa lower probably due to residual porosities of of 1 – 3 %.

LPS-SiC and SiC- Si_3N_4 Particle Reinforced Materials Prepared by Axial Hot-Pressing

Since isopressed SiC- Si_3N_4 powder mixtures did not densify using the standard gas pressure sintering cycle and 5 vol-% YAG-based sintering aids, axial hot-pressing was chosen as densification technique. As expected, this technique enabled to achieve not only full density, but also a submicron-sized-microstructure in the sintered bodies. From Table1 it can be seen that hot-pressed SiC- Si_3N_4 composite materials are characterized by a very high toughness in the range of 5.5 – 6.2 MPa $\cdot$ m$^{1/2}$ and a very high hardness (23 – 26 GPa). X-ray diffraction of the composites revealed the presence of α-SiC, ß-Si_3N_4 and Si. Since YAG was not detected, it is suggested that an in-situ reaction of some Si_3N_4 with YAG at 1900°C generated elemental Si ($d_{50} < 1$ μm) and an amorphous Y-Si-Al-O-N binder-phase. The lower sintering temperature (1900°C) as well as the pinning of the moving grain-interfaces by the fine needle-like ß-Si_3N_4 crystals and the small Si grains are suspected to be the major reasons for the very fine microstructure ($d_{50} = 0.5$ μm, see Figure 5). High fracture toughness is obtained in analogy to [13] by having in-situ formed ß-Si_3N_4 whiskers that can bridge cracks; furthermore high hardness is obtained by having a submicron grained microstructure (grain size effect, analogous to sintered Al_2O_3 [14]). However with 20 wt-% Si_3N_4 a drop of 3 MPa in Vickers hardness is observed which can be related to the high volume fraction of the less hard Si_3N_4 phase. Hot-pressed EKasic®T with 5 vol-% YAG-based binder phase is as hard as the hot-pressed 90 SiC-10 Si_3N_4 (wt-%) and 95 SiC-5 Si_3N_4 (wt-%) composites and has also a submicron grained microstructure ($d_{50} = 0.6$ μm, see Figure 6) . It should be noted, that hardness of these hot-pressed materials is higher than that of solid state sintered SiC with a mean grain size of about 2.4 μm (see Table 1). Surprisingly Ray et al. [15] observed a much lower hardness for submicron grain sized, hot-pressed Y_2O_3-Al doped LPS-SiC materials. Probably the AlN containing segregation film at the SiC-SiC grain boundaries in our YAG-AlN doped LPS-SiC materials may prevent grain boundary deformation during hardness measurement (AlN contributing to a higher interface strength). Toughness of hot-pressed EKasic®T (~ 5 MPa $\cdot$ m$^{1/2}$) is similar to in-situ platelet reinforced LPS-SiC and the SiC-SiC / SiC-TiB_2 particle reinforced materials. The toughness development in hot-pressed EKasic®T is believed to result from an almost 100 % intercrystalline fracture mode due to crack deflection and last not least from microcracking at SiC/YAG interfaces, since the number of interfaces increases significantly with decreasing grain size. It is also noteworthy that quantification of the different fracture modes (inter/transgranular) measured along the crack propagation paths shows that percentage of transgranular fracture is increasing, starting from 0.6 % for hot-pressed (0.6 μm grain size, globular grains) going over 2.2 % for gas pressure sintered (2.2 μm grain size, globular grains) and finally ending at 6.5 % for platelet reinforced LPS-SiC materials.

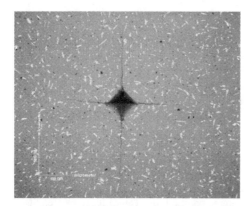

Figure 4.
Light microscopic photomicrograph of a polished and unetched SiC-TiB$_2$ composite (light areas = TiB$_2$, grey matrix = SiC) featuring a Vickers HV-10 indentation, bar length = 100 μm.

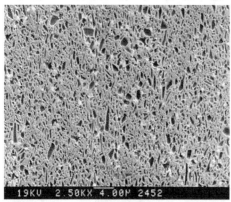

Figure 5.
SEM micrograph of a polished and plasma-etched SiC-Si$_3$N$_4$ composite (black areas = in-situ ß-Si$_3$N$_4$ whiskers and SiC cores, white = elemental silicon), bar length = 4 μm.

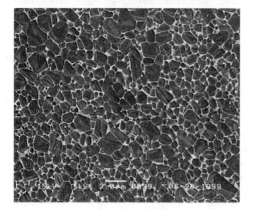

Figure 6.
SEM image showing the plasma-etched microstructure of axially hot pressed EKasic®T (5 vol-% YAG-AlN binder), bar length = 2 μm.

586

In comparison to gas-pressure sintered LPS-SiC materials, axially hot-pressed EKasic®T exhibits highest hardness (~ 26 GPa) at the same high level of toughness (~ 5.0 MPa $\cdot$ m$^{\frac{1}{2}}$).

Comparison Axial Hot-Pressing and Gas-Pressure Sintering
The axial hot-pressing process requires large amounts of energy and expensive mold material and can normally be used only for few parts with simple, uncomplicated shapes. Owing to this limitation, precision parts can only be produced by expensive diamond machining.

The striking cost- and production-related advantages associated with the gas-pressure sintering process [16] are
- shaping is done in a separate step via pressure-slip casting, automatic dry-pressing or isostatic dry-pressing followed by green machining, which allows manufacture of more complex shaped parts without the need for expensive final machining,
- a large number of stacked parts (50 to 100) can be densified in one sintering cycle using a production-scale furnace (high economy of operation),
- costs by wear of kiln furniture is only 10 % of costs necessary for mold materials in axially hot-pressing.

CONCLUSIONS
Improved LPS-SiC materials as well as liquid phase sintered SiC/SiC, SiC/TiB$_2$ and SiC/Si$_3$N$_4$ composite materials with improved toughness (up to ~ 6 MPa $\cdot$ m$^{\frac{1}{2}}$) and high hardness (up to ~ 26 GPa) were developed. Toughness improvement was achieved in various ways,
- by using a debindering treatment of pressed parts in air prior to sintering ("decarburization", ≤ 0.1 wt-% free carbon),
- by inducing growth of elongated SiC grains ("in-situ platelet reinforcement") through subsequent annealing of gas-pressure sintered specimens,
- by dispersion of larger ceramic particles (SiC, TiB$_2$, Si$_3$N$_4$) in the fine-grained matrix.
Hardness improvement occurred only in materials with a submicron-grained microstructure (mean grain size < 1.0 µm) formed at reduced sintering temperatures.

By further ballistic testing of such improved SiC ceramics it should be determined to which extent the toughness and hardness improvements lead to improved protection against armor piercing bullets.

ACKNOWLEDGEMENT

The authors thank Elmar Strassburger (Ernst-Mach-Institute) for coordination and performance of preliminary ballistic tests at EMI-Freiburg, Germany and Eva Maria Hengeler, ESK Ceramics, for preparing the manuscript.

REFERENCES

[1] L.S. Sigl and H.J. Kleebe, "Core/rim Structure of LPS-SiC", *J. Am. Ceram. Soc.*, **76** 773-776 (1993).

[2] K.A. Schwetz, E. Schäfer, and R. Telle, "Influence of Powder Properties on in-situ Platelet Reinforcement of LPS-SiC", *cfi/Ber.DKG*, **80** [3] E40 – 45 (2003).

[3] F. Meschke and C. Nitsche, "New SiC Materials with a Broadened Range of Performance", *cfi/Ber.DKG*, **78** [10] E20 – 22 (2001).

[4] K.A. Schwetz and J. Hassler, "Stability of High Technology Ceramics Against Liquid Corrosion", *cfi/Ber.DKG*, **79** [11] D14 – 19 (2002).

[5] J.C. LaSalvia, "A Physically-Based Model for the Effect of Microstructure and Mechanic. Properties on Ballistic Performance", *Ceram. Sci. and Eng. Proceedings*, **23** [3] 213 - 220 (2002)

[6] K.A. Schwetz, L.S. Sigl, G. Victor, and T. Kempf, "Liquid Phase Sintered SiC Shaped Bodies having improved Fracture Toughness as well as High Electrical Resistance, and Method of Making them", European Patent 1070686B1 (2001), US Pat. 6531423 (March 11, 2003).

[7] D.W. Saldsieder, "Improvement of Fracture Toughness in LPS-SiC Based Ceramics", *PhD Thesis (in German)*, RWTH Aachen (2002).

[8] K.A. Schwetz, H. Werheit and E. Nold, "Sintered and Monocrystalline Black and Green Silicon Carbide – Chemical Compositions and Optical Properties", *cfi/Ber.DKG*, **80** [12] E37 – 44 (2003).

[9] ASTM C 1427-99, "Standard Test Methods for Vickers Indentation Hardness of Advance Ceramics", pp. 480 – 487 in 1999 *Annual Book of Standards* (ASTM, Philadelphia, PA 1999).

[10] W. Kollenberg, B. Mössner, and K.A. Schwetz, "Temperature – and Load Dependance o Vickers Hardness in Hot-Pressed SiC Materials", *VDI-Berichte*, No. **804**, 347 – 358 (1990).

[11] K.A. Schwetz, L.S. Sigl, and L. Pfau, "Mechanical Properites of Injection Molded B_4C-C Ceramics", *Journal of Solid State Chemistry*, **133** 68 - 76 (1997).

[12] K.S. Cho, H.J. Choi, J.G. Lee, and Y.W. Kim, "In-Situ Enhancement of Toughness of SiC-TiB_2 Composites", *J. Mat. Sci.* **33**, 211 - 214 (1998).

[13] C.J. Shih and A. Ezis, "SiC Matrix Composites using Liquid Phase Sintering Techniques", in *Processing and Fabrication of Advanced Materials III*, Proc. Symp. 3rd 1993 (V. A. Ravi et al., eds.), the Minerals, Metals and Materials Society, 99 - 112 (1994).

[14] A. Krell, "A New Look at the Influences of Load, Grain Size, and Grain Boundaries on the Room Temperature Hardness of Ceramics", *Int. J. Refract. Mater.*, **16** 331 - 335 (1998).

[15] D. Ray, M. Flinders, A. Anderson, and R.A. Cutler, "Hardness/Toughness Relationship for SiC Armor", *Ceram. Eng. Sci. Proc.*, **24** Issue 3, 401 – 410 (2003).

[16] H.U. Kessel and W.P. Engel, "Gas Pressure Sintering with Controlled Densification", *cfi/Ber.DKG*, **66** [5/6] 227 - 234 (1989).

588

INDENTATION TESTING OF ARMOR CERAMICS

Eugene Medvedovski*　　　　　　　Partho Sarkar
Ceramic Protection Corporation　　　Alberta Research Council
3905 – 32nd St. N.E.　　　　　　　250 Karl Clark Rd.,
Calgary, AB, T1Y 7C1, Canada　　　Edmonton, AB, T6N 1E4, Canada

ABSTRACT

Hardness and fracture toughness are important factors typically used for characterizing armor ceramics. Their values depend on composition and microstructure of the ceramics, on the features of testing methods (e.g. on the type of indentation technique and an indentation load) and on the sample preparation procedure. Vickers and Rockwell indentation commonly used for hardness testing and fracture toughness K_{Ic} determined by the indentation method have been studied for different armor ceramics. The sample preparation procedure for the Vickers hardness testing has been optimized in order to increase the accuracy of measurement and to reduce the stresses occurring at indentation. The influence of microstructure for different alumina and carbide-based armor ceramics, as well as the influence of an indentation load, on hardness and fracture toughness have been analyzed. It should be noted, that for dense alumina armor ceramics Vickers hardness can be successfully tested at high indentation loads, such as 10-50 kg, but dense carbide-based armor ceramics may be tested only at indentation loads not greater than 1 kg. Brittleness of ceramics that depends on the hardness to fracture toughness ratio, presence of strong cleavage plane and microstructure should all be considered during selecting a proper hardness testing procedure and the indentation load and also for characterization of armor ceramics. Rockwell hardness testing is recommended as the most suitable method for characterization of heterogeneous materials, such as carbide-based ceramics, where relatively coarser grains are bonded by a fine crystalline-glassy matrix, as well as for ceramic-metal-matrix composites.

INTRODUCTION

Hardness and fracture toughness are important properties, which affect the performance of ceramic armor. Armor ceramics should withstand the attack of hard materials acting with various energies, velocities and angles, hence they should demonstrate a remarkable level of hardness and fracture toughness values depending on the particular application conditions. These properties help to understand and characterize the ability of ceramics to resist deformation and crack forming and propagation.

Hardness values are significantly distinguished if different testing methods or various testing conditions are used. A similar difference is also observed between fracture toughness values determined either by an indentation technique or from the notched beam (or chevron-notched beam) breakage method. Hence, hardness and fracture toughness cannot be considered as independent physical parameters, but as the response to certain conditions of a test.

EXPERIMENTAL

Some armor ceramic materials manufactured by CPC studied in the present paper have been described previously[1, 2]. Boron carbide-based ceramics prepared in the framework of the special

---

joint CPC-ARC project in the systems of B_4C-TiB_2, B_4C-$MoSi_2$ and some others were manufactured by pressing and were pressureless sintered. Some other silicon carbide- and boron carbide-based ceramics from other companies were manufactured based on their proprietary procedures using pressureless sintering, hot pressing or reaction sintering.

Test samples were cut from specially prepared test tiles with approximate dimensions of 100x100x(7-10) or 50x50x(7-10) mm or from actual products using a diamond saw and then ground and polished in accordance with a specially developed procedure.

Microstructure was studied using scanning electron microscopy (SEM). Vickers hardness of dense ceramics was tested in accordance with ASTM C1327 at different indentation loads ranging from 0.1 to 50 kg. Vickers hardness was calculated using the formula: $HV = 1.8544P/d^2$, where P is the indentation load (kg), d is the arithmetic mean of the length of the two diagonals d_1 and d_2 of the indentation replica (mm). A load of 10 kg was selected as a "standard" load for dense alumina ceramics based on the customers' requirements and on our own experience. Rockwell hardness of heterogeneous silicon carbide-based ceramics, as well as some dense homogeneous alumina and carbide-based ceramics, was tested in accordance with ASTM E18 at indentation loads of 60 kg (HRA) and 150 kg (HRC) without specially used sample preparation. Fracture toughness (critical stress intensity factor) K_{Ic} was determined using the indentation technique for the samples prepared for Vickers hardness testing and was calculated using the well-known formula: $K_{Ic} = 0.941(P.c^{-3/2})$, where P is the indentation load (N) and c is the crack length (m) measured using a microscope. Young's modulus was determined using the ultrasonic technique, measuring the longitudinal ultrasonic velocity in accordance with ASTM C769.

RESULTS AND DISCUSSION
Influence of Phase Composition and Microstructure

The values of hardness and fracture toughness as indentation properties depend on the factors related ceramics, mostly on their phase composition and microstructure, and on the factors related indentation techniques and the samples surface quality. As to the "structural" factors, the following have an influence:

- Nature and quantity of the major crystalline phases;
- Composition and quantity of a bonding (glassy or crystalline-glassy) phase; presence, nature and quantity of secondary crystalline phases reinforcing a glassy phase; composition and quantity of a glassy phase;
- Nature, quantity and distribution of the matrix bonding the major crystalline phases;
- Grain size of the major phases, grain size distribution and grain compaction;
- Porosity, a size of pores and pore size distribution.

The hardness and fracture toughness test results for armor alumina ceramics with different phase compositions and microstructures, manufactured by CPC, are presented in Table 1. The highest hardness among the studied alumina ceramics is observed for AL99.7 ceramics as the materials with a higher corundum content, more uniform micro-crystalline structure (an average size of the isometric grains is 1-3 μm), and a very low glassy phase content. Opposite to others, a glassy phase of AL99.7 ceramics was formed due to a presence of oxides-impurities in the starting alumina materials; this glassy phase is reinforced by the tiny spinel grains. AL98.5 ceramics can be ranked as the next hard material in the listed ceramics. It has superior hardness in comparison with AL98 ceramics ($HV10$ is 1380-1420 kg/mm^2 vs. 1300-1350 kg/mm^2 for slip cast ceramics and 1320-1380 kg/mm^2 vs. 1250-1300 kg/mm^2 for uniaxially pressed ceramics) although they have similar Al_2O_3 contents. It is explained through the use of a higher content of a finer starting alumina powder in the AL98.5 ceramic batch composition that results in the formation of a finer and more uniform microstructure. Among the studied alumina ceramics, AL97ML material possesses lower hardness value due to relatively coarser corundum grains and an elevated content of a aluminosilicate glassy phase. A presence of tiny mullite crystals as a secondary crystalline phase reinforcing a glassy phase of AL97ML ceramics may be considered as a positive factor affecting hardness values. The difference in the hardness values for slip cast and pressed ceramics may be explained by the formation of a more uniform microstructure with lower microcracking for slip cast ceramics. The microcracks in pressed ceramics are formed mostly due to pressing stresses and binder burnout stresses (pressed ceramics have a higher content of temporary organic binder and plasticizer). Fracture toughness of the studied alumina ceramics has similar values (K_{Ic} is in the range of 3.0-3.3 MPa.m$^{1/2}$), especially for the ceramics with a high Al_2O_3 content, that may be explained by the same nature and a general similarity of their microstructures.

The quality of raw materials (presence of impurities, particle size distribution, etc.) affects hardness values due to its influence on the microstructure of the ceramics. Finer crystal sizes of starting powders promote the formation of finer ceramic microstructures and, as a result, increased hardness values. However, the particle size distribution of the powders may be adjusted during milling process of ceramic manufacturing. Purity of the starting materials affects the crystal growth during sintering and the defects occurring in ceramics that may affect hardness values. For example, the study of the influence of the raw materials quality on Vickers hardness for AL99.7 ceramics made from the same batch composition and by the same technology showed that the ceramics based on the starting powders with higher impurities contents, especially with a presence of alkali oxides (especially, Na_2O) and TiO_2 and Fe_2O_3 and some others, have more uneven microstructure with elevated crystal sizes that results in lower Vickers hardness values. Indentation fracture toughness values do not have valuable differences for high alumina ceramics made from starting materials with different contents of impurities.

Vickers hardness of dense microcrystalline carbide-based ceramics measured at relatively low indentation loads (0.3-1 kg) has significantly higher values (Table 2) than studied oxide-based ceramics. Hot-pressed B_4C ceramics have greater hardness than pressureless-sintered ($HV0.3$ is 3200-3500 vs. 2800-3000, $HV1$ is 2430-2500 vs. 2300-2400 kg/mm^2) that is caused by a higher level of densification (99+% and 95-96% of theoretical density, respectively) and fewer defects in the microstructure. Pressureless-sintered ceramics based on the systems B_4C-TiB_2, B_4C-TiB_2-ZrB_2, B_4C-$MoSi_2$ and some others have a rather high level of hardness ($HV0.3$ is 2900-3100 kg/mm^2 for the B_4C-TiB_2-based compositions and 2600-2650 kg/mm^2 for B_4C-$MoSi_2$-based compositions). However, it is still lower than that of hot-pressed B_4C-ceramics, which is explained by a residual closed porosity and the presence of micro-cracks on the boundaries of the grains with different natures. All B_4C ceramics demonstrate a presence of strong cleavage plane at the Vickers hardness testing, that is especially noted at "elevated" loads and in the case of a relatively coarse-grain structure. Pressureless-sintered SiC ceramics with a homogeneous microstructure formed by SiC grains of a few microns size also have a high level of hardness (values of $HV0.3$-$HV1$ are in the range of 2200-2600 kg/mm^2). Fracture toughness K_{Ic} values of the mentioned dense carbide-based ceramics are in the range of 2.5-3.5 MPa.m$^{1/2}$.

If carbide-based ceramics have a heterogeneous structure, hardness should be measured for the fine-crystalline matrix and for the larger primary grains. For example, dense reaction-bonded silicon carbide (RBSC) ceramics formed by SiC grains of sizes up to 50-70 μm and greater bonded by a fine-crystalline SiC-based matrix have high hardness for the primary SiC grains

($HV1$ is in the range of 2350-2450 kg/mm^2) and significantly lower hardness values for the matrix ($HV1$ is in the range of 1600-1850 kg/mm^2). Lower hardness of the matrix may be a consequence of the presence of residual silicon (approximately 10-12%) and incomplete formation of silicon carbide. Measurement of hardness for the matrix can be used as a method for evaluation of the level of the completeness of the reaction-bonding processes in this type of ceramics. As an example, RBSC ceramics with a lower "level" of silicon carbide formation in the matrix had lower matrix hardness ($HV0.5$ is less than 1000 kg/mm^2) and an intense shattering of a matrix at the applied load. This elevated shattering results in a high standard deviation in the Vickers hardness measurement, and it does not allow the proper determination of fracture toughness. As a sequence, RBSC ceramics with weakened matrix has lower ballistic performance. Due to some limitations dealt with silicon infiltration and reaction-bonding process for RBSC ceramics, the hardness values tend to decrease with increase of a products thickness. For example, the samples cut from the plates with a thickness of 9.2, 7.8 and 6 mm have $HV1$ values (for large grains) of 2330, 2350 and 2415 kg/mm^2, respectively.

Porous (even with low porosity) heterogeneous ceramics formed by hard primary grains with increased sizes bonded by a matrix with lower hardness cannot be tested using a sharp pyramid indenter. The sharp indenter destroys the "weak" matrix demonstrating low hardness values even at small indentation loads. In this case, the Rockwell hardness testing can be successfully used. For example, the ceramics AS and ASN^2 formed by silicon carbide grains with a size ranging from 30 to 120 μm bonded by an aluminosilicate crystalline-glassy matrix have been tested using the Rockwell hardness technique at different high loads. These ceramics demonstrated the values of HRA and HRC of 55-75 and 35-55, respectively, depending on the composition (mineral and particle size distribution), on the ratio between relatively large grains and a crystalline-glassy matrix, and on the bonding between the grains and the matrix. A specially selected particle size distribution provided a remarkable level of grain compaction promoted elevated values of Rockwell hardness. Silicon nitride and sialon constituents presenting in the matrix of the studied ceramics ASN^2 promote a higher level of bonding, lower porosity, less fracturing at the testing loads, and, as a result, elevated hardness.

Dense alumina and carbide-based ceramics demonstrate a high level of Rockwell hardness (HRA is 85-92 for alumina ceramics and 90-94 for silicon carbide and boron carbide ceramics). The testing of these ceramics at a higher load (i.e. HRC of 150 kg-load) is not desirable because it may result in damage of the indenter. However, Rockwell hardness testing does not allow determining the accurate correlation of microstructure features and hardness for hard dense ceramics. Rockwell hardness testing may be recommended also for characterization of ceramic-ceramic- and ceramic-metal-matrix composites, which also have heterogeneous structures with different natures and properties of a matrix and reinforcing constituents.

Sample Preparation

Good surface quality of ceramic samples is important for hardness testing using a pyramid-shaped indenter, e.g. for hardness values, for accuracy of measurement and for standard deviation. Especially, it is important if high loads are applied (e.g. for 10-kg or greater loads). Often the hardness values determined for as-fired samples are varied significantly. The friction effect between an indenter and a hard ceramic surface induces error during measurement. Moreover, a rough surface causes problems of accurate reading of the replica and cracks sizes. An indenter applied on the rough surface creates additional cracks to occur that causes errors during hardness and fracture toughness determination. Another important requirement of the testing is that a surface of samples should be perpendicular to the indenter axis.

ASTM C1327, which is used for Vickers hardness testing, does not indicate a sample preparation procedure. The influence of a sample preparation technique and a surface quality on Vickers hardness was studied for the samples of various ceramics as-fired and after grinding and polishing using different techniques with and without subsequent annealing at 1100-1200°C. Based on our study, the following steps are recommended:

1) smooth and slow cutting of a sample using a saw with fine diamond grit;
2) polishing of a sample with 30 μm-diamond grit until a sample is flat;
3) then polishing with 6 μm-diamond grit;
4) and, as a last step, polishing with 0.06 μm colloidal silica.

Microscopic observation of the polished surface for the alumina ceramic samples prepared using the recommended steps shows that the suitable surface for the hardness measurement is achieved after at least 2 hours of polishing with colloidal silica (i.e., 2-2.5 hours of final polishing is recommended). The optimal time of the final polishing step is selected depending on properties of ceramics. Each polishing step is continued until all the grinding marks from the previous step are removed. Annealing is effective when the samples are polished using coarse diamond grit. In the case of the "stepped" polishing process with the use of colloidal silica as the final step, when the stresses connected with the mechanical treatment are significantly reduced, the annealing is already not effective, and, hence, is not recommended.

The influence of surface quality (smoothness) on hardness measurement is exemplified by the testing of the AL98 alumina armor tiles. As-pressed and as-fired tiles had an average surface roughness of 0.2 μm (sometimes even rougher). The same samples after grinding and polishing using the developed procedure had an average surface roughness less than 0.1 μm (actually, 0.05 μm or even finer) (Fig.1). As a result, *HV10* values for the samples as-fired and for the samples after grinding and polishing have considerable difference; the standard deviation for the polished samples is significantly lower (from +/- 27 to +/-75 for as-fired samples and from +/-17 to +/-45 for polished samples).

Rockwell hardness testing does not require special sample preparation, and samples are tested as-fired. However, if a ceramic surface is very rough, local surface damage occurs with cracks initiation that results in a hardness values decrease. The samples with no porosity and higher smoothness had less standard deviation than the samples with a rougher surface. Thus, dense oxide and carbide ceramics demonstrated lower standard deviation than the SiC-based ceramics[2] with an originally rougher surface.

Influence of Indentation Load on Hardness

The influence of indentation load on hardness and fracture toughness is not common[3-5]. The size of a replica from the indentation increases with an increase of the applied load. This is noted in testing using different methods, e.g., at the testing of Vickers and Rockwell hardness, studied in the present work.

Studying Vickers hardness as a function of indentation load and plotting a graph of hardness vs. indentation load, minimum two areas are observed on the graph. Usually, under small loads, the hardness values decrease drastically with a load increase. Then, under higher loads, hardness values exhibit little change. J. Quinn and G. Quinn[5], studying hardness of SiC ceramics vs. indentation load under the load range of 0.1-1 kg, have noted two areas with a critical point of approximately 0.5 kg. These results correlate well with the hardness study for various dense B_4C

and SiC ceramics conducted by the authors of the present paper and by some other researchers. Our study conducted for different alumina ceramics manufactured by CPC showed that these materials can be tested at 10, 20 and even at 50 kg without serious damages (depending on their general ability to withstand heavy indentations). As an example, a graph of the dependence of Vickers hardness vs. indentation load for the armor AL98.5 ceramics commercially produced by CPC is shown on Fig. 2. It can be seen that at very small loads (up to 1 kg) hardness values decrease significantly with a load increase (from 1540 kg/mm^2 at 0.1 kg to 1370 kg/mm^2 at 1 kg). Then, at a load increase up to 10 kg, the hardness values remain at the same level. At higher loads, hardness slightly decreases, and this trend may be assumed at higher loads.

In order to read properly the indentation replica size at the applied load, a ceramic sample should not shatter under this load. Brittleness of ceramics (B) is an important property for ceramic characterization, especially for armor applications. It may be expressed by a formula[5]: $B=HV.E/K_{Ic}^2$, where HV is Vickers hardness (GPa), E is Young's modulus (GPa), K_{Ic} is fracture toughness (MPa.m$^{1/2}$). A higher B value denotes a more brittle material. This approach helps to understand, why alumina and some other oxide ceramics can be successfully tested without damage at the loads of 10-50 kg (macro-hardness testing). Opposite to oxide ceramics, dense carbide-based ceramics can be tested only at the loads less than 1 kg (in some cases, even at 0.5 kg). At higher loads, these samples start cracking and shattering; it is observed for both hot-pressed and pressureless sintered ceramics. This effect can be explained by generally high hardness and relatively not very high fracture toughness of these ceramics, i.e. their low brittleness (higher values of B). The values of the brittleness factor calculated using the aforementioned formula are shown in Tables 1 and 2. However, for accurate calculation and comparison of the brittleness factor for different ceramics, the parameters affecting brittleness should be measured using the same procedure and under the same conditions. As practical experience shows, the ceramics with high brittleness factor (e.g. pressureless sintered SiC and B$_4$C- based ceramics) do not perform well in multi-hit ballistic testing situations due to increased shattering even after the first ballistic impact.

Testing of Rockwell hardness of various oxide and carbide ceramics (dense and porous) under different indentation loads (HRA and HRC in the present study) also demonstrates the influence of the indentation load. The higher the load, the lower the Rockwell hardness values. Standard deviation usually has greater values in the case of testing at higher loads, especially for porous heterogeneous materials, such as silicon carbide-based ceramics[2].

Laminated ceramic composites may be tested by both Vickers and Rockwell techniques; however, nature of laminates should be taken into consideration at the selection of indentation loads. In the case of laminated composites, lower loads are recommended.

Considering the formula describing brittleness of ceramics and the semi-phenomenological expression for ballistic impact energy dissipation ability $D = 0.36(HV.E.c)/K_{Ic}^2$, where c is sonic velocity (m/s) [6], the relationship between these two factors may be found. D factor may be expressed as $D = 0.36(B.c)$. Ballistic energy dissipation also depends on structure of ceramics, so, more accurate, it may be expressed as $D = B.c.S$, where S is a "structural" factor relating the features of structure and phase composition of ceramics.

SUMMARY

Extensive study of Vickers and Rockwell hardness and indentation fracture toughness of various armor alumina, silicon carbide- and boron carbide-based ceramics prepared using different manufacturing processes showed a strong influence of phase composition and microstructure on these properties. The influence of the sample preparation technique and the surface quality on hardness has been investigated, and a sample preparation procedure including several steps to reduce stress in ceramics has been developed and recommended. The study of the influence of an indentation load on Vickers hardness conducted for different ceramics has been shown that ceramics can be tested only up to some level of a load depending on the brittleness of the ceramics. The relationship between brittleness of ceramics and their ballistic energy dissipation ability is shown.

REFERENCES
[1]E. Medvedovski, "Alumina Ceramics for Ballistic Protection", *Amer. Ceram. Soc. Bull.*, Part 1, **81** [3] 27-32 (2002); Part 2, **81** [4] 45-50 (2002)
[2]E. Medvedovski, "Silicon Carbide-Based Armor Ceramics", pp. 365-374 in *Ceram. Eng. Sci. Proc.*, issue 3, **24** (2003)
[3]H. Li, R.C. Bradt, "The Effect of Indentation Induced Cracking on the Microhardness", *J. Mater. Sci.*, **31** 1065-70 (1996)
[4]J. Gong, J. Wu, Z. Guan, "Analysis of the Indentation Size Effect on the Apparent Hardness for Ceramics", *Mater. Letters*, **38** 197-201 (1999)
[5]J.B. Quinn, G.D. Quinn, "On the Hardness and Brittleness of Ceramics"; pp. 460-463 in *Key Engineering Materials*, **132-136** (1997); 1997 Trans Tech Publications, Switzerland
[6]V.C. Neshpor, G.P. Zaitsev, E.J. Dovgal, et al., "Armour Ceramics Ballistic Efficiency Evaluation"; pp.2395-2401 in *Ceramics: Charting the Future*. Proc. 8th CIMTEC Florence, 28 June-4 July 1994; Ed. by P. Vincenzini, Techna Srl., 1995

a)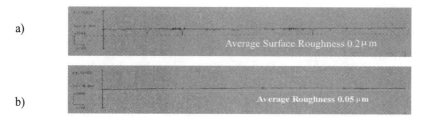

b)

Fig.1. Surface Roughness of Alumina Specimens
a) as received specimen b) prepared in accordance with the recommended procedure

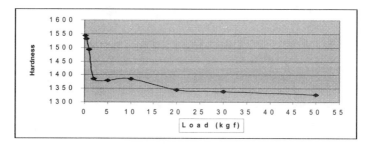

Fig. 2. Vickers Hardness vs. Indentation Load for Alumina AL98.5 Ceramics

Table. 1. Indentation Properties of the Studied CPC-Alumina Ceramics

Property	AL97ML	AL98	AL98.5	AL99.7
Vickers Hardness $HV10$,				
kg/mm²	1230-1260	1250-1320	1320-1420	1540-1570
GPa	12.0-12.4	12.2-12.9	12.9-14.3	15.1-15.4
Rockwell Hardness				
HRA	80-84	88-92	91-93	92-93
HRC	70-73	76-79	78-80	80-82
Fracture Toughness K_{Ic}, MPa.m$^{1/2}$	3.0-3.3	3.2-3.3	3.2-3.4	3.1-3.3
Young's Modulus, GPa	280-300	320-360	330-370	340-380
Brittleness $Bx10^{-6}$, m^{-1}	340-380	370-430	420-460	525-545

Table 2. Indentation Properties of the Studied Carbide-Based Ceramics

Property	B₄C (hp)	B₄C (pls)	SiC (pls)	RBSC large grains	matrix	AS-ASN
Vickers Hardness $HV1$						
kg/mm²	2430-2500	2300-2350	2200-2250	2380-2550	1650-1850	-
GPa	24-24.5	22-23	19.5-22	23.5-25	16-18.2	
Rockwell Hardness						
HRA	92.5	91.5	91	90-91		65-78
HRC	-	-	-			40-55
Fracture Toughness K_{Ic}, MPa.m$^{1/2}$	2.4-2.6	3.1-3.3	3.1-3.3	2.2-2.8	3.4-4.3	-
Young's Modulus, GPa	450-470	450-470	360-390	330-400		240-310
Brittleness $Bx10^{-6}$, m^{-1}	2280-2330	1000-1030	1050-1100	-		-

hp – hot pressed, pls – pressureless sintered

METALLIC BONDING OF CERAMIC ARMOR USING REACTIVE MULTILAYER FOILS

A. Duckham, M. Brown, E. Besnoin, D. van Heerden, , O.M. Knio, and T.P. Weihs
Reactive NanoTechnologies
111 Lake Front Drive
Hunt Valley, MD 21030

ABSTRACT
    The room temperature brazing and soldering of ceramic armor to metallic plates is described. The novel metallic bonding process uses reactive multilayer foils (NanoFoil^{TM}) as local heat sources to melt braze or solder layers placed between the ceramic and metal components. By replacing furnace cycles with local heating this NanoBond^{TM} process eliminates significant heating of the components and thus large area components can be bonded relatively stress-free. Examples of bonding Al_2O_3 to aluminum and of bonding SiC to titanium are presented and several configurations involving different braze and solder layers have been investigated in detail for the bonding of SiC to titanium. Shear strengths have been measured and are found to be significantly higher compared to those measured for similar samples bonded by epoxy adhesives. Numerical predictions of heat flow are also described for the reactive joining process and are validated by IR measurements. Lastly, bond areas over 100 cm^2 are demonstrated.

INTRODUCTION
    The large differences in the coefficients of thermal expansion (CTE) between ceramics and metals limit the size of ceramic to metal joints that can be formed by conventional soldering and brazing processes that utilize moderate to high temperature furnace cycles. In such conventional processing, the entire joint assembly is heated above the melting temperature of the solder or braze. On cooling, the excessive contraction of the metal component relative to the ceramic component results in severe residual stresses within the components and thus limits the size of joint areas to less than 7 cm^2 (for brazing). Ceramic armor tiles typically measure approximately 100 cm^2 and thus suffer delaminating and cracking following conventional furnace cycles. Currently these large ceramic tiles are joined to metal substrates by adhesives such as epoxy. This solution is not ideal since the strength of these joints is limited by the low strength of the epoxy and furthermore the organic epoxy results in a large elastic impedance mismatch with both the ceramic and the metal. Both the limited epoxy joint strength and the impedance mismatch are believed to limit the effectiveness of the ceramic armor in ballistic protection.
    Nanobond^{TM} is a new joining technology that enables soldering or brazing without significantly heating the components being bonded. The reactive multilayer foils (NanoFoil^{TM}) are magnetron sputtered and consist of thousands of alternating nanoscale layers of Ni and Al [1-16]. The layers react exothermically when atomic diffusion between the layers is initiated by an external energy impulse (Fig. 1), and release a rapid burst of heat. If the foils are sandwiched between layers of solder or braze, the heat released by the foils can be harnessed to melt these layers (see Fig. 2). By controlling the properties of the foils the exact amount of heat released by the foils can be tuned so that while there is sufficient heat to melt the solder or braze layers the components will not be heated significantly. Nanobond^{TM} is thus a room temperature method for achieving metallic joints between ceramics and metals. The components are not heated significantly, and there is minimal contraction of the components when the solder or braze cools

after melting. This means that large area, metallic joints can be formed between ceramic armor and metal plates with virtually no residual stresses within the components.

In this paper we demonstrate the NanoBond[TM] of two ceramic armor systems: Al_2O_3 to aluminum and SiC to titanium. Heat transfer into the components is numerically modeled for both these systems and is validated by infrared (IR) imaging. Several configurations involving different brazes and solders are investigated further for the SiC to titanium system. The success of bonding is evaluated by compression shear testing.

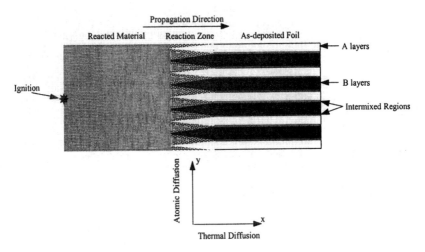

Figure 1: Cross-sectional schematic of atomic and thermal diffusion during reaction propagation in NanoFoil[TM]. Note that atoms diffuse normal to the layers while heat diffuses parallel to the layers.

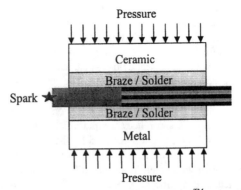

Figure 2: A cross-sectional schematic showing the NanoBond[TM] process of joining ceramic and metal components using NanoFoil[TM] and braze or solder layers.

HEAT FLOW DURING THE NANOBOND™ PROCESS

We performed numerical simulations using a previously developed computational model, in which the reaction in NanoFoil™ is described in terms of a self-propagating front with known velocity and heat release rates. Experimentally determined propagation velocities and heats of reaction are used for this purpose. Based on this description, the melting of solder or braze layers and the temperature evolution within the bonded components are determined by integration of the following energy conservation equation:

$$\rho \frac{\partial h}{\partial t} = \nabla \bullet \mathbf{q} + \dot{Q} \tag{1}$$

where h is the enthalpy, ρ is the density, t is time, $\mathbf{q}$ is the heat flux vector, and $\dot{Q}$ is the heat release rate. The temperature, T, is related to the enthalpy, h, through a complex relationship that involves the latent heats. The model also accounts for the effects of thermal contact resistance, and provides for variation of the thermal contact resistance as melting occurs along various interfaces. Simulations are performed using a numerical scheme that is based on a finite-difference discretization of Eqn. 1.

Predicted temperature distributions during the NanoBond™ process calculated by the above method are presented in Fig. 3 for Al$_2$O$_3$ to aluminum bonding and for SiC to titanium bonding. The results indicate that thermal transport occurs in an asymmetric fashion on each side of the foil due mainly to the relative thermal conductivities of the solder, braze and components. Most importantly though, it is apparent that the any significant heating of the components is limited in terms of time after the foil reaction and in terms of the distance from the braze or solder interface. For instance, the temperatures of the Al$_2$O$_3$ and the aluminum near the solder interfaces drop to about 200 °C (473 K) after only 16 ms following the foil reaction (Fig. 3(a)), while the titanium component is only heated by a few degrees at a distance of 1 mm from the braze interface (Fig. 3(b)).

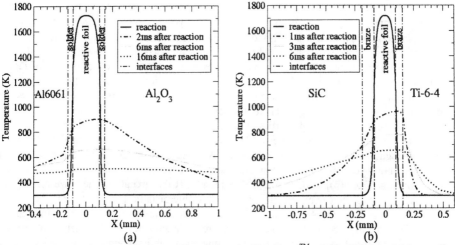

Figure 3: Predicted temperature profiles during the NanoBond™ of (a) Al$_2$O$_3$ to aluminum alloy 6101 melting a AgSn solder and (b) SiC to Ti-6Al-4V, melting a AgCuInTi braze (Incusil ABA).

We validated the model predictions of the heat flow into the components during the NanoBond™ process by temperature measurements using an IR camera with a spatial resolution of 108 μm and a temporal resolution of 100 ms. Prior to joining we polished the sides of the component samples and painted them white to ensure uniform emissivity. Figure 4(a) shows IR images of SiC being bonded to Ti-6-4 at the time of foil reaction and Figure 4(b) shows IR images at 240 ms after reaction. Note that the SiC component absorbs most of the heat, as predicted, but its average temperature does not rise above 100°C following foil reaction.

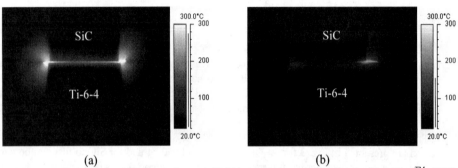

(a)   (b)

Figure 4: IR images at (a) foil reaction and (b) 240 ms after reaction during the NanoBond™ of SiC to Ti-6-4, melting a AgCuInTi braze (Incusil ABA).

REACTIVE JOINING OF SIC TO TITANIUM

We investigated various joint configurations using different solder and braze layers for the NanoBond™ of SiC to titanium (Ti-6Al-4V). Two of these configurations are illustrated in Fig. 5. The configuration shown in Fig. 5(a) involved a pre-deposition of a AgCuInTi braze (Incusil ABA – Morgan Advanced Ceramics) onto both the SiC and Ti-6Al-4V. We achieved this by first applying the braze in paste form and then performing a vacuum heat treatment above the melting temperature of the braze. This braze layer thickness on the components was approximately 50 μm thick. The NanoFoil™ used was coated with 1 μm of the same braze. This coating was sputter deposited just prior and just after depositing the Al/Ni foil NanoFoil™ without breaking vacuum. The configuration shown in Fig. 5(b) is identical to that of Fig. 5(a) except that we pre-deposited a layer of SnAgSb solder (Indium Corp.) over the braze layers on the components. In this configuration (Fig. 5(b)) the solder layers are the melting layers while the braze layers are the melting layers for the configuration in Fig. 5(a).

We determined the shear properties of these two joining configurations by compressive shear lap testing of 1.3 cm cube specimens. The testing geometry is illustrated in Fig. 6(a) and the shear strengths are summarized in Fig. 6(b). We measured the average shear strength of the braze melting layer configuration (Fig. 5(a)) to be 60 MPa and that of the solder melting layer configuration (Fig. 5(b)) to be 67 MPa. We found that the strength of the braze melting layer configuration was limited by fracture at the brittle intermetallic layer that forms between the Incusil braze and the Ti-6Al-4V during the heat treatment prior to the reactive joining process [18]. This type of fracture also explains the appreciable amount of scatter that we observed for this configuration. Fracture for the solder melting layer configuration was most often observed to occur within the solder layer. Also plotted in Fig. 6(b) are shear strengths measured for SiC bonded to Ti-6Al-4V by epoxy adhesive [17]. The average strength of these joints was 26 MPa. Thus the strengths of the metallic bonds formed by NanoBond™ in this study are more than double those of epoxy joints.

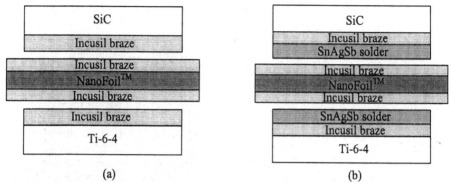

(a)   (b)

Figure 5: Two different reactive joining configurations investigated for the metallic bonding of SiC to Ti-6Al-4V by our NanoBond™ process. In (a) a AgCuInTi braze (Incusil ABA) is the melting layer while in (b) a SnAgSb braze is the melting layer.

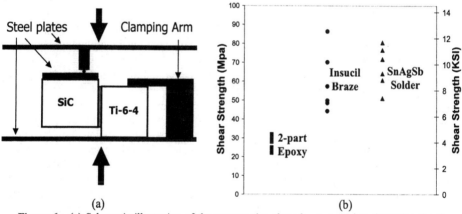

(a)   (b)

Figure 6: (a) Schematic illustration of the compressive shear lap test used to determine the shear strength of NanoBond™ joints. (b) Shear strengths for SiC/Ti-6Al-4V NanoBond™ joints for the two configurations shown in Fig. 5 (a) Incusil braze melting layer and (b) SnAgSb solder melting layer. Also shown for comparison are strengths for joints bonded by epoxy [17].

We also attempted large area (10 cm x 10 cm) SiC/Ti-6Al-4V NanoBond™ joints. Fig. 7 shows such a joint, where we used the configuration of Fig. 5(b). We evaluated the success of such joining by cutting these larger area bonds into smaller specimens (1.3 cm x 1.3 cm bond area) for shear strength testing. We measured the average shear strength of the sectioned specimens from the joint shown in Fig. 7 to be 67 MPa, which agrees very favorably with the results obtained for the smaller scale specimens (Fig. 6(b)).

Figure 7: Photograph of a large area, 10 cm x 10 cm (note the ruler is graduated in inches) SiC tile joined to a Ti-6Al-4V plate by our NanoBond™ process.

CONCLUSIONS

The feasibility of bonding ceramics to metals by using NanoFoil™ as localized heat sources to melt solder and braze layers has been demonstrated. The heat flow during this NanoBond™ process has been numerically predicted for the bonding of Al_2O_3 to aluminum and for the bonding of SiC to titanium. These model predictions show that the components are not significantly heated and that heating is of a very short duration. Direct IR imaging of components during reactive joining verify these predictions. This implies that that large area stress-free bonding of ceramics to metals is possible using our NanoBond™ process. Several configurations involving different solder and braze melting layers were investigated experimentally for bonding SiC to titanium. Average shear strengths of up to 67 MPa were measured for small-scale (1.3 cm cubed) samples. Measured shear strengths were well above those for epoxy bonded samples for two of the configurations investigated. Large-scale (100 cm^2) solder bonds were also fabricated by our NanoBond™ process. When these samples were sectioned into smaller specimens for shear strength testing, they not only survived the cutting, but the measured shear strengths compared very favorably with the smaller scale bonded specimens.

ACKNOWLEDGEMENTS

The authors wish to acknowledge the financial support of the Army Research Laboratory through Award No. DAAD17-03-C-0052, and the National Science Foundation through Award No. 0232395

REFERENCES
[1] T.P. Weihs, M.E. Reiss, D.Van Heerden, and Omar Knio, "Reactive Joining using Multilayer Materials", Provisional U.S. Patent Application filed on May 2, 2000. Three US and PCT Patent Applications were filed on May 1, 2001.
[2] T. P. Weihs, "Self-Propagating Reactions in Multilayer Materials", chapter in Handbook of Thin Film Process Technology, edited by D.A. Glocker and S.I. Shah, IOP Publishing, 1998.

[3.] D. M. Makowiecki and R.M. Bionta, "Low Temperature Reactive Bonding", US Patent 05381944, 1995.

[4.] T.W. Barbee, Jr. and T.P. Weihs, Ignitable, "Heterogeneous, Stratified Structures for the Propagation of an Internal Exothermic, Chemical reaction along an Expanding Wavefront", U.S. Patent 5,538,795, issued July 23, 1996.

[5.] T.W. Barbee, Jr. and T.P. Weihs, "Method for Fabricating an Ignitable, Heterogeneous, Stratified Structure", U.S. Patent 5,547,715, August 20, 1996.

[6.] U. Anselmi-Tamburni and Z.A. Munir, "The Propagation of a Solid-State Combustion Wave in Ni-Al Foils", *J. Appl. Phys.*, 66 (1989) 5039.

[7.] E. Ma, C.V. Thompson, L.A. Clevenger, and K.N. Tu, "Self-propagating Explosive Reactions in Al/Ni Multilayer Thin Films", *Appl. Phys. Lett.*, 57 (1990) 1262.

[8.] F. Bordeaux and A.R. Yavari, "Ultra Rapid Heating by Spontaneous Mixing Reactions in Metal-Metal Multilayer Composites", *J. Mater. Res.*, **5**, 1656 (1990).

[9.] T.S. Dyer, Z.A. Munir, and V. Ruth, "The Combustion Synthesis of Multilayer NiAl Systems", *Scripta Met.*, **30**, 1281 (1994).

[10.] T. P. Weihs, A. Gavens, M.E. Reiss, D. van Heerden, A. Draffin, and D. Stanfield, "Self-propagating Exothermic Reactions in Nanoscale Multilayer Materials", *TMS Proceedings*, Feb., 1997 Meeting, Orlando, FL.

[11.] D. Van Heerden, A.J. Gavens, A.B. Mann, and T.P. Weihs, "Metastable Phase Formation and Microstructural Evolution during Self-Propagating Reactions in Al/Ni and Al/Monel Multilayers", Mat. *Res. Soc. Symp. Proceedings*, Vol. 481, 533-8, edited by M. Atzmon, E. Ma, P. Bellon, and R. Trivedi, Fall 1997.

[12.] T.P. Weihs, P. Searson, M. Nikilova, and K. Blobaum, "Reactive Particle Dispersed Films and Process for the Electrochemical Deposition of Reactive Foils and Reactive Particle-Dispersed Films", U.S. Patent Application and PCT Application filed July 31, 1998.

[13.] A.J. Gavens, D.Van Heerden, A.B. Mann, M.E. Reiss, and T.P. Weihs, "Effect of Intermixing on Self-Propagating Exothermic Reactions in Al/Ni Nanolaminate Foils", *J. Appl. Phys.*, **87**, 1255-1263 (2000).

[14.] K.J. Blobaum and T.P. Weihs, "Self-propagating thermite reactions in Al.CuO$_x$ Multilayer foils", *submitted to App. Phys. Let.* (2002).

[15.] J. Wang, E. Besnoin, A. Duckham, S.J. Spey, M.E. Reiss, O.M. Knio, M. Powers, M. Whitener, and T.P. Weihs, "Room-temperature soldering with nanostructured foils", *Appl. Phys. Lett.*, **83** 3987-3989 (2003).

[16.] J. Wang, E. Besnoin, A. Duckham, S.J. Spey, M.E. Reiss, O.M. Knio, and T.P. Weihs, "Joining of stainless steel specimens with nanostuctured foils Al/Ni foils", *J. Appl. Phys.*, **95** 248-256 (2004).

[17.] K.J. Doherty, "Active Soldering Silicon Carbide to Ti-6Al-4V", *Soldering and Brazing Conference* (2003)

[18.] A. Duckham, E. Besnoin, S.J. Spey, J. Wang, M.E. Reiss, O.M. Knio, T.P. Weihs, "Reactive Nanostructured Foil used as a Heat Source for Joining Titanium", *submitted to J. Appl. Phys.* (2003).

STRAIN RATE EFFECTS ON FRAGMENT SIZE OF BRITTLE MATERIALS

Fenghua Zhou, Jean-Francois Molinari and KT Ramesh
Department of Mechanical Engineering
Johns Hopkins University
3400 N. Charles Street, Baltimore, MD 21218

ABSTRACT

A one-dimensional (1D) analytic model was developed to study the fragmentation phenomenon of ceramics under dynamic loading. The model integrates the elastic stress wave propagation with the cohesive crack initiation and growth process. Both the average fragment size and the fragment size distribution can be calculated using this model. Rate-dependent fragment size was calculated using the model. A scaling law was obtained to describe the size-strain-rate relationship, which explains the available experimental observations.

INTRODUCTION

Ceramics are used in armor systems because of their excellent specific strength and hardness. Under explosive loading, numerous cracks nucleate, propagate and coalesce in the material. As a result, the ceramic armor breaks into many pieces (fragments). Understanding the characteristics of the fragments (fragment size and fragment size distribution) is crucial to the design and evaluation of a ceramic armor system. For example, one important issue in ceramic armor design is obtaining maximum crack density in the fragmented zone (the comminuted zone), in order to maximally dissipate the penetrator's kinetic energy.

Many analytical models have been developed [1, 2, 7-10]. Most of them use energy criteria to relate material properties and strain-rate to the average fragment size $\bar{s}$. Assuming that the local kinetic energy is responsible for the creation of new fractured surfaces, Grady [1] obtained:

$$\bar{s} = \left(\frac{24G_c}{\rho\dot{\varepsilon}^2}\right)^{1/3} \tag{1}$$

where G_c is the material's fracture energy, ρ density and $\dot{\varepsilon}$ strain-rate. Fracture energy is calculated through formula $G_c = K_{Ic}^2/E$, with K_{Ic} denoting fracture toughness and E Young's modulus.

Grady [1] predicts that $\bar{s}$ decreases with increasing $\dot{\varepsilon}$ to the power of $-2/3$. According to Equation 1, $\bar{s}$ is infinitely large when $\dot{\varepsilon}$ is very small. Glenn and Chudnovsky [2] introduced a correction term accounting for the stored elastic energy before failure. The fragment size is:

$$\bar{s} = 4\sqrt{\frac{\alpha}{3}}\sinh\left(\frac{\phi}{3}\right) \tag{2}$$

where $\alpha = \frac{3\sigma_b^2}{\rho E \dot{\varepsilon}_0^2}$, σ_b is the breaking strength of the bar. The quantity ϕ is expressed as

$\phi = \sinh^{-1}\left[\beta\left(\frac{3}{\alpha}\right)^{3/2}\right]$, where $\beta = \frac{3}{2}\frac{G_c}{\rho\dot{\varepsilon}_0^2}$. Glenn and Chudnovsky's theory (called G&C theory hereafter) is identical to Grady's in the high strain-rate region where the local kinetic energy term dominates, while in the very low strain-rate (quasi static) region the theory predicts a fragment size independent of the strain-rates, as:

$$\bar{s}_{quasistatic}\Big|_{G\&C} = \frac{4\beta}{\alpha} = \frac{2EG_c}{\sigma_b^{\,2}} \tag{3}$$

These energy models are straightforward and reflect qualitatively the decreasing trend of fragment size with increasing strain-rate. Nevertheless, many experimental observations show that in the high strain-rate region, the energy model estimation (Equation 1) over estimates the fragment size. The experimental fragment sizes of Lankford and Blanchard [3] are half of the theoretical estimation. The experimental values of Shih et al. [4] are one order smaller. Wang and Ramesh [5] also reported experimental measurements that are much smaller than Grady [1] estimation.

Grady and Kipp [6] conducted experiments where E51200 steel balls were shot into targets at high velocity. They measured the fragment sizes of the steel balls after experiments, and applied Equation 1 conversely to estimate the dynamic fracture toughness of the steel:

$$K_{Ic} = \left(\frac{\rho E \dot{\varepsilon}^2 \bar{s}^3}{24} \right)^{\frac{1}{2}} = \frac{\rho c \dot{\varepsilon}\, \bar{s}^{3/2}}{\sqrt{24}} \tag{4}$$

where $c = \sqrt{E/\rho}$ is the elastic uniaxial stress wave speed. In Grady and Kipp's experiments the strain-rate is 10^4-10^5 s^{-1}. The obtained dynamic K_{Ic} values were one quarter of the quasistatic fracture toughness. Since common materials exhibit higher fracture toughness under impact loading, Grady and Kipp [6] considered their estimated dynamic K_{Ic} values unreliable, because the fragment size theory (either Equation 1 or 4) is not fully developed.

Besides the energy models, numerical and analytic models have been proposed to investigate the dynamic process of fragmentation [7-9]. Miller et al. [7] used finite element method to simulate the fragmentation process. In their simulation, the crack growth process is modeled by using cohesive elements. Miller et al. found that the calculated fragment size is approximately one order of magnitude smaller than the energy models. Drugan [8] developed an analytical model to investigate the fragmentation process of a bar under uniform tensile strain-rate. Drugan's fragment sizes are close to Miller et al.'s in the intermediate strain-rate region, which are much smaller than the energy model estimation. Recent work of Shenoy and Kim [9] gives similar conclusions.

The limitation of an energy model originates from its simple assumption of energy conservation. During the fragmentation process, not only energy, but also momentum is conserved. A quantitative estimation of fragment size necessitates detailed consideration of the dynamic fracture process.

In this paper we present a model to study the fragmentation process of a brittle bar. The model incorporates the mechanisms of stress wave propagation with the local crack nucleation and growth process. The model renders the average fragmentation size and the fragment size distribution, both dependent on the strain-rates. The calculations were compared to the energy models. It is seen that at the high strain-rate region, the average fragment size is significantly smaller than the energy model estimations; while at the low strain-rate region it is lager. The physical mechanisms explaining these deviations are provided.

ANALYTIC MODEL

Figure 1 illustrates the model. A long bar lying along X-axis is being pulled with a constant tensile strain-rate $\dot{\varepsilon}$. At time $t=0$ a linear velocity distribution exists along the bar: $v(X,0) = \dot{\varepsilon}_0 X$, so that the deformation of the bar is uniform. Before fracture, the stress of the bar increases linearly with time:

$$\sigma(X,t) = E\dot{\varepsilon}_0 \qquad (5)$$

Cracks can be initiated at any internal point when the local stress is greater then the local tensile strength. The crack initiation criterion is:

$$\sigma(X,t) \geq \sigma_c(X) \qquad (6)$$

where $\sigma_c(X)$ is the local strength of the bar, which is identical to the bar's breaking strength σ_b as long as the bar is homogeneous.

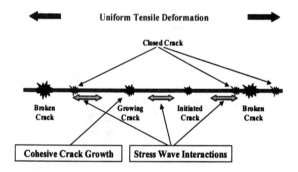

Figure 1 A brittle bar breaks into many pieces under constant strain-rate loading

An initiated crack may grow or close depending on the local stress state. It may contact again if the local stress is compressive. When the local stress is tensile, the crack growth behavior obeys a cohesive law, which links the crack opening force σ_{coh} to the crack opening distance δ_{coh}. A modified initial-rigid, linear-decaying, irreversible (damage dependent) cohesive law is used. The law is illustrated in Figure 2. As long as the cracked node does not contact, the cohesive law takes the form:

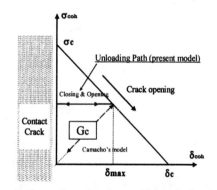

Figure 2 Linear-decaying, irreversible cohesive law

$$\frac{\sigma_{coh}}{\sigma_c} = 1 - \frac{\delta_{coh}}{\delta_c} \qquad (\dot{\delta}_{coh} > 0, \delta_{coh} = \delta_{max}, D < 1.0)$$

$$\frac{\sigma_{coh}}{\sigma_c} = 1 - \frac{\delta_{max}}{\delta_c} \qquad (\delta_{coh} < \delta_{max}, D < 1.0) \qquad (7)$$

where $\sigma_c = \sigma_c(X)$ is the local strength, $\delta_c = 2G_c/\sigma_c$ the critical opening distance, and $D = \min(\delta_{max}/\delta_c, 1.0)$ the local damage number.

Some initiated cracks grow up to the critical opening distance, representing complete breakages. Others are closed due to the unloading wave released from the neighboring crack. The fully broken points separate the bar into many pieces, each with a specific length. When the bar is completely unloaded, the total number of fragments and the length of each fragment are obtained.

The fragmentation process is simulated as an integration of nodal variables along the time axis. Fundamental equations of stress wave propagation are used to evaluate the local velocity and stress at an intact or contact node. Wave equations incorporating the cohesive law are used to solve the cracked node. In Reference [10] we gave detailed information about the numerical scheme.

CACULATION RESULTS

We consider a 50mm long bar made of model ceramic. Basic material parameters are assumed as: $\rho = 2.75 \times 10^3$ kg/m^3, $E = 275$ GPa, $c = 10^4$ m/s. We assume that the bar is completely homogeneous, with $\sigma_c = 300$ MPa and $G_c = 100$ N/m. The fragmentation process is analyzed in each designated strain-rates. Figure 3a shows the history curves of the average stress and the fragment numbers for the case $\dot{\varepsilon} = 500\,s^{-1}$. The stress in the bar increases linearly according to Equation 5, and reaches the peak value (300 MPa) at failure point when Equation 6 is satisfied. The fragment number increases some time *after* the bar fails. The formation of each fragment is not simultaneous. Most fragments form during the time the stress drops from peak value to zero. Some fragments form after the bar stress drops to zero (with some oscillations). This observation agrees with the experimental observations of Wang and Ramesh [5], which show that apparent cracks formed within the specimen *after* the failure point. The final number of fragments is constant. Figure 3b shows a part of the damage (D) distribution after fragmentation. The points where $D=1$ separate each fragment. The length of each fragment is not necessarily identical. Some damage remains within each fragment.

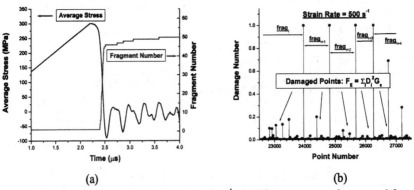

(a)   (b)

Figure 3 Fragmentation of a bar under $\dot{\varepsilon} = 500\,s^{-1}$: (a) History curves of stress and fragment numbers; (b) Damage distribution of the fragmented bar

The average fragment length ($\bar{s}$) is calculated from the final fragment number N_f, through the formula $\bar{s} = 50/N_f \ (mm)$. In the $\dot{\varepsilon} = 500\,\text{s}^{-1}$ case, $N_f=50$, the average fragment size is 1 mm.

The cases where strain-rate ranges from 1 to 5×10^6 s^{-1} are simulated. The obtained $\bar{s}$ vs. $\dot{\varepsilon}$ relationship is shown in Figure 4. Grady's curve (Equation 1) and G&C's curve (Equation 2) are also plotted for comparison. Qualitatively our curve is similar to G&C's curve. Namely, in the low $\dot{\varepsilon}$ region the fragment size approaches to a constant value, while in the high $\dot{\varepsilon}$ region fragment size decreases as $\dot{\varepsilon}$ increases. However, the calculated fragment sizes quantitatively differ from G&C's theory in two aspects. Firstly, our low strain-rate fragment size is about 2 to 3 times larger than G&C's quasistatic value (Equation 3). The reason is that the sudden failure of the bar causes the stress wave to propagate within the fragments. The stress waves trapped within the fragments account 50% to 70% of the stored strain energy, only about 30% to 50% strain energy are used to create fracture. Our estimation of the quasistatic fragment size is:

$$\bar{s}_{quasistatic} = (2 \text{ to } 3)\bar{s}_{quasistatic}\big|_{G\&C} = \frac{(4 \text{ to } 6)EG_c}{\sigma_c^{\ 2}} \tag{8}$$

Secondly, in the high strain-rate region our fragment size is about five times smaller than the energy model estimation (Equation 1). The intrinsic mechanisms of this phenomenon are: (a) during the cohesive fracture process, global kinetic energy and external work are consistently being converted into energy available for failure; (b) the internal impacts between fragments facilitate this energy conversion process, making large fragment to break further; (c) the size distributions may also play a role in the fragmentation process.

Our calculations intersect with G&C curve at strain rate strain-rate $\dot{\varepsilon} \approx 2500\,\text{s}^{-1}$. In many dynamic tests the strain-rate is close to or larger than this point. The difference between our model and Grady's model in this strain-rate region agrees with other numerical and theoretical analyses [7-9] in which the dynamic fragmentation process is considered. It explains the discrepancies between the experimental measurements and the energy estimations [3-5].

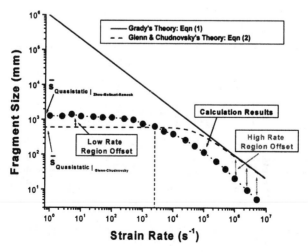

Figure 4 Average fragment size ($\bar{s}$) vs. strain-rate ($\dot{\varepsilon}$) curves

609

GENERALIZED FORMULA FOR FRAGMENT SIZE ESTIMATION

Parametric studies have been conducted where different values of material parameter E, c, G_c, σ_c are considered [11]. It has been shown that the normalized fragment size is an exclusive function of the normalized strain-rate. The function is fitted as:

$$\frac{\bar{s}}{\bar{s}_0} = 5.0\left[1 + 1.47\left(\frac{\dot{\varepsilon}}{\dot{\varepsilon}_0}\right)^{0.243} + 4.77\left(\frac{\dot{\varepsilon}}{\dot{\varepsilon}_0}\right)^{0.763}\right]^{-1} \qquad (9)$$

where $\bar{s}_0 = EG_c/\sigma_c^2$, $\dot{c}_0 = c\sigma_c^3/E^2 G_c$ are the characteristic fragment size and the characteristic strain-rate.

Equation 9 applies in a broad strain-rates range from quasistatic $(\dot{\varepsilon}/\dot{\varepsilon}_0 = 0)$ to $\dot{\varepsilon}/\dot{\varepsilon}_0 = 1000$. Different material parameters and strain-rate can be substituted into (9) to obtain a fragment size estimation. Experimental data of Grady and Kipp [6] are revisited using Equation 9. For the ANSI E52100 high carbon chromium steel: E=207 GPa, c=5135 m/s. We take yielding stress σ_y=2.03 GPa as the cohesive critical stress σ_c, and assume that the material parameters E, c, σ_c do not change under high strain-rates. Literature [6] reported test results of four shots. The effective strain-rates are 1.6×10^4, 5.4×10^4, 6.1×10^4, and 8.5×10^4 s^{-1}. The nominal fragment sizes are 1.45, 0.72, 0.96, and 0.71 mm, respectively (Plotted in the lower part of Figure 5). Substituting the experimental data and the E, c, σ_c values into (9), the fracture energy G_c can be calculated. These values are converted into the fracture toughness K_{Ic} via equation $K_{Ic} = \sqrt{EG_c}$.

The estimated K_{Ic} values are plotted in Figure 5, in comparison with the values estimated by Grady and Kipp [6]. Grady and Kipp's early dynamic K_{Ic} values are 7.5 to 15 MPa m$^{1/2}$, much smaller than the quasistatic value (30 to 40 MPa m$^{1/2}$). Contrarily, our estimated K_{Ic} values changes from 70 MPa m$^{1/2}$ to 105 MPa m$^{1/2}$ when the strain-rates increases from 1.6×10^4 s^{-1} to 8.5×10^4 s^{-1}. This appears to be a reasonable range of K_{Ic} under intense impact loading.

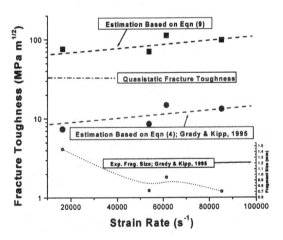

Figure 5 Estimated K_{Ic} values based on the experimental $\bar{s}(\dot{\varepsilon})$ data of [6]

CONCLUSIONS

An analytical approach has been developed to investigate the dynamic fragmentation process of brittle materials. The approach renders a relationship between the average fragment size $\bar{s}$ and the strain-rate $\dot{\varepsilon}$. The $\bar{s}$ vs $\dot{\varepsilon}$ curve differs from those given in energy models, specifically in the high strain-rate region where our model gives an estimation of fragment size 5 times smaller than conventional energy model. Our results agree with the available experimental observations [3-5] and dynamic models [7-9]. A quantitative scaling law Applying the new fragmentation theory to reanalyze the test data of Grady and Kipp [6], we obtain a more reasonable dynamic fracture toughness of high-strength steel.

ACKNOWLEDGEMENT

This work was performed under the auspices of the Center for Advanced Metallic and Ceramic Systems at Johns Hopkins. The research was sponsored by the Army Research Laboratory (ARMAC-RTP) and was accomplished under the ARMAC-RTP Cooperative Agreement Number DAAD19-01-2-0003.

REFERENCES

[1]D.E. Grady, "Local inertial effects in dynamic fragmentation", *Journal of Applied Physics*, **53**, 322-325 (1982)

[2]L.A. Glenn and A. Chudnovsky, "Strain-energy effects on dynamic fragmentation", *Journal of Applied Physics*, **59**, 1379-1380 (1986)

[3]J. Lankford and C.R. Blanchard, "Fragmentation of brittle materials at high rates of loading", *Journal of Material Science*, **26**, 3067-3072 (1991)

[4]C.J. Shih, M.A. Meyers, V.F. Nesterenko and S.J. Chen, "Damage evolution in dynamic deformation of silicon carbide", *Acta Materialia*, **48**, 2399-2420 (2000)

[5]H. Wang and K.T. Ramesh, "Dynamic strength and fragmentation of hot-pressed silicon carbide under uniaxial compression", *Acta Materialia*, **52**, 355-367 (2004)

[6]D.E. Grady and M.E. Kipp, "Experimental measurement of dynamic failure and fragmentation properties of metals", *Int. J. Solids Structures*, **32**, 2779-2991 (1995)

[7]O. Miller, L.B. Freund and A. Needleman, "Modeling and simulation of dynamic fragmentation in brittle materials" *Int. J. Fracture*, **96**, 101-125 (1999)

[8]W.J. Drugan, "Dynamic fragmentation of brittle materials: analytical mechanics-based models", *J. Mech. Phys. Solids*, **49**, 1181-1208 (2001)

[9]V.B. Shenoy and K.-S. Kim, "Disorder effects in dynamic fragmentation of brittle materials", *J. Mech. Phys. Solids*, **51**, 2023-2035 (2003)

[10]F. Zhou, J-F. Molinari and K.T Ramesh, "One-dimensional fragmentation analysis: cohesive model, explicit dynamics and defect distributions", submitted to *J. Mech. Phys. Solids* (2004)

[11]F. Zhou, J-F. Molinari and K.T Ramesh, "Effects of Material Parameters and Strain Rate on Fragmentation of Ceramics", in preparation (2004)

Author Index